Operator Theory
Advances and Applications
Vol. 74

Editor
I. Gohberg

Spectral Theory of Approximation Methods for Convolution Equations

Roland Hagen
Steffen Roch
Bernd Silbermann

Birkhäuser Verlag
Basel · Boston · Berlin

Authors' address

Technische Universität Chemnitz-Zwickau
Fakultät für Mathematik
PSF 964
D-09009 Chemnitz

A CIP catalogue record for this book is available from the Library of Congress, Washington D.C., USA

Deutsche Bibliothek Cataloging-in-Publication Data
Hagen, Roland:
Spectral theory of approximation methods for convolution
equations / Roland Hagen ; Steffen Roch ; Bernd Silbermann. –
Basel ; Boston ; Berlin : Birkhäuser, 1995
   (Operator theory ; Vol. 74)
   ISBN 3-7643-5112-8 (Basel ...)
   ISBN 0-8176-5112-8 (Boston)
NE: Roch Steffen:; Silbermann, Bernd:; GT

© 1995 Birkhäuser Verlag, P.O. Box 133, CH-4010 Basel, Switzerland
Camera-ready copy prepared by the editors
Printed on acid-free paper produced from chlorine-free pulp
Cover design: Heinz Hiltbrunner, Basel
Printed in Germany
ISBN 3-7643-5112-8
ISBN 0-8176-5112-8

9 8 7 6 5 4 3 2 1

Contents

Introduction   VII

1 Invertibility in Banach algebras
  1.1   Banach algebras and C^*-algebras   1
  1.2   Linear operators   4
  1.3   Stability of operator sequences   7
  1.4   Local principles   12
  1.5   The finite section method for Toeplitz operators   17
  1.6   A general invertibility scheme   24
  1.7   Norm-preserving localization   32
  1.8   Exercises   36
  1.9   Comments and references   39

2 Spline spaces and Toeplitz operators
  2.1   Singular integral operators-constant coefficients   46
  2.2   Piecewise constant splines   54
  2.3   Algebras of Toeplitz operators (Basic facts)   59
  2.4   Discretized Mellin convolutions   64
  2.5   Algebras of Toeplitz operators (Fredholmness)   72
  2.6   General spline spaces   79
  2.7   Spline projections   83
  2.8   Canonical prebases   91
  2.9   Concrete spline spaces   96
  2.10 Concrete spline projections   102
  2.11 Approximation of singular integral operators   105
  2.12 Proofs   126
  2.13 Exercises   138
  2.14 Comments and references   142

3 Algebras of approximation sequences
  3.1   Algebras of singular integral operators   146
  3.2   Approximation using piecewise constant splines   153
  3.3   Approximation of homogeneous operators   163

3.4	The stability theorem	171
3.5	Basic properties of approximation sequences	175
3.6	Proof of the stability theorem	189
3.7	Sequences of local type	197
3.8	Concrete approximation methods	205
3.9	Exercises	218
3.10	Comments and references	222

4 Singularities

4.1	Approximation of operators in Toeplitz algebras	228
4.2	Multiindiced approximation methods	240
4.3	Approximation of singular integral operators	255
4.4	Approximation of compound Mellin operators	260
4.5	Approximation over unbounded domains	270
4.6	Exercises	277
4.7	Comments and references	279

5 Manifolds

5.1	Algebras of singular integral operators	281
5.2	Splines over homogeneous curves	289
5.3	Splines over composed curves	299
5.4	The stability theorem	306
5.5	A Galerkin method	320
5.6	Exercises	335
5.7	Comments and references	336

6 Finite sections

6.1	Finite sections of singular integrals	337
6.2	Finite sections of discrete convolutions	344
6.3	Around spline approximation methods	350

Bibliography	356
Index	371

Introduction

The topic of this book is the different kinds of invertibility of linear -in particular, convolution- operators. The common or *classical invertibility* of a linear bounded operator A on a Banach space X is immediately related with the solvability of the equations $Ax = y$ where x and y are in X. These equations are uniquely solvable for all right sides y if and only if the operator A is classically invertible, and their solutions x can be found by applying the inverse operator A^{-1} to y. So far so good- but, given a concrete operator, it is in general a highly non-trivial problem to decide whether it is invertible and to determine its inverse practically. For that reason, besides this classical invertibility, other invertibility concepts are in discussion.

One of these concepts is that of *essential invertibility*. The operator A is essentially invertible (or a Fredholm, Noether, or Φ-operator) if its kernel has finite dimension (in contrast to $\mathrm{Ker} A = \{0\}$ in case of classical invertibility), and if its image possesses a finite dimensional complement in X (whereas $\mathrm{Im} A = X$ for classical invertibility). Clearly, it should be much simpler to study essential invertibility of an operator rather than its classical one. That this simplification is also of practical significance will be illustrated by the following example.

Let us start with recalling that both kinds of invertibility correspond to invertibility problems in associated Banach algebras. Indeed, by a theorem of Banach, an operator A is classically invertible if and only if it is invertible in the Banach algebra $L(X)$ of all linear and bounded operators on X. Further, A is essentially invertible if and only if it has an inverse modulo the ideal $K(X)$ of all linear compact operators on X or, equivalently, if the coset $A + K(X)$ is invertible in the so-called Calkin algebra $L(X)/K(X)$. So one is led to the study of invertibility of elements in the Banach algebras $L(X)$ and $L(X)/K(X)$. In general, these algebras prove to be too large to tackle invertibility problems successfully, and therefore one has to restrict oneself to looking for inverses not in all of $L(X)$ or $L(X)/K(X)$ but in suitably chosen subalgebras. (This is a well known method: In order to determine, e.g., the inverse of the operator of the Sturm boundary value problem, one does not look for this inverse among all possible operators but only among the integral operators with continuous kernel function - the Green function of the operator.) Of course, here one has to take care to guarantee that invertibility in the larger algebra is equivalent to invertibility in the subalgebra.

Let now A be the operator $aI + bS$ where S is the singular integral operator on the real axis,

$$(Sf)(t) = \frac{1}{\pi i} \int_{-\infty}^{\infty} \frac{f(s)}{s - t} \, ds \quad , \quad t \in \mathbf{R},$$

and where the coefficients a and b are supposed to be continuous on $\mathbf{R} \cup \{\infty\} =: \dot{\mathbf{R}}$ (continuity of a function b at ∞ means that both limits $\lim_{s \to +\infty} b(s)$ and $\lim_{s \to -\infty} b(s)$ exist and coincide). This operator is bounded on $L^2 = L^2(\mathbf{R})$. In place of the comprehensive algebras $L(L^2)$ and $L(L^2)/K(L^2)$ we consider their subalgebras $\mathcal{A}$ and $\mathcal{A}/K(L^2)$, respectively, where $\mathcal{A}$ stands for the smallest closed subalgebra of $L(L^2)$ containing all operators $aI + bS$ with a and b in $C(\dot{\mathbf{R}})$.

The difficulties arising with classical invertibility find their algebraic expression in the fact that the algebra $\mathcal{A}$ is irreducible. Roughly speaking, this means that it is principally impossible to reduce invertibility in $\mathcal{A}$ to a simpler problem. On the other hand, one can show that the quotient algebra $\mathcal{A}/K(L^2)$ is even commutative!

The invertibility of elements of a commutative Banach algebra is a matter of Gelfand's spectral theory. This theory associates with each commutative Banach algebra a compact set T, and with each element of this algebra a continuous complex-valued function on T with the property that the element is invertible if and only if this function does not vanish on T. In case of the algebra $\mathcal{A}/K(L^2)$ one finds that

$$T = \dot{\mathbf{R}} \times \{-1, 1\} ,$$

and the function corresponding to the coset $aI + bS + K(L^2)$ is just given by

$$(t, n) \mapsto a(t) + nb(t) . \tag{1}$$

Hence, the singular integral operator $aI + bS$ is essentially invertible if and only if the function (1) has no zero, and this condition is of course also necessary for the classical invertibility of $aI + bS$. But in this special situation, a famous theorem by Coburn states some additional conditions which, together with the essential invertibility, imply the classical one. More precisely, the following theorem holds.

Theorem 0.1 *(a) The singular integral operator $aI + bS$ is essentially invertible if and only if the function (1) does not vanish.*
(b) This operator is classically invertible if and only if it is essentially invertible and if the winding number of the curve $\{(a(t)+b(t))/(a(t)-b(t)) , t \in \dot{\mathbf{R}}\}$ is zero.

This theorem shows, in particular, that essential invertibility is a *local* property (it only depends on the values of a function), whereas classical invertibility is of *global* nature (the winding number depends on a function as a whole).

The function (1) is called the *symbol* of the singular integral operator $aI + bS$ (related with its Fredholm property).

Now we are going to discuss the practical computation of the inverse of an operator $A \in L(X)$. Usually, A^{-1} is determined by a certain approximation method. Accordingly, the operator A is called to be *approximately invertible* by a sequence (A_n) of approximation operators which converges strongly to A (i.e. $A_n x \to Ax$ as $n \to \infty$ for all $x \in X$) if and only if there is a number n_0 so that the equations $A_n x_n = y$ are uniquely solvable for all y and $n \geq n_0$, and if the sequence (x_n) of their solutions converges to a solution x of the equation $Ax = y$.

It is evident that this notion of invertibility is stronger than the classical one. As the latter, approximate invertibility is equivalent to invertibility in a certain

Banach algebra again. In order to see this, recall that an operator A is approximately invertible by a sequence (A_n) converging strongly to A if and only if the operator A is (classically) invertible and if the sequence (A_n) is stable. Stability of a sequence (A_n) means that there is an index n_0 such that the operators A_n are invertible for all $n \geq n_0$ and that $\sup_{n \geq n_0} \|A_n^{-1}\| < \infty$.

To get the desired formulation of stability in algebraic language, consider the collection $\mathcal{F}$ of all bounded sequences. On defining operations by $(A_n) + (B_n) := (A_n + B_n)$ and $(A_n)(B_n) := (A_n B_n)$ and a norm by $\|(A_n)\| := \sup_n \|A_n\|$, the set $\mathcal{F}$ even becomes a Banach algebra. Further, the set $\mathcal{G}$ of all sequences tending to zero in the norm forms a closed two-sided ideal in $\mathcal{F}$. Now, it is not hard to see that a bounded sequence (A_n) is stable if and only if its coset $(A_n) + \mathcal{G}$ is invertible in the quotient algebra $\mathcal{F}/\mathcal{G}$. (Indeed, if $(A_n) + \mathcal{G}$ is invertible then there are a bounded sequence (B_n) and sequences (C_n) and (D_n) tending to zero such that $A_n B_n = I + C_n$ and $B_n A_n = I + D_n$ whence, via a Neumann series argument, stability of (A_n) follows.)

The role of the algebra $\mathcal{F}/\mathcal{G}$ in approximate invertibility is the same as that of $L(X)$ in classical invertibility. Moreover, similarly to the classical invertibility, it turns out that the algebra $\mathcal{F}/\mathcal{G}$ is too large to work in successfully, and that its relevant subalgebras are irreducible again. The experiences from Fredholm theory suggest to look for an *"essential approximate invertibility"* which should allow to study first a simpler (possibly, local) problem, and then (under additional conditions) to turn back to proper approximate invertibility.

It was A. Kozak who first succeeded in developing a local theory of approximate invertibility (in case of the finite section method for two-dimensional Toeplitz operators). But, as one can show, Kozak's approach is restricted to rather special classes of operators (e.g. Toeplitz operators with continuous generating function). Later on, one of the authors (see [S 2]) proposed another approach with an essentially wider area of applications. He discovered a canonical procedure to construct ideals $\mathcal{J}$ of subalgebras of $\mathcal{F}$ with the following properties:

- essential approximate invertibility is invertibility modulo the ideal $\mathcal{J}$,

- invertibility modulo $\mathcal{J}$ is of local nature,

- essential approximate invertibility implies essential invertibility

and, in contrast to Fredholm theory, where Coburn's theorem appears as a lucky circumstance only,

- there are *well-defined* conditions which, together with essential approximate invertibility, imply approximate invertibility itself.

Let us illustrate the efficiency of this "lifting mechanism" by consideration of a simple spline Galerkin method for singular integral operators. We choose the spline space S_n of all piecewise constant functions over the partition $\frac{1}{n}\mathbf{Z}$ both as trial and test space, and we let L_n stand for the Galerkin projection of $L(L^2)$ onto S_n.

In order to study the approximate invertibility of $aI + bS$ by the spline Galerkin method, we introduce the smallest closed subalgebra $\mathcal{B}$ of $\mathcal{F}$ which contains all approximation sequences $(L_n(aI + bS)|S_n)$ with $a, b \in C(\dot{\mathbf{R}})$. As already mentioned, both the algebra $\mathcal{B}$ and its quotient algebra $\mathcal{B}/\mathcal{G}$ are irreducible. Let now a subset $\mathcal{J}$ of $\mathcal{B}$ be defined by

$$\mathcal{J} = \{(L_n K|S_n) + (C_n) \quad \text{with} \quad K \in K(L^2) \quad \text{and} \quad (C_n) \in \mathcal{G}\}.$$

This set (which is the analogue to the ideal $K(X)$ in Fredholm theory) has exactly the properties we are interested in:

- $\mathcal{J}$ *is an ideal in* $\mathcal{B}$.

To verify that $\mathcal{J}$ is, e.g., a left ideal, let (A_n) be a sequence in $\mathcal{B}$ with strong limit A. Then

$$(A_n)(L_n K|S_n + C_n) = ((A_n L_n - L_n A)K|S_n + A_n C_n + L_n AK|S_n),$$

and the sequence on the right hand side is in $\mathcal{J}$ since $A_n L_n - L_n A \to 0$ strongly as $n \to \infty$ which implies that the sequence $((A_n L_n - L_n A)K|S_n + A_n C_n)$ is in $\mathcal{G}$, and since AK is compact again.

- *The algebra* $\mathcal{B}/\mathcal{J}$ *is commutative.*

This follows readily from the commutator relation (due to Arnold/Wendland and Prößdorf) $\|L_n fI - fL_n\| \to 0$ holding for each $f \in C(\dot{\mathbf{R}})$.

- *The ideal* $\mathcal{J}$ *can be lifted.*

Indeed, let, for example, the coset $(A_n) + \mathcal{J}$ be invertible from the right, that is, there are sequences $(B_n) \in \mathcal{B}$ and $(L_n K|S_n) + (C_n) \in \mathcal{J}$ such that

$$A_n B_n = I + L_n K|S_n + C_n.$$

Now suppose additionally the strong limit A of the sequence $(A_n L_n)$ to be invertible. Then set $B'_n = B_n - L_n A^{-1} K|S_n$ to obtain

$$
\begin{aligned}
A_n B'_n &= A_n B_n - A_n L_n A^{-1} K|S_n \\
&= I + L_n K|S_n + C_n - (A_n L_n - L_n A)A^{-1} K|S_n - L_n K|S_n \\
&= I + C'_n
\end{aligned}
$$

with $(C'_n) = (C_n) - ((A_n L_n - L_n A)A^{-1} K|S_n) \in \mathcal{G}$.

Thus, the sequence (A_n) is right invertible modulo $\mathcal{G}$. Similarly one verifies its invertibility from the left.

The Gelfand theory yields the following assertions now. The compact set associated with $\mathcal{B}/\mathcal{J}$ is $T = \dot{\mathbf{R}} \times [-1, 1]$, and the function on T corresponding to the coset $(L_n(aI + bS)|S_n)$ is given by

$$(t, z) \mapsto a(t) + z\, b(t). \tag{2}$$

Summarizing these results we get the following theorem about the applicabilty of the Galerkin method.

Theorem 0.2 *(a) The singular integral operator $aI + bS$ is essentially approximately invertible by the spline Galerkin method if and only if the function (2) does not vanish.*
(b) This operator is approximately invertible by the spline Galerkin method (in other words: this method is applicable) if and only if it is essentially approximately invertible and if the operator itself is classically invertible.

A comparison with the function (1) and its role in Theorem 1 suggests to interprete the function (2) as the *symbol* of the operator $aI + bS$ *which is related with the applicability of the Galerkin method.*

   We hope that the reader now has got a certain idea of what this book is all about. Let us turn over to a brief description of its contents.

   In the first chapter, we start with summarizing some necessary facts from operator theory. Then we analyze the so-called finite section method for Toeplitz operators with continuous generating function. From this method, we finally extract a general heuristic scheme for investigating invertibility problems in Banach algebras which will serve as a guide throughout the whole book.

   The second chapter contains a detailed treatise on algebras generated by Toeplitz operators with piecewise continuous generating function with special emphasis on local and global symbol constructions for Fredholmness. Further we present some important aspects from the algebraic theory of spline spaces where the notion *spline* is used in a wide sense. Splines are functions satisfying certain axioms. Specifying the axioms yields the classical smoothest piecewise polynomial splines (B-splines), or piecewise polynomial splines with defect (based on Bernstein polynomials), or wavelets.

   Here we focus our attention on the construction of spline spaces with a finite number of generating (or mother-) splines, and to the existence and main properties of projection operators onto these spline spaces. Finally we examine the structure of the approximation operators which arise from the application of spline projections to pure singular integral and Mellin operators.

   The third chapter is the heart of the whole book. Here we introduce and investigate a comprehensive Banach algebra of approximation sequences coming from spline Galerkin, collocation, qualocation, but also from certain quadrature and other fully discretized methods. Complete necessary and sufficient conditions for the stability of the involved sequences are derived. The definition of this algebra is motivated by two observations: the special structure of approximation sequences for homogeneous operators (as pointed out at the end of the second chapter), and the shift invariance of the algebra of all singular integral operators with piecewise continuous coefficients on the real axis (which is briefly examined in the first sections of Chapter 3).

   In contrast to the algebra $\mathcal{B}/\mathcal{J}$ introduced above, this algebra proves to be highly non-commutative. In particular, the classical Gelfand theory is no longer an adequate tool to study the related quotient algebras, and it has to be replaced by more general local principles.

The topic of the forth chapter is spline approximation methods based on modified spline spaces. These spaces result from the original ones by omitting basic spline functions in neighbourhoods of singular points (such as discontinuity points of the coefficient functions or corner points of the underlying curve; compare Chapter 5) in order to improve the stability behaviour. We extend the algebra introduced in the third chapter by including certain cutting-off-sequences and derive necessary and sufficient conditions for the stability of the modified spline projection methods.

In the fifth chapter, we employ the results of Chapters 3 and 4 for approximation methods for singular integral operators on arbitrary Lyapunov curves (having, possibly, corners, intersections, end points). In this context, the approximation sequences considered in the previous chapters serve as local models for the sequences over curves.

Let us emphasize once more that all stability results obtained in Chapters II–V for singular integral operators and Mellin operators over the real axis or more complicated curves hold both for classic smoothest piecewise polynomial splines, for defect splines, and for compactly supported wavelets as well.

The concluding sixth chapter highlights some further applications of the Banach algebra approach to approximation methods for convolution operators.

Acknowledgements

We wish to express our sincere appreciation to our colleagues and students Dietmar Berthold, Victor Didenko, Torsten Ehrhardt, Tilo Finck and Wolfram Hoppe who read the bulk of the manuscript very carefully and eliminated not only a large number of mistakes and errors but helped with their constructive criticizm to essentially improve the book. Special thank is due to Dietmar Berthold who did the computations presented in Section 5.5.6.

Finally, we are pleased to express our gratitude to Professor Israel Gohberg for his encouragement and support in writing this book.

Chapter 1

Invertibility in Banach algebras

We start with summarizing some necessary auxiliary material from functional analysis and operator theory with special emphasis on local invertibility theories in Banach algebras. Then to give the reader a first idea of the usefulness of abstract local principles, we examine the so-called finite section method for Toeplitz operators. From this application we are finally going to extract a general scheme of using Banach algebra techniques which allows to tackle a lot of different invertibility problems (including the Fredholmness of Toeplitz and Wiener-Hopf operators and the stability of spline approximation methods for singular integral and Mellin operators) from a common point of view.

We renounce (with one exception) the proofs of the standard material in Sections 1.1 and 1.2 but we shall give precise references at the end of the chapter.

1.1 Banach algebras and C^*-algebras

1.1.1 Banach algebras

A *Banach algebra* is a complex Banach space $\mathcal{A}$ which is also an associative ring where the ring multiplication is connected with the Banach space norm by the inequality

$\|ab\| \leq \|a\| \, \|b\| \, , \quad a \, , b \in \mathcal{A} \, .$

The Banach algebra $\mathcal{A}$ is *commutative* if $ab = ba$ for all elements a and b of $\mathcal{A}$, and the algebra $\mathcal{A}$ is called *unital* if there is an identity element e in $\mathcal{A}$ such that $ae = ea = a$ for all $a \in \mathcal{A}$ and $\|e\| = 1$.

Let $\mathcal{A}$ and $\mathcal{B}$ be Banach algebras. A bounded linear operator $W : \mathcal{A} \to \mathcal{B}$ is said to be an *algebra homomorphism* if

$$W(ab) = W(a)\,W(b)\,, \quad a, b \in \mathcal{A}.$$

An *isomorphism* is an invertible homomorphism, and a homomorphism W is *unital* if $\mathcal{A}$ and $\mathcal{B}$ are unital Banach algebras and if W maps the identity element of $\mathcal{A}$ onto that of $\mathcal{B}$.

1.1.2 Invertibility

Let $\mathcal{A}$ be a unital Banach algebra. An element $a \in \mathcal{A}$ is *left* (*right, resp.two-sided*) *invertible* if there exists an element $b \in \mathcal{A}$ such that $ba = e$ ($ab = e$, resp. $ab = ba = e$). Two-sided invertible elements will be called invertible. Obviously, the two-sided inverse of an element $a \in \mathcal{A}$ is uniquely determined, and we denote it by a^{-1}. The collection of all invertible elements of a Banach algebra $\mathcal{A}$ forms an open subset of $\mathcal{A}$.

The set of all complex numbers λ for which $a - \lambda e$ is not invertible is called the *spectrum* of a and will be denoted by $\sigma_{\mathcal{A}}(a)$ or simply by $\sigma(a)$. The spectrum of an element is a non-empty and compact subset of the complex plane. If $W : \mathcal{A} \to \mathcal{B}$ is a unital homomorphism then, evidently,

$$\sigma_{\mathcal{B}}(W(a)) \subseteq \sigma_{\mathcal{A}}(a)\,, \quad a \in \mathcal{A}.$$

If, moreover, $\sigma_{\mathcal{B}}(W(a)) = \sigma_{\mathcal{A}}(a)$ for all $a \in \mathcal{A}$ then the homomorphism W is called a *symbol mapping* for the algebra $\mathcal{A}$, and $W(a)$ is referred to as the *symbol* of a. Evidently, an element a is invertible if and only if its symbol $W(a)$ is invertible. Notice, however, that a Banach algebra may possess several symbol mappings.

If $\mathcal{B}$ is a subalgebra of the unital Banach algebra $\mathcal{A}$ which contains the identity element then

$$\sigma_{\mathcal{A}}(b) \subseteq \sigma_{\mathcal{B}}(b)\,, \quad b \in \mathcal{B}\,.$$

The subalgebra $\mathcal{B}$ is said to be *inverse closed in* $\mathcal{A}$ if $\sigma_{\mathcal{A}}(b) = \sigma_{\mathcal{B}}(b)$ for all $b \in \mathcal{B}$ or, equivalently, if the embedding homomorphism of $\mathcal{B}$ into $\mathcal{A}$ is a symbol mapping for $\mathcal{B}$. The following result keeps some information about the difference between the spectra of $b \in \mathcal{B}$ in $\mathcal{B}$ and $\mathcal{A}$.

Theorem 1.1 *Let $\mathcal{B}$ be a subalgebra of $\mathcal{A}$ and $e \in \mathcal{B}$. If $b \in \mathcal{B}$ then $\sigma_{\mathcal{B}}(b)$ is the union of $\sigma_{\mathcal{A}}(b)$ and a (possibly empty) collection of bounded connected components of the complement of $\sigma_{\mathcal{A}}(b)$.*

Corollary 1.1 *If $\sigma_{\mathcal{B}}(b)$ has no interior points for all elements b of a dense subalgebra of $\mathcal{B}$ then $\mathcal{B}$ is inverse closed in $\mathcal{A}$.*

1.1.3 Ideals

A closed subalgebra $\mathcal{J}$ of the Banach algebra $\mathcal{A}$ is called a *left* (resp. *right*) *ideal* of $\mathcal{A}$ if $aj \in \mathcal{J}$ (resp. $ja \in \mathcal{J}$) for all $a \in \mathcal{A}$ and $j \in \mathcal{J}$. A *two-sided ideal* is a closed subalgebra which is both a left and a right ideal. An ideal $\mathcal{J}$ of $\mathcal{A}$ is *proper* if $\mathcal{J} \neq \mathcal{A}$.

A proper left (right, resp. two-sided) ideal is called a *maximal* left (right, resp. two-sided) ideal if it is not properly contained in any other proper left (right, resp. two-sided) ideal. In a Banach algebra with identity, every proper left (right, resp. two-sided) ideal is contained in some maximal left (right, resp. two-sided) ideal.

The *radical* of $\mathcal{A}$ is the intersection of all maximal left ideals of $\mathcal{A}$. The radical of a unital Banach algebra $\mathcal{A}$ is a two-sided ideal of $\mathcal{A}$, and it coincides with the intersection of all maximal right ideals. An element $k \in \mathcal{A}$ belongs to the radical of $\mathcal{A}$ if and only if $\sigma_{\mathcal{A}}(a) = \sigma_{\mathcal{A}}(a + k)$ for all a in $\mathcal{A}$. A unital Banach algebra is called *semi-simple* if its radical consists of the zero element only.

If $\mathcal{A}$ is a Banach algebra and $\mathcal{J}$ is a proper two-sided ideal then the quotient space $\mathcal{A}/\mathcal{J}$ becomes a Banach algebra on defining the norm by

$$\|a + \mathcal{J}\| := \inf_{j \in \mathcal{J}} \|a + j\|$$

and a multiplication by $(a + \mathcal{J})(b + \mathcal{J}) = ab + \mathcal{J}$. The homomorphism

$$\mathcal{A} \to \mathcal{A}/\mathcal{J} \quad , \quad a \mapsto a + \mathcal{J}$$

will be referred to as the *canonical homomorphism* from $\mathcal{A}$ onto $\mathcal{A}/\mathcal{J}$. If $\mathcal{A}$ is unital then this homomorphism is unital, and its norm is equal to one. In case $\mathcal{J}$ is an ideal in the radical of $\mathcal{A}$, the canonical homomorphism is even a symbol mapping for $\mathcal{A}$.

Let $\mathcal{A}$ and $\mathcal{B}$ be Banach algebras, $W : \mathcal{A} \to \mathcal{B}$ be a homomorphism, and let $\mathcal{J}$ be a two-sided ideal of $\mathcal{A}$ which belongs to the kernel of W. Then $W(a) = W(a + j)$ for all $a \in \mathcal{A}$ and $j \in \mathcal{J}$, and, thus, $W(a)$ depends on the coset $a + \mathcal{J}$ only. So, the mapping

$$\mathcal{A}/\mathcal{J} \to \mathcal{B} \quad , \quad a + \mathcal{J} \mapsto W(a)$$

is correctly defined, and it is called the *quotient homomorphism* of W by $\mathcal{J}$. The quotient homomorphism is continuous (unital, resp. a symbol mapping) whenever W is continuous (unital, resp. a symbol mapping).

1.1.4 C*-algebras

A mapping $a \mapsto a^*$ of a Banach algebra $\mathcal{A}$ onto itself is called an involution if $(a^*)^* = a$, $(a + b)^* = a^* + b^*$, $(ab)^* = b^* a^*$, $(\lambda a)^* = \bar{\lambda} a^*$, and $\|a\| = \|a^*\|$ for all $a, b \in \mathcal{A}$ and $\lambda \in \mathbf{C}$. A *C*-algebra* is a Banach algebra with involution that satisfies $\|aa^*\| = \|a\|^2$ for all of its elements a.

Here are some basic properties of C^*-algebras:

(a) If $\mathcal{A}$ is a C^*-algebra and $\mathcal{J}$ is a two-sided ideal of $\mathcal{A}$ then $\mathcal{J}$ is self-adjoint (i.e. $\mathcal{J}^* = \mathcal{J}$), and $\mathcal{A}/\mathcal{J}$ provided with the involution $(a + \mathcal{J})^* = a^* + \mathcal{J}$ is a C^*-algebra.

(b) If $\mathcal{A}$ is a C^*-algebra, $\mathcal{B}$ is a C^*-subalgebra of $\mathcal{A}$, and $\mathcal{J}$ is a two-sided ideal of $\mathcal{A}$ then $\mathcal{B}+\mathcal{J}$ is a C^*-subalgebra of $\mathcal{A}$, and the C^*-algebras $(\mathcal{B}+\mathcal{J})/\mathcal{J}$ and $\mathcal{B}/(\mathcal{B} \cap \mathcal{J})$ are isometrically isomorphic.

(c) If $\mathcal{A}$ is a unital C^*-algebra and $\mathcal{B}$ is a C^*-subalgebra of $\mathcal{A}$ containing the identity element then $\mathcal{B}$ is inverse closed in $\mathcal{A}$.

(d) C^*-algebras are semi-simple.

1.2   Linear operators

1.2.1   Bounded and compact operators

Given Banach spaces X and Y we denote by $L(X,Y)$ the linear space of all bounded and linear operators from X to Y and by $K(X,Y)$ the class of all compact operators from X into Y. In case $Y = X$ we simply write $L(X)$ and $K(X)$ in place of $L(X, X)$ and $K(X, X)$, respectively. On defining a norm on $L(X,Y)$ by

$$\|A\| = \sup_{\|x\|_X \leq 1} \|Ax\|_Y \,,$$

$L(X,Y)$ becomes a Banach space and $K(X,Y)$ a closed subspace of $L(X,Y)$. Further, $L(X)$ is a unital Banach algebra, and $K(X)$ is a two-sided ideal of $L(X)$. In case X is a Hilbert space, $L(X)$ proves even to be a C^*-algebra. The quotient algebra $L(X)/K(X)$ will be referred to as the *Calkin algebra* of the Banach space X, and the canonical homomorphism from $L(X)$ onto $L(X)/K(X)$ will be abbreviated by π throughout this book.

A sequence (A_n) of operators $A_n \in L(X,Y)$ is said to *converge* to an operator $A \in L(X,Y)$

- *weakly*, if $f(A_n x) \to f(Ax)$ for each $x \in X$ and each functional $f \in Y^*$ (= the dual space of Y);

- *strongly*, if $\|A_n x - Ax\|_Y \to 0$ for each $x \in X$;

- *uniformly*, if $\|A_n - A\| \to 0$.

Here are some properties of convergent operator sequences:

(a) Let X, Y, Z be Banach spaces. If $A_n \in L(X,Y), K \in K(Y, Z)$, and $A_n \to A$ weakly, then $KA_n \to KA$ strongly.

(b) Let X, Y, Z, W be Banach spaces. If $A_n \in L(X,Y), K \in K(Y,Z), B_n \in L(Z,W)$ and $A_n^* \to A^*$ strongly, $B_n \to B$ strongly then $B_n K A_n \to BKA$ uniformly (as usual, the star marks the adjoint operators).

(c) If $A \in L(X,Y)$, if $\|A_n x - Ax\|_Y \to 0$ for all x belonging to a dense subset of X, and if $\sup_n \|A_n\| < \infty$ then $A_n \to A$ strongly.

(d) (Banach-Steinhaus theorem) If $(A_n x)$ is a convergent sequence in Y for each $x \in X$ then $\sup_n \|A_n\| < \infty$, the operator A defined by $Ax = \lim_{n \to \infty} A_n x$ is bounded, and $\|A\| \leq \liminf_{n \to \infty} \|A_n\|$.

Proof of (a) and (b). Clearly, $K_n A \to KA$ weakly. Suppose that the sequence $(K_n A)$ is not strongly converging to KA. Then there are an $x \in X$, a number $\epsilon > 0$, and a sequence (n_k) of positive integers such that

$$\|KA_{n_k} x - KAx\| \geq \epsilon . \tag{1}$$

Since $(A_n x)$ converges weakly to Ax by hypothesis, the norms $\|A_n x\|$ are uniformly bounded (see [Yo], Chap. V, Theorem 3, or [ReS], Vol. I, Chap. IV.5, Prop. (a)). Thus, the sequence $(KA_{n_k} x - KAx)_{k \geq 1}$ is bounded with respect to the norm in Y, and the compactness of K implies that there is a subsequence $(KA_{n_{k_l}} x - KAx)_{l \geq 1}$ which converges in the norm. Because of (1), the limit of this sequence is not 0, hence, its weak limit is also different from 0. But this is a contradiction.

For a proof of (b) we restrict ourselves to the case when A_n is the identity operator. The general case can be reduced to the considered one in an obvious manner. Set $S := \{y \in Y : \|y\| \leq 1\}$. Since K is compact, the image of S under K is a relatively compact subset of Z. Thus, given $\epsilon > 0$, there are elements $y_1, \ldots, y_k \in S$ such that $Ky_1, \ldots, Ky_k \in S$ form a finite ϵ-net in $K(S)$. Now let y be an arbitrary element in S and choose an y_j which satisfies $\|Ky - Ky_j\| < \epsilon$. Then

$$\begin{aligned}
\|(B_n K - BK)y\| &\leq \|(B_n K - BK)(y - y_j)\| + \|(B_n K - BK)y_j\| \\
&\leq \|B_n - B\|\epsilon + \max_{1 \leq j \leq k} \|(B_n K - BK)y_j\| .
\end{aligned}$$

The uniform boundedness principle entails that $\sup_n \|B_n - B\| < \infty$, hence the first item on the right-hand side is less than a constant times ϵ. The second item goes to zero as $n \to \infty$ due to the strong convergence of $B_n K$ to BK. ∎

1.2.2 Fredholm operators

Let X and Y be Banach spaces. To each operator $A \in L(X,Y)$ one associates two linear subspaces of X and Y, the *kernel* and the *image* of A, by

$$\begin{aligned}
\text{Ker } A &= \{x \in X : Ax = 0\} \quad \text{and} \\
\text{Im } A &= \{y \in Y : \text{there exists an } x \in X \text{ with } y = Ax\} .
\end{aligned}$$

The operator A is said to be *normally solvable* if Im A is closed in Y. A normally solvable operator $A \in L(X,Y)$ is called a $\Phi_+ - operator$ if dim Ker $A < \infty$, and A is a $\Phi_- - operator$ if dim Coker $A := \dim(Y/\text{Im } A) < \infty$. If A is both a Φ_+- and a Φ_-−operator then it is called a *Fredholm operator* (or Φ-operator or Noetherian operator), and the integer

ind $A := \dim \text{Ker } A - \dim \text{Coker } A$

is referred to as the *index* of A. The collection of all Fredholm operators will be denoted by $\Phi(X,Y)$, and $\Phi(X,X)$ will be abbreviated to $\Phi(X)$.

Some basic properties of Fredholm operators are:

(a) (Small perturbations) $\Phi(X,Y)$ is an open subset of $L(X,Y)$ and the mapping ind : $\Phi(X,Y) \to \mathbf{Z}$ is constant on each of the connected components of $\Phi(X,Y)$. Moreover, if $A \in \Phi(X,Y)$ and if $C \in L(X,Y)$ has a sufficiently small norm then

$$\dim \text{Ker } (A + C) \le \dim \text{Ker } A , \quad \dim \text{Coker } (A + C) \le \dim \text{Coker } A.$$

(b) (Compact perturbations) If $A \in \Phi(X,Y)$ and $K \in K(X,Y)$ then $A + K \in \Phi(X,Y)$ and ind $(A + K) = $ ind A.

(c) (Atkinson's theorem) Let X, Y, Z be Banach spaces and $A \in \Phi(X,Y), B \in \Phi(Y,Z)$. Then $BA \in \Phi(X,Z)$ and ind $BA = $ ind $B + $ ind A.

(d) (Adjoint operators) If $A \in \Phi(X,Y)$ then $A^* \in \Phi(Y^*, X^*)$ and

$$\dim \text{Ker } A^* = \dim \text{Coker } A , \ \dim \text{Coker } A^* = \dim \text{Ker } A$$

whence ind $A^* = -$ind A.

(e) (Equivalent definitions) The following are equivalent:

(i) $A \in \Phi(X,Y)$.

(ii) There exist operators $R, L \in L(Y, X)$ such that the operators $AR - I$ and $LA - I$ are compact.

(iii) There exists an operator $B \in L(Y, X)$ such that the operators $AB - I$ and $BA - I$ have a finite dimensional image.

If $X = Y$ then $A \in \Phi(X)$ if and only if the coset $\pi(A)$ is invertible in the Calkin algebra of X.

An operator $A \in L(X,Y)$ is referred to as an *operator of regular type* if there exists a positive constant C such that

$$\|Ax\| \ge C \|x\| \quad \text{for} \quad x \in X .$$

An operator is of regular type if and only if it is normally solvable and if its kernel consits of the zero element only. If both A and A^* are of regular type then A is invertible.

1.2.3 Projection operators

Let X be a Banach space. A bounded linear operator P on X is called a projection operator if $P^2 = P$. If P is a projection operator then the operator $I - P$ is also a projection on X. Thereby,

$$\text{Ker}\, P = \text{Im}\, (I - P) \quad \text{and} \quad \text{Im}\, P = \text{Ker}\, (I - P)\,,$$

whence in particular follows that the kernel and the image of a projection are closed subspaces of X. Moreover, the Banach space X is the direct sum of these subspaces. Conversely, let the Banach space X be the direct sum $X_1 \dotplus X_2$ of its closed subspaces, that is, $X_1 \cap X_2 = \{0\}$, $X_1 + X_2 = X$ algebraically, and there is a constant C such that $\|x_1\| \leq C\|x\|$ and $\|x_2\| \leq C\|x\|$ whenever $x_1 \in X_1$, $x_2 \in X_2$, and $x = x_1 + x_2$. Then there is a projection P on X with $X_1 = \text{Im}\, P$ and $X_2 = \text{Ker}\, P$.

1.3 Stability of operator sequences

1.3.1 Approximate solution of operator equations

Let X and Y be Banach spaces and $A \in L(X, Y)$. A broad variety of methods for the approximate solution of the operator equation

$$Ax = y\,, \quad y \in Y\,, \tag{1}$$

is comprehended by the following scheme: One chooses two sequences $(P_n) \subseteq L(X)$ and $(R_n) \subseteq L(Y)$ of projection operators such that $P_n \to I \in L(X)$ and $R_n \to I \in L(Y)$ strongly as $n \to \infty$, as well as a sequence (A_n) of bounded operators $A_n : \text{Im}\, P_n \to \text{Im}\, R_n$, and considers instead of (1) the sequence of equations

$$A_n x_n = R_n y\,, \quad y \in Y\,, \tag{2}$$

with the x_n being sought in $\text{Im}\, P_n$.

Definition The approximation method (2) *applies* to the operator A if

(i) there is an n_0 such that the equation (2) has a unique solution $x_n \in \text{Im}\, P_n$ for all $y \in Y$ and for all $n \geq n_0$,

(ii) the sequence $(x_n)_{n \geq n_0}$ of these solutions converges in the norm of X to a solution of the equation (1).

Before stating a simple criterion for the applicability of an approximation method we introduce the notion of a *stable* operator sequence.

Definition The sequence (A_n) of operators $A_n : \text{Im}\, P_n \to \text{Im}\, R_n$ is *stable* if there exists an n_0 such that all operators A_n with $n \geq n_0$ are invertible and if

$$\sup_{n \geq n_0} \|A_n^{-1} R_n\| < \infty\,.$$

The following proposition shows that stability investigations form the corner stone of any numerical analysis.

Proposition 1.1 *Suppose $A_n P_n \to A$ strongly. Then the method (2) applies to the operator A if and only if the operator A is invertible and the sequence (A_n) is stable.*

Proof   Let (2) be applicable to A. Then, by definition, the operators (A_n) are invertible for large n, say for $n \geq n_0$, and the sequence $(A_n^{-1} R_n y)_{n \geq n_0}$ converges for each $y \in Y$. The Banach-Steinhaus theorem involves the stability of the sequence (A_n). For the invertibility of A we consider the estimate

$$\|P_n x - A_n^{-1} R_n A x\| \leq \|A_n^{-1} R_n\| \ \ \|A_n P_n x - Ax\| .$$

Since $A_n P_n \to A$ strongly and since the sequence (A_n) is stable this estimate shows that $\|P_n x - A_n^{-1} R_n A x\| \to 0$ as $n \to \infty$ for each $x \in X$. In particular, if $x \in \text{Ker } A$ then $\|P_n x\| \to 0$ whence $x = 0$. Thus, the kernel of A is trivial, and since the image of A coincides with all of Y by definition, the operator A is invertible.

Now let the operator A be invertible and the sequence (A_n) be stable. Then the operators A_n are invertible for large n, and so it suffices to verify the strong convergence of the sequence $(A_n^{-1} R_n)_{n \geq n_0}$ to A^{-1}. But this is a consequence of the estimate

$$
\begin{aligned}
\|A_n^{-1} R_n y - A^{-1} y\| &\leq \ \|A_n^{-1} R_n A x - P_n x\| + \|P_n x - x\| \\
&\leq \ \|A_n^{-1} R_n\| \, \|A x - A_n P_n x\| + \|P_n x - x\|
\end{aligned}
$$

in combination with $P_n \to I$ and $A_n P_n \to A$ strongly. ∎

1.3.2   Algebraization

Our next goal is an equivalent formulation of the stability of an operator sequence in a Banach algebra language. For, we suppose that $X = Y$ and $\text{Im } P_n = \text{Im } R_n$. Let $\mathcal{E}$ denote the collection of all bounded sequences (A_n) of operators $A_n : \text{Im } P_n \to \text{Im } P_n$. On defining operations on $\mathcal{E}$ by $(A_n) + (B_n) = (A_n + B_n)$ and $(A_n)(B_n) = (A_n B_n)$, and a norm by $\|(A_n)\| = \sup_n \|A_n P_n\|$, the set $\mathcal{E}$ becomes a normed algebra and, as one can easily check, even a Banach algebra. Further it is easy to see that the subset $\mathcal{G}$ of $\mathcal{E}$ consisting of all sequences (A_n) with $\|A_n P_n\| \to 0$ as $n \to \infty$ forms a closed two-sided ideal of $\mathcal{E}$.

Proposition 1.2 *Let $(A_n) \in \mathcal{E}$. The sequence (A_n) is stable if and only if the coset $(A_n) + \mathcal{G}$ is invertible in the quotient algebra $\mathcal{E}/\mathcal{G}$.*

Proof   Let (A_n) be stable, i.e. A_n is invertible for $n \geq n_0$. Put $B_n := P_n$ for $n < n_0$ and $B_n := A_n^{-1}$ for $n \geq n_0$. Then the sequence (B_n) is in $\mathcal{E}$ and $(A_n)(B_n) - (P_n) \in \mathcal{G}$ as well as $(B_n)(A_n) - (P_n) \in \mathcal{G}$. Thus, $(A_n) + \mathcal{G}$ is invertible in $\mathcal{E}/\mathcal{G}$.

Conversely, let $(A_n) + \mathcal{G}$ be invertible in $\mathcal{E}/\mathcal{G}$, and let $(B_n) + \mathcal{G}$ be the inverse of $(A_n) + \mathcal{G}$. Then there are sequences (G_n) and (H_n) in $\mathcal{G}$ such that $A_n B_n = P_n + G_n$ and $B_n A_n = P_n + H_n$. Let n_0 be an integer such that $\|G_n\| < 1/2$ and $\|H_n\| < 1/2$ for all $n \geq n_0$. Then the operators $P_n + G_n$ and $P_n + H_n$ are invertible whenever $n \geq n_0$, and the norms of their inverses are uniformly bounded by $1/(1 - 1/2) = 2$. This implies that the operators A_n are two-sided invertible if $n \geq n_0$ and that $\sup_{n \geq n_0} \|A_n^{-1} P_n\| \leq 2 \sup_{n \geq n_0} \|B_n P_n\| < \infty$. Thus, (A_n) is a stable sequence. ∎

1.3.3 Essentialization

It is a plain truth that it is much easier to study Fredholmness of an operator than to study its invertibility. Similarly, it is often a hard thing to attack the invertibility of a coset in the quotient algebra $\mathcal{E}/\mathcal{G}$ directly, but the problem becomes manageable if the algebra will be factorized by a larger ideal than the ideal $\mathcal{G}$. Next we present a simple version of a mechanism to produce large ideals and to reduce the invertibility problem in $\mathcal{E}/\mathcal{G}$ to its essential part. An extension of these ideas will be given in Section 1.6.2.

Let $\mathcal{F}$ stand for the subset of the algebra $\mathcal{E}$ consisting of all sequences $(A_n P_n)$ and $(A_n^* P_n^*)$ converging strongly as $n \to \infty$. This implies that $\operatorname*{s-lim}_{n \to \infty} A_n^* P_n^* = (\operatorname*{s-lim}_{n \to \infty} A_n P_n)^*$. Further we write $\mathcal{J}$ for the set of all sequences $(P_n T P_n + G_n)$ with T running through the compact operators and $\|G_n P_n\| \to 0$ as $n \to \infty$. Clearly, $\mathcal{F}$ is a closed subalgebra of $\mathcal{E}$, and $\mathcal{G}$ is a two-sided ideal in $\mathcal{F}$. Finally, we suppose that $P_n^* \to I^* \in L(X^*)$ strongly as $n \to \infty$.

Proposition 1.3 *The set $\mathcal{J}$ is a closed two-sided ideal in the algebra $\mathcal{F}$.*

Proof The assumptions guarantee that the sequences $(P_n T P_n + G_n)$ and $((P_n T P_n + G_n)^*) = (P_n^* T^* P_n^* + G_n^*)$ are strongly convergent to T and T^*, respectively. Thus, $\mathcal{J} \subseteq \mathcal{F}$. Let $(A_n), (B_n) \in \mathcal{F}$ and $A_n P_n \to A$, $B_n P_n \to B$ strongly. Then $(A_n)(P_n T P_n + G_n)(B_n) = (P_n A T B P_n + G_n')$ with

$$G_n' = A_n G_n B_n + (A_n P_n - P_n A) T P_n B_n + P_n A T (P_n B_n - B P_n) \,.$$

Taking into account that $A_n P_n - P_n A \to 0$ and $(P_n B_n P_n - B P_n)^* \to 0$ strongly the property (b) in 1.2.1 yields $\|G_n' P_n\| \to 0$, and, since the operator ATB is compact we conclude that $(A_n)(P_n T P_n + G_n)(B_n) \in \mathcal{J}$. ∎

Theorem 1.2 *Let $(A_n) \in \mathcal{F}$ be a sequence with strong limit A. Then the following assertions are equivalent:*

(i) *The method $A_n x_n = P_n y$ applies to the operator A.*

(ii) *The coset $(A_n) + \mathcal{G}$ is invertible in the quotient algebra $\mathcal{F}/\mathcal{G}$.*

(iii) *The operator A is invertible, and the coset $(A_n) + \mathcal{J}$ is invertible in the quotient algebra $\mathcal{F}/\mathcal{J}$.*

Proof   (i) $\Rightarrow$ (ii): By Propositions 1.1 and 1.2, the coset $(A_n) + \mathcal{G}$ is invertible in the algebra $\mathcal{E}/\mathcal{G}$. It remains to show that the sequence (B_n) with $B_n = P_n$ if $n < n_0$ and $B_n = A_n^{-1}$ if $n \geq n_0$ as well as the conjugate sequence $(B_n)^*$ converge strongly. According to Proposition 1.1 the operator A is invertible. We claim that $B_n P_n \to A^{-1}$ strongly. Indeed, for $n \geq n_0$,

$$
\begin{aligned}
\|B_n P_n y - A^{-1} y\| &= \|A_n^{-1} P_n y - A^{-1} y\| \\
&\leq \|A_n^{-1} P_n y - P_n A^{-1} y\| + \|P_n A^{-1} y - A^{-1} y\| \\
&\leq \|A_n^{-1} P_n\| \, \|y - A_n P_n A^{-1} y\| + \|P_n A^{-1} y - A^{-1} y\| \,,
\end{aligned}
$$

and since $\sup \|A_n^{-1} P_n\| < \infty$, $P_n \to I$, and $A_n P_n \to A$, the desired convergence follows. For the adjoint sequence, the proof is analogous.

(ii) $\Rightarrow$ (iii): Since $\mathcal{G} \subseteq \mathcal{J}$, the invertibility of the coset $(A_n) + \mathcal{G}$ in $\mathcal{F}/\mathcal{G}$ evidently implies the invertibility of the coset $(A_n) + \mathcal{J}$ in $\mathcal{F}/\mathcal{J}$. Further, let $(B_n) + \mathcal{G}$ be the inverse of $(A_n) + \mathcal{G}$. Then there are sequences (G_n) and (H_n) in $\mathcal{G}$ such that $A_n B_n = P_n + G_n$ and $B_n A_n = P_n + H_n$. Passage to the limit $n \to \infty$ yields $AB = I$ and $BA = I$ with B referring to the strong limit of (B_n), whence the invertibility of A follows.

(iii) $\Rightarrow$ (i): since $(A_n) + \mathcal{J}$ is invertible there is a sequence $(B_n) \in \mathcal{F}$, a sequence $(G_n) \in \mathcal{G}$, and a compact operator T such that $A_n B_n = P_n + P_n T P_n + G_n$. Put $B_n' = B_n - P_n A^{-1} T P_n$. Then $(B_n) - (B_n') \in \mathcal{J}$ and $A_n B_n' = P_n + G_n'$ with

$$
G_n' = G_n + P_n T P_n - A_n P_n A^{-1} T P_n = G_n + (P_n - A_n P_n A^{-1}) T P_n \,.
$$

Since $P_n - A_n P_n A^{-1} \to 0$ strongly we conclude that $(G_n') \in \mathcal{G}$. Thus, the coset $(A_n) + \mathcal{G}$ is right invertible in $\mathcal{F}/\mathcal{G}$, and analogously one verifies its left invertibility. Now Proposition 1.2 ensures the stability of the sequence (A_n). This fact in connection with the invertibility of A yields the assertion via Proposition 1.1. ∎

1.3.4   Perturbations

As a consequence of Theorem 1.2 we derive the following corollary characterizing the influence of small and of compact perturbations to the applicability of approximation methods.

Corollary 1.2 *Let $(A_n) \in \mathcal{F}$ and suppose the method $A_n x_n = P_n y$ to apply to the operator A.*

(i) *If $(A_n') \in \mathcal{F}$, $A_n' P_n \to A$ strongly, and if $\limsup_{n \to \infty} \|A_n P_n - A_n' P_n\|$ is sufficiently small then the method $A_n' x_n = P_n y$ applies to A , too.*

(ii) *If K is a compact operator and $A + K$ is invertible then the method $(A_n + P_n K P_n) x_n = P_n x$ applies to $A + K$.*

Proof   (i) By Theorem 1.2 we have to verify the invertibility of the coset $(A_n') + \mathcal{G}$ in $\mathcal{F}/\mathcal{G}$. Since the set of the invertible elements in a Banach algebra is open, and since $(A_n) + \mathcal{G}$ is invertible, it suffices to check that the norm of $(A_n - A'_n) + \mathcal{G}$ is small enough. But this is just the assumption.

(ii) Following Theorem 1.2 we have to prove the invertibility of $(A_n + P_n K P_n) + \mathcal{J}$ in $\mathcal{F}/\mathcal{J}$. Since $(A_n + P_n K P_n) - (A_n) \in \mathcal{J}$ this invertibility is a consequence of that of $(A_n) + \mathcal{J}$. ∎

1.3.5 Invertibility of bounded operators

Up to now we have dealt with four notions of invertibility of a linear bounded operator A acting on a Banach space X:

- the *usual* invertibility of A ,

- the *essential* invertibility of A , viz. the invertibility of $\pi(A)$ in the Calkin algebra or, equivalently, the Fredholmness of A,

- the *approximate* invertibility of A by a sequence $(A_n) \in \mathcal{F}$ which means the applicability of the approximation method $A_n x_n = P_n y$ to the operator A , and

- the *essential approximate* invertibility of A by $(A_n) \in \mathcal{F}$ which signifies the invertibility of the coset $(A_n) + \mathcal{J}$ in the algebra $\mathcal{F}/\mathcal{J}$ (compare Theorem 1.2).

Thereby, the following implications are valid.

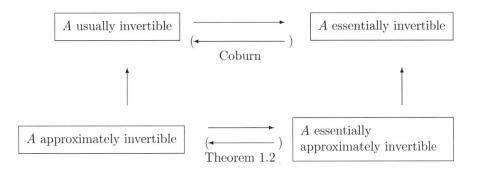

The verification of the second vertical implication is left as an exercise. The other implications either have been already shown, or they are trivial.

Sometimes, and under additional conditions, the horizontal implications can be reversed. Let, for instance, A be a Toeplitz operator or a one-dimensional singular integral operator. Then a theorem of Coburn (see 2.2) states that A is invertible if A is essentially invertible and ind $A = 0$. Similarly, in view of Theorem 1.2, an operator A is approximately invertible if A is essentially approximately invertible and if A is invertible in the usual sense.

Remember that the four kinds of invertibility are equivalent to invertibility problems in accordingly constructed Banach algebras (namely in the algebras $L(X)$, $L(X)/K(X), \mathcal{F}/\mathcal{G}$, and $\mathcal{F}/\mathcal{J}$, respectively). So we have to look for generally applicable invertibility theories in Banach algebras. In case the considered algebras are commutative, a candidate for such a theory is the well-known Gelfand theory which will be shortly explained in the following section.

During the last 25 years several generalizations of Gelfand's theory to non-commutative Banach algebras have been created which apply to algebras being, in a certain sense, sufficiently close to the category of commutative Banach algebras. Our main favourite for a non-commutative Gelfand theory is *Allan's local principle* which works in Banach algebras having a large center (see 1.4.4). Due to its elegance and its adaptability to the invertibility problems we shall be concerned with, this principle will be used throughout this book.

When trying to apply Allan's local principle to the algebras $L(X), L(X)/K(X)$, $\mathcal{F}/\mathcal{G}$, $\mathcal{F}/\mathcal{J}$, or to some of their interesting subalgebras, it soon turns out that the algebras $L(X)$ and $\mathcal{F}/\mathcal{G}$ (corresponding to the 'direct' invertibilities) are too far away from the commutativity, whereas we shall construct subalgebras of the Calkin algebra as well as of the algebra $\mathcal{F}/\mathcal{J}$ (corresponding to the 'essential' invertibilities) which, on the one hand, contain all objects we are interested in but, on the other hand, whose centers are large enough for a successful analysis by means of Allan's principle. This suggests the following philosophy for solving usual or approximate invertibility problems: First solve the corresponding *essential* invertibility problem by a local principle and then try to lift the results taking advantage of Coburn's theorem (see 2.2), Theorem 1.2, or of related theorems.

Remark Notice that the definition of the essential approximate invertibility is less unique than that of the essential invertibility. This is due to the fact that, if X is a separable infinite-dimensional Hilbert space, the ideal $K(X)$ is the only non-trivial closed two-sided ideal in $L(X)$ and, thus, the Calkin algebra is the only relevant quotient algebra of $L(X)$. On the other hand, the algebra $\mathcal{F}$ contains a variety of closed two-sided ideals between $\mathcal{J}$ and $\mathcal{F}$, and many of them can be lifted by a Theorem 1.2-type theorem. So, to each of these ideals there corresponds another 'essential approximate invertibility'. Indeed, soon we shall meet situations where the ideal $\mathcal{J} = \{P_n T P_n + G_n\}$ introduced in Section 1.3.3 proves to be too small.

1.4   Local principles

1.4.1   The maximal ideal space

Let $\mathcal{B}$ be a commutative unital and complex Banach algebra. If x is a maximal ideal of $\mathcal{B}$ then the quotient algebra $\mathcal{B}/x$ proves to be isomorphic to the complex field $\mathbf{C}$. Thus, to each maximal ideal x of $\mathcal{B}$ and to each $b \in \mathcal{B}$, a complex number, $\Phi_x(b)$, is associated which is the image of the coset $b+x$ under the above mentioned isomorphism. The mapping $\Phi_x \,:\, b \mapsto \Phi_x(b)$ is a multiplicative linear functional on $\mathcal{B}$ and, conversely, one can prove that each non-trivial multiplicative linear functional on $\mathcal{B}$ is of this form. So there is a one-to-one correspondence between the non-trivial multiplicative linear functionals and the maximal ideals of $\mathcal{B}$: the kernel of every non-trivial multiplicative linear functional is a maximal ideal, and every maximal ideal is the kernel of a uniquely determined non-trivial multiplicative linear functional.

Let $M(\mathcal{B})$ refer to the set of all maximal ideals of the Banach algebra $\mathcal{B}$. Given an element $b \in \mathcal{B}$ we introduce the complex valued function $\Phi(b)$ on $M(\mathcal{B})$ by

$$\Phi(b) \,:\, M(\mathcal{B}) \rightarrow \mathbf{C}\,,\; x \mapsto \Phi_x(b)\,,$$

and call this function the *Gelfand transform* of $b \in \mathcal{B}$. The set $M(\mathcal{B})$ can be made into a topological space in a natural way: The Gelfand topology on $M(\mathcal{B})$ is the coarsest topology on $M(\mathcal{B})$ that makes all Gelfand transforms $\Phi(b)\,,\, b \in \mathcal{B}$, continuous. The set $M(\mathcal{B})$ provided with the Gelfand topology will be referred to as the *maximal ideal space* of the Banach algebra $\mathcal{B}$. Finally, the mapping

$$\Phi \,:\, \mathcal{B} \rightarrow C(M(\mathcal{B}))\,,\, b \mapsto \Phi(b)$$

is called the *Gelfand transform* . Herein, $C(M)$ stands for the algebra of all continuous functions on the topological space M. If M is a Hausdorff compact we endow $C(M)$ with the supremum norm to make $C(M)$ into a Banach algebra.

1.4.2 Classic Gelfand theory

Theorem 1.3 *(i) The element $b \in \mathcal{B}$ is invertible if and only if $\Phi_x(b) \neq 0$ for all $x \in M(\mathcal{B})$.*

(ii) The maximal ideal space $M(\mathcal{B})$ is a Hausdorff compact, and the set $\Phi(\mathcal{B})$ separates the points of $M(\mathcal{B})$.

(iii) The Gelfand transform Φ is a continuous algebra homomorphism from $\mathcal{B}$ into the Banach algebra $C(M(\mathcal{B}))$, and the norm of Φ is 1.

(iv) The kernel of Φ is the radical of $\mathcal{B}$.

The assertions (i) and (iii) are equivalent to the following: *The Gelfand transform $\Phi : \mathcal{B} \rightarrow C(M(\mathcal{B}))$ is a symbol mapping for $\mathcal{B}$.* Further, (iii) reveals that, for $b \in \mathcal{B}$,

$$\|\Phi(b)\| \,=\, \sup\{|\Phi_x(b)|\,,\; x \in M(\mathcal{B})\} \leq \|b\|\,,$$

but notice that, generally, neither equality holds in this estimate nor the mapping Φ is injective or surjective. Only in the C^*-case we have the following famous result.

Theorem 1.4 *(Gelfand-Naimark) Let $\mathcal{B}$ be a commutative unital C^*-algebra. Then the Gelfand transform is an isometrical isomorphism from $\mathcal{B}$ onto the algebra $C(M(\mathcal{B}))$.*

This theorem justifies to think of elements of commutative C^*-algebras as continuous functions on a Hausdorff compact, and we shall do this henceforth without further comment.

1.4.3 Central subalgebras

Allan's local principle is a generalization of the (commutative) Gelfand theory to Banach algebras which are close to the commutative ones in the sense that their centers are non-trivial. Remember that the *center* of an algebra $\mathcal{A}$ consists of all elements $a \in \mathcal{A}$ with the property that $ab = ba$ for all $b \in \mathcal{A}$. Clearly, the center of a unital Banach algebra $\mathcal{A}$ is a closed, unital, commutative, and inverse closed subalgebra of $\mathcal{A}$.

Let $\mathcal{A}$ be a unital Banach algebra. By a *central subalgebra* $\mathcal{B}$ of $\mathcal{A}$ we mean a closed subalgebra of the center of $\mathcal{A}$ which contains the identity element. Obviously, $\mathcal{B}$ is a commutative Banach algebra, and we denote its maximal ideal space by $M(\mathcal{B})$ again. To each maximal ideal x of $\mathcal{B}$ we associate the smallest closed two-sided ideal I_x of $\mathcal{A}$ which contains x, and we let Φ_x refer to the canonical homomorphism from $\mathcal{A}$ onto the quotient algebra $\mathcal{A}/I_x$. Notice that, in contrast to the commutative case, the quotient algebras $\mathcal{A}/I_x$ can differ from each other in dependence of $x \in M(\mathcal{B})$. In particular, it may happen that $I_x = \mathcal{A}$ for some points x. In this case we *define* that $\Phi_x(a)$ is invertible in $\mathcal{A}/I_x$ and that $\|\Phi_x(a)\| = 0$ for each $a \in \mathcal{A}$.

1.4.4 Allan's local principle

Theorem 1.5 *Let $\mathcal{B}$ be a central subalgebra of $\mathcal{A}$. Then*

    (i) an element $a \in \mathcal{A}$ is invertible if and only if the cosets $\Phi_x(a)$ are invertible in $\mathcal{A}/I_x$ for every $x \in M(\mathcal{B})$,

    (ii) the mapping $M(\mathcal{B}) \to \mathbf{R}^+$, $x \mapsto \|\Phi_x(a)\|$ is upper semi-continuous for each $a \in \mathcal{A}$.

The proof of this theorem is based on the following observation.

Proposition 1.4 *Let $\mathcal{B}$ be a central subalgebra of $\mathcal{A}$. If L is a maximal left, right, or two-sided ideal of $\mathcal{A}$ then $L \cap \mathcal{B}$ is a (two-sided) maximal ideal of $\mathcal{B}$.*

Proof Suppose that L is a maximal left ideal. It is clear that $L \cap \mathcal{B}$ is a proper ($e \notin L$) closed two-sided ideal of $\mathcal{B}$ and so we are left with the proof of its maximality. Let $z \in \mathcal{B} \setminus L$. Then $I_z := \{l + az : l \in L, a \in \mathcal{A}\}$ is a left ideal of $\mathcal{A}$ containing L properly ($z \notin L$). The maximality of L implies that $I_z = \mathcal{A}$, hence $e \in I_z$, and it follows that z has an inverse modulo L (note that $z \in \text{Cen}\,\mathcal{A}$). Furthermore, $K_z := \{a \in \mathcal{A} : az \in L\}$ is a proper ($e \notin K_z$) left ideal of $\mathcal{A}$ containing L. Since L is maximal, we have $K_z = L$. In particular, if y_1, y_2 are both inverses modulo L of

z, then $y_1 - y_2 \in L$. Thus, the inverses modulo L of z determine a unique element of the quotient space $\mathcal{A}/L$.

Suppose $z - \lambda e \notin L$ for all $\lambda \in \mathbf{C}$. Let $y^\pi(\lambda)$ denote the (uniquely determined) coset of $\mathcal{A}/L$ containing the inverses modulo L of $z - \lambda e$. We claim that $y^\pi : \mathbf{C} \to \mathcal{A}/L$ is an analytic function. To see this, let $\lambda_0 \in \mathbf{C}$ and let $y_0 \in y^\pi(\lambda_0)$ be an inverse modulo L of $z - \lambda_0 e$. Then, for $|\lambda - \lambda_0| < 1/\|y_0\|$, the element $e - (\lambda - \lambda_0)y_0$ is invertible in $\mathcal{A}$ and it is readily verified that $y_0[e - (\lambda - \lambda_0)y_0]^{-1}$ is an inverse modulo L of $z - \lambda e$. Thus, for $|\lambda - \lambda_0| < 1/\|y_0\|$,

$$y^\pi(\lambda) = y_0 \left[e - (\lambda - \lambda_0)y_0\right]^{-1} + L \, ,$$

which implies the asseted analyticity of y^π. If $|\lambda| > \|z\|$, then $z - \lambda e$ is actually invertible in $\mathcal{A}$ and, as $|\lambda| \to \infty$,

$$\|y^\pi(\lambda)\| \le \|(z - \lambda e)^{-1}\| = (1/|\lambda|)\|\sum_{n \ge 0} z^n/\lambda^n\| = o(1) \, .$$

Therefore, by Liouville's theorem, $y^\pi(\lambda) = 0$ for all $\lambda \in \mathbf{C}$, contrary to the assumption that L is a proper ideal of $\mathcal{A}$ (for $y^\pi(0) = 0$ would imply that there is a $y_0 \in L$ with $y_0 z - e \in L$, whence $e \in L$).

Hence, there is some $\lambda \in \mathbf{C}$ such that $z - \lambda e \in L$ and, since $z \notin L$, we have $\lambda \ne 0$. It follows that $e = \lambda^{-1} z + l$ for some $l \in L \cap \mathcal{B}$.

Assume there is a two-sided ideal I of $\mathcal{B}$ such that $L \cap \mathcal{B} \subset I$ and $L \cap \mathcal{B} \ne I$. Then there is a $z \in I \setminus (L \cap \mathcal{B}) \subset \mathcal{B} \setminus L$, and by what has been proved above, there exist $\lambda \in \mathbf{C} \setminus \{0\}$ and $l \in L \cap \mathcal{B}$ with $e = \lambda^{-1} z + l$. But this implies that $e \in I$ and, hence, $I = \mathcal{B}$, which proves the maximality of $L \cap \mathcal{B}$. $\blacksquare$

Proof of the theorem. (i) We show that $a \in \mathcal{A}$ is left invertible if and only if $\Phi_x(a)$ is left invertible for all $x \in M(\mathcal{B})$. The proof for the right invertibility is analogous.

Clearly, $\Phi_x(a)$ is left invertible if a is so. To verify the reverse implication assume the contrary, i.e. suppose $\Phi_x(a)$ to be left invertible in $\mathcal{A}/I_x$ for all $x \in M(\mathcal{B})$ but let a have no left inverse in $\mathcal{A}$. Denote by L a maximal left ideal of $\mathcal{A}$ containing the set $I := \{ba : b \in \mathcal{A}\}$ (note that $e \notin I$ and recall 1.1.3). Put $x = L \cap \mathcal{B}$. By the above proposition, x is a maximal ideal of $\mathcal{B}$. We claim that $I_x \subseteq L$. Indeed, if $l = \sum_{k=1}^n a_k x_k b_k$ where $x_k \in x$ and $a_k, b_k \in \mathcal{A}$ then $l = \sum_{k=1}^n a_k b_k x_k$ (because $\mathcal{B}$ is central), and hence $l \in L$ (because L is a left ideal). Thus, $I_x \subseteq L$. By our assumption, $\Phi_x(a)$ is left invertible in $\mathcal{A}/I_x$, that is, there exist a $b \in \mathcal{A}$ with $ba - e \in I_x$, and since $I_x \subseteq L$ we have $ba - e \in L$. On the other hand, $ba \in I \subseteq L$. This implies that $e \in L$ which contradicts the maximality of L.

(ii) Let $x \in M(\mathcal{B})$ and $\epsilon > 0$, and choose elements $a_1, ..., a_n \in \mathcal{A}$ and $x_1, ..., x_n \in x$ such that

$$\|a + \sum_{j=1}^n a_j x_j\| < \|\Phi_x(a)\| + \epsilon/2 \, . \tag{1}$$

Define an open neighbourhood $U \subset M(\mathcal{B})$ of x by

$$U = \{y \in M(\mathcal{B}) : |\Phi_y(x_j)| < \epsilon (2\sum_{i=1}^{n} \|a_i\| + 1)^{-1} \text{ for } j = 1, ..., n\},$$

and put $y_j = x_j - \Phi_y(x_j)e$ (note that the x_j's are elements of the *commutative* Banach algebra $\mathcal{B}$, and so the cosets $\Phi_y(x_j)$ may be identified with complex numbers). Then $y_j \in y$ (since $\Phi_y(y_j) = \Phi_y(x_j - \Phi_y(x_j)e) = 0$), and thus

$$\|\Phi_y(a)\| \leq \|a + \sum a_j y_j\|. \tag{2}$$

Now, if $y \in U$ then (1) and (2) give

$$\begin{aligned}\|\Phi_y(a)\| - \|\Phi_x(a)\| &\leq \|a + \sum a_j y_j\| - \|a + \sum a_j x_j\| + \epsilon/2 \\ &\leq \|\sum a_j(y_j - x_j)\| + \epsilon/2 \\ &= \|\sum \Phi_y(x_j)a_j\| + \epsilon/2 < \epsilon,\end{aligned}$$

which proves the upper semi-continuity of $y \mapsto \|\Phi_y(a)\|$ at x. ∎

As an important consequence of the upper semi-continuity we remark the following property of the local invertibility.

1.4.5 Local invertibility

Proposition 1.5 *Let the situation be as in Theorem 1.5. If $\Phi_x(a)$ is invertible in $\mathcal{A}/I_x$ then $\Phi_y(a)$ is invertible in $\mathcal{A}/I_y$ for all y in some open neighbourhood of x.*

Proof Suppose $\Phi_x(a)$ to be invertible. Then there exists an element $b \in \mathcal{A}$ such that $\Phi_x(ab - e) = 0$ and $\Phi_x(ba - e) = 0$. By what has just been proved, the mappings

$$y \mapsto \|\Phi_y(ab - e)\| \quad \text{and} \quad y \mapsto \|\Phi_y(ba - e)\|$$

are upper semi-continuous. Hence,

$$\|\Phi_y(ab - e)\| < 1/2 \quad \text{and} \quad \|\Phi_y(ba - e)\| < 1/2$$

for all maximal ideals y in a certain neighbourhood U of x. Since

$$\Phi_y(a)\Phi_y(b) = \Phi_y(e) + \Phi_y(ab - e), \quad \Phi_y(b)\Phi_y(a) = \Phi_y(e) + \Phi_y(ba - e),$$

and since $\Phi_y(e)$ is the identity element in $\mathcal{A}/I_y$ this implies the invertibility of $\Phi_y(a)$ for all $y \in U$. ∎

1.4.6   Douglas' local principle

In case $\mathcal{A}$ is a C^*-algebra and $\mathcal{B}$ is a central C^*-subalgebra of $\mathcal{A}$ the Theorem 1.5 is usually referred to as *Douglas' local principle*. In this setting the assertions of Theorem 1.5 can be completed by the *norm-preserving property* of the localization:

$$\|a\| \;=\; \max_{x\in M(\mathcal{B})} \|\Phi_x(a)\| \quad \text{for} \quad \text{all} \quad a \in \mathcal{A}. \tag{1}$$

We shall not prove this formula here since we shall later on introduce a more general concept of norm-preserving localization. Notice further that the maximum in formula (1) is justified since each upper semi-continuous function attains its supremum.

1.5   The finite section method for Toeplitz operators

1.5.1   Matrix operators

In this section we are going to illustrate the application of Allan's local principle to study the stability of the so-called finite section method for block Toeplitz operators. Let us first introduce some notations.

Given a linear space X denote by X_N the linear space of column-vectors of length N with components from X and let $X_{N\times N}$ stand for the linear space of $N \times N$ matrices with entries from X. If X is a Banach space, X_N can be made into a Banach space on defining a norm in X_N by

$$\|(x_1, ..., x_N)^T\|_{X_N} \;:=\; \|x_1\|_X + ... + \|x_N\|_X$$

or on choosing any norm in X_N equivalent to that one. Every operator $A \in L(X_N)$ may then be written as an operator matrix $A = (A_{ij})_{i,j=1}^N$ where $A_{ij} \in L(X)$, that is, $L(X_N)$ can be identified with $(L(X))_{N\times N}$.

1.5.2   Toeplitz operators

Let l^2 denote the Hilbert space of all sequences $(x_n)_{n=0}^\infty$ of complex numbers endowed with the norm

$$\|(x_n)\| \;=\; \Big(\sum_{n=0}^\infty |x_n|^2\Big)^{1/2}$$

and, given a function $a \in L^\infty(\mathbf{T})$ (the space of all bounded measurable functions on the unit circle $\mathbf{T}$), we let $(a_n)_{n=-\infty}^\infty$ stand for the sequence of its Fourier coefficients. As usual,

$$a_n \;=\; \int_0^1 a(e^{2\pi it})e^{-2\pi int}\, dt \quad .$$

Then the mapping $T(a)$ which acts on sequences $(x_n) \in l^2$ with finite support by

$$T(a)(x_n) = (y_n), \ y_n = \sum_{k=0}^{\infty} a_{n-k} x_k \quad , \quad n = 0, 1, \ldots \quad ,$$

extends by continuity to a linear bounded operator on l^2 referred to as the *Toeplitz operator with generating function* a and denoted by $T(a)$ again.

Similarly, if $a \in L^{\infty}(\mathbf{T})$ again then the *Hankel operator* $H(a)$ can be defined as the continuous extension of the mapping $H(a)$ acting on sequences $(x_n) \in l^2$ with finite support by

$$H(a)(x_n) = (y_n) \quad , \quad y_n = \sum_{k=0}^{n} a_{n+k+1} x_k \quad , \quad n = 0, 1, \ldots \quad .$$

Let $a_{ij} \in L^{\infty}(\mathbf{T})$ $(i, j = 1, \ldots, N)$. Then the operator $A = (T(a_{ij}))_{i,j=1}^{N}$, thought of as acting on the space l_N^2, is called a *block Toeplitz operator* with generating function $a = (a_{ij})_{i,j=1}^{N} \in L^{\infty}(\mathbf{T})_{N \times N}$, and we write $A = T(a)$ for brevity. Analogously, block Hankel operators will be considered.

Here are three important properties of Toeplitz operators needed below to study their finite sections in case of continuous generating functions:

(a) *If $a, b \in L^{\infty}(\mathbf{T})$ then*

$$T(ab) = T(a) T(b) + H(a) H(\tilde{b})$$

with the function $\tilde{b}$ being defined by $\tilde{b}(t) = b(1/t)$.

(b) *If $a \in C(\mathbf{T})$ then the Hankel operator $H(a)$ is compact.*

(c) *If $a \in C(\mathbf{T})_{N \times N}$ then the block Toeplitz operator $T(a)$ is Fredholm if and only if the matrix function a is invertible.*

The proofs of (a) and (b) are not hard and will be left as an exercise. The proof of property (c) can be performed by using local principles (compare the arguments in the proof of Theorem 1.6 below).

1.5.3   Finite sections

Let $a \in L^{\infty}(\mathbf{T})$ and consider the operator equation $T(a)x = y$ or, in matrix representation,

$$\begin{pmatrix} a_0 & a_{-1} & a_{-2} & \cdots \\ a_1 & a_0 & a_{-1} & \cdots \\ a_2 & a_1 & a_0 & \cdots \\ \vdots & \vdots & \vdots & \cdots \end{pmatrix} \begin{pmatrix} x_0 \\ x_1 \\ x_2 \\ \vdots \end{pmatrix} = \begin{pmatrix} y_0 \\ y_1 \\ y_2 \\ \vdots \end{pmatrix} . \tag{1}$$

For the approximate solution of this equation by the *finite section method* we consider the finite linear equations

$$A_n \, x^{(n)} \; = \; y^{(n)} \tag{2}$$

where $y^{(n)} = (y_0, ..., y_n)$ and where the A_n are finite sections of the Toeplitz matrix in (1):

$$A_0 = (a_0), \; A_1 = \begin{pmatrix} a_0 & a_{-1} \\ a_1 & a_0 \end{pmatrix}, \; A_2 = \begin{pmatrix} a_0 & a_{-1} & a_{-2} \\ a_1 & a_0 & a_{-1} \\ a_2 & a_1 & a_0 \end{pmatrix}, \; ...$$

Defining operators P_n by

$$P_n \; : \; (x_0, x_1, ...) \; \mapsto \; (x_0, x_1, ..., x_n, 0, 0, ...)$$

and identifying the image of P_n with the $n+1$ dimensional linear space $\mathbf{C}_{n+1}$, one can rewrite the equations (2) in the form

$$P_n \, T(a) \, P_n \, x^{(n)} \; = \; P_n y$$

which is in accordance with (2) in 1.3.1. Notice further that $P_n \to I$ strongly and $P_n^* \to I$ strongly.

If $T(a) = (T(a_{ij}))_{i,j=1}^N$ is a block Toeplitz operator then the finite section method for solving the equation $T(a)x = y$, $x, y \in l_N^2$, requires to solve the approximation equations

$$\begin{pmatrix} P_n & \\ & \ddots \\ & & P_n \end{pmatrix} \begin{pmatrix} T(a_{11}) & \cdots & T(a_{1N}) \\ \vdots & \vdots & \vdots \\ T(a_{N1}) & \cdots & T(a_{NN}) \end{pmatrix} \begin{pmatrix} P_n & \\ & \ddots \\ & & P_n \end{pmatrix} x^{(n)} =$$

$$= \begin{pmatrix} P_n & \\ & \ddots \\ & & P_n \end{pmatrix} y$$

or, abbreviating the diagonal operator $\mathrm{diag}(P_n \ldots P_n)$ by P_n again,

$$P_n \, T(a) \, P_n \, x^{(n)} \; = \; P_n \, y \quad . \tag{3}$$

Theorem 1.6 *Let $T(a)$ be a block Toeplitz operator with continuous generating function, i.e. $a \in C(\mathbf{T})_{N \times N}$. Then the finite section method (3) applies to $T(a)$ if and only if the operators $T(a)$ and $T(\tilde{a})$ with $\tilde{a}(t) = a(1/t)$ are invertible.*

The *proof* will be split into several steps. First we *algebraize* the assertion by translating it into an invertibility problem in a Banach algebra. Following Theorem 1.2, the applicability of the finite section method is equivalent to the invertibility of the coset $(P_n T(a) P_n) + \mathcal{G}$ in the quotient algebra $\mathcal{F}/\mathcal{G}$ with accordingly specified $\mathcal{F}$ and $\mathcal{G}$.

Since the algebra $\mathcal{F}$ contains "all" approximation methods it is clearly too large for a successfull analysis. Thus we shall consider the smallest closed subalgebra $\mathcal{A}_N$ of $\mathcal{F}$ which contains all sequences $(P_n T(a) P_n)$ with $a \in C(\mathbf{T})_{N \times N}$ and try to work in $\mathcal{A}_N$ instead of in $\mathcal{F}$. One can prove that the ideal $\mathcal{G}$ is also contained in the algebra $\mathcal{A}_N$ (see the proposition below). Thus we can form the quotient algebra $\mathcal{A}_N / \mathcal{G}$, and since $(P_n T(a) P_n)^* = (P_n T(\tilde{a}) P_n)$ is in $\mathcal{A}_N$ again, the algebra $\mathcal{A}_N / \mathcal{G}$ is a C^*-subalgebra of the C^*-algebra $\mathcal{F}/\mathcal{G}$. Thus, property (c) in 1.1.4 entails that the finite section method applies if and only if the coset $(P_n T(a) P_n) + \mathcal{G}$ is invertible in $\mathcal{A}_N / \mathcal{G}$.

In the next step we *essentialize* the problem, i.e. we shall look for an ideal $\mathcal{J}_N$ of $\mathcal{A}_N$ which can be lifted in the sense of the remark in Section 1.3.5 and makes the quotient algebra $\mathcal{A}_N / \mathcal{J}_N$ to possess a managable center. The experiences gathered in Fredholm theory further suggest to construct the ideal $\mathcal{J}_N$ in such a manner that the cosets diag $((P_n T(f) P_n), ..., (P_n T(f) P_n)) + \mathcal{J}_N$ belong to the center of $\mathcal{A}_N / \mathcal{J}_N$ for each continuous function f. Trying to find such an ideal one soon realizes that neither the ideal $\mathcal{G}$ nor the ideal $\mathcal{J}$ defined in Theorem 1.2 fits for our purposes. Only the following simple identity will strikingly make obvious how to construct the ideal $\mathcal{J}_N$. For its presentation we define operators $W_n : l^2 \to l^2$ by

$$W_n \; : \; (x_0, x_1, ... \,) \mapsto (x_n, x_{n-1}, ..., x_1, x_0, 0, 0, ... \,) .$$

Proposition 1.6 *If $a, b \in L^\infty(\mathbf{T})$ then*

$$P_n T(ab) P_n = P_n T(a) P_n T(b) P_n + P_n H(a) H(\tilde{b}) P_n + W_n H(\tilde{a}) H(b) W_n .$$

The proof is left as an exercise.

Now, given $a \in L^\infty(\mathbf{T})$ and $f \in C(\mathbf{T})$, Proposition 1.6 involves that

$$
\begin{aligned}
P_n T(a) P_n T(f) P_n \; &- \; P_n T(f) P_n T(a) P_n \; = \\
&= \; P_n (H(f) H(\tilde{a}) - H(a) H(\tilde{f})) P_n + W_n (H(\tilde{f}) H(a) - H(\tilde{a}) H(f)) W_n , \quad (4)
\end{aligned}
$$

and since the function $\tilde{f}$ is continuous again we conclude from property (b) in 1.5.2 that the operators $H(f) H(\tilde{a}) - H(a) H(\tilde{f})$ and $H(\tilde{f}) H(a) - H(\tilde{a}) H(f)$ are compact. Hence, the ideal $\mathcal{J}_N$ should be of the form $\mathcal{J}_N = (\mathcal{J}_1)_{N \times N}$ with

$$
\begin{aligned}
\mathcal{J}_1 \; = \; &\{ (A_n) \in \mathcal{F} : A_n = P_n T_1 P_n + W_n T_2 W_n + G_n \\
&\text{with } T_1, T_2 \text{ compact and } (G_n) \in \mathcal{G} \} .
\end{aligned}
$$

The following proposition reveals that the so-defined set $\mathcal{J}_N$ indeed provides us with the desired ideal.

Proposition 1.7   *(i) (Ideal property) The set $\mathcal{J}_1$ belongs to the algebra $\mathcal{A}_1$ and forms a closed two-sided ideal of $\mathcal{A}_1$.*

(ii) *(Lifting property)* If $(A_n) \in \mathcal{A}_N$ then the strong limit

$$\underset{n \to \infty}{\text{s-lim}}\ W_n A_n W_n\ =:\ W(A_n)$$

exists, and the mapping $(A_n) \mapsto W(A_n)$ is an algebra homomorphism. Further, if A denotes the strong limit of the sequence (A_n) then the coset $(A_n)+\mathcal{G}$ is invertible in $\mathcal{A}_N/\mathcal{G}$ (that is, the sequence (A_n) is stable!) if and only if the operators A and $W(A_n)$ are invertible and if the coset $(A_n) + \mathcal{J}_N$ is invertible in $\mathcal{A}_N/\mathcal{J}_N$.

(iii) *(Commutator property)* Let $f \in C(\mathbf{T})$. Then the coset
diag $((P_n T(f) P_n), ..., (P_n T(f) P_n)) + \mathcal{J}_N$ is in the center of $\mathcal{A}_N/\mathcal{J}_N$.

Proof   Our first goal is to explain the existence of the strong limits $W(A_n)$. If $A_n = P_n T(a) P_n$ then the assertion follows from the identity $W_n P_n T(a) P_n W_n = P_n T(\tilde{a}) P_n$ and from $P_n \to I$. In particular, $W(P_n T(a) P_n) = T(\tilde{a})$. In case

$$(A_n)\ =\ \sum_i \prod_j (P_n T(a_{ij}) P_n) \tag{5}$$

we write $P_n = W_n W_n$ to find that $W(A_n)$ exists and that $W(A_n) = \sum_i \prod_j T(\tilde{a}_{ij})$. Further, since $\|W_n\| = 1$, the Banach-Steinhaus theorem involves that $\|W(A_n)\| \le \|(A_n)\|$ for all sequences (A_n) of the form (5). Thus the operator W acts as a continuous algebra homomorphism on the collection $\mathcal{A}'$ of all sequences (5), and since $\mathcal{A}'$ is a dense subalgebra of $\mathcal{A}_1$ this gives our goal.

(i) Since the proof of the inclusion $\mathcal{J}_1 \subseteq \mathcal{A}_1$ is not relevant here we refer to ([BS 2], 7.27) or to Section 3.4.2 for a similar proof. Let us only verify that $\mathcal{J}_1$ is an ideal in $\mathcal{A}_1$. If (A_n) and (G_n) are sequences in $\mathcal{A}_1$ and $\mathcal{G}$, respectively, if T_1 and T_2 are compact operators then

$$(A_n)(P_n T_1 P_n + W_n T_2 W_n + G_n) = (P_n A T_1 P_n + W_n W(A_n) T_2 W_n + G_n')$$

with

$$G_n' = P_n(A_n P_n - A) T_1 P_n + W_n(W_n A_n W_n - W(A_n)) T_2 W_n + G_n\ .$$

In view of $A_n P_n - A \to 0$ and $W_n A_n W_n - W(A_n) \to 0$ strongly, and of 1.2.1(b), we conclude that $(G_n') \in \mathcal{G}$ and, hence, $\mathcal{J}_1$ is a left ideal. Analogously, $\mathcal{J}_1$ proves to be a right ideal in $\mathcal{A}_1$. The proof of the closedness of $\mathcal{J}_1$ is standard.

(ii) If $(A_n) + \mathcal{G}$ is invertible in $\mathcal{A}_N/\mathcal{G}$ then there are sequences (B_n) in $\mathcal{A}_N$ and $(G_n), (G_n')$ in $\mathcal{G}$ such that

$$A_n B_n\ =\ P_n + G_n \quad \text{and} \quad B_n A_n\ =\ P_n + G_n'\ . \tag{6}$$

Now apply the homomorphisms $(A_n) \mapsto A$, $(A_n) \mapsto W(A_n)$, and $(A_n) + \mathcal{G} \mapsto (A_n) + \mathcal{J}_N$ to the equalities (6) to obtain the invertibility of $A, W(A_n)$ and $(A_n) + \mathcal{J}_N$.

Conversely, let $A, W(A_n)$ and $(A_n) + \mathcal{J}_N$ be invertible. Then there exist a sequence $(B_n) \in \mathcal{A}_N$, a sequence $(G_n) \in \mathcal{G}$, and compact operators T_1, T_2 such that

$$A_n B_n = P_n + P_n T_1 P_n + W_n T_2 W_n + G_n \ .$$

Put $B'_n = B_n - P_n A^{-1} T_1 P_n - W_n W(A_n)^{-1} T_2 W_n$. Then $(B_n) - (B'_n) \in \mathcal{J}_N$ and $A_n B'_n = P_n + G'_n$ with

$$G'_n = G_n + (P_n - A_n P_n A^{-1}) T_1 P_n + W_n (P_n - W_n A_n W_n W(A_n)^{-1}) T_2 W_n \ .$$

Since $P_n - A_n P_n A^{-1} \to 0$ and $P_n - W_n A_n W_n W(A_n)^{-1} \to 0$ strongly we derive that $(G'_n) \in \mathcal{G}$ (remember 1.2.1(b)). Thus, the coset $(A_n) + \mathcal{G}$ is right invertible in $\mathcal{A}_N/\mathcal{G}$. The proof of its left invertibility proceeds analogously.

(iii) It is evidently sufficient to check whether the commutator

$$\operatorname{diag}((P_n T(f) P_n), ..., (P_n T(f) P_n)) \left((P_n T(a_{ij}) P_n)\right)_{i,j=1}^N -$$
$$- ((P_n T(a_{ij}) P_n))_{i,j=1}^N \operatorname{diag}((P_n T(f) P_n), ..., (P_n T(f) P_n))$$
$$= ((P_n T(f) P_n T(a_{ij}) P_n - P_n T(a_{ij}) P_n T(f) P_n))_{i,j=1}^N$$

belongs to the ideal $\mathcal{J}_N$. But this is just a consequence of the identity (4) and the definition of $\mathcal{J}_N$. ∎

Next we want to explain how Allan's principle applies to *localize* the study of the invertibility in $\mathcal{A}_N/\mathcal{J}_N$. The preceding proposition suggests to let the algebra $\mathcal{B}_N$, spanned by all sequences

$$\operatorname{diag}\left((P_n T(f) P_n), ..., (P_n T(f) P_n)\right) + \mathcal{J}_N \ , \ f \in C(\mathbf{T}) \ , \tag{7}$$

in $\mathcal{A}_N/\mathcal{J}_N$, play the role of the central subalgebra in the local principle. For we have to determine the maximal ideal space of this algebra.

First we observe that by property (d)

$$(P_n T(fg) P_n) - (P_n T(f) P_n)(P_n T(g) P_n) \in \mathcal{J}_1 \tag{8}$$

for all continuous functions f and g. Thus, every element of $\mathcal{B}_N$ is of the form (7). In particular, if the coset (7) is invertible in $\mathcal{B}_N$ then there are a continuous function g, compact operators T_1 and T_2, and a sequence $(G_n) \in \mathcal{G}$ such that

$$P_n T(g) P_n T(f) P_n = P_n + P_n T_1 P_n + W_n T_2 W_n + G_n \ .$$

Letting n go to infinity this yields $T(g)T(f) = I + T_1$ (notice that $W_n \to 0$ weakly whence $W_n T_2 W_n \to 0$ strongly by 1.2.1(a)). Analogously, $T(f)T(g) = I + T'_1$ with a compact operator T'_1. Hence, the invertibility of (7) implies that $T(f)$ is a Fredholm operator, and now property 1.5.2(c) entails us the invertibility of the function f.

Conversely, if f is an invertible continuous function then $(P_n) - (P_n T(f) P_n)(P_n T(f^{-1}) P_n) \in \mathcal{J}_1$ by (8), i.e. the coset (7) is invertible. These considerations show that the algebras $\mathcal{B}_N$ and $C(\mathbf{T})$ are isomorphic and hence, the maximal ideal space $M(\mathcal{B}_N)$ is homeomorphic to the unit circle $\mathbf{T}$. More precisely, the maximal ideal of $\mathcal{B}_N$ corresponding to $x \in \mathbf{T}$ is

$$\{ \operatorname{diag}((P_nT(f)P_n), ..., (P_nT(f)P_n)) + \mathcal{J}_N \, , \, f \in C(\mathbf{T}) \, , \, f(x) = 0 \} \, . \tag{9}$$

According to the abstract local principle we let I_x denote the smallest closed two-sided ideal in $\mathcal{A}_N/\mathcal{J}_N$ containing the maximal ideal (9), and we write Φ_x^J for the canonical homomorphism from $\mathcal{A}_N$ onto the quotient algebra $(\mathcal{A}_N/\mathcal{J}_N)/I_x$. Thus, localizing our problem, the coset $(P_nT(a)P_n) + \mathcal{J}_N$ is invertible if and only if all cosets $\Phi_x^J(P_nT(a)P_n)$ are invertible.

In the next step we shall *identify* these local cosets with the cosets of certain simpler objects. For this goal, we fix the point $x \in \mathbf{T}$ and consider the matrix valued function $\mathbf{T} \to \mathbf{C}_{N \times N}$, $y \mapsto a(y) - a(x)$. We claim that $\Phi_x^J(P_n(T(a) - a(x)I)P_n) = 0$. Indeed, given $\epsilon > 0$ we find a continuous matrix-valued function $b = (b_{ij})$ and a neighbourhood U of x such that $\|a - a(x) - b\| < \epsilon$ and $b_{ij} = 0$ on U. Choose a continuous scalar-valued function f which vanishes on $\mathbf{T} \setminus U$ and which takes the value 1 at x. Then, since $\Phi_x^J(P_n\operatorname{diag}(T(f), \dots, T(f))P_n)$ is the identity element of $(\mathcal{A}_N/\mathcal{J}_N)/I_x$,

$$\begin{aligned}
\|\Phi_x^J(P_n(T(a) - a(x))P_n)\| &\leq \epsilon + \|\Phi_x^J(P_nT(b)P_n)\| \\
&= \epsilon + \|\Phi_x^J(P_nT(b)P_n \operatorname{diag}(T(f), \dots T(f))P_n\| \\
&= \epsilon + \|\Phi_x^J((P_nT(b_{ij})P_nT(f)P_n)_{i,j=1}^N)\| \, ,
\end{aligned}$$

and remember 1.5.3(d) to find that $(P_nT(b_{ij})P_nT(f)P_n) \in \mathcal{J}_1$ whence $\|\Phi_x^J(P_n(T(a) - a(x)I)P_n)\| < \epsilon$. Letting ϵ go to zero we get our claim.

Now we are going to complete the proof of the theorem 1.6. By part (ii) of the above proposition it remains to show that the invertibility of $T(a)$ $(= \underset{n \to \infty}{\text{s-lim}} P_nT(a)P_n)$ and of $T(\tilde{a})$ $(= \underset{n \to \infty}{\text{s-lim}} W_nT(a)W_n)$ imply the invertibility of the coset $(P_nT(a)P_n) + \mathcal{J}_N$ or, equivalently, of all of the cosets $\Phi_x^J(P_na(x)P_n)$, $x \in \mathbf{T}$. But this is easily seen: if $T(a)$ is Fredholm then the property 1.5.2(c) gives the invertibility of the matrices $a(x) \in \mathbf{C}_{N \times N}$ for all $x \in \mathbf{T}$, and since $P_na(x)P_n \, P_na(x)^{-1}P_n = P_n$ we conclude that

$$\Phi_x^J(P_na(x)P_n) \, \Phi_x^J(P_na(x)^{-1}P_n) \; = \; \Phi_x^J(P_n)$$

as desired. ∎

1.5.4 Compact perturbations

As we have just seen, the finite section method applies to the block Toeplitz operator $T(a)$, $a \in C(\mathbf{T})_{N \times N}$, if and only if the operators $T(a)$ and $T(\tilde{a})$ are invertible. Now we ask for necessary and sufficient conditions for the finite section method

$$P_n(T(a) + K)P_n \, x^{(n)} \; = \; P_n y \, , \; K \in K(l_N^2) \, ,$$

to apply to the compactly perturbed block Toeplitz operator $T(a) + K$. A first answer to this question results from the Corollary 1.2: If the finite section method applies to $T(a)$ and if $T(a) + K$ is invertible then the finite section method applies to $T(a) + K$. Since we have already characterized the applicability of the finite section method to Toeplitz operators we get the following result first obtained by Gohberg and Feldman:

Proposition 1.8 *Let $a \in C(\mathbf{T})_{N \times N}$. If $T(a), T(\tilde{a})$ and $T(a) + K$ are invertible then the finite section method applies to $T(a) + K$.*

It has been an open problem for a long time whether the reverse implication holds. The Proposition 1.7(ii) only shows that the applicability of the finite section method to $T(a) + K$ involves the invertibility of $T(a) + K$ and of $T(\tilde{a})$ (notice that the sequence $(A_n) = (P_n(T(a) + K)P_n)$ is in $\mathcal{A}_N$ again by part (i) of the very same proposition, and that $W(A_n) = T(\tilde{a})$ due to the weak convergence $W_n \to 0$), but there seems to be no way to derive the invertibility of $T(a)$.

The Banach algebra approach (in particular, Proposition 1.7(ii) in combination with Allan's principle) opened a new and elegant way to tackle this problem. Ten years after Gohberg and Feldman's approach one of the authors obtained the following result:

Theorem 1.7 *Let $a \in C(\mathbf{T})_{N \times N}$ and K be compact. The finite section method applies to the operator $T(a) + K$ if and only if the operators $T(a) + K$ and $T(\tilde{a})$ are invertible.*

Since the cosets $(P_n(T(a) + K)P_n) + \mathcal{J}_N$ and $(P_n T(a) P_n) + \mathcal{J}_N$ evidently coincide, the proof of this theorem is verbatim the same as that of Theorem 1.6. For another elementary proof see Exercises 1.14 and 1.15. ∎

In the next section we analyse the proof of Theorem 1.6 from the methodological point of view in order to extract a general scheme for tackling invertibility problems (in the wide sense of Section 1.3.5) by Banach algebra methods.

1.6 A general scheme for solving invertibility problems

1.6.1 1st step: Algebraization

Find a unital Banach algebra $\mathcal{F}$ and a two-sided ideal $\mathcal{K}$ in $\mathcal{F}$ such that the original problem becomes equivalent to an invertibility problem in the quotient algebra $\mathcal{F}/\mathcal{K}$.

It should be clear from the discussion in 1.3.5 what is meant: The Fredholmness of a bounded linear operator A is equivalent to the invertibility of the coset $A + K(X)$ in the Calkin algebra $L(X)/K(X)$; the applicability of an approximation method $A_n x_n = P_n y$ is equivalent to the invertibility of the coset $(A_n) + \mathcal{G}$ in the quotient algebra $\mathcal{F}/\mathcal{G}$ defined in 1.3.3, and so on.

Generally, the algebra $\mathcal{F}$ will prove to be too large, and/or the ideal $\mathcal{K}$ will be too small to have a rich center in the quotient algebra $\mathcal{F}/\mathcal{K}$ which would offer the applicability of Allan's local principle. In these situations, we add the following step.

1.6.2   2nd step: Essentialization

Find a unital subalgebra $\mathcal{A}$ of $\mathcal{F}$ and an ideal $\mathcal{G} \subseteq \mathcal{A} \cap \mathcal{K}$ such that, for all $A \in \mathcal{A}$, the coset $A + \mathcal{K}$ is invertible in $\mathcal{F}/\mathcal{K}$ if and only if the coset $A + \mathcal{G}$ is invertible in $\mathcal{A}/\mathcal{G}$, and find an ideal $\mathcal{J}$ of $\mathcal{A}$, which contains $\mathcal{G}$ and which can be lifted, such that the quotient algebra possesses a sufficiently large center.

"Lifting an ideal $\mathcal{J}$" means that, under certain additional conditions, the invertibility of the coset $A + \mathcal{J}$ in $\mathcal{A}/\mathcal{J}$ implies the invertibility of the coset $A + \mathcal{G}$ in $\mathcal{A}/\mathcal{G}$. We have already become acquainted with two lifting theorems: Theorem 1.2 and Proposition 1.7(ii). Here is a general version of these results:

Theorem 1.8 (Lifting theorem) *Let $\mathcal{G} \subseteq \mathcal{A} \cap \mathcal{K}$ be a closed two-sided ideal in $\mathcal{A}$, Ω an index set, and suppose that, for each $t \in \Omega$, we are given a unital Banach algebra $\mathcal{C}^t$, a unital homomorphism $W^t : \mathcal{A} \to \mathcal{C}^t$, and closed two-sided ideals $\mathcal{J}^t$ of $\mathcal{A}/\mathcal{G}$ and $\mathcal{R}^t$ of $\mathcal{C}^t$ such that*

(i) $\mathcal{G} \subseteq \operatorname{Ker} W^t$,

(ii) the restriction of the quotient homomorphism

$$\mathcal{A}/\mathcal{G} \to \mathcal{C}^t , \ A + \mathcal{G} \mapsto W^t(A)$$

(being correctly defined by (i) and denoted by W^t again) onto the ideal $\mathcal{J}^t$ is an isomorphism between $\mathcal{J}^t$ and the ideal $\mathcal{R}^t$.

Let $\mathcal{J}$ stand for the smallest closed two-sided ideal of $\mathcal{A}$ containing all elements K for which the coset $K + \mathcal{G}$ belongs to one of the ideals $\mathcal{J}^t$, and let $A \in \mathcal{A}$. Then the coset $A + \mathcal{G}$ is left (right, resp. two-sided) invertible in $\mathcal{A}/\mathcal{G}$ if and only if the elements $W^t(A)$ are left (right, resp. two-sided) invertible for all $t \in \Omega$ and if the coset $A + \mathcal{J}$ is left (right, resp. two-sided) invertible in $\mathcal{A}/\mathcal{J}$.

Example The Proposition 1.7(ii) results from the lifting theorem by choosing $\Omega = \{0, 1\}$ and identifying $\mathcal{C}^0 = \mathcal{C}^1 = L(l_N^2)$, $\mathcal{R}^0 = \mathcal{R}^1 = K(l_N^2)$, $\mathcal{A} = \mathcal{A}_N$,

$$\mathcal{J}^0 = \{(P_n T P_n + G_n) \ \text{ with } T \text{ compact } \text{ and } (G_n) \in \mathcal{G} \},$$

$$\mathcal{J}^1 = \{(W_n T W_n + G_n) \ \text{ with } T \text{ compact } \text{ and } (G_n) \in \mathcal{G} \}$$

and $W^0(A_n) := \operatorname*{s\text{-}lim}_{n \to \infty} A_n P_n$, $W^1(A_n) := \operatorname*{s\text{-}lim}_{n \to \infty} W_n A_n W_n$. ∎

Proof of the theorem. The proof will be given for left invertibility only. Clearly, the left invertibility of the coset $A + \mathcal{G}$ involves that of $W^t(A)$ in $\mathcal{C}^t$ and of $A + \mathcal{J}$ in

$\mathcal{A}/\mathcal{J}$. For the reverse implication, let $W^t(a)$ $(t \in \Omega)$ and $A + \mathcal{J}$ be left invertible. Then there are elements $B \in \mathcal{A}$ and $K \in \mathcal{J}$ such that $BA = I + K$ with I the identity in $\mathcal{A}$. By the definition of the ideal $\mathcal{J}$ one can choose elements $K^{t_i} \in \mathcal{A}$ ($i = 1, \ldots, n$), $G \in \mathcal{G}$, and $K' \in \mathcal{J}$ such that $K^{t_i} + G \in \mathcal{J}^{t_i}$, $K' = K^{t_1} + \ldots + K^{t_n} + G$, and $\|K - K'\| < 1/2$.

Let $W^{t_i}(A)^{-1}$ stand for one left inverse of $W^{t_i}(A)$. Then, since $W^{t_i}(K^{t_i})W^{t_i}(A)^{-1}$ belongs to the ideal $\mathcal{R}^{t_i}$, there exist by (ii) uniquely determined elements $B^{t_i} \in \mathcal{A}$ with $B^{t_i} + G \in \mathcal{J}^{t_i}$ such that $W^{t_i}(B^{t_i}) = W^{t_i}(K^{t_i})W^{t_i}(A)^{-1}$. Put $B' = B - B^{t_1} - \ldots - B^{t_n}$. Then $B + \mathcal{J} = B' + \mathcal{J}$ and

$$B'A = I + K^{t_1} - B^{t_1}A + \ldots + K^{t_n} - B^{t_n}A + K - K' + G.$$

Since, by definition, $W^{t_i}(K^{t_i} - B^{t_i}A) = 0$, and $K^{t_i} - B^{t_i}A + G \in \mathcal{J}^{t_i}$, we conclude again from (ii) that $K^{t_i} - B^{t_i}A \in \mathcal{G}$ and, thus,

$$B'A = I + K - K' + G + G'$$

with a certain $G' \in \mathcal{G}$. Due to $\|K - K'\| < 1/2$ this equality yields the left invertibility of $A + \mathcal{G}$ in $\mathcal{A}/\mathcal{G}$. ∎

Concerning the first part of this essentialization step we only notice that invertibility of $A+\mathcal{G}$ in $\mathcal{A}/\mathcal{G}$ evidently implies invertibility of $A+\mathcal{K}$ in $\mathcal{F}/\mathcal{K}$. The verification of the reverse direction is often rather subtle. In the 5 th step we shall discuss one approach which works in many situations to solve this problem.

1.6.3   3rd step: Localization

Find a manageable large subalgebra $\mathcal{B}$ of the center of $\mathcal{A}/\mathcal{J}$, determine its maximal ideal space $M(\mathcal{B})$, and localize $\mathcal{A}/\mathcal{J}$ over $M(\mathcal{B})$ via Allan.

The outcome of this step is 'local' algebras $(\mathcal{A}/\mathcal{J})/I_x =: \mathcal{A}_x^J$, and we are required to study invertibility in these algebras. Let Φ_x^J refer to the canonical homomorphism from $\mathcal{A}$ onto $\mathcal{A}_x^J$.

1.6.4   4th step: Identification

Identify the local algebras.

This comprises such things as the construction of (possibly, matrix- or operator-valued) symbol mappings for $\mathcal{A}_x^J$, the discovery of the structure of the algebras $\mathcal{A}_x^J$, and/or the identification of $\mathcal{A}_x^J$ with a 'well known' algebra up to isomorphy. For instance, let $\mathcal{A}/\mathcal{J}$ be commutative itself. Then we can set $\mathcal{B} = \mathcal{A}/\mathcal{J}$, the local algebras $\mathcal{A}_x^J$ prove to be isomorphic to the complex field $\mathbf{C}$, and the mapping assigning to $a \in \mathcal{A}/\mathcal{J}$ the value of the Gelfand transform of a at x yields a symbol mapping for $\mathcal{A}_x^J$. For another example we refer the reader to the proof of Theorem 1.6 where we have de facto seen that the local algebras occuring there are isomorphic to $\mathbf{C}_{N \times N}$.

In the literature two methods of identifying local algebras seem to be favoured: the utilization of the special structure of the algebra, and the construction of locally equivalent representations.

Concerning the first method we emphasize that localizing often produces local algebras with a very clear structure: they are singly generated, or generated by two idempotents. For these simply structured algebras general symbol mappings are known. Although these symbol mappings will not be used in the sequel, we decided to present them here explicitly (but without proof) because of their importance in concrete operator theory. Several applications will be discussed in the exercises to Chapters 2 and 3. Before stating these results let us give a precise definition of what is meant by an algebra generated by some of its elements.

Let $\mathcal{H}$ be a Banach algebra and $h_1, \ldots, h_n \in \mathcal{H}$. Then the smallest closed sub-algebra of $\mathcal{H}$ containing the elements $h_1, \ldots, h_n$ is called the *algebra generated by* $h_1, \ldots, h_n$ and will be denoted by $alg_{\mathcal{H}}(h_1, \ldots, h_n)$. In case $alg_{\mathcal{H}}(h_1, \ldots, h_n) = \mathcal{H}$ we simply write $\mathcal{H} = alg(h_1, \ldots, h_n)$.

Theorem 1.9 *Let $\mathcal{H} = alg(e, h)$, that is, $\mathcal{H}$ is a singly generated (by h) algebra with identity e. Then $\mathcal{H}$ is commutative, its maximal ideal space is homeomorphic to the spectrum of h in $\mathcal{H}$, and the Gelfand transform (which is a symbol map for $\mathcal{H}$) acts on the dense subalgebra of $\mathcal{H}$ consisting of all polynomials in h as*

$$\Phi(\sum_{i=0}^{n} a_i h^i) \; : \; \sigma_{\mathcal{H}}(h) \to \mathbf{C} \, , \, x \mapsto \sum_{i=0}^{n} a_i x^i \quad .$$

Theorem 1.10 *Let $\mathcal{H} = alg(e, p, q)$ with identity element e and idempotents p and $q : p^2 = p$, $q^2 = q$. Suppose further that $\{0, 1\} = \sigma(p) = \sigma(q) \subseteq \sigma(pqp)$, and that both the spectrum of pqp as its complement in $\mathbf{C}$ are connected. Then there exists a symbol map ψ for $\mathcal{H}$ which assigns to each h in $\mathcal{H}$ a 2×2 matrix-valued function $\psi(h)$ defined on $\sigma(pqp)$. In particular, the symbols of the generating elements of $\mathcal{H}$ look as follows:*

$$\psi(e) : x \mapsto \begin{pmatrix} 1 & 0 \\ 0 & 1 \end{pmatrix}, \psi(p) : x \mapsto \begin{pmatrix} 1 & 0 \\ 0 & 0 \end{pmatrix}, \psi(q) : x \mapsto \begin{pmatrix} x & w(x) \\ w(x) & 1-x \end{pmatrix}$$

where $x \in \sigma(pqp)$ and where $w : \mathbf{C} \to \mathbf{C}$ is an arbitrary function satisfying the identity $w(x)^2 = x(1-x)$.

If the local algebras $\mathcal{A}_x^J$ are of the special structure described in the preceding theorems then these theorems provide us with symbol mappings, and so we solve the invertibility problems in these algebras. On the other hand, these symbol mappings need not to be isomorphisms in general, thus, they do not assert anything about the structure of the local algebras. In what follows we describe a mechanism which produces *isomorphic* symbol mappings for the local algebras which will be called *locally equivalent representations*. To each maximal ideal x of the central subalgebra $\mathcal{B}$ of $\mathcal{A}/\mathcal{J}$ we associate a unital homomorphism W_x of $\mathcal{A}$ into a unital Banach algebra $\mathcal{C}_x$ such that

(iii) $\mathcal{J} \subseteq \operatorname{Ker} W_x$, and

(iv) if $B \in \mathcal{A}$ is an element with $B + \mathcal{J} \in \mathcal{B}$ and $\Phi_x^J(B) = 0$ then $B \in \operatorname{Ker} W_x$.

These assumptions ensure that, for each $A \in \mathcal{A}$, the element $W_x(A)$ only depends on the coset $\Phi_x^J(A)$. Thus, the quotient homomorphism $\mathcal{A}_x^J \to \mathcal{C}_x$, $\Phi_x^J(A) \mapsto W_x(A)$, is correctly defined, and we denote it by W_x again.

The homomorphism W_x is said to be locally equivalent if

(v) the image of W_x in $\mathcal{C}_x$ is an inverse closed and closed subalgebra of $\mathcal{C}_x$, and

(vi) there exists a unital homomorphism W_x' mapping $\operatorname{Im} W_x$ into $\mathcal{A}_x^J$ such that

$$\Phi_x^J(A) = W_x'(W_x(A)) \quad \text{for all } A \in \mathcal{A}.$$

Theorem 1.11 *Let (i)-(vi) from Sections 1.6.2 and 1.6.4 be fulfilled. Then,*
(a) for each $A \in \mathcal{A}$, the coset $\Phi_x^J(A)$ is invertible in $\mathcal{A}_x^J$ if and only if $W_x(A)$ is invertible in $\mathcal{C}_x$,
(b) for each $A \in \mathcal{A}$, the coset $A + \mathcal{J}$ is invertible in $\mathcal{A}/\mathcal{J}$ if and only if $W_x(A)$ is invertible for all $x \in M(\mathcal{B})$,
(c) for each $A \in \mathcal{A}$, the coset $A + \mathcal{G}$ is invertible in $\mathcal{A}/\mathcal{G}$ if and only if $W_x(A)$ and $W^t(A)$ are invertible for all $x \in M(\mathcal{B})$ and $t \in \Omega$,
(d) the algebras $\mathcal{A}_x^J$ and $\operatorname{Im} W_x$ are topologically isomorphic. They are isometrically isomorphic if the norms of $W_x : \mathcal{A} \to \mathcal{C}_x$ and $W_x' : \operatorname{Im} W_x \to \mathcal{A}_x^J$ are 1.

Proof Since the mapping $\Phi_x^J(A) \mapsto W_x(A)$ is a unital homomorphism, the invertibility of $\Phi_x^J(A)$ in $\mathcal{A}_x^J$ implies the invertibility of $W_x(A)$ in $\mathcal{C}_x$. Conversely, if $W_x(A)$ is invertible in $\mathcal{C}_x$ then, by (v), it is invertible in $\operatorname{Im} W_x$ and consequently, $W_x'(W_x(A))$ is invertible in $\mathcal{A}_x^J$. Now (vi) gives the invertibility of $\Phi_x^J(A)$ which proves part (a).

Assertions (b) and (c) are immediate consequences of Allan's local principle and of the lifting theorem 1.8. For (d) notice that the kernel of the homomorphism $\Phi_x^J(A) \mapsto W_x(A)$ is trivial by (vi), thus, this homomorphism is actually an isomorphism. Finally, let C_1 and C_2 stand for the norms of the homomorphisms $W_x : \mathcal{A} \to \mathcal{C}_x$ and $W_x' : \operatorname{Im} W_x \to \mathcal{A}_x^J$. If $A, B \in \mathcal{A}$ and $B + \mathcal{J} \in I_x$ then

$$\|W_x(A)\| = \|W_x(A + B)\| \leq C_1 \|A + B\| .$$

Taking the infimum over all B we conclude that $\|W_x(A)\| \leq C_1 \|\Phi_x^J(A)\|$. On the other hand,

$$\|\Phi_x^J(A)\| = \|W_x'(W_x(A))\| \leq C_2 \|W_x(A)\|$$

which completes the proof. ∎

1.6.5 5th step: Inverse closedness

Show that, for each $A \in \mathcal{A}$, the invertibility of $A+\mathcal{K}$ in $\mathcal{F}/\mathcal{K}$ implies the invertibility of $A+\mathcal{G}$ in $\mathcal{A}/\mathcal{G}$.

Let us refer to this implication by (vii) for brevity. By Theorems 1.8 and 1.11, (vii) is equivalent to each of the following:

- If $A+\mathcal{K}$ is invertible in $\mathcal{F}/\mathcal{K}$ then $A+\mathcal{J}$ is invertible in $\mathcal{A}/\mathcal{J}$ and $W^t(A)$ is invertible for all $t \in \Omega$.

- If $A+\mathcal{K}$ is invertible in $\mathcal{F}/\mathcal{K}$ then $W_x(A)$ and $W^t(A)$ are invertible for all $x \in M(\mathcal{B})$ and $t \in \Omega$.

Here are some situations where (vii) can be guaranteed:

Situation I: If $\mathcal{G} = \mathcal{K}$ (resp. $\mathcal{J} = \mathcal{K}$) then $\mathcal{A}/\mathcal{G}$ (resp. $\mathcal{A}/\mathcal{J}$) is a subalgebra of $\mathcal{F}/\mathcal{K}$, and (vii) is satisfied if and only if this subalgebra is inverse closed. Criteria for inverse closedness have been stated in 1.1.4(c) ($\mathcal{F}$ and $\mathcal{A}$ are C^*-algebras) and in Corollary 1.1.

Situation II: If $\mathcal{G} = \mathcal{A} \cap \mathcal{K}$ and if $\mathcal{F}$ and $\mathcal{A}$ are C^*-algebras then, by 1.1.4(b),

$$\mathcal{A}/\mathcal{G} = \mathcal{A}/\mathcal{A} \cap \mathcal{K} = (\mathcal{A}+\mathcal{K})/\mathcal{K},$$

whence (vii) follows since $(\mathcal{A}+\mathcal{K})/\mathcal{K}$ is a C^*-subalgebra of $\mathcal{F}/\mathcal{K}$ and hence, inverse closed.

Situation III: Let there exist a subalgebra $\mathcal{D}$ of $\mathcal{F}$ containing $\mathcal{A}$ and $\mathcal{K}$ such that $\mathcal{D}/\mathcal{K}$ is inverse closed in $\mathcal{F}/\mathcal{K}$, that all homomorphisms W_x and W^t can be extended to continuous homomorphisms from $\mathcal{D}$ into $\mathcal{C}_x$ and $\mathcal{C}^t$, respectively, and that $\mathcal{K} \subseteq \operatorname{Ker} W_x$ ($x \in M(\mathcal{B})$) and $\mathcal{K} \subseteq \operatorname{Ker} W^t$ ($t \in \Omega$). Then, if $A+\mathcal{K}$ is invertible in $\mathcal{F}/\mathcal{K}$, it is also invertible in $\mathcal{D}/\mathcal{K}$, and since the quotient homomorphisms

$$\mathcal{D}/\mathcal{K} \to \mathcal{C}_x \ , \ A+\mathcal{K} \mapsto W_x(A) \text{ and } \mathcal{D}/\mathcal{K} \to \mathcal{C}^t \ , \ A+\mathcal{K} \mapsto W^t(A)$$

are correctly defined, we arrive at the invertibility of $W_x(A)$ and $W^t(A)$ for all x and t.

Next we are going to explain a simple mechanism which works if the homomorphisms W_x and W^t are of a special form and which then provides us with an algebra $\mathcal{D}$ as above.

Let $t \in \Omega$ be fixed. The homomorphism W^t is called *strong-limit-homomorphism* if $\mathcal{F}$ and $\mathcal{C}^t$ are subalgebras of the algebras $L(X)$ and $L(X^t)$ of all linear bounded operators on Banach spaces X and X^t, respectively, and if there exist bounded sequences $(F_n^t)_{n\geq 0}$ and $(F_{-n}^t)_{n\geq 0}$ of linear operators $F_n^t : X^t \to X$ and $F_{-n}^t : X \to X^t$ such that

(viii) $W^t(A) = \underset{n\to\infty}{\text{s-lim}} \ F_{-n}^t A F_n^t \quad \text{and} \quad W^t(A)^* = \underset{n\to\infty}{\text{s-lim}} \ (F_{-n}^t A F_n^t)^*$
     for all $A \in \mathcal{A}$,

(ix) $F_{-n}^t A F_n^t F_{-n}^t B F_n^t = F_{-n}^t A B F_n^t \quad \text{for all} \quad A, B \in \mathcal{F} \text{ and all } n \in \mathbf{N},$

(x) $\lim_{n\to\infty} \|F^t_{-n} K F^t_n\| = 0$ for all $K \in \mathcal{K}$.

Analogously the property of W_x to be a strong-limit-homomorphism can be defined.

Example Notice that in case of the finite section method studied in sections 1.5.3-1.5.4 these assumptions are just satisfied: Indeed, $\mathcal{F} \subseteq L(X)$ where X is the Banach space of all sequences $(x_k)_{k\geq 0}$ with $x_k \in \mathrm{Im}P_k$ provided with elementwise operations and the norm

$$\|(x_k)_{k\geq 0}\| = (\sum_{k=0}^{\infty} \|x_k\|^2)^{1/2} \,,$$

and $\mathcal{C}^0 = \mathcal{C}^1 = L(l^2_N)$. Further we identify

F^0_n with $l^2_N \ni x \mapsto \underbrace{(0,\ldots,0}_{n}, P_n x, 0, \ldots) \in X$,

F^0_{-n} with $X \ni (x_k) \mapsto x_n \in l^2_N$

in case of the homomorphisms $W^0(A_n) = \underset{n\to\infty}{\mathrm{s\text{-}lim}}\, A_n P_n$, and

F^1_n with $l^2_N \ni x \mapsto (0,\ldots,0, W_n x, 0, \ldots) \in X$,

F^1_{-n} with $X \ni (x_k) \mapsto W_n x_n \in l^2_N$

in case of the homomorphisms $W^1(A_n) = \underset{n\to\infty}{\mathrm{s\text{-}lim}}\, W_n A_n W_n$. The conditions (viii)-(x) are readily verified. ∎

Let us return to the general situation. By $\mathcal{D}^t$ we denote the collection of all elements A of $\mathcal{F}$ for which the sequences $(F^t_{-n} A F^t_n)$ and $((F^t_{-n} A F^t_n)^*)$ are strongly convergent on X^t and on its dual $(X^t)^*$, respectively. The corresponding strong limits will be denoted by $W^t(A)$ and $W^t(A)^*$ again. Clearly, by (viii) and (x), $\mathcal{A} \subseteq \mathcal{D}^t$ and $\mathcal{K} \subseteq \mathcal{D}^t$.

Proposition 1.9 (a) $\mathcal{D}^t$ is a closed subalgebra of $\mathcal{F}$, and $W^t : \mathcal{D}^t \to L(X^t)$ is a unital homomorphism.
(b) If $A \in \mathcal{D}^t$ and $A + \mathcal{K}$ is invertible in $\mathcal{F}/\mathcal{K}$ then $W^t(A)$ is invertible.
(c) The algebra $\mathcal{D}^t$ is inverse closed in $\mathcal{F}$, and the algebra $\mathcal{D}^t/\mathcal{K}$ is inverse closed in $\mathcal{F}/\mathcal{K}$.

Proof (a) One easily derives from (ix) that $\mathcal{D}^t$ is an algebra and that $W^t : \mathcal{D}^t \to L(X^t)$ is an algebra homomorphism which is even unital since $W^t : \mathcal{A} \to L(X^t)$ is unital. The closedness of $\mathcal{D}^t$ in $\mathcal{F}$ is a standard consequence of the boundedness of the sequences (F^t_n) and (F^t_{-n}).
(b) Setting $A = B = I$ in (ix) one concludes that the operators $\Pi^t_n := F^t_{-n} F^t_n : L(X^t) \to L(X^t)$ are projections which, moreover, converge strongly to the identity

since the homomorphisms W^t are supposed to be unital. Further, choosing $A = I$ or $B = I$ in (ix) one gets

$$F^t_{-n}AF^t_n \;=\; F^t_{-n}AF^t_n\Pi^t_n \;=\; \Pi^t_n F^t_{-n}AF^t_n$$

for all $A \in \mathcal{F}$, i.e. one can think of $F^t_{-n}AF^t_n$ as acting from $\operatorname{Im}\Pi^t_n$ into $\operatorname{Im}\Pi^t_n$.

Let now $A \in \mathcal{F}$ and $A + \mathcal{K}$ be invertible in $\mathcal{F}/\mathcal{K}$. Then there is an operator $B \in \mathcal{F}$ such that $BA - I =: K \in \mathcal{K}$. Thus,

$$F^t_{-n}BAF^t_n \;=\; F^t_{-n}(I + K)F^t_n$$

and, by (ix),

$$F^t_{-n}BF^t_n\,F^t_{-n}AF^t_n \;=\; \Pi^t_n \,+\, F^t_{-n}KF^t_n\;.$$

Condition (x) involves that the operators $F^t_{-n}AF^t_n : \operatorname{Im}\Pi^t_n \to \operatorname{Im}\Pi^t_n$ are invertible from the left for n large enough. Analogously, their invertibility from the right follows. Hence, these operators are invertible from both sides for large n and the norms of their inverses are uniformly bounded. So there is a constant $C > 0$ such that

$$\|\Pi^t_n x\| \;\le\; C\,\|F^t_{-n}AF^t_n x\|$$

for all $x \in X^t$, and letting n go to infinity one gets that

$$\|x\| \;\le\; C\,\|W^t(A)\,x\|\,,$$

i.e. $W^t(A)$ is an operator of regular type. Similarly, $W^t(A)^*$ is of regular type, too, which shows the invertibility of $W^t(A)$.

(c) To prove the inverse closedness of $\mathcal{D}^t$ it remains to check whether the sequences $(F^t_{-n}BF^t_n)$ and $((F^t_{-n}BF^t_n)^*)$ are strongly convergent. Let us do this for the first one. Given $x \in X^t$ set $y := W^t(A)^{-1}x$. Then

$$
\begin{aligned}
&\|F^t_{-n}BF^t_n x - W^t(A)^{-1}x\|\\
&\quad\le\; \|F^t_{-n}BF^t_n x - F^t_{-n}BAF^t_n W^t(A)^{-1}x - F^t_{-n}KF^t_n W^t(A)^{-1}x + F^t_{-n}F^t_n y - y\|\\
&\quad\le\; \|F^t_{-n}BF^t_n\|\,\|x - F^t_{-n}AF^t_n W^t(A)^{-1}x\| \,+\\
&\qquad +\, \|F^t_{-n}KF^t_n W^t(A)^{-1}x\| + \|F^t_{-n}F^t_n y - y\|\\
&\quad=\; \|F^t_{-n}BF^t_n\|\,\|W^t(A)y - F^t_{-n}AF^t_n y\| + \|F^t_{-n}KF^t_n y\| + \|F^t_{-n}F^t_n y - y\| \to 0
\end{aligned}
$$

as $n \to \infty$, i.e. $B \in \mathcal{D}^t$ and $W^t(B) = W^t(A)^{-1}$. $\blacksquare$

Now suppose all homomorphisms W^t and W_x ($t \in \Omega$ and $x \in M(\mathcal{B})$) to be strong-limit-homomorphisms and let $\mathcal{D}^t$ and $\mathcal{D}_x$ refer to the associated operator algebras. Then clearly, the intersection $\mathcal{D} := (\cap_{t\in\Omega}\mathcal{D}^t) \cap (\cap_{x\in M(\mathcal{B})}\mathcal{D}_x)$ is a subalgebra of $\mathcal{F}$ which just satisfies all requirements met in situation III. Thus, in this setting we have:

Theorem 1.12 *Let (i)-(vi) be satisfied and suppose W^t and W_x to be strong-limit-homomorphisms for all $t \in \Omega$ and $x \in M(\mathcal{B})$. Then the following assertions are equivalent for all $A \in \mathcal{A}$:*
(i) $A + \mathcal{K}$ is invertible in $\mathcal{F}/\mathcal{K}$.
(ii) $A + \mathcal{G}$ is invertible in $\mathcal{A}/\mathcal{G}$.
(iii) $W^t(A)$ and $W_x(A)$ are invertible for all $t \in \Omega$ and $x \in M(\mathcal{B})$.

1.7   Norm-preserving localization

1.7.1   The KMS-property

Let the situation be as in Theorem 1.5, i.e. $\mathcal{A}$ is a unital Banach algebra and $\mathcal{B}$ a central subalgebra of $\mathcal{A}$. Then Allan's local principle reduces the question whether an element $a \in \mathcal{A}$ is invertible to a whole variety of "simpler" invertibility problems in the local algebras $\mathcal{A}/I_x$. So, localization by Allan is sufficient to study invertibility, but, if one wants to infer something about the algebraical or topological structure of the algebra $\mathcal{A}$ then one will soon meet situations where this localization fails. This is caused by the circumstance that the intersection $\cap_{x \in M(\mathcal{B})} I_x$ may contain non-zero elements in which case some information gets lost when localizing. In this section we establish an effective criterion for this intersection to be trivial.

The Banach algebra $\mathcal{A}$ is said to be *KMS with respect to* $\mathcal{B}$ if

(a) $\mathcal{B}$ is a C^*-algebra under a C^*-norm $\|.\|_*$, and the norms $\|.\|_\mathcal{A}$ and $\|.\|_*$ are equivalent, more precisely: there is an involution $b \mapsto \bar{b}$ in $\mathcal{B}$ and a norm $\|.\|_*$ on $\mathcal{B}$ which makes $\mathcal{B}$ into a C^*-algebra, and there is a constant $C_1 > 0$ such that

$$\|b\|_* \leq \|b\|_\mathcal{A} \leq C_1 \|b\|_* \quad \text{for} \quad \text{all} \quad b \in \mathcal{B} ; \tag{1}$$

(the first inequality holds since, for b belonging to the commutative C^*-algebra $\mathcal{B}$, $\|b\|_*$ coincides with the spectral radius of b), and if

(b) there is a constant $C_2 > 0$ such that

$$\left\| a \sum_{i=1}^{s} b_i \right\| \leq C_2 \max_{1 \leq i \leq s} \{ \|ab_i\| \} \tag{2}$$

for all $s \in \mathbf{N}$, $a \in \mathcal{A}$, and $b_1, \dots, b_s \in \mathcal{B}$ with $\operatorname{supp} b_i \cap \operatorname{supp} b_j = \emptyset$ whenever $i \neq j$.

By the *support* $\operatorname{supp} b$ of $b \in \mathcal{B}$ we mean the support of the Gelfand transform of b which is a complex-valued function. Let us further mention that in case $C_2 = 1$ the estimate (2) can be replaced by the following one:

$$\|a(b_1 + b_2)\| \leq \max\{\|ab_1\|, \|ab_2\| \}$$

whenever $\operatorname{supp} b_1 \cap \operatorname{supp} b_2 = \emptyset$.

1.7.2   Local norm estimates

Theorem 1.13 *Let $\mathcal{A}$ be KMS with respect to $\mathcal{B}$. Then*

$$\|a\|_{loc} := \max_{x \in M(\mathcal{B})} \|\Phi_x(a)\|$$

defines a norm on $\mathcal{A}$ which is equivalent to the original norm $\|.\|_{\mathcal{A}}$ on $\mathcal{A}$. In particular,

$$\|a\|_{loc} \leq \|a\|_{\mathcal{A}} \leq C_1^2 C_2 \|a\|_{loc}$$

with C_1 and C_2 referring to the constants in (1) and (2).

For the proof we need a technical proposition. Given $a \in \mathcal{A}$ and $x \in M(\mathcal{B})$ define

$$\|a\|_x := \inf \{ \|ba\|_{\mathcal{A}} : b \in \mathcal{B}, 0 \leq b \leq 1 ,$$
$$b \text{ is identically 1 in some open neighbourhood of } x \} .$$

Proposition 1.10 *Let $\mathcal{B}$ be a central subalgebra of $\mathcal{A}$ satisfying (a). Then $\|.\|_x$ is an equivalent norm on $\mathcal{A}/I_x$; precisely:*

$$\|\Phi_x(a)\| \leq \|a\|_x \leq C_1 \|\Phi_x(a)\| \quad \text{with} \quad C_1 \quad \text{as} \quad \text{in} \quad (1).$$

Proof   If $b \in \mathcal{B}$ is identically 1 in some open neighbourhood of x then $(b-1)a$ belongs to I_x, and so $\|\Phi_x(a)\| \leq \|a + (b-1)a\|_{\mathcal{A}} = \|ba\|_{\mathcal{A}}$.

To show the reverse inequality, choose $b_1, \ldots, b_n \in \mathcal{B}$ satisfying $\Phi_x(b_1) = \ldots = \Phi_x(b_n) = 0$ and elements $c_1, \ldots, c_n \in \mathcal{A}$ such that $\|a + b_1 c_1 + \ldots + b_n c_n\|_{\mathcal{A}} < \|\Phi_x(a)\| + \epsilon$. If b is any function in $C(M(\mathcal{B}))$ such that $0 \leq b \leq 1$ and b is identically 1 in some open neighbourhood of x then

$$
\begin{aligned}
\|ba\|_{\mathcal{A}} &\leq \|b(a + \sum_i b_i c_i)\|_{\mathcal{A}} + \|\sum_i b b_i c_i\|_{\mathcal{A}} \\
&\leq C_1 \|a + \sum_i b_i c_i\|_{\mathcal{A}} + \sum_i \|b b_i\|_{\mathcal{A}} \|c_i\|_{\mathcal{A}} \\
&\leq C_1 \|\Phi_x(a)\| + C_1 \epsilon + \sum_i \|b b_i\|_{\mathcal{A}} \|c_i\|_{\mathcal{A}} ;
\end{aligned}
$$

and this is smaller than $C_1 \|\Phi_x(a)\| + 2C_1 \epsilon$ if only b is chosen so that $\|b b_i\|_* < \epsilon/(n \|c_i\|_{\mathcal{A}})$ for all i. Hence, $\|a\|_x \leq C_1 \|\Phi_x(a)\|$. ∎

Proof   of the Theorem 1.13. Let $\mathcal{A}$ be KMS with respect to $\mathcal{B}$. Then, by definition of the quotient norm, we have $\|a\|_{loc} \leq \|a\|_{\mathcal{A}}$, and so we are left with the inequality $\|a\|_{\mathcal{A}} \leq C_1^2 C_2 \|a\|_{loc}$. Let x be in $M(\mathcal{B})$. As in the proof of Proposition 1.10, there is a b_0 in $C(M(\mathcal{B}))$ such that $0 \leq b_0 \leq 1$, b_0 is identically 1 in some open neighbourhood $U(x)$ of x, and $\|b_0 a\| < C_1 \|\Phi_x(a)\| + \epsilon$ ($\epsilon > 0$ arbitrarily given). Hence, if b is any function in $C(M(\mathcal{B}))$ with $0 \leq b \leq 1$ and $\operatorname{supp} b \subset U(x)$ then

$$\|ba\| = \|b b_0 a\| \leq \|b\| \|b_0 a\| < C_1^2 \|\Phi_x(a)\| + C_1 \epsilon \leq C_1^2 \|a\|_{loc} + C_1 \epsilon.$$

In other words, each x in $M(\mathcal{B})$ has an open neighbourhood $U(x)$ such that $\|ba\| < C_1^2\|a\|_{loc} + C_1\epsilon$ whenever $b \in C(M(\mathcal{B}))$, $0 \leq b \leq 1$, and supp $b \subseteq U(x)$. Choose a finite number $U_1, \ldots, U_n$ of such neighbourhoods which cover $M(\mathcal{B})$, fix any (large) positive integer m, and let $k \in \{1, \ldots, m\}$. Further let $1 = f_1 + \ldots + f_n$ be a partition of unity subordinate to the covering $M(\mathcal{B}) = \cup_{i=1}^n U_i$ and put

$$V_{ki}^m := \{x \in M(\mathcal{B}) : f_i(x) \geq \frac{k+1}{n(m+1)} ; \quad f_{i+1}(x) \leq \frac{k}{n(m+1)}, \ldots$$

$$\ldots, f_n(x) \leq \frac{k}{n(m+1)}\} \quad \text{for} \quad i = 1, \ldots, n-1,$$

$$V_{kn}^m := \{x \in M(\mathcal{B}) : f_n(x) \geq \frac{k+1}{n(m+1)}\}.$$

A straightforward check shows that the sets $V_{k1}^m, \ldots, V_{kn}^m$ are closed and pairwise disjoint, that $V_{ki}^m \subset U_i$ for $i = 1, \ldots, n$, and that each x in $M(\mathcal{B})$ belongs to at most n of the sets $G_k^m := M(\mathcal{B}) \setminus \cup_{i=1}^n V_{ki}^m$. Now let $g_{k1}^m, \ldots, g_{kn}^m$ be any functions in $C(M(\mathcal{B}))$ such that $g_{ki}^m \mid_{V_{ki}^m} = 1$, supp $g_{ki}^m \cap$ supp $g_{kj}^m = \emptyset$ whenever $i \neq j$, supp $g_{ki}^m \subset U_i$, and $0 \leq g_{ki}^m \leq 1$. Finally, put $g_k^m = g_{k1}^m + \ldots + g_{kn}^m$. Because $\mathcal{A}$ is KMS with respect to $\mathcal{B}$, we have

$$\|g_k^m a\| = \|(g_{k1}^m + \ldots + g_{kn}^m)a\| \leq C_2 \max_i \|g_{ki}^m a\| < C_1^2 C_2 \|a\|_{loc} + C_1 C_2 \epsilon$$

(for the last '<' recall that supp $g_{ki}^m \subset U_i$), and hence

$$\|(g_1^m + \ldots + g_m^m)a\| \leq m(C_1^2 C_2 \|a\|_{loc} + C_1 C_2 \epsilon). \tag{1}$$

Now put $h_k^m = 1 - g_k^m$. Then $0 \leq h_k^m \leq 1$ and supp $h_k^m \subseteq G_k^m$. We have

$$\|(g_1^m + \ldots + g_m^m)a\| = \|ma - (h_1^m + \ldots + h_m^m)a\| \geq m\|a\| - \|h_1^m + \ldots + h_m^m\| \|a\|,$$

and because supp $(h_1^m + \ldots + h_m^m) \subset \cup_{k=1}^m G_k^m$ and each $x \in M(\mathcal{B})$ belongs to at most n of the sets G_k^m, it follows that $h_1^m(x) + \ldots + h_m^m(x) \leq n$ for all x in $M(\mathcal{B})$, whence

$$\|(g_1^m + \ldots + g_m^m)a\| \geq (m - C_1 n)\|a\|. \tag{2}$$

Combining (1) and (2) we arrive at the inequality

$$\|a\| \leq \frac{m}{m - C_1 n}(C_1^2 C_2 \|a\|_{loc} + C_1 C_2 \epsilon),$$

and letting m go to infinity and ϵ go to zero we obtain the desired inequality. ∎

1.7.3   Local inclusion theorems

As an application of Theorem 1.13 we shall derive the following *local inclusion theorem*. We recommend the reader to compare this theorem with Glicksberg's theorem (quantitative version) and with the Shilov-Bishop theorem (qualitative version) in the theory of uniform commutative Banach algebras (see, e.g., [BS 1], 1.21 and 1.22).

Given an element a of a Banach algebra as well as a subset $\mathcal{C}$ of the same algebra we define the *distance between a and $\mathcal{C}$* by

$$\text{dist}\,(a,\mathcal{C}) \;:=\; \inf_{c\in\mathcal{C}} \|a - c\|\,.$$

In the context of Allan's localization we further introduce the *local distance between a and $\mathcal{C}$* by

$$\text{dist}_{loc}\,(a,\mathcal{C}) \;:=\; \sup_{x\in M(\mathcal{B})}\;\text{dist}\,(\Phi_x(a),\Phi_x(\mathcal{C}))\,.$$

Theorem 1.14 *(a) (Quantitative version). Let $\mathcal{A}$ be KMS with respect to $\mathcal{B}$ and let $\mathcal{C} \subseteq \mathcal{A}$ be a (not necessarily closed) $\mathcal{B}$-module of $\mathcal{A}$ (that is, $bc \in \mathcal{C}$ if $b \in \mathcal{B}$ and $c \in \mathcal{C}$). Then, for all elements a in $\mathcal{A}$,*

$$\text{dist}_{loc}\,(a,\mathcal{C}) \;\le\; \text{dist}\,(a,\,\mathcal{C}) \;\le\; C_1^2 C_2\,\text{dist}_{loc}\,(a,\mathcal{C})\,.$$

(b) (Qualitative version). Let, moreover, $\mathcal{C}$ be closed in $\mathcal{A}$, and suppose $\Phi_x(a) \in \Phi_x(\mathcal{C})$ for all $x \in M(\mathcal{B})$. Then $a \in \mathcal{C}$.

Proof   (a) Since $\|\Phi_x(a) - \Phi_x(c)\| \le \|a - c\|$ we have $\text{dist}\,(\Phi_x(a),\Phi_x(\mathcal{C})) \le \text{dist}\,(a,\mathcal{C})$ whence follows $\text{dist}_{loc}(a,\mathcal{C}) \le \text{dist}\,(a,\mathcal{C})$.

For a proof of the second inequality we start from the obvious estimate $\text{dist}\,(\Phi_x(a),\Phi_x(\mathcal{C})) \le \text{dist}_{loc}(a,\mathcal{C})$. Thus, given $x \in M(\mathcal{B})$ and $\epsilon > 0$ there exists an element $c_x \in \mathcal{C}$ such that

$$\|\Phi_x(a) - \Phi_x(c_x)\| \le \text{dist}\,(\Phi_x(a),\Phi_x(\mathcal{C})) + \epsilon \le \text{dist}_{loc}(a,\mathcal{C}) + \epsilon\,.$$

From the upper semi-continuity of the local norms (Theorem 1.5(ii)) we conclude that

$$\|\Phi_y(a - c_x)\| \;\le\; \text{dist}_{loc}(a,\mathcal{C}) + \epsilon$$

for all y belonging to a certain neighbourhood $U(x)$ of x. Choose a finite number $U(x_1),\ldots,U(x_n)$ of such neighbourhoods which cover $M(\mathcal{B})$, and let $1 = f_1 + \ldots + f_n$ ($f_i \in \mathcal{B}$) be a partition of unity subordinate to the covering $M(\mathcal{B}) = \cup U(x_i)$. Put $b_\epsilon := \sum_{i=1}^{n} f_i c_{x_i}$. Then, by Theorem 1.13,

$$\|a - b_\epsilon\| \;=\; \Big\|\sum_i f_i(a - c_{x_i})\Big\|$$

$$\le\; C_1^2 C_2 \sup_{y\in M(\mathcal{B})} \Big\|\Phi_y\Big(\sum_i f_i(a - c_{x_i})\Big)\Big\| = C_1^2 C_2 \sup_{y\in M(\mathcal{B})} \Big\|\sum_i f_i(y)\Phi_y(a - c_{x_i})\Big\|$$

$$=\; C_1^2 C_2 \sup_{y\in M(\mathcal{B})} \Big\|\sum_{i:\,y\in U(x_i)} f_i(y)\Phi_y(a - c_{x_i})\Big\|$$

$$<\; C_1^2 C_2 \sup_{y\in M(\mathcal{B})} \sum_{i:\,y\in U(x_i)} f_i(y)(\text{dist}_{loc}(a,\mathcal{C}) + \epsilon) = C_1^2 C_2\,(\text{dist}_{loc}(a,\mathcal{C}) + \epsilon)\,.$$

Letting ϵ go to zero we arrive at the desired inequality.

(b) If $\Phi_x(a) \in \Phi_x(\mathcal{C})$ then $\text{dist}_{loc}(a, \mathcal{C}) = 0$. By (a), this implies that $\text{dist}(a, \mathcal{C}) = 0$ and, since $\mathcal{C}$ is closed, $a \in \mathcal{C}$. ■

Later on we shall employ this result for showing that certain algebras contain certain elements.

1.8   Exercises

E 1.1 Let X, Y be Banach spaces and $A \in L(X, Y)$. Prove that A is normally solvable if and only if A^* is normally solvable.

(Hint: The proof is easy if $X = Y$ is a Hilbert space. For the general setting see [Yo]).

E 1.2 Let X, Y, A be as in the first exercise. Prove that A is a Φ_+-(resp. Φ_--) operator if and only if A^* is a Φ_--(resp. Φ_+-) operator.

E 1.3 Let X, Y, A be as in the first exercise. Prove that A is a Φ_+-operator if and only if there are a constant $C > 0$, a positive integer r, Banach spaces $Z_1, \ldots, Z_r$, and compact operators $T_1 : X \to Z_1, \ldots, T_r : X \to Z_r$, such that

$$\|x\| \leq C \left(\|Ax\| + \sum_{j=1}^{r} \|T_j x\| \right)$$

for all $x \in X$.

(See [MP], Chapter 1, Lemma2.1 for a proof.)

E 1.4 Let (P_n), (R_n), (A_n) be operator sequences as in 1.3.1 and suppose that $\sup \|A_n\| < \infty$. Show that, if the method 1.3.1(2) applies to the operator A then

1. the sequence $\left(A_n^{-1} R_n \right)_{n \geq n_0}$ converges strongly to an operator $B \in L(Y, X)$.

2. $AB = I$ (i.e. A is right invertible).

3. $A_n P_n BA \to A$ strongly.

E 1.5 Let (P_n) and (R_n) be as above and suppose the method 1.3.1(2) with A_n being specified as $R_n A P_n$ to apply to the operator A. Prove the following estimate for the speed of convergence:

$$\|x - x_n\| \leq \left(1 + \|A\| \, \|A_n^{-1} R_n\| \right) \inf_{z \in X} \|P_n z - A^{-1} y\| .$$

E 1.6 Let (P_n) and (R_n) be as above. Show that the collection of all operators $A \in L(X, Y)$ to which the method $R_n A P_n x_n = R_n y$ applies is open in $L(X, Y)$.

E 1.7 Verify the second vertical implication in the scheme in 1.3.5.

E 1.8 Let the situation be as in Theorem 1.5.

1. Prove that if $\mathcal{A}$ is semi-simple then $\cap_{x \in M(\mathcal{B})} I_x = \{0\}$ (see Theorem 1.34 in [BS]).

2. Let $\mathcal{A}$ be a C^*-algebra and $\mathcal{B}$ be a central $*$-subalgebra of $\mathcal{A}$. Verify that then $I_x \neq \mathcal{A}$ for all $x \in M(\mathcal{B})$.

3. Specify $\mathcal{A}$ to be the algebra $C(\mathbf{T})$ of all continuous functions on the unit circle $\mathbf{T}$ and take the disk algebra for $\mathcal{B}$. Show that the maximal ideal space of $\mathcal{B}$ is homeomorphic to the closed unit disk clos $\mathbf{D}$, and that $I_x \neq C(\mathbf{T})$ if and only if $x \in \mathbf{T}$.

E 1.9 Let $\mathcal{A}$ and $\mathcal{B}$ be unital C^*-algebras and let $W : \mathcal{A} \rightarrow \mathcal{B}$ be a $*$-symbol mapping (i.e. a symbol mapping satisfying $W(a^*) = W(a)^*$ for all $a \in \mathcal{A}$). Show that W is an isometry.

(Hint: Prove that $\operatorname{Ker} W$ is trivial and use [Kh], Proposition 7.62 (II), or [Dix], 1.8.1.).

E 1.10 Prove (a), (b) in 1.5.2 and Proposition 1.6.
Hints: For (a), consider the operators

$$T^0(a): \quad (x_n)_{n \in Z} \mapsto (y_n)_{n \in Z}, \quad y_n = \sum_{k=-\infty}^{\infty} a_{n-k} x_k,$$

$$P: \quad (x_n)_{n \in Z} \mapsto (y_n)_{n \in Z}, \quad y_n = \begin{cases} x_n & n \geq 0 \\ 0 & n < 0, \end{cases}$$

$$J: \quad (x_n)_{n \in Z} \mapsto (y_n)_{n \in Z}, \quad y_n = x_{n-1}$$

on the two-sided l^2-space and employ the identity

$$
\begin{aligned}
PT^0(ab)P &= PT^0(a)T^0(b)P \\
&= PT^0(a)PT^0(b)P + PT^0(a)(I - P)T^0(b)P \\
&= PT^0(a)PT^0(b)P + PT^0(a)J\,PJT^0(b)P
\end{aligned}
$$

by identifying the operators $PT^0(a)P$, $PT^0(a)JP$, and $PJT^0(b)P$ with $T(a)$, $H(a)$, and $H(\tilde{b})$, respectively.

For (b) first analyze the situation when a is a trigonometric polynomial on $\mathbf{T}$ and then use the density of the polynomials in $C(\mathbf{T})$.

For the proof of Proposition 1.6 introduce the one-sided shift operators V_n and V_{-n} by

$$V_n : (x_0, x_1, \ldots) \mapsto (\underbrace{0, 0, \ldots, 0}_{n \text{ zeros}}, x_0, x_1, \ldots)$$

and $V_{-n} : (x_0, x_1, \dots) \mapsto (x_n, x_{n+1,\dots})$, write

$$P_n T(ab) P_n =$$
$$P_n T(a) P_n T(b) P_n + P_n T(a)(I - P_n) T(b) P_n \ + \ P_n \, H(a) H(\tilde{b}) P_n$$

and make use of the identities $P_n = W_n^2$, $I - P_n = V_n V_{-n}$, and $W_n T(a) V_n = P_n H(\tilde{a})$, $V_{-n} T(b) W_n = H(b) P_n$.

E 1.11 Let $\mathcal{K}$ be a unital algebra, $\mathcal{G}$ a unital subalgebra of $\mathcal{K}$, and $\mathcal{J} \subseteq \mathcal{G}$ a two-sided ideal of $\mathcal{K}$. Prove that if the quotient algebra $\mathcal{G}/\mathcal{J}$ is inverse closed in $\mathcal{K}/\mathcal{J}$ then $\mathcal{G}$ is inverse closed in $\mathcal{K}$.

E 1.12 Let $\mathcal{K}$ be a unital algebra, $\mathcal{G}$ an inverse closed subalgebra of $\mathcal{K}$, and p an idempotent in $\mathcal{G}$.

 (a) Show that $p\mathcal{K}p = \{pkp, \ k \in \mathcal{K}\}$ is an algebra with identity p.

 (b) Show that $p\mathcal{G}p$ is inverse closed in $p\mathcal{K}p$.

E 1.13 Let $\mathcal{K}$ be a unital algebra, $p \in \mathcal{K}$ an idempotent, $q = e - p$, and $a, b, c \in \mathcal{K}$ with c being invertible.

 (a) Show that $e + ab$ is invertible if and only if $e + ba$ is invertible. (Hint: Verify

$$(e \ + \ ab)^{-1} \ = \ e \ - \ a(e \ + \ ba)^{-1} b \ .)$$

 (b) Show that pcp is invertible in $p\mathcal{K}p$ if and only if $qc^{-1}q$ is invertible in $q\mathcal{K}q$. (Hint: Verify *Kozak's identity*

$$(pcp)^{-1} \ = \ pc^{-1}p \ - \ pc^{-1}q(qc^{-1}q)^{-1} qc^{-1}p \ .)$$

E 1.14 Let $a \in C(\mathbf{T})_{N \times N}$ and suppose $T(a)$ to be invertible. Use Kozak's identity to verify that the finite section method applies to $T(a)^{-1}$.

E 1.15 Use Exercise 1.14 to prove Theorem 1.7.
  (Hint: Prove that $T(a) = T(a^{-1})^{-1} + C$ with a compact operator C and apply Corollary 1.2.)

E 1.16 Let $\mathcal{K}$ be an algebra with identity e and let $\mathcal{K}_{k,l}$ stand for the linear space (the algebra if $k = l$) of all $k \times l$ matrices with entries in $\mathcal{K}$. Show that the $k \times k$ matrix

$$\begin{pmatrix} e & & & \\ & \ddots & & 0 \\ & & e & \\ \hline A & & & B \end{pmatrix}, \ B \in \mathcal{K}_{l,l}, \ A \in \mathcal{K}_{l,k-l},$$

is invertible in $\mathcal{K}_{k,k}$ if and only if B is invertible in $\mathcal{K}_{l,l}$.

E 1.17 Prove the reverse to Theorem 1.13, that is, if the norms $\|.\|_{\mathcal{A}}$ and $\|.\|_{loc}$ are equivalent then $\mathcal{A}$ is KMS with respect to $\mathcal{B}$.

E 1.18 Show that a C^*-algebra is KMS with respect to every C^*-subalgebra of its center, and that the constants C_1 and C_2 in 1.7.1(1) and (2) are equal to one.

E 1.19 Let X be a compact Hausdorff space with a non-negative (not necessarily finite) Borel measure and denote by $L^p(X)$ ($1 \le p \le \infty$) the Lebesgue space over X. To each function $a \in L^\infty(X)$ we associate the operator aI on $L^p(X)$ of multiplication by a. An operator A in $L(L^p(X))$ is said to be of *local type* if $AaI - aIA$ is compact on $L^p(X)$ for all a in $C(X)$. The collection of all operators of local type on $L^p(X)$ is clearly a Banach algebra, and it will be denoted by $\Lambda(L^p(X))$. Prove that the quotient algebra $\Lambda(L^p(X)) / K(L^p(X))$ is KMS with respect to its central subalgebra $\{aI + K(L^p(X)),\ a \in C(X)\}$, and that the constants in 1.7.1(1) and (2) are equal to 1 (see ([BKS], Proposition 5.5) or the proof of Theorem 3.6).

1.9 Comments and references

1.1.1-1.2.3 These facts are well known and can be found in almost each book on Banach- or C^*-algebras. In particular, Theorem 1.1 is in [Ru] (Theorem 10.18), properties (a)-(d) in 1.1.4 can be found in [Dix] (Proposition 1.8.2, Corollary 1.8.4, Proposition 1.3.10 and Remark 1.9.15, respectively), the Banach-Steinhaus theorem is in [Ru] (Theorem 2.7), and the elementary properties of Fredholm operators can be looked up in Sections 1-3 of Chapter 1 in [MP] or in Sections 4.1-4.7 of [GK]. Section 4.15 of the latter monograph also contains some results on operators of regular type.

The idea of a *symbol* was brought forth for the first time in 1936 by Michlin who associated with each two-dimensional singular integral operator a certain function (see [M 1], [M 2]). Gohberg [Go] shaved in 1952 that Michlin's symbol actually coincides with the Gelfand transform applied to the algebra of singular integral operators with continuous coefficients, factored by the compact operators. Since that time, the distinguished role of the symbol concept for several important classes of operators has been pointed out by many mathematicians (Kohn/Nirenberg [KN], Hörmander [Ho]).

Nevertheless, it was rather surprising that one can also assign a symbol function to certain approximation methods which plays a fruitful and important role in stability analysis: In 1983, Böttcher and Silbermann succeeded in establishing a symbol calculus for certain approximation sequences for quarter-plane Toeplitz operators ([BS 3]). (In implicit form, the symbol for the finite section method for one-dimensional Toeplitz operators is already contained in Silbermann's paper [S 2].) Nowadays, symbol calculi are available for broad varities of approximation methods for pseudodifferential operators (Hagen, Silbermann [HS 2], Hagen, Roch, Silbermann [HRS], Prößdorf, Schneider [PSch 2], Roch, Silbermann [RS 2], [RS 3], [RS 6] and Silbermann [S 7]). The definition of a symbol mapping given in 1.1.2 is a natural generalization of these concrete symbol calculi. (It is interesting to observe that both pioneering papers, [M 1] and [BS 3], deal with the two-dimensional setting!).

1.3.1-1.3.4 The elementary theory of abstract approximation methods can be found, e.g., in the textbooks Prößdorf, Silbermann [PS 1], [PS 2], Böttcher, Silbermann [BS 2], Gohberg, Feldmann [GF 1], and Mikhlin, Prößdorf [MP].

Polskij [Po] was the first who realized the meaning of stability of an approximation sequence. The observation that the stability of a sequence is equivalent to an invertibility problem in a suitable chosen Banach algebra was made by Kozak [Ko]. Finally, the idea of essentialization goes back to Silbermann [S 2] (see also the notes and comments to Section 1.6.2 for detailed references). The reader is also recommended to see Silbermann's review paper [S 1].

1.3.5-1.4.4 The simplest local principle is the classic Gelfand theory which states that invertibility in commutative Banach algebras is local in nature (see, e.g., Rudin [Ru], 11.8-11.9). The first step in establishing generalizations of the Gelfand theory into the context of non-commutative Banach algebras was done by Simonenko [Si]. He both realized the local nature of Fredholmness of convolution and related operators and at the same time created a powerful machinery (his local principle) for tackling successfully a whole series of problems. Simonenko's local principle was generalized by Kozak [Ko] to arbitrary Banach algebras. Also see Chapter XV of Mikhlin, Prößdorf [MP] and Böttcher, Krupnik, Silbermann [BKS]. Another modification of Simonenko's local principle, which is distinguished for its simplicity on the one hand and for its wide arc of applications on the other hand, is due to Gohberg and Krupnik (see, e.g., Chapter 5 in volume 1 of [GK] or Chapter 1 in [BS 2]).

In the present monograph we prefer Allan's local principle [A] due to its elegance and its well-adaptability to our purposes. Independently, Douglas [D 1] stated Theorem 1.5 for C^*-algebras (see 1.4.4), and he was the first to realize its importance for operator theory, in particular for the investigation of Toeplitz operators ([D 2]). In this connection, it seems to be remarkable that Allan did not have operator theoretic applications in mind when he created his local principle. For a consequent thorough use of this principle see the monograph Böttcher/Silbermann [BS 2]. From this monograph, we also took the proofs of Proposition 1.4 and Theorem 1.5.

1.4.5-1.4.6 See, e.g., [BS 1], [BS 2] and [D 1].

1.5.1-1.5.4 The finite section method is the oldest and most obvious way to solve infinite systems of linear equations with infinite number of unknowns practically. Already Fourier used this method to compute the coefficients in the trigonometric series $\sum_{n=0}^{\infty} a_n \cos nx = f(x)$ for a given function f (see [Die], Section 8.5). But it was no doubt for Fourier that this method really converges. Only at the end of the 19 th century, Hill, Poincare, and H.v.Koch started with a detailed investigation and foundation of the finite section method. Nevertheless the question whether the finite section method applies to *each* infinite system which possesses a unique solution for each right-hand side remained open for a long time. That the answer is "No" can be seen by the following simple examples which were communicated to us by I.Gohberg: Let $U : (x_k) \mapsto (x_{k+1})$ denote the shift operator acting on the Hilbert space $l_{\mathbf{Z}}^2(0)$, and consider the operator $A = 2I - U$. Since $\|U\| = 1$, and $A = 2(I - \frac{1}{2}U)$, this operator is invertible and, thus, the infinite system

$$2x_n - x_{n-1} = a_n \quad , \quad n \in \mathbf{Z} ,$$

is uniquely solvable for each right-hand side.

Moreover, using Neumann series, one gets $A^{-1} = \frac{1}{2}(I + \frac{1}{2}U + \frac{1}{4}U^2 + \ldots)$ or, equivalently,

$$
A^{-1} = \frac{1}{2}
\begin{pmatrix}
\ddots & \ddots & \ddots & \ddots & & & \\
\ldots & 0 & 1 & 1/2 & 1/4 & \ldots & \\
& \ldots & 0 & 1 & 1/2 & 1/4 & \ldots \\
& & \ldots & 0 & 1 & 1/2 & 1/4 & \ldots \\
& & & \ddots & \ddots & \ddots & \ddots
\end{pmatrix}. \tag{1}
$$

The finite section matrices

$$
A_n =
\begin{pmatrix}
2 & -1 & 0 & \ldots & 0 \\
0 & 2 & -1 & \ddots & 0 \\
0 & 0 & 2 & \ddots & 0 \\
\vdots & \vdots & \ddots & \ddots & \vdots \\
0 & 0 & 0 & \ldots & 2
\end{pmatrix}_{(n+1)\times(n+1)}
$$

are invertible for all n, and

$$
A_n^{-1} = \frac{1}{2}
\begin{pmatrix}
1 & 1/2 & 1/4 & \ldots & 1/2^n \\
0 & 1 & 1/2 & \ldots & 1/2^{n-1} \\
0 & 0 & 1 & \ldots & 1/2^{n-2} \\
\vdots & \vdots & \ddots & \ddots & \vdots \\
0 & 0 & 0 & \ldots & 1
\end{pmatrix}. \tag{2}
$$

Comparing (1) with (2) one clearly gets applicability of the finite section method for A. Now we modify the operator A a little bit by changing its coefficients, that is, we consider the operator $B = -I + 2U = 2U(I - \frac{1}{2}U^{-1})$. The Neumann series entails again the invertibility of B, and it gives moreover

$$
B^{-1} = \frac{1}{2}U^{-1}(I + \frac{1}{2}U^{-1} + \frac{1}{4}U^{-2} + \ldots)
$$

or

$$
B^{-1} = \frac{1}{2}
\begin{pmatrix}
\ddots & \vdots & & & & \\
\ddots & 0 & \vdots & & & \\
\ddots & 1 & 0 & \vdots & & \\
\ddots & 1/2 & 1 & 0 & & \\
& 1/4 & 1/2 & 1 & \ddots & \\
& \vdots & 1/4 & 1/2 & \ddots & \\
& & \vdots & 1/4 & \ddots & \\
& & & \vdots & \ddots
\end{pmatrix},
$$

whereas

$$
B_n^{-1} = \begin{pmatrix}
1 & 2 & 4 & \cdots & 2^n \\
0 & 1 & 2 & \cdots & 2^{n-1} \\
0 & 0 & 1 & \cdots & 2^{n-2} \\
\vdots & \vdots & \ddots & \ddots & \vdots \\
0 & 0 & 0 & \cdots & 1
\end{pmatrix} .
$$

Thus, B^{-1} is a lower triangular matrix, but the sequence (B_n^{-1}) consists of upper triangular matrices and is, moreover, unbounded. This shows that the finite section method
cannot solve the uniquely solvable infinite system with system operator B ! (Let us remark that there are modifications of the finite section method which work also in this
situation, compare Chapter III in [GF 1]).

Concerning the history of the finite section method for Toeplitz operators one could
distinguish between two periods. The first one is characterized by using non-local factorization techniques, and it was opened by Baxter [Ba] and Reich [Re]. They showed that
the finite section method applies to the Toeplitz operator $T(a)$, acting on l^1 and being
generated by a function in the Wiener algebra, if this function does not vanish on the unit
circle and if it has winding number zero. Gohberg and Feldman [GF 2] then proved the
nessecity of this condition, and they were also able to show that this result remains valid
for Toeplitz operators $T(a)$ acting on arbitrary l^p-spaces if a is an arbitrary continuous
function in the class $C_{p,0}$ introduced in Section 2.3.1 below. The development culminated
with Gohberg and Feldman's book [GF 1], in which a first systematic and comprehensive
theory of projection methods for convolution equations was given and which is a basic
reference on this topic till now. In this monograph, they also successfully tackled the case
of continuous matrix valued coefficients. (See Chapter VIII, Section 5 of this monograph
for the proof of Theorem 1.6.)

Pertaining to compact pertubations we included some historical remarks into the text.
Theorem 1.7 is Silbermann's [S 2]. It makes essential use of Widom's [W 1] formula stated
in Proposition 1.6.

The second period of investigating the finite section method for Toeplitz operators
started at the beginning of the seventies when, under the impression of Simonenko's local Fredholm theory, V.B.Dybin brought forth the idea that local methods ought to be
applicable to projection methods, too. This idea is by no means evident because applicability of an approximation method is global in its nature. Kozak, a student of Dybin's,
was the first to carry out this programme. He algebraized and essentialized the stability (and did so factorize out the "globalness"), and then he developed a self-contained
local theory of the finite section method for one- and higher-dimensional operators with
continuous symbols [Ko]. The problem whether the finite section method for Toeplitz
operators with piecewise continuous generating function is also local in Kozak's sense
had been open for a long time. That the answer is negative had been shown by one of the
authors in 1988 (see 8.59 in [BS 2]). The method presented in 1.5.3-1.5.4 is Silbermann's
[S 2]. His method is originally intended to operators with discontinuous coefficients. For
that reason we interrupt our brief history of the finite section method here, and we are
going to continue it after dealing with finite sections of Toeplitz operators with piecewise
continuous generating functions in Chapter 4.

A further important advantage of Silbermann's approach is its applicability not only to
finite section methods (or other Galerkin approximation methods) but also to collocation

methods. The pioneering paper into this direction is Junghanns/Silbermann [JS] where the trigonometric collocation for systems of singular integral operators with piecewise continuous coefficients is studied.

1.6.1-1.6.5 The general scheme presented in these sections is a summary of techniques and ideas which were developed to apply local principles to invertibility problems. It cannot be a rigid algorithm, but we hope that it could serve as a guide into the world of applications of non-commutative Gelfand theories to concrete problems both from operator theory and numerical analysis.

The first version of the lifting theorem goes back to Silbermann [S 2]. Inspired by Widom's formula in Proposition 1.6, he formulated and proved the result which we re-stated in Proposition 1.7(ii). In Roch, Silbermann [RS 2] there is a generalization of Silbermann's lifting theorem to the case of an arbitrary family of lifting homomorphisms, and the formulation presented in Theorem 1.8 appeared in Roch's thesis for the first time.

Theorem 1.9 is a well-known result (see any book on commutative Banach algebras), and the two-projections-theorem (Theorem 1.10) in the Banach algebra setting goes back to Roch, Silbermann [RS 7]; for the history of the C^*-version of this theorem we refer to the first chapter of the monograph [PS 2].

1.7.1-1.7.3 Theorem 1.13 is Böttcher, Krupnik and Silbermann's [BKS] (with a minor modification: in [BKS], there is $C_1 = C_2 = 1$). The quantitative version of the local inclusion theorem 1.14 was derived by Roch, Silbermann ([RS 1], Theorem 5.3). Part (a) of this theorem is new. The local inclusion theorem is also a key in understanding why in Simonenko's theory of operators of local type (compare Exercise 1.19) the existence of the envelope can be guaranteed. In this concrete setting, the inclusion theorem plays the role of a Stone-Weierstraß-theorem (see Simonenko/Chin Ngok Minh [SC] for the notion of an envelope and Roch/Silbermann [RS 8]).

Chapter 2

Spline spaces and Toeplitz operators

This chapter presents the basic theory of Toeplitz operators on l^p-spaces as well as of spline approximation methods for singular integral operators with constant coefficients, and it highlights the fruitful interplay between these two fields. The first fundamental result is de Boor's estimate which shows that spline spaces can be viewed as embeddings of the l^p-spaces into L^p-spaces. This allows to think of Toeplitz operators previously acting on l^p-spaces as acting on subspaces of L^p, and it offers the applicability of (continuous) Fourier and Mellin techniques to the (discrete) Toeplitz operators which will be done in the first part of the chapter. In its second part we shall employ the theory of Toeplitz operators to derive results on spline approximation methods. For instance, we shall explain how the local behaviour of Toeplitz operators determines such things as the convergence of certain approximation sequences in the strong operator topology. The basic observations in this chapter are four theorems (Lemma 2.1, Theorem 2.5, Proposition 2.17, Theorem 2.7) whose proofs are unfortunately rather technical and not very instructive. For that reason we have separated these proofs from the other material and present them in a separate section which can be omitted in the first reading without loss.

To start with, we repeat some auxiliary material from the theory of singular integral operators with constant coefficients.

2.1   Singular integral operators with constant coefficients

2.1.1   Singular integrals

Throughout this chapter we let p and α denote fixed real numbers satisfying the inequalities $p > 1$ and $0 < 1/p + \alpha < 1$ and, given a (possibly unbounded) subinterval $\mathbf{I}$ of the real axis $\mathbf{R}$, we let $L^p_I(\alpha)$ refer to the weighted Lebesgue space endowed with the norm

$$\|f\| = \left(\int_I |f(t)|^p \, |t|^{\alpha p} \, dt \right)^{1/p} \ . \tag{1}$$

Clearly, $L^p_I(\alpha)$ can be viewed as a closed subspace of $L^p_R(\alpha)$, and this point of view allows, for example, to identify the identity operator on $L^p_I(\alpha)$ with the operator $\chi_I I$ of multiplication by the characteristic function χ_I of $\mathbf{I}$ acting on $L^p_R(\alpha)$. Analogously, a linear bounded operator A on $L^p_I(\alpha)$ can be identified with the operator $\chi_I A \chi_I I$ on $L^p_R(\alpha)$. In the sequel we shall often use this identification without comment. Further we shall always identify the dual space $(L^p_I(\alpha))^*$ with the space $L^q_I(-\alpha)$, where $1/p + 1/q = 1$, and we let $(.,.)$ refer to the sesquilinear form $(u, v) = \int_I u(t)\overline{v(t)} \, dt$.

The *singular integral operator* S_I on $\mathbf{I}$ is defined by

$$(S_I f)(t) = \frac{1}{\pi i} \int_I \frac{f(s)}{s - t} \, ds \ , \ t \in \mathbf{I} \ .$$

The kernel singularity as $s \to t$ involves that this integral does not exist in the usual sense, but it exists as a Cauchy principal value integral. The operator S_I is bounded on $L^p_I(\alpha)$ if the conditions imposed on p and α are fulfilled. Here are two important properties of the singular integral operator S_R:

- It is its own inverse, that is

$$S^2_R = I \ . \tag{2}$$

- It is a Fourier convolution operator.

To make the latter more precise, let F denote the *Fourier transform* acting on the Schwartz space $S(\mathbf{R})$ via

$$(F f)(z) = \int_R e^{-2\pi i x z} f(x) \, dx \ , \ z \in \mathbf{R} \ ,$$

and write F^{-1} for the inverse Fourier transform,

$$(F^{-1} f)(x) = \int_R e^{2\pi i x z} f(z) \, dz \ , \ x \in \mathbf{R} \ .$$

The Fourier transform extends by continuity to a unitary operator on the Hilbert space $L_R^2(0)$ which is denoted by F again. Now one can show that the restriction of the singular integral operator S_R onto $L_R^p(\alpha) \cap L_R^2(0)$ coincides with the restriction of the *Fourier convolution operator* $F^{-1}aF$ with generating function $a(z) = \operatorname{sgn} z$ (the sign function) onto the same space.

More general, if b is a bounded function then the operator $F^{-1}bF$ is well-defined on $L_R^p(\alpha) \cap L_R^2(0)$. If this operator extends boundedly onto all of $L_R^p(\alpha)$ then we call this extension Fourier convolution operator and denote it by $W^0(b)$, and the function b is called an $L_R^p(\alpha)$-Fourier multiplier. Each $L_R^p(\alpha)$-multiplier is a bounded function, and

$$\| b \|_\infty \leq \| W^0(b) \|_{L(L_R^p(\alpha))} . \tag{3}$$

Conversely, if b is a bounded function with finite total variation $V(b)$, then b is an $L_R^p(\alpha)$-multiplier for each $p > 1$ and $\alpha \in (-1/p, 1 - 1/p)$ and, moreover

$$\| W^0(b) \|_{L(L_R^p(\alpha))} \leq C_{p,\alpha} (\| b \|_\infty + V(b)) \tag{4}$$

(Stechkin's inequality, see [BS 2], 9.3(e)). Further, the set of $L_R^2(0)$-multipliers coincides with the algebra L_R^∞ of all essentially bounded and measurable functions, and

$$\| W^0(a) \|_{L(L_R^2(0))} = \| a \|_\infty .$$

To the singular integral operators S_{R+} and S_R, we associate two Banach algebras: we let $\Sigma^p(\alpha)$ stand for the smallest closed subalgebra of $L(L_{R+}^p(\alpha))$ containing the operator S_{R+} and the identity operator, and let $\Sigma_R^p(\alpha)$ refer to the smallest closed subalgebra of $L(L_R^p(\alpha))$ containing the operator S_R, the operator $\chi_{R+}I$ of multiplication by the characteristic function of the semi-axis $\mathbf{R}^+$, and the identity operator.

2.1.2 The algebra $\Sigma^p(\alpha)$

There is an alternative description of the algebra $\Sigma^p(\alpha)$ which dominates the whole theory of this algebra. For, we define the *Mellin transform M* (depending on p and α) by

$$(M f)(z) = \int_0^\infty x^{1/p+\alpha-zi-1} f(x) \, dx , \ z \in \mathbf{R} ,$$

and the inverse Mellin transform M^{-1} by

$$(M^{-1} f)(x) = \frac{1}{2\pi} \int_{-\infty}^\infty x^{zi-1/p-\alpha} f(z) \, dz , \ x \in \mathbf{R}^+ .$$

Further, we introduce operators $E_{p,\alpha}$ by

$$(E_{p,\alpha}f)(x) = \frac{1}{2\pi} f(\frac{1}{2\pi} \ln x) \, x^{-1/p-\alpha} , \ x \in \mathbf{R}^+ .$$

One easily checks that

$$\|E_{p,\alpha}f\|_{L^p_{R+}(\alpha)} = (2\pi)^{1/p-1} \|f\|_{L^p_R(0)} ,$$

and that $M E_{p,\alpha} = F$. If b is an $L^p_R(0)$-multiplier then the operator

$$M^0(b) := E_{p,\alpha}W^0(b) E^{-1}_{p,\alpha}$$

is bounded on $L^p_{R+}(\alpha)$, and we call $M^0(b)$ the Mellin convolution by b. For another representation of Mellin convolutions see 2.1.4. As a consequence of 2.1.1(3) one gets that

$$\| b \|_\infty \leq \|M^0(b)\|_{L(L^p_{R+}(\alpha))} \tag{1}$$

for each multiplier, and 2.1.1(4) implies the Stechkin inequality for Mellin convolutions:

$$\|M^0(b)\|_{L(L^p_{R+}(\alpha))} \leq C \left(\| b \|_\infty + V(b) \right) \tag{2}$$

whenever b is a bounded function with finite total variation. Moreover,

$$\|M^0(b)\|_{L(L^2_{R+}(0))} = \| b \|_\infty . \tag{3}$$

The algebra $\Sigma^p(\alpha)$ can be characterized as follows.

(a) (Simonenko/Chin Ngok Minh) $\Sigma^p(\alpha)$ is the smallest closed subalgebra of $L(L^p_{R+}(\alpha))$ which contains all Mellin convolution operators $M^0(b)$ where b is a continuous function of finite total variation which possesses finite limits at $\pm\infty$.

Proof Let $\mathcal{A}$ denote the smallest closed subalgebra of $L(L^p_{R+}(\alpha))$ which contains all Mellin convolutions $M^0(b)$ with b being as in assertion (a). Let us first show that $S_{R+} \in \mathcal{A}$ and, hence, $\Sigma^p(\alpha) \subseteq \mathcal{A}$. Formulas 3.238.1 and 3.238.2 of [GR] yield that the Mellin transform of the kernel function $x \mapsto \frac{1}{\pi i}(1 - x)^{-1}$ of S_{R+} is equal to

$$s(z) = \frac{1}{\pi i} \int_0^\infty x^{1/p+\alpha-zi-1} \frac{1}{1 - x}\, dx = \coth \pi(z + i(1/p + \alpha)). \tag{4}$$

Thus, $S_{R+} = M^0(s)$, and since s is continuous, of finite total variation, and with limits ± 1 at infinity, we get the desired inclusion.

For the reverse inclusion, set $K := s(\overline{\mathbf{R}})$ and define the operator $Q : C(K) \to C(\overline{\mathbf{R}})$ by

$$(Qf)(z) = f(s(t)) , \quad t \in \overline{\mathbf{R}} .$$

Clearly, Q is an isometrical isomorphism which does not change the total variation. Our first claim is the inclusion $M^0(b) \in \Sigma^p(\alpha)$ for all $b \in Q(C^1(K))$. (As usual, $C^1(K)$ refers to the Banach space of all continuously differentiable functions on the compact K.) Indeed, let $f = Q^{-1}b$, and approximate f by a sequence (f_n) of polynomials in the norm of $C^1(K)$. Then $\|f - f_n\|_\infty + V(f - f_n) \to 0$ and

$$\|b - Qf_n\|_\infty + V(b - Qf_n) \to 0$$

whence via Stechkin's inequality (2) follows

$$\| M^0(b) - M^0(Qf_n) \|_{L(L^p_{R+}(\alpha))} \to 0 .$$

Since $M^0(Qf_n) \in \Sigma^p(\alpha)$ (observe that Qf_n is a polynomial in s and $M^0(Qf_n)$ is a polynomial in S_{R+}) we get our claim.

Let now b be an arbitrary function in $C(\overline{\mathbf{R}})$ with finite total variation. Then there is a sequence (b_n) of functions in $Q(C^1(K))$ which approximates b in the norm of $C(\overline{\mathbf{R}})$ and with $\sup_n V(b_n) < \infty$. By what has already shown, $M^0(b_n) \in \Sigma^p(\alpha)$.

From (3) we derive that $\|M^0(b_n) - M^0(b)\|_{L(L^2_{R+}(0))} \to 0$, and from (2) it is obvious that $\sup_n \|M^0(b_n) - M^0(b)\|_{L(L^{p'}_{R+}(\alpha'))} < \infty$ for all $p' > 1$ and $\alpha' \in (-1/p', 1 - 1/p')$. Choose p' and α' so that $1/p = (1-\theta)/2 + \theta/p'$ and $\alpha = \alpha'\theta/p'$ with some $\theta \in (0,1)$. Then the Stein-Weiss interpolation theorem (compare [BL], Corollary 5.5.4) yields that $\|M^0(b_n) - M^0(b)\|_{L(L^p_{R+}(\alpha))} \to 0$ as $n \to \infty$ and, consequently, $\mathcal{A} \subseteq \Sigma^p(\alpha)$. ∎

Let us mention some operators whose affiliation to $\Sigma^p(\alpha)$ is a consequence of the above description. For each complex β with $0 < \operatorname{Re}\beta < 2\pi$, the *generalized Hankel operator* N_β,

$$(N_\beta f)(t) = \frac{1}{\pi i} \int_{R+} \frac{f(s)}{s - e^{i\beta}t}\,ds , \ t \in \mathbf{R}^+ ,$$

belongs to $\Sigma^p(\alpha)$. It is a Mellin convolution

$$N_\beta = M^0(n_\beta) \quad \text{with} \quad n_\beta(z) = \frac{e^{(z+i(1/p+\alpha))(\pi-\beta)}}{\sinh \pi(z + i(1/p + \alpha))} , \tag{5}$$

(see [GR], 3.194.4) and the operators S_{R+} and N_β are related by the identity

$$S^2_{R+} - I = N_\beta N_{2\pi-\beta} . \tag{6}$$

Further, if $0 < \operatorname{Re}(1/p + \alpha + \gamma) < 1$, then the *weighted singular integral operator* $S_{R+,\gamma}$,

$$(S_{R+,\gamma}f)(t) = \frac{1}{\pi i} \int_{R+} \left(\frac{t}{s}\right)^\gamma \frac{f(s)}{s - t}\,ds , \ t \in \mathbf{R}^+ ,$$

and, if moreover $0 < \operatorname{Re}\beta < 2\pi$, the *weighted generalized Hankel operator* $N_{\beta,\gamma}$,

$$(N_{\beta,\gamma}f)(t) = \frac{1}{\pi i} \int_{R+} \left(\frac{t}{s}\right)^\gamma \frac{f(s)}{s - e^{i\beta}t}\,ds , \ t \in \mathbf{R}^+ ,$$

lie in $\Sigma^p(\alpha)$, and

$$S_{R^+,\gamma} = M^0(s_\gamma) \quad \text{with} \quad s_\gamma(z) = \coth \pi(z + i(1/p + \alpha + \gamma)) , \tag{7}$$

$$N_{\beta,\gamma} = M^0(n_{\beta,\gamma}) \quad \text{with} \quad n_{\beta,\gamma}(z) = \frac{e^{(z+i(1/p+\alpha+\gamma))(\pi-\beta)}}{\sinh \pi(z + i(1/p + \alpha + \gamma))} . \tag{8}$$

The weighted operators $S_{R^+,\beta}$ and $N_{\pi,\gamma}$ are connected by

$$S_{R^+,\beta} S_{R^+,\gamma} - \cos \pi(\beta - \gamma) N_{\pi,\beta} N_{\pi,\gamma} = I . \tag{9}$$

Denote the operator N_π simply by N, and let $N^p(\alpha)$ refer to the smallest closed two-sided ideal of $\Sigma^p(\alpha)$ which contains N. Then we have

(b) *An operator $M^0(b) \in \Sigma^p(\alpha)$ belongs to the ideal $N^p(\alpha)$ if and only if $b(\pm\infty) = 0$.*

(c)   (i) *$N^p(\alpha)$ is the smallest closed two-sided ideal of $\Sigma^p(\alpha)$ which contains any operator N_β with $0 < \operatorname{Re} \beta < 2\pi$.*

    (ii) *$N^p(\alpha)$ is the smallest closed two-sided ideal of $\Sigma^p(\alpha)$ which contains N^2.*

    (iii) *$N^p(\alpha)$ is the smallest closed subalgebra of $\Sigma^p(\alpha)$ which contains the operators N^2 and $S_{R^+}N^2$.*

    (iv) *$N^p(\alpha)$ is the smallest closed subalgebra of $\Sigma^p(\alpha)$ which contains the operators N and $S_{R^+}N$.*

(d) *(Costabel's decomposition of $\Sigma^p(\alpha)$). The algebra $\Sigma^p(\alpha)$ decomposes into the direct sum $\mathbf{C}\, I \dotplus \mathbf{C}\, S_{R^+} \dotplus N^p(\alpha)$.*

(e) *The maximal ideal space of the commutative Banach algebra $\Sigma^p(\alpha)$ is homeomorphic to the two-point compactification $\overline{\mathbf{R}}$ of the real axis. In particular, the Mellin convolution operator $M^0(b) \in \Sigma^p(\alpha)$ is invertible in $\Sigma^p(\alpha)$ if and only if $b(z) \neq 0$ for all $z \in \mathbf{R}$ and if $b(\pm\infty) \neq 0$. Thus, $\Sigma^p(\alpha)$ is an inverse closed subalgebra of $L(L^p_{R^+}(\alpha))$.*

(f) *The algebras $\Sigma^p(\alpha)$ and $\Sigma^p(0)$ are isometrically isomorphic. The isomorphism sends the singular integral operator $S_{R^+} \in \Sigma^p(\alpha)$ into the weighted singular integral operator $S_{R^+,\alpha} \in \Sigma^p(0)$.*

Proof   From (a) we know that each operator in $\Sigma^p(\alpha)$ is of the form $M^0(b)$ with a continuous function b. Thus, $M^0(b)$ can be written as

$$M^0(b) = \alpha I + \beta S_{R^+} + M^0(c) , \tag{10}$$

where c is a continuous function vanishing at $\pm\infty$, and $\alpha, \beta \in \mathbf{C}$. Moreover, the decomposition (10) is unique, and one has

$$\alpha = \frac{1}{2}(b(+\infty) + b(-\infty)) , \quad \beta = \frac{1}{2}(b(+\infty) - b(-\infty)) .$$

These identities in combination with (1) imply that α, β depend continuously on $M^0(b)$. Thus, $M^0(c)$ depends continuously on $M^0(b)$ (with respect to the norm in $L(L^p_{R^+}(\alpha))$).

In order to prove (b), let $M^0(b) \in \Sigma(\alpha)$ with $b(\pm\infty) = 0$. By definition, $M^0(b)$ can be approximated by a sequence of polynomials

$$M^0(p_n(s)) = p_n(M^0(s)) = p_n(S_{R^+}) \quad \text{in} \quad S_{R^+} \; .$$

By the continuous dependence stated above, we can suppose that $p_n(s)(\pm\infty) = 0$. Hence, $p_n(s) = (1 - s^2) q_n(s)$ with other polynomials q_n. Identity (6) yields that $I - S^2_{R^+} = -N^2$ and, consequently,

$$M^0(p_n(s)) = (I - S^2_{R^+})M^0(q_n(s)) = -N^2 M^0(q_n(s)) \in N^2\Sigma^p(\alpha) \subseteq N^p(\alpha), \quad (11)$$

which implies that $M^0(b) \in N^p(\alpha)$. Conversely, if $M^0(b) \in N^p(\alpha)$ then, evidently, $b(\pm\infty) = 0$.

The inclusion (11) shows moreover that (c), (ii)-(iv) hold, and for (c), (i) one employs identity (6) in its general form and shows (in full analogy to the proof of (a)) that the weighted singular integral operator $S_{R^+,\gamma}$ also generates $\Sigma^p(\alpha)$.

Assertion (d) is now obvious, and for (e) we return once more to (1). This identity shows that, for each $\tau \in \overline{R}$, the mapping $M^0(b) \mapsto b(\tau)$ is a bounded multiplicative functional of the commutative algebra $\Sigma^p(\alpha)$ and, conversely, it is not hard to check that each bounded multiplicative functional is of this form, and that the Gelfand topology on $\overline{R}$ coincides with the usual one. Theorem 1.3 gives the assertion (for details see also [BS 2], Proposition 2.46).

For a proof of (f), set $w(t) = t^\alpha$. Obviously, the mapping $T : A \mapsto wAw^{-1}I$ is an isometrical isomorphism of $L(L^p_{R^+}(\alpha))$ onto $L(L^p_{R^+}(0))$ which sends S_{R^+} into $S_{R^+,\alpha}$. Since $S_{R^+,\alpha}$ together with the identity operator also generate $\Sigma^p(0)$ (compare the above remark), one easily gets that $T(\Sigma^p(\alpha)) = \Sigma^p(0)$. ∎

2.1.3 The algebra $\Sigma^p_R(\alpha)$

Let $L^p_{R^+}(\alpha)_2$ stand for the Banach space of all pairs $(f_1, f_2)^T$ of functions $f_1, f_2 \in L^p_{R^+}(\alpha)$ endowed with the norm

$$\| (f_1, f_2)^T \| := (\|f_1\|^p + \|f_2\|^p)^{1/p} \; .$$

The mapping $\eta : f \mapsto (f_1, f_2)^T$ with $f_1(t) = f(t)$ and $f_2(t) = f(-t)$ for all $t \in R^+$ is a linear isometry from $L^p_R(\alpha)$ onto $L^p_{R^+}(\alpha)_2$, and the mapping $A \mapsto \eta A\eta^{-1}$ is an isometric algebra isomorphism from $L(L^p_R(\alpha))$ onto $L(L^p_{R^+}(\alpha)_2)$. The following proposition describes the image of the algebra $\Sigma^p_R(\alpha)$ under this isomorphism.

Proposition 2.1 $\eta\Sigma^p_R(\alpha)\eta^{-1}$ *is the algebra of all operators*
$\begin{pmatrix} A & B \\ C & D \end{pmatrix} \in L(L^p_{R^+}(\alpha)_2)$ *with* $A, D \in \Sigma^p(\alpha)$ *and* $B, C \in N^p(\alpha)$.

Proof   Denote, for a moment, by $\mathcal{A}$ the set of all operators $\begin{pmatrix} A & B \\ C & D \end{pmatrix} \in$
$L(L^p_{R+}(\alpha)_2)$ with $A, D \in \Sigma^p(\alpha)$ and $B, C \in N^p(\alpha)$. It is easy to see that $\mathcal{A}$ is
actually an algebra. A straightforward computation shows that

$$\eta \chi_{R+} \eta^{-1} = \begin{pmatrix} I & 0 \\ 0 & 0 \end{pmatrix} \quad \text{and} \quad \eta S_R \eta^{-1} = \begin{pmatrix} S_{R+} & -N \\ N & -S_{R+} \end{pmatrix}$$

and, thus, $\eta \Sigma^p_R(\alpha) \eta^{-1} \subseteq \mathcal{A}$. It remains to verify that the mapping $A \mapsto \eta A \eta^{-1}$
from $\Sigma^p_R(\alpha)$ into $\mathcal{A}$ is even onto. But this is a simple consequence of the identities

$$\eta \chi_{R+} S_R \chi_{R+} \eta^{-1} = \begin{pmatrix} S_{R+} & 0 \\ 0 & 0 \end{pmatrix} \quad \text{and} \quad -\eta \chi_{R+} S_R \chi_{R-} \eta^{-1} = \begin{pmatrix} 0 & N \\ 0 & 0 \end{pmatrix}. \quad \blacksquare$$

This proposition furnishes some important consequences: Since the entries of a
matrix $\begin{pmatrix} A & B \\ C & D \end{pmatrix} \in \eta \Sigma^p_R(\alpha) \eta^{-1}$ commute with each other one has an effective
invertibility criterion for the operators in $\Sigma^p_R(\alpha) : E \in \Sigma^p_R(\alpha)$ is invertible if and
only if $\eta E \eta^{-1} = \begin{pmatrix} A & B \\ C & D \end{pmatrix}$ is invertible or if and only if

$$\det \begin{pmatrix} A & B \\ C & D \end{pmatrix} = AD - BC \in \Sigma^p(\alpha)$$

is invertible, and now it remains to apply 2.1.2(e). Moreover, $\Sigma^p_R(\alpha)$ is inverse
closed in $L(L^p_R(\alpha))$.

Property 2.1.2(f) entails via the above proposition that the algebras $\eta \Sigma^p_R(\alpha) \eta^{-1}$
and $\eta \Sigma^p_R(0) \eta^{-1}$ are isomorphic; hence, the algebras $\Sigma^p_R(\alpha)$ and $\Sigma^p_R(0)$ coincide up
to isomorphy. Finally, we have the following analogue of Costabel's decomposition
of $\Sigma^p(\alpha)$:

$$\begin{aligned} \Sigma^p_R(\alpha) \;=\;& \mathbf{C}\chi_{R+} I \dotplus \mathbf{C}\chi_{R-} I \dotplus \mathbf{C}S_{R+} \dotplus \mathbf{C}S_{R-} \dotplus \\ &+\; N^p(\alpha) \dotplus JN^p(\alpha) \dotplus N^p(\alpha)J \dotplus JN^p(\alpha)J \,, \end{aligned}$$

where $S_{R-} = \chi_{R-} S_R \chi_{R-} I$ is the singular integral operator along the negative
semi-axis $\mathbf{R}^-$, and J stands for the flip operator $(Jf)(t) = f(-t), t \in \mathbf{R}$. It is
interesting to observe that, although the operator J does not belong to $\Sigma^p_R(\alpha)$, all
operators in $JN^p(\alpha)$ and $N^p(\alpha)J$ are contained in $\Sigma^p_R(\alpha)$ (see Exercise 2.3).

Example 2.1 *Let* $0 < \mathrm{Re}\,(1/p + \alpha + \gamma) < 1$. *Then the weighted singular integral
operator on* $\mathbf{R}$,

$$(S_{R,\gamma}f)(t) = \frac{1}{\pi i} \int_R \left|\frac{t}{s}\right|^\gamma \frac{f(s)}{s-t} \, ds \,, \quad t \in \mathbf{R} \,,$$

satisfies the identity

$$\eta S_{R,\gamma} \eta^{-1} = \begin{pmatrix} S_{R+,\gamma} & -N_{\pi,\gamma} \\ N_{\pi,\gamma} & -S_{R+,\gamma} \end{pmatrix} \tag{1}$$

whence follows that $S_{R,\gamma}$ belongs to $\Sigma_R^p(\alpha)$ and that the spectrum of $S_{R,\gamma}$ both in $L(L_R^p(\alpha))$ as in $\Sigma_R^p(\alpha)$ equals $\{-1,1\}$. Finally, Costabel's decomposition of $S_{R,\gamma}$ is given by

$$S_{R,\gamma} = S_{R^+} + S_{R^-} + M^0(b) + JN_{\pi,\gamma} - N_{\pi,\gamma}J - JM^0(b)J \,,$$

where $b = s_\gamma - s$.

2.1.4 Integral representations

Let $\mathcal{N}$ denote the smallest (not necessarily closed) subalgebra of the algebra of all continuous functions on $\mathbf{R}$ containing all functions b which are, together with their Mellin preimages $k = M^{-1}b$, continuously differentiable and for which there exist constants M_1 and M_2 and polynomials P_1 and P_2 such that

$$\sup_{z\in R} |b(z)(1+|z|)| \leq M_1 \,, \quad \sup_{z\in R} |b'(z)(1+|z|)^2| \leq M_2 \tag{1}$$

and

$$|k(x)(1+x)| \leq P_1(|\ln x|) \quad , \quad |k'(x)(1+x)^2| \leq P_2(|\ln x|) \,. \tag{2}$$

Proposition 2.2 *If $b \in \mathcal{N}$ then $\lim_{z\to\pm\infty} b(z) = 0$, the function b has a finite total variation, and the set $N_0^p(\alpha) = M^{-1}\mathcal{N}M$ is a dense subalgebra of the ideal $N^p(\alpha)$.*

Proof   The first assertion is immediately from (1). Let $V_x^y(b)$ denote the total variation of b over the interval $[x,y]$. Noticing that $V_{-\infty}^\infty(b) < \infty$ if only $V_{-\infty}^\infty(\mathrm{Re}\,b) < \infty$ and $V_{-\infty}^\infty(\mathrm{Im}\,b) < \infty$ and that $\mathrm{Re}\,b$ and $\mathrm{Im}\,b$ belong to $\mathcal{N}$ whenever $b \in \mathcal{N}$, we can suppose the function b to be real-valued without loss. In this case,

$$V_{-\infty}^\infty(b) = \sum_{k\in Z} V_k^{k+1}(b) = \sum_{k\in Z} \int_k^{k+1} |b'(z)|\,dz \leq \sum_{k\in Z} \frac{M_2}{(1+|k|)^2} < \infty$$

which proves the second assertion. Thus, by 2.1.2(a) and (b), the functions in $\mathcal{N}$ are generating functions of Mellin convolution operators which belong to the ideal $N^p(\alpha)$. Since $\mathcal{N}$ is an algebra, the set $M^{-1}\mathcal{N}M \subseteq N^p(\alpha)$ is also an algebra, and the density of $N_0^p(\alpha)$ in $N^p(\alpha)$ is a simple consequence of the fact that the functions n_π and sn_π given by 2.1.2(4) and 2.1.2(5) belong to $\mathcal{N}$ (their Mellin preimages are the functions $x \mapsto (\pi i(1+x))^{-1}$ and $x \mapsto \ln x (\pi^2(1+x))^{-1}$, respectively), and that the operators $N = M^0(n_\pi)$ and $S_{R^+}N = M^0(sn_\pi)$ generate a dense subalgebra of $N^p(\alpha)$ by 2.1.2(c)(IV). ∎

Obviously, the set $M^{-1}\mathcal{N}$ is an algebra with respect to the *Mellin convolution*

$$(k_1 * k_2)(x) = \int_{R^+} k_1\left(\frac{x}{s}\right) k_2(s) \frac{ds}{s} \,, \quad x \in \mathbf{R}^+ , \tag{3}$$

as multiplication, and a function k belongs to $M^{-1}\mathcal{N}$ if and only if k is the kernel function of a Mellin convolution operator $M^0(b)$ with $b \in \mathcal{N}$, that is

$$(M^0(b)f)(x) = \int_{R^+} k\left(\frac{x}{s}\right) f(s) \frac{ds}{s} \ , \ x \in \mathbf{R}^+ . \tag{4}$$

The operators in $N_0^p(\alpha)$ are distinguished by a lot of remarkable properties (see Sections 2.4.1, 2.4.2 and 5.2.6 below) which will be exhibited at those places where they are really needed.

2.2   Piecewise constant splines

2.2.1   Definitions

Given numbers $n \in \mathbf{Z}^+$ and $k \in \mathbf{Z}$ we define functions ϕ_{kn} by

$$\phi_{kn}(t) = \chi_{[0,1)}(nt - k) .$$

The condition $0 < 1/p + \alpha < 1$ imposed on α guarantees that all functions ϕ_{kn} belong both to the space $L_R^p(\alpha)$ and to its dual $L_R^q(-\alpha)$ with $1/p + 1/q = 1$. So it makes sense to consider the smallest closed subspace S_n of $L_R^p(\alpha)$ which contains all functions ϕ_{kn} with k ranging over the integers. The space S_n is the simplest example of a *spline space*. Its elements, the *piecewise constant* splines, are just those functions in $L_R^p(\alpha)$ which are constant on each interval $[k/n, (k+1)/n)$. There is a natural projection operator L_n mapping $L_R^p(\alpha)$ onto its subspace S_n, the so-called *Galerkin projection*, which sends the function $u \in L_R^p(\alpha)$ into the function

$$L_n u = n \sum_{k \in Z}(u, \phi_{kn}) \phi_{kn} \tag{1}$$

with $(.,.)$ defined as in 2.1.1.

Proposition 2.3 *The operators L_n defined by (1) are bounded projections from $L_R^p(\alpha)$ onto S_n, and $L_n \to I$ and $L_n^* \to I$ strongly as $n \to \infty$.*

In Sections 2.8.2 and 2.9.1 we shall prove a more general result. That's why we renounce to give a separate proof of this assertion.

2.2.2   De Boor's estimates

Given a subset $\mathbf{I}$ of $\mathbf{Z}$ we let $l_I^p(\alpha)$ stand for the discrete analogue of the $L_I^p(\alpha)$-spaces, viz. for the space of all sequences $(x_k)_{k \in I}$ of complex numbers provided with the norm

$$\|(x_k)\| = \left(\sum_{k \in I}|x_k|^p(|k|+1)^{\alpha p}\right)^{1/p} .$$

Again, we shall think of $l_I^p(\alpha)$ as being embedded into $l_Z^p(\alpha)$, and we shall always identify the dual of $l_I^p(\alpha)$ with $l_I^q(-\alpha), 1/p + 1/q = 1$.

The following estimates entail that the spaces $l_Z^p(\alpha)$ and $S_n (\subseteq L_R^p(\alpha))$ can be identified up to isomorphy. Estimates of this kind are usually attributed to de Boor.

Proposition 2.4 (a) If $(x_k) \in l_Z^p(\alpha)$ then the series $\sum_{k \in Z} x_k \phi_{kn}$ converges in $L_R^p(\alpha)$, and there is a constant C_1 such that

$$\left\| \sum_k x_k \phi_{kn} \right\| \leq C_1 \, n^{-(1/p+\alpha)} \left\| (x_k) \right\| .$$

(b) If the series $\sum_{k \in Z} x_k \phi_{kn}$ converges in $L_R^p(\alpha)$ then the sequence (x_k) is in $l_Z^p(\alpha)$, and there is a constant C_2 with

$$\left\| (x_k) \right\| \leq C_2 \, n^{1/p+\alpha} \left\| \sum_k x_k \phi_{kn} \right\| .$$

Again, the proofs are not hard but we refer to Sections 2.6.2 and 2.7.4 where more general estimates will be verified. As a consequence of de Boor's estimates we obtain that the mappings

$$E_n \; : \; l_Z^p(\alpha) \to S_n(\subseteq L_R^p(\alpha)) \; , \quad (x_k) \mapsto \sum_k x_k \phi_{kn} \; ,$$

and

$$E_{-n} \; : \; S_n \to l_Z^p(\alpha) \; , \quad \sum_k x_k \phi_{kn} \mapsto (x_k) \; ,$$

are linear and bounded, that $\|E_n\| \leq C_1 \, n^{-(1/p+\alpha)}$ and $\|E_{-n}\| \leq C_2 \, n^{1/p+\alpha}$ and, consequently, $\sup_n \|E_n\| \, \|E_{-n}\| < \infty$.

Thus, every number $n \in \mathbf{Z}^+$ gives rise to a continuous embedding E_n of the space $l_Z^p(\alpha)$ into $L_R^p(\alpha)$. In particular, each operator $A \in L(l_Z^p(\alpha))$ corresponds to the operator $E_n A E_{-n}$ which acts boundedly on the space S_n and, conversely, if $B \in L(S_n)$ then $E_{-n} B E_n \in L(l_Z^p(\alpha))$.

Next we explain how this correspondence can be used to study a special, but typical, spline approximation method for singular integral operators.

2.2.3 The Galerkin method for singular integral operators on the axis

Consider the singular integral equation

$$(aI + bS_R)x \; = \; u \quad , \; u \in L_R^p(\alpha)$$

with a and b being complex constants. The S_n-*Galerkin method* with S_n both as trial and as test space seeks for an approximate solution $x_n \in S_n$ of this equation satisfying

$$((aI + bS_R) x_n \, , \, \phi_{kn}) \; = \; (u \, , \, \phi_{kn}) \tag{1}$$

for all $k \in \mathbf{Z}$. It is obvious from the definition of the projections L_n in 2.2.1 that the approximating equations (1) can be rewritten in the form

$$L_n (aI + bS_R) x_n \; = \; L_n u \, . \tag{2}$$

By Proposition 1.1, and due to the strong convergence of the operators $L_n(aI + bS_R)L_n$ to $aI + bS_R$, the Galerkin method (2) applies to the operator $aI + bS_R$ if and only if this operator is invertible and if the sequence $(L_n(aI + bS_R)|S_n)$ is stable. Taking into account 2.1.1(2) or Proposition 2.1, the invertibility of $aI + bS_R$ is easy to check: it is invertible if and only if $a + b \neq 0$ and $a - b \neq 0$. So we are left with the stability problem. Clearly, the sequence $(L_n(aI + bS_R)|S_n)$ is stable if and only if the corresponding sequence $(E_{-n} L_n(aI + bS_R)E_n)$ of operators on $l^p_Z(\alpha)$ is stable (remember that $\sup_n \|E_n\| \, \|E_{-n}\| < \infty$!).

Let $(\sigma^{(n)}_{kl})_{k,l \in Z}$ stand for the matrix representation of the operator $E_{-n} L_n(aI + bS_R)E_n$ with respect to the standard basis of $l^p_Z(\alpha)$. By the definitions of $E_{\pm n}$ and L_n, we have

$$\sigma^{(n)}_{kl} = n((aI + bS_R)\phi_{ln}, \phi_{kn}) = \int_R ((aI + bS_R)\chi_{[0,1)})(t + k - l)\chi_{[0,1)}(t) \, dt \, . \tag{3}$$

Hence, the operators $E_{-n} L_n(aI + bS_R)E_n$ are independent of n whence follows that the sequence $(L_n(aI + bS_R)|S_n)$ is stable if and only if the operator $E_{-1} L_1(aI + bS_R)E_1$ is invertible.

Moreover, (3) shows that the entries $\sigma^{(1)}_{kl}$ only depend on the differences $k - l$. Thus, the matrix $E_{-1} L_1(aI + bS_R)E_1$ is a two-sided infinite Toeplitz matrix, and our next goal will be the determination of the generating function of this matrix. That is, we look for a function on the unit circle whose k th Fourier coefficient equals $\sigma^{(1)}_{k0}$.

Since $aI + bS_R = F^{-1}cF$ with $c(z) = a + b \operatorname{sgn} z$ (see Section 2.1.1), and using the convolution theorem, we find

$$\sigma^{(1)}_{k0} = \int_R (F^{-1}cF \chi_{[0,1)})(t + k) \, \chi_{[0,1)}(t) \, dt = \int_R (FcF\chi_{[0,1)})(-k - t) \, \chi_{[0,1)}(t) \, dt$$

$$= ((FcF\chi_{[0,1)}) * \chi_{[0,1)})(-k) = (F(cF\chi_{[0,1)})(F^{-1}\chi_{[0,1)}))(-k)$$

$$= \int_R e^{2\pi i x k} c(x)(F\chi_{[0,1)})(x)(F^{-1}\chi_{[0,1)})(x) \, dx$$

$$= \int_R e^{-2\pi i k x} c(-x)(F\chi_{[0,1)})(-x)(F^{-1}\chi_{[0,1)})(-x) \, dx$$

$$= \int_R e^{-2\pi i x k} c(-x)\overline{(F\chi_{[0,1)})(x)} \, (F\chi_{[0,1)})(x) \, dx \, .$$

Introduce new variables $m \in \mathbf{Z}$ and $y \in [0, 1)$ with $x = m + y$ to obtain

$$\sigma_{k0}^{(1)} = \sum_{m \in Z} \int_0^1 e^{-2\pi i y k} c(-m - y) |(F\chi_{[0,1)})(m + y)|^2 \, dy$$

whence, taking into account the identity

$$(F\chi_{[0,1)})(z) = \begin{cases} e^{i\pi z} \frac{\sin(\pi z)}{\pi z} & \text{if} \quad z \neq 0 \\ 1 & \text{if} \quad z = 0 \end{cases},$$

follows

$$\sigma_{k0}^{(1)} = \sum_{m \in Z} \int_0^1 e^{-2\pi i y k} c(-m - y) \frac{\sin^2 \pi(m + y)}{\pi^2 (m + y)^2} \, dy. \tag{4}$$

Changing the summation and integration in (4), taking into account that $\sum_{m \in Z} \frac{\sin^2 \pi(m+y)}{\pi^2(m+y)^2} = 1$ (see, for example, [Re], 11.23), and abbreviating the function

$$y \mapsto -\frac{\sin^2 \pi y}{\pi^2} \sum_{m \in Z} \frac{\text{sgn}\,(m + 1/2)}{(y + m)^2} = \frac{\sin^2 \pi y}{\pi^2} \sum_{m \in Z} \frac{\text{sgn}\,(-y - m)}{(y + m)^2}$$

for $y \in (0, 1)$ by $\hat{\sigma}$, we obtain

$$\sigma_{k0}^{(1)} = \int_0^1 e^{-2\pi i y k} (a + b\,\hat{\sigma}(y)) \, dy. \tag{5}$$

Thus, if we let σ refer to the function on the unit circle $\mathbf{T}$

$$\sigma(e^{2\pi i y}) := \hat{\sigma}(y), \quad y \in [0, 1) \tag{6}$$

then (5) shows that $\sigma_{k0}^{(1)}$ is actually the k th Fourier coefficient of the function $a + b\sigma$ defined by (6). We denote the operator $E_{-1}L_1(aI + bS_R)E_1$ henceforth by $T^0(a + b\sigma)$. It is not hard to verify that the function σ possesses the one-sided limits $\sigma(1 + 0) = -1$ and $\sigma(1 - 0) = +1$ at $1 \in \mathbf{T}$ and that $\hat{\sigma}$ is real-valued and continuously and monotonically increasing on $(0, 1)$. Hence, the operator $T^0(a + b\sigma)$ is invertible if and only if the point 0 does not belong to the line segment $[a - b, a + b]$ joining $a - b$ to $a + b$ (see ([BS 2], 6.28) or 2.3.1(e) below). Summarizing these facts we get:

Proposition 2.5 *The S_n-Galerkin method applies to the operator $aI + bS_R$ if and only if the operator $T^0(a + b\sigma)$ is invertible or, equivalently, if $0 \notin [a - b, a + b]$.*

2.2.4   The Galerkin method for singular integral operators on the semi-axis

Now we are going to treat the analogous question for the integral operator $aI + bS_{R^+}$ acting on $L^p_{R^+}(\alpha)$. By 2.1.2(e), this operator is invertible if and only if the point 0 is not contained in the circular arc

$$\{a + b \coth \pi(z + i(1/p + \alpha)) , \quad z \in \overline{\mathbf{R}}\} \tag{1}$$

joining $a - b$ to $a + b$. By the same reasoning as in the previous section, we further conclude that the sequence of approximating operators involved by the equations

$$((aI + bS_{R^+})x_n , \phi_{kn}) = (u, \phi_{kn}) , \quad k \geq 0 ,$$

is stable if and only if the (one-sided) Toeplitz operator $T(a + b\sigma)$ is invertible. This operator is invertible if and only if the point 0 is not contained in the region bounded by the line segment $[a - b, a + b]$ and the circular arc (1) (see Section 2.5.3 and compare [BS 2], 6.32 and 2.40).

Proposition 2.6 *The S_n-Galerkin method applies to the operator $aI + bS_{R^+}$ if and only if the Toeplitz operator $T(a + b\sigma)$ is invertible.*

2.2.5   The Galerkin method for singular integral operators on the interval $[0, 1]$

Finally, we consider the Galerkin method

$$((aI + bS_{[0,1]})x_n , \phi_{kn}) = (u, \phi_{kn}) , \quad k = 0, \ldots, n - 1 , \tag{1}$$

with $x_n \in \text{span}\{\phi_{0n}, \ldots, \phi_{n-1,n}\}$ for the singular integral operator $aI + bS_{[0,1]}$ on $L^p_{[0,1]}(\alpha)$. This operator is invertible if and only if the origin 0 lies outside the region which is bounded by the two circular arcs

$$\{a + b \coth \pi(z + i(1/p + \alpha)) , \quad z \in \overline{\mathbf{R}}\}$$

and

$$\{a - b \coth \pi(z + i/p) , \quad z \in \overline{\mathbf{R}}\}$$

(see Section 3.1.7 and compare [Du 1], Th.8.1 or [GK 1], Ch.IX, Th.5.1). As in the preceding sections, we translate the approximating operators determined by (1) into operators on $l^p_Z(\alpha)$ to study stability. The resulting sequence

$$(\sigma_0) \quad , \quad \begin{pmatrix} \sigma_0 & \sigma_{-1} \\ \sigma_1 & \sigma_0 \end{pmatrix} \quad , \quad \begin{pmatrix} \sigma_0 & \sigma_{-1} & \sigma_{-2} \\ \sigma_1 & \sigma_0 & \sigma_{-1} \\ \sigma_2 & \sigma_1 & \sigma_0 \end{pmatrix} , \ldots \tag{2}$$

is just the sequence of the finite section method applied to the Toeplitz operator $T(a + b\sigma)$ (compare Section 1.5.3).

Proposition 2.7 *The S_n-Galerkin method (1) applies to the operator $aI + bS_{[0,1]}$ if and only if the operator itself is invertible and if the finite section method (2) applies to the operator $T(a + b\sigma)$.*

Since the function σ is discontinuous, Theorem 1.6 is not competent to tell us something about the finite sections of $T(a+b\sigma)$. In Section 4.1.1 we shall establish necessary and sufficient conditions for the convergence of the finite section method in a more general situation.

2.2.6 The Galerkin method for Mellin convolutions

The Sections 2.2.3-2.2.5 suggest that the Galerkin method for singular integral operators is in some way related to Toeplitz operators. Proceeding with the Mellin convolution equation

$$M^0(b)x = u, \quad b \in \mathcal{N}, \ u \in L^p_{R^+}(\alpha)$$

or, equivalently, with

$$\int_0^\infty k\left(\frac{t}{s}\right) x(s)\, \frac{ds}{s} = u(t), \quad Mk = b,$$

in an analogous manner, we arrive at an operator on $l^p_{Z^+}(\alpha)$ with matrix representation

$$\left(\int_l^{l+1} \int_m^{m+1} k\left(\frac{t}{s}\right) \frac{ds}{s}\, dt\right)_{l,m=0}^\infty. \tag{1}$$

Clearly, this is not a "Toeplitz-like matrix" but, as we shall point out later on, the corresponding operator is an *element of the Banach algebra generated by Toeplitz operators* with piecewise continuous generating functions. In the following sections we give a detailed treatise on this algebra with special emphasis on operators of the form (1).

2.3 Algebras of Toeplitz operators (Basic facts)

2.3.1 Piecewise continuous multipliers

Given a function $a \in L^\infty(\mathbf{T})$ with Fourier coefficient sequence $(a_k)_{k\in Z}$, we let $T^0(a)$ denote the operator acting on finitely supported sequences as follows:

$$T^0(a)(x_l) = (y_k) \quad \text{with} \quad y_k = \sum_{l\in Z} a_{k-l} x_l, \ k \in \mathbf{Z}.$$

If this operator extends to a bounded operator on all of $l^p_Z(\alpha)$ then we denote it by $T^0(a)$ again, and we call the function a an $l^p(\alpha)$-*multiplier*. Since the operator

$E_{-1}L_1S_RE_1$ is bounded by de Boor and by Proposition 2.3, the function σ defined in 2.2.3 (6) is an example of an $l^p(\alpha)$-multiplier. Here we list some basic properties of $l^p(\alpha)$-multipliers without proof:

(a) *The class of all $l^p(\alpha)$-multipliers forms a commutative Banach algebra under the norm* $\|a\| := \|T^0(a)\|_{L(l_Z^p(\alpha))}$. *In particular,* $T^0(ab) = T^0(a)T^0(b)$.

(b) *Let a be an $l^p(\alpha)$-multiplier. Then $\bar{a}$ (the conjugate complex of a) is also an $l^p(\alpha)$-multiplier, and a and $\bar{a}$ are $l^q(-\alpha)$-multipliers (1/p+1/q=1). Moreover,*

$$\|T^0(a)\|_{L(l_Z^p(\alpha))} = \|T^0(\bar{a})\|_{L(l_Z^p(\alpha))} = \|T^0(a)\|_{L(l_Z^q(-\alpha))} = \|T^0(\bar{a})\|_{L(l_Z^q(-\alpha))}$$

and

$$\|a\|_\infty \leq \|T^0(a)\|_{L(l_Z^p(\alpha))} .$$

(c) *Let $a \in L^\infty(\mathbf{T})$ be a function with finite total variation $V(a)$. Then a is an $l^p(\alpha)$-multiplier, and*

$$\|T^0(a)\|_{L(l_Z^p(\alpha))} \leq C \left(\|a\|_{L^\infty(\mathbf{T})} + V(a) \right)$$

with a constant C depending only on p and α.

If $p = 2$ and $\alpha = 0$ then each $L^\infty(\mathbf{T})$-function provides us with an $l^2(0)$-multiplier. On the other hand, in case $p \neq 2$, $L^\infty(\mathbf{T})$-functions which are not $l^p(\alpha)$-multipliers can be found even among the continuous functions (see [BS 2],6.15). This justifies the following definitions. By $PC_{p,\alpha}$ we denote the closure in the $l^p(\alpha)$-multiplier-norm of the set of all piecewise continuous functions on $\mathbf{T}$ having only finitely many jump discontinuities and possessing a finite total variation, and we write $C_{p,\alpha}$ for the intersection $PC_{p,\alpha} \cap C(\mathbf{T})$. Henceforth, we shall simply refer to the functions in $PC_{p,\alpha}$ and in $C_{p,\alpha}$ as *piecewise continuous* and *continuous multipliers*, respectively, but we draw the reader's attention to the fact that there exist multipliers in $C(\mathbf{T})$ which are *not* in $C_{p,\alpha}$.

(d) *The class $C_{p,\alpha}$ is a Banach algebra, and its maximal ideal space is homeomorphic to the unit circle $\mathbf{T}$. The multiplicative functional associated with $t \in \mathbf{T}$ is given by*

$$C_{p,\alpha} \ni a \mapsto a(t) \in \mathbf{C} .$$

(e) *The class $PC_{p,\alpha}$ is a Banach algebra, and its maximal ideal space is homeomorphic to $\mathbf{T} \times \{0,1\}$. The multiplicative functional associated with $(t,j) \in \mathbf{T} \times \{0,1\}$ is given by*

$$PC_{p,\alpha} \ni a \mapsto \begin{cases} a(t+0) & \text{if} \quad j = 0 \\ a(t-0) & \text{if} \quad j = 1 \end{cases} .$$

(f) *Both the class of all trigonometric polynomials and the class of all continuous functions with finite total variation are dense in $C_{p,\alpha}$. The collection $\mathbf{PC}$ of all piecewise constant functions is dense in $PC_{p,\alpha}$.*

2.3.2 Commutativity in Toeplitz algebras

Define the projection operator $P : l_Z^p(\alpha) \to l_Z^p(\alpha)$ by

$$P(x_k) = (y_k), \quad y_k = \begin{cases} 0 & \text{if} \quad k < 0 \\ x_k & \text{if} \quad k \geq 0 \end{cases},$$

and let $\mathrm{alg}\,(T^0(PC_{p,\alpha}), P)$ stand for the smallest closed subalgebraP of $L(l_Z^p(\alpha))$ which contains all operators $T^0(a)$ with $a \in PC_{p,\alpha}$ and the projection P. Further we let $\mathrm{alg}\,T(PC_{p,\alpha})$ refer to the smallest closed subalgebra of $\mathrm{alg}\,(T^0(PC_{p,\alpha}), P)$ which contains all operators $PT^0(a)P$ with $a \in PC_{p,\alpha}$. If we think of $l_{Z+}^p(\alpha)$ as a subspace of $l_Z^p(\alpha)$ we can identify the operators $PT^0(a)P$ with the usual Toeplitz operators $T(a)$ acting on $l_{Z+}^p(\alpha)$. Thus, $\mathrm{alg}\,T(PC_{p,\alpha})$ can be thought of as the smallest closed subalgebra of $L(l_{Z+}^p(\alpha))$ which encloses all Toeplitz operators $T(a)$ with $a \in PC_{p,\alpha}$.

Proposition 2.8 *(a) If $f \in C_{p,\alpha}$ then the commutator $T^0(f)A - AT^0(f)$ is compact for every operator $A \in \mathrm{alg}\,(T^0(PC_{p,\alpha}), P)$.*
(b) If $f \in C_{p,\alpha}$ then the commutator $T(f)B - BT(f)$ is a compact operator for every $B \in \mathrm{alg}\,T(PC_{p,\alpha})$.
(c) If $f \in C_{p,\alpha}$ and $a \in PC_{p,\alpha}$ then the quasicommutator $T(a)T(f) - T(af)$ is compact.

Proof (a) The operator $T^0(f)$ commutes with each operator $T^0(a), a \in PC_{p,\alpha}$ by 2.3.1(a). So it remains to show the compactness of the commutators $T^0(f)P - PT^0(f)$. By 2.3.1(f) we can suppose without loss that f is a polynomial, $f(t) = \sum_{r=-n}^{n} f_r t^r$. Abbreviating the operators $T^0(t^r)$ to V_r and taking into account that $V_r V_s = V_{r+s}$ and $V_r^{-1} = V_{-r}$ for all $r, s \in \mathbf{Z}$ it remains to check the compactness of $PV_1 - V_1 P$. The matrix representation $(a_{jk})_{j,k \in Z}$ of this operator with respect to the standard basis of $l_Z^p(\alpha)$ contains only one non-vanishing entry, viz. $a_{0,-1} = 1$. Hence, $PV_1 - V_1 P$ has rank one.
(b) Immediate from (a) or (c) below.
(c) The identity $T(af) = T(a)T(f) + H(a)H(\tilde{f})$ (see 1.10) as well as the evident compactness of the Hankel operator $H(\tilde{f})$ for f being a trigonometric polynomial yield the assertion. ∎

Proposition 2.9 *(a) The algebra $\mathrm{alg}\,(T^0(C_{p,\alpha}), P)$ encloses the ideal $K(l_Z^p(\alpha))$ of all compact operators.*
(b) The algebra $\mathrm{alg}\,T(C_{p,\alpha})$ encloses the ideal $K(l_{Z+}^p(\alpha))$.

Proof Each compact operator K on $l_Z^p(\alpha)$ can be approximated as closely as desired by an operator whose matrix representation (a_{jk}) contains only finitely many non-vanishing entries. Thus it remains to verify that $K \in \mathrm{alg}\,(T^0(C_{p,\alpha}), P)$ for each operator of this special kind. Clearly,

$$K = \sum_{j,k \in Z} a_{jk} V_{-j}(PV_1 - V_1 P)V_{k+1} \qquad (1)$$

which proves our claim (a). Assertion (b) is an immediate consequence of (a). ∎

In the sequel, we let $\mathrm{alg}\,^\pi(T^0(PC_{p,\alpha}), P)$ and $\mathrm{alg}\,^\pi T(PC_{p,\alpha})$ refer to the quotient algebras $\mathrm{alg}\,(T^0(PC_{p,\alpha}), P)/K(l_Z^p(\alpha))$ and $\mathrm{alg}\,T(PC_{p,\alpha})/K(l_{Z_+}^p(\alpha))$, respectively. We write π both for the canonical homomorphism from $\mathrm{alg}\,(T^0(PC_{p,\alpha}), P)$ onto $\mathrm{alg}\,^\pi(T^0(PC_{p,\alpha}), P)$ and for the canonical homomorphism from $\mathrm{alg}\,T(PC_{p,\alpha})$ onto $\mathrm{alg}\,^\pi T(PC_{p,\alpha})$. Now, the Proposition 2.8 can be restated as follows:

The algebra $\pi(T^0(C_{p,\alpha}))$ lies in the center of $\mathrm{alg}\,^\pi(T^0(PC_{p,\alpha}), P)$, and the algebra $\pi(T(C_{p,\alpha}))$ lies in the center of $\mathrm{alg}\,^\pi T(PC_{p,\alpha})$.

2.3.3   Homogenization of operators in Toeplitz algebras

The following proposition shows that there is a natural homomorphism which sends each element of the Toeplitz algebra $\mathrm{alg}\,(T^0(PC_{p,\alpha}), P)$ to a homogeneous operator on $L_R^p(\alpha)$, that is, to an operator $B \in L(L_R^p(\alpha))$ satisfying $Z_t^{-1} B Z_t = B$ for all $t > 0$, where the operator Z_t is defined by

$$(Z_t f)(s) \;=\; f(s/t)\,.$$

This homomorphism will be denoted by 'smb' since we shall see later on that it is a symbol mapping in the sense of 1.1.2.

Proposition 2.10 *(a) If $A \in \mathrm{alg}\,(T^0(PC_{p,\alpha}), P)$ then the strong limit*

$$\mathrm{smb}\,A \;:=\; \operatorname*{s-lim}_{n\to\infty} E_n A E_{-n} L_n$$

exists and belongs to $\Sigma_R^p(\alpha)$, the mapping $\mathrm{smb} : \mathrm{alg}\,(T^0(PC_{p,\alpha}), P) \to \Sigma_R^p(\alpha)$ is a continuous algebra homomorphism, and this homomorphism acts on the generating operators of the algebra $\mathrm{alg}\,(T^0(PC_{p,\alpha}), P)$ as follows:

$$\mathrm{smb}\,T^0(a) \;=\; a(1+0)(I - S_R)/2 + a(1-0)(I + S_R)/2\,, \tag{1}$$
$$\mathrm{smb}\,P \;=\; \chi_{R^+} I\,, \tag{2}$$
$$\mathrm{smb}\,K \;=\; 0 \quad if \quad K \quad is \quad compact\,. \tag{3}$$

(b) The restriction of the mapping smb onto the subalgebra $\mathrm{alg}\,T(PC_{p,\alpha})$ of $\mathrm{alg}\,(T^0(PC_{p,\alpha}), P)$ is a continuous algebra homomorphism into the algebra $\Sigma^p(\alpha)$.

Proof (a) Since $E_n A E_{-n} L_n\, E_n B E_{-n} L_n = E_n A B E_{-n} L_n$, the mapping smb is obviously an algebra homomorphism, and its continuity is a consequence of de Boor's estimate. In particular, the existence of the limit $\mathrm{smb}\,A$ for all $A \in \mathrm{alg}\,(T^0(PC_{p,\alpha}), P)$ follows once we have verified its existence for the generating operators of this algebra, viz. for the operators $T^0(a)(a \in PC_{p,\alpha})$ and P. Thus, we are left on verifying the identities (1)-(3).

For a proof of (1) we start with the function $a(t) = t$. Then $T^0(a)$ is just the (discrete) shift operator V_1 and, clearly,

$$E_n T^0(a) E_{-n} L_n \;=\; E_n V_1 E_{-n} L_n \;=\; U_{1/n} L_n \tag{4}$$

where the U_s stands for the continuous analogue of the shift operator, $(U_s g)(t) = g(t - s)$. In case $\alpha = 0$ we have the strong convergences $U_{1/n} \to I$ and $L_n \to I$ as $n \to \infty$ and thus, smb $V_1 = I$. In case $\alpha \neq 0$ some more care is in order, since the operators $U_{1/n}$ are unbounded on $L_R^p(\alpha)$ (it is easy to find functions $g \in L_R^p(\alpha)$ with $U_{1/n} g \notin L_R^p(\alpha)$). But from (4) it is obvious that the operators $U_{1/n} L_n$ are even uniformly bounded on $L_R^p(\alpha)$. We are going to show that $U_{1/n} L_n \to I$ strongly or, equivalently, $U_{1/n} L_n - L_n \to 0$ strongly. By 1.2.1(c) it suffices to check that $\|U_{1/n} L_n g - L_n g\| \to 0$ for each piecewise constant function g with compact support. For these g,

$$U_{1/n} L_n g - L_n g =$$
$$n \sum_k (g, \phi_{kn}) \phi_{k+1,n} - n \sum_k (g, \phi_{kn}) \phi_{kn} = n \sum_k ((g, \phi_{k-1,n}) - (g, \phi_{kn})) \phi_{kn}, \quad (5)$$

and if g has N discontinuities then at most $2N$ items in (5) are non-vanishing. For instance, if g is discontinuous at $r \in \mathbf{R}$ the corresponding non-vanishing items in (5) are

$$n((g, \phi_{[rn]-1,n}) - (g, \phi_{[rn],n})) \phi_{[rn],n} \quad \text{and} \quad n((g, \phi_{[rn],n}) - (g, \phi_{[rn]+1,n})) \phi_{[rn]+1,n} \quad (6)$$

with [.] refering to the entire function, that is, $[x]$ is the integer in $(x - 1, x]$. A straightforward computation yields that the $L_R^p(\alpha)$-norms of the functions (6) tend to zero as $n \to \infty$, whence follows smb$V_1 = I$ in the weighted case, too. Analogously, smb$V_{-1} = I$.

Now let a be a polynomial, $a(t) = \sum_k a_k t^k$. Since smb is an algebra homomorphism,

$$\text{smb} \, T^0(a) = \text{smb} \sum_k a_k V_k = \sum_k a_k (\text{smb} V_1)^k = \sum_k a_k I \ ,$$

that is

$$\text{smb} \, T^0(a) \ = \ a(1) I \ . \quad (7)$$

By 2.3.1(f), the polynomials are dense in $C_{p,\alpha}$. Hence, a density argument reveals that (7) remains valid for $a \in C_{p,\alpha}$.

For the next step, let a be a piecewise continuous multiplier which is continuous on a neighborhood U of the point $1 \in \mathbf{T}$ and which has only finitely many discontinuities. Choose a continuous multiplier b which coincides with a on U as well as a continuous multiplier c with $c(1) = 0$ and $c(t) = 1$ whenever $t \in \mathbf{T} \setminus U$. Then $a = (a - b)c + b$. As we have already seen, smb$T^0(b) = b(1)I = a(1)I$, and for $(a - b)c$ we find

$$E_n T^0(a - b) T^0(c) E_{-n} L_n \ = \ E_n T^0(a - b) E_{-n} L_n E_n T^0(c) E_{-n} L_n \ .$$

Since the operators $E_n T^0(a - b) E_{-n} L_n$ are uniformly bounded, and since $E_n T^0(c) E_{-n} L_n \to c(1)I = 0$ by (7) we conclude that smb$T^0(a - b)T^0(c) = 0$ and, consequently, smb$T^0(a) = a(1)I$.

Now let a be an arbitrary piecewise continuous multiplier with only finitely many discontinuities. Write

$$a = a(1+0)(1-\sigma)/2 + a(1-0)(1+\sigma)/2 + d$$

with σ standing for the function 2.2.3(6). The function d is continuous in a neighborhood of 1, and $d(1) = 0$. Thus, $\mathrm{smb}T^0(d) = 0$, and the identity

$$E_n T^0(\sigma) E_{-n} L_n = L_n S_R L_n$$

established in Section 2.2.3 involves

$$\mathrm{smb}\,T^0(a) = a(1+0)(I - S_R)/2 + a(1-0)(I + S_R)/2 . \tag{8}$$

Finally, by 2.3.1(f), we can approximate an arbitrary piecewise continuous multiplier by a piecewise continuous multiplier with finitely many jumps as closely as desired. Thus, (8) remains true for all $a \in PC_{p,\alpha}$.

The remaining assertions of the proposition are readily verified: (2) follows from the identity $E_n P E_{-n} L_n = L_n \chi_{R^+} L_n$ and from the strong convergence $L_n \to I$, and for (3) we approximate K by a linear combination of the operators $V_{-j}(PV_1 - V_1 P)V_{k+1}$ (cf. 2.3.2(1)) whence follows

$$\mathrm{smb}\,V_{-j}(PV_1 - V_1 P)V_{k+1} = 0 \quad\text{and}\quad \mathrm{smb}\,K = 0$$

by what has already been proved.
(b) Immediate from (a). ∎

The Propositions 2.8, 2.9 and 2.10 in combination with Lemma 2.1 below are the basic observations we need for the study of the discretized Mellin convolutions as well as for the derivation of Fredholm criteria for elements in the Toeplitz algebra in the following sections.

2.4   Discretized Mellin convolutions

2.4.1   A class of kernel functions

Let b be a continuous function on $\mathbf{R}$ such that the Mellin convolution operator $M^0(b)$ belongs to the ideal $N^p(\alpha)$. As we have seen in Section 2.2.6, the operators $E_{-n} L_n M^0(b) E_n$ are independent of n. We denote (one of) them by $G(b)$ and call these operators *discretized Mellin convolutions*. In particular, if $b \in \mathcal{N}$ then, by the representation 2.2.6 (1),

$$G(b) = \left(\int_m^{m+1} \int_n^{n+1} k\left(\frac{t}{s}\right) \frac{ds}{s} \, dt \right)_{m,n=0}^{\infty} \tag{1}$$

with $k = M^{-1}b$. For the following technical lemma we fix a positive integer d and denote by $\mathcal{M}(d)$ or, shortly, by $\mathcal{M}$, the collection of all continuous functions k on $\mathbf{R}^+$ such that

(i) $k \in \cap_{p,\alpha} L^p_{R^+}(\alpha)$ where the intersection is taken over all p with $p > 1$ and all α with $0 < 1/p + \alpha < 1$, and

(ii) $\sum_{m=0}^{\infty} (m+1)^{\alpha p} \left(\sum_{n=0}^{\infty} (n+1)^{-\beta v} \left| k\left(\frac{t_{mn}}{s_{nm}}\right) \frac{1}{s_{nm}} - k\left(\frac{m+1}{n+1}\right) \frac{1}{n+1} \right|^v \right)^{p/v} < \infty$

for all p, v, α, β with $p, v > 1$ and $0 < 1/p + \alpha, 1/v - \beta < 1$, and for all choices of non-zero numbers $t_{mn} \in [m, m+d]$ and $s_{nm} \in [n, n+d]$.

Lemma 2.1 *The class $\mathcal{M}(d)$ is an algebra with respect to the pointwise addition and the Mellin convolution (cf. 2.1.4 (3)) as multiplication, and $M^{-1}\mathcal{N}$ (see 2.1.4) is a subalgebra of $\mathcal{M}(d)$.*

The proof will be given in Section 2.12.1.

In case $p = v = 2$ and $\alpha = \beta = 0$ the somewhat mysterious condition (ii) reduces to

$$\sum_{m=0}^{\infty} \sum_{n=0}^{\infty} \left| k\left(\frac{t_{mn}}{s_{nm}}\right) \frac{1}{s_{nm}} - k\left(\frac{m+1}{n+1}\right) \frac{1}{n+1} \right|^2 < \infty ,$$

i.e. the operator K with mn th entry $k\left(\frac{t_{mn}}{s_{nm}}\right) \frac{1}{s_{nm}} - k\left(\frac{m+1}{n+1}\right) \frac{1}{n+1}$ is Hilbert-Schmidt and, thus, compact. Similarly, condition (ii) involves in the general case (where operator K is usually referred to as Hille-Tamarkin-operator)

Corollary 2.1 *If k is in $\mathcal{M}$ and t_{mn} and s_{nm} are as above then the operator K with mn th entry*

$$k\left(\frac{t_{mn}}{s_{nm}}\right) \frac{1}{s_{nm}} - k\left(\frac{m+1}{n+1}\right) \frac{1}{n+1}$$

is compact from $l^s_{Z^+}(\beta)$ into $l^p_{Z^+}(\alpha)$ for every choice of numbers $p, s > 1$ and α, β with $0 < 1/p + \alpha, 1/s + \beta < 1$.

2.4.2 Properties of discretized Mellin convolutions

Next we establish some properties of discretized Mellin convolutions which are related to their commutativity with Toeplitz operators.

Proposition 2.11 *Let b, b_1, b_2 be continuous functions on $\mathbf{R}$ such that $M^0(b)$, $M^0(b_1), M^0(b_2) \in N^p(\alpha)$, and let $a \in PC_{p,\alpha}$ be a piecewise continuous multiplier. Then*

(a) *for $b \in \mathcal{N}$, the operator $G(b) - \left(k\left(\frac{m+1}{n+1}\right) \frac{1}{n+1} \right)_{m,n=0}^{\infty}$ with $k = M^{-1}b$ is compact on $l^p_{Z^+}(\alpha)$.*

(b) *$G(b_1)G(b_2) - G(b_1 b_2)$ is compact on $l^p_{Z^+}(\alpha)$.*

(c) $G(b_1)G(b_2) - G(b_2)G(b_1)$ is compact on $l^p_{Z^+}(\alpha)$.

(d) $T(a)G(b) - G\left(\left(a(1-0)\frac{1+s}{2} + a(1+0)\frac{1-s}{2}\right)b\right)$ is compact on $l^p_{Z^+}(\alpha)$. In particular, $T(a)G(b) - a(1)G(b)$ is compact whenever $a \in C_{p,\alpha}$.

Remember that the function s was defined in 2.1.2(4) as the Mellin symbol of the singular integral operator S_{R^+}.

Proof　(a) Assume without loss k to be a real-valued function. Then the representation 2.4.1(1) of the operator $G(b)$ in combination with the mean value theorem yields

$$G(b) = \left(k\left(\frac{t_{mn}}{s_{nm}}\right)\frac{1}{s_{nm}}\right)^{\infty}_{m,n=0}$$

with numbers $t_{mn} \in (m, m+1)$ and $s_{nm} \in (n, n+1)$. Now invoke Corollary 2.1 to get the assertion.

(b) The set $N^p_0(\alpha) = M^{-1}\mathcal{N}M$ is dense in $N^p(\alpha)$ by Proposition 2.2. So we can supppose $b_1, b_2 \in \mathcal{N}$ without loss. Put $k_1 = M^{-1}b_1$ and $k_2 = M^{-1}b_2$. Then, by (a),

$$G(b_1)G(b_2) = \left(\sum_{r=0}^{\infty} k_1\left(\frac{m+1}{r+1}\right)\frac{1}{r+1} k_2\left(\frac{r+1}{n+1}\right)\frac{1}{n+1}\right)^{\infty}_{m,n=0} + K_1$$

and

$$
\begin{aligned}
G(b_1 b_2) &= \left(\int_0^{\infty} k_1\left(\frac{m+1}{x}\right) k_2\left(\frac{x}{n+1}\right)\frac{1}{n+1}\frac{dx}{x}\right)^{\infty}_{m,n=0} + K_2 \\
&= \left(\sum_{r=0}^{\infty}\int_r^{r+1} k_1\left(\frac{m+1}{x}\right) k_2\left(\frac{x}{n+1}\right)\frac{1}{n+1}\frac{dx}{x}\right)^{\infty}_{m,n=0} + K_2 \\
&= \left(\sum_{r=0}^{\infty} k_1\left(\frac{m+1}{x_r}\right) k_2\left(\frac{x_r}{n+1}\right)\frac{1}{n+1}\frac{1}{x_r}\right)^{\infty}_{m,n=0} + K_2
\end{aligned}
$$

with compact operators K_1 and K_2 and numbers $x_r \in (r, r+1)$ depending on m and n. Thus, with $K_3 := K_1 - K_2$,

$$
\begin{aligned}
&G(b_1)G(b_2) - G(b_1 b_2) - K_3 = \\
&= \left(\sum_r \left(k_1\left(\frac{m+1}{r+1}\right)\frac{1}{r+1} k_2\left(\frac{r+1}{n+1}\right)\frac{1}{n+1} - \right.\right. \\
&\qquad \left.\left. -k_1\left(\frac{m+1}{x_r}\right)\frac{1}{x_r} k_2\left(\frac{x_r}{n+1}\right)\frac{1}{n+1}\right)\right)^{\infty}_{m,n=0} \\
&= \left(\sum_r \left(k_1\left(\frac{m+1}{r+1}\right)\frac{1}{r+1} - k_1\left(\frac{m+1}{x_r}\right)\frac{1}{x_r}\right) k_2\left(\frac{r+1}{n+1}\right)\frac{1}{n+1}\right)^{\infty}_{m,n=0} +
\end{aligned}
$$

$$+ \left(\sum_r k_1\left(\frac{m+1}{x_r}\right) \frac{1}{x_r} \left(k_2\left(\frac{r+1}{n+1}\right) \frac{1}{n+1} - k_2\left(\frac{x_r}{n+1}\right) \frac{1}{n+1} \right) \right)_{m,n=0}^{\infty}$$

$$= K_4(G(b_2) + K_5) + (G(b_1) + K_6)K_7 ,$$

where the operators

$$K_4 = \left(k_1\left(\frac{m+1}{r+1}\right) \frac{1}{r+1} - k_1\left(\frac{m+1}{x_r}\right) \frac{1}{x_r} \right)_{m,r=0}^{\infty} ,$$

$$K_5 = \left(k_2\left(\frac{r+1}{n+1}\right) \frac{1}{n+1} \right)_{r,n=0}^{\infty} - G(b_2) ,$$

$$K_6 = \left(k_1\left(\frac{m+1}{x_r}\right) \frac{1}{x_r} \right)_{m,r=0}^{\infty} - G(b_1) ,$$

$$K_7 = \left(k_2\left(\frac{r+1}{n+1}\right) \frac{1}{n+1} - k_2\left(\frac{x_r}{n+1}\right) \frac{1}{n+1} \right)_{r,n=0}^{\infty}$$

are compact by (a) and by Corollary 2.1. Thus, $G(b_1)G(b_2) - G(b_1 b_2) \in K(l_{Z+}^p(\alpha))$.
(c) Immediate from (b).
(d) First let $a \in C_{p,\alpha}$. By Proposition 2.2 and by 2.3.1(f) we can suppose $b \in \mathcal{N}$
and $T^0(a) = T^0(t^r) = V_r$ without loss of generality. Let, for instance, $r > 0$. Then
the identity $V_r G(b) - G(b) = \sum_{i=0}^{r-1} V_i(V_1 G(b) - G(b))$ shows that we can further
assume that $r = 1$. The mn th entry of the operator $V_1 G(b) - G(b)$ equals

$$- \int_0^1 \int_n^{n+1} k\left(\frac{t}{s}\right) \frac{ds}{s} \, dt \quad \text{if} \quad m = 0$$

and

$$\int_{m-1}^m \int_n^{n+1} k\left(\frac{t}{s}\right) \frac{ds}{s} \, dt - \int_m^{m+1} \int_n^{n+1} k\left(\frac{t}{s}\right) \frac{ds}{s} \, dt \quad \text{if} \quad m > 0 . \qquad (1)$$

Again by the mean value theorem, (1) can be written as

$$k\left(\frac{t_{mn}}{s_{nm}}\right) \frac{1}{s_{nm}} - k\left(\frac{t'_{mn}}{s'_{nm}}\right) \frac{1}{s'_{nm}}$$

with certain numbers $t_{mn} \in (m-1,m) \subseteq (m-1,m+1)$, $t'_{mn} \in (m,m+1) \subseteq$
$(m-1,m+1)$, $s_{nm}, s'_{nm} \in (n,n+1) \subseteq (n,n+2)$. Thus, Corollary 2.1 applies
with $d = 2$ to give the compactness of $V_1 G(b) - G(b)$. So we get for $a \in C_{p,\alpha}$ the
compactness of $T(a)G(b) - a(1)G(b)$.

If a is an arbitrary piecewise continuous multiplier, we approximate it by multipliers having only finite many jump discontinuities. But such a multiplier can be
represented as in the proof of Proposition 2.10 by the function

$$a(1+0)\frac{1-\sigma}{2} + a(1-0)\frac{1+\sigma}{2} + a'c + d$$

where $a' \in PC_{p,\alpha}$, $c,d \in C_{p,\alpha}$ with $c(1) = d(1) = 0$, and where σ refers to the function 2.2.3(6). Via Proposition 2.8(c) we conclude that

$$
\begin{aligned}
T(a)G(b) &= a(1+0)\frac{1-T(\sigma)}{2}G(b) + a(1-0)\frac{1+T(\sigma)}{2}G(b) + \\
&\quad + T(a')T(c)G(b) + T(d)G(b) + \text{compact}
\end{aligned}
$$

whence by the already proved part follows

$$
T(a)G(b) = a(1+0)\frac{1-T(\sigma)}{2}G(b) + a(1-0)\frac{1+T(\sigma)}{2}G(b) + \text{compact} .
$$

So it remains to verify the compactness of $T(\sigma)G(b) - G(sb)$. It is evident from the definition of the ideal $N^p(\alpha)$ that we can assume $b = n_\pi c$ with a continuous function c such that $M^0(c) \in N^p(\alpha)$ (compare 2.1.2(c)(iii)) without loss of generality. Then part (b) states that

$$
\pi\left(T(\sigma)G(n_\pi c) - G(n_\pi sc)\right) = \pi(T(\sigma)G(n_\pi) - G(sn_\pi))\,\pi\left(G(c)\right) ,
$$

and we are left on verifying the compactness of $T(\sigma)G(n_\pi) - G(sn_\pi)$.

Consider the function

$$
f : e^{ix} \mapsto -x/\pi + 1 \quad , \quad x \in [0, 2\pi) . \tag{2}
$$

This function is in $PC_{p,\alpha}$, and its n th Fourier coefficient f_n is given by

$$
f_n = \left\{ \begin{array}{lcl} 0 & \text{if} & n = 0 \\ \frac{1}{\pi i n} & \text{if} & n \neq 0 \end{array} \right. .
$$

Further we have the identities

$$
\pi\left(G(n_\pi)\right) = \pi\left(H(f)\right) \tag{3}
$$

and

$$
\pi\left(G(sn_\pi)\right) = \pi\left(-T(f)H(f)\right) \tag{4}
$$

which proofs are left to the reader as exercises. Thus,

$$
\begin{aligned}
\pi\left(T(\sigma)G(n_\pi) - G(sn_\pi)\right) &= \\
&= \pi\left(T(\sigma+f)G(n_\pi)\right) - \pi\left(T(f)G(n_\pi) + G(sn_\pi)\right) \\
&= -\pi\left(T(f)H(f) - T(f)H(f)\right) = 0
\end{aligned}
$$

since the function $\sigma + f$ is continuous on $\mathbf{T}$ and $(\sigma + f)(1) = 0$. ∎

2.4.3   Discretized Mellin convolutions in Toeplitz algebras

Theorem 2.1 *If* $M^0(b) \in N^p(\alpha)$ *then the operator* $G(b)$ *defined by 2.4.1(1) belongs to the algebra* $\operatorname{alg} T(PC_{p,\alpha})$.

Proof   It is obviously sufficient to prove that $G(b) \in \operatorname{alg} T(PC_{p,\alpha})$ whenever $b \in \mathcal{N}$. Let $\operatorname{alg}(T(PC_{p,\alpha}), G(\mathcal{N}))$ stand for the smallest closed subalgebra of $L(l^p_{Z+}(\alpha))$ which contains all operators $T(a)$ and $G(b)$ with $a \in PC_{p,\alpha}$ and $b \in \mathcal{N}$. This algebra encloses all compact operators (Proposition 2.9(b)). The quotient algebras $\operatorname{alg} T(PC_{p,\alpha})/K(l^p_{Z+}(\alpha))$ and $\operatorname{alg}(T(PC_{p,\alpha}), G(\mathcal{N}))/K(l^p_{Z+}(\alpha))$ will be abbreviated henceforth to $\mathcal{A}$ and $\mathcal{B}$, respectively. Clearly, $\mathcal{A}$ is a subalgebra of $\mathcal{B}$. We claim that $\mathcal{A} = \mathcal{B}$ which would prove our assertion.

1 st step. Given an operator $A \in L(l^p_{Z+}(\alpha))$ we let $\pi(A)$ denote the coset $A + K(l^p_{Z+}(\alpha))$. Further we write $\mathcal{J}$ for the smallest closed two-sided ideal of the algebra $\mathcal{A}$ which contains all cosets $\pi(T(f))$ where $f \in C_{p,\alpha}$ and $f(1) = 0$. Since $\pi(T(f))$ lies in the center of $\mathcal{A}$ by Proposition 2.8(b), and since $\pi(T(f)G(b)) = 0$ for each $G(b)$ by Proposition 2.11(d) we conclude that $\mathcal{J}$ is also a closed two-sided ideal of the algebra $\mathcal{B}$. Thus, it is sufficient for getting our claim to verify that $\mathcal{A}/\mathcal{J} = \mathcal{B}/\mathcal{J}$.

2 nd step. We proceed with constructing a homomorphism from the quotient algebra $\mathcal{B}/\mathcal{J}$ into the algebra $\Sigma^p(\alpha)$. Let $A \in \operatorname{alg}(T(PC_{p,\alpha}), G(\mathcal{N}))$ and define the operator $\operatorname{smb} A \in L(L^p_{R+}(\alpha))$ by

$$\operatorname{smb} A = \operatorname*{s\text{-}lim}_{n \to \infty} E_n A E_{-n} L_n . \tag{1}$$

Taking into account that

$$E_n G(b) E_{-n} L_n = L_n M^0(b) L_n \to M^0(b) \quad \text{strongly} \tag{2}$$

(see 2.4.1) and that $M^0(b) \in \Sigma^p(\alpha)$ by 2.1.2(a), one gets as in Proposition 2.10 that the limits (1) indeed exist for each operator $A \in \operatorname{alg}(T(PC_{p,\alpha}), G(\mathcal{N}))$, that the mapping $\operatorname{smb} : A \mapsto \operatorname{smb} A$ is a continuous algebra homomorphism from $\operatorname{alg}(T(PC_{p,\alpha}), G(\mathcal{N}))$ into $\Sigma^p(\alpha)$, and that this homomorphism coincides on $\operatorname{alg} T(PC_{p,\alpha})$ with the homomorphism smb defined in 2.3.3.

Further, 2.3.3(3) shows that all compact operators lie in the kernel of the homomorphism smb. Thus, $\operatorname{smb} A$ depends actually only on the coset $\pi(A) = A + K(l^p_{Z+}(\alpha))$, and we can think of smb as acting on the quotient algebra $\mathcal{B}$ into $\Sigma^p(\alpha)$. We denote this quotient homomorphism by smb again. Moreover, the identity 2.3.3(1) yields that $\pi(T(f)) \in \operatorname{Ker} \operatorname{smb}$ whenever $f \in C_{p,\alpha}$ and $f(1) = 0$. Thus, the whole ideal $\mathcal{J}$ belongs to the kernel of smb or, in other words, if we let Φ stand for the canonical homomorphism from $\operatorname{alg}(T(PC_{p,\alpha}), G(\mathcal{N}))$ onto $\mathcal{B}/\mathcal{J}$ then $\operatorname{smb} A$ depends on $\Phi(A)$ only. The quotient homomorphism

$$\mathcal{B}/\mathcal{J} \to \Sigma^p(\alpha) , \quad \Phi(A) \mapsto \operatorname{smb} A \tag{3}$$

will be denoted by smb again.

3 rd step. Here we prove that the homomorphism smb from (3) is an isomorphism. Consider the mapping

$$\text{smb}' \; : \; \Sigma^p(\alpha) \to \mathcal{B}/\mathcal{J} \, , \; B \mapsto \Phi(E_n^{-1} L_n B E_n) \, .$$

Clearly, this definition is correct: to see this write B as $\alpha I + \beta S_{R^+} + M^0(b)$ with an operator $M^0(b) \in N^p(\alpha)$ which is possible by Costabel's decomposition 2.1.2(d), and then apply 2.2.4 and 2.2.6 to obtain that the operator

$$E_{-n} L_n B E_n \; = \; E_{-n} L_n (\alpha I + \beta S_{R^+}) E_n + E_{-n} L_n M^0(b) E_n$$

is independent of n and belongs to the algebra $\text{alg}\,(T(PC_{p,\alpha}), G(\mathcal{N}))$. Further, smb$'$ is obviously a continuous linear mapping. We claim that $\text{smb} \circ \text{smb}' = I$ and smb$' \circ \text{smb} = I$. These two identities yield that smb is bijective and, hence, an isomorphism.

The first one of this identities is readily verified: If $B = \alpha I + \beta S_{R^+} + M^0(b) \in \Sigma^p(\alpha)$ then smb$' B = \Phi(\alpha I + \beta\, T(\sigma)) + \Phi(G(b))$ by 2.2.4 and 2.2.6, and Proposition 2.10 in combination with (2) yields

$$\text{smb}(\text{smb}' B) = \text{smb}\, \Phi(\alpha I + \beta\, T(\sigma)) + \text{smb}\, \Phi(G(b)) = \alpha I + \beta S_{R^+} + M^0(b) = B \, .$$

For the second identity we start with showing that smb$'$ is an algebra homomorphism. Let $B_1 = \alpha_1 I + \beta_1 S_{R^+} + M^0(b_1)$ and $B_2 = \alpha_2 I + \beta_2 S_{R^+} + M^0(b_2)$ be arbitrary operators in $\Sigma^p(\alpha)$. Then

$$\begin{aligned} \text{smb}' B_1 &= \Phi(\alpha_1 I + \beta_1 T(\sigma) + G(b_1)) \, , \\ \text{smb}' B_2 &= \Phi(\alpha_2 I + \beta_2 T(\sigma) + G(b_2)) \end{aligned} \qquad (4)$$

and, since Costabel's decomposition of $B_1 B_2$ equals by 2.1.2(9)

$$(\alpha_1\alpha_2 + \beta_1\beta_2)I + (\alpha_1\beta_2 + \alpha_2\beta_1)S_{R^+} + $$
$$+ M^0(\alpha_1 b_2 + \beta_1 s b_2 + \alpha_2 b_1 + \beta_2 s b_1 + \beta_1\beta_2 n_\pi^2 + b_1 b_2)$$

with functions s, n_π defined in 2.1.2(4) and 2.1.2(5), we have

$$\begin{aligned} \text{smb}' B_1 B_1 &= \Phi((\alpha_1\alpha_2 + \beta_1\beta_2)I + (\alpha_1\beta_2 + \alpha_2\beta_1)T(\sigma) + \\ &+ \; G(\alpha_1 b_2 + \beta_1 s b_2 + \alpha_2 b_1 + \beta_2 s b_1 + \beta_1\beta_2 n_\pi^2 + b_1 b_2)) \, . \end{aligned} \qquad (5)$$

Comparing (4) with (5) it remains to show that

$$\Phi(T(\sigma)^2) \; = \; \Phi(I + G(n_\pi^2)) \qquad (6)$$

and

$$\Phi(T(\sigma)G(b_i)) \; = \; \Phi(G(s b_i)) \quad (i = 1, 2) \, .$$

The latter equality is an immediate consequence of Proposition 2.11(d). To prove
(6) let f refer to the function 2.4.2(2), and define $\tilde{f}$ by $\tilde{f}(t) = f(1/t)$. The identity
1.5.2(a) (which holds for multipliers, too) gives

$$-T(f)\,T(\tilde{f}) \ = \ -T(f\tilde{f}) \ + \ H(f)^2 \ . \tag{7}$$

Since $f\tilde{f}$ is a continuous function with $(f\tilde{f})(1) = -1$ we have $\Phi(-T(f\tilde{f})) = \Phi(I)$,
and since $f + \sigma$ is a continuous function with $(f + \sigma)(1) = 0$ we have $\Phi(T(f)) =$
$-\Phi(T(\sigma))$ and $\Phi(T(\tilde{f})) = \Phi(T(\sigma))$. Moreover, $H(f)$ coincides with $G(n_\pi)$ up to
a compact summand. So we conclude from (7) that $\Phi(T(\sigma)^2) = \Phi(I + G(n_\pi)^2)$,
and this is just (6) by Proposition 2.11(b). Hence, smb$'$ is a continuous algebra
homomorphism.

In particular, in order to prove the identity smb$'$ $\circ$ smb $= I$ it suffices to show
that (smb$'$ $\circ$ smb)$(\Phi(A)) = \Phi(A)$ for the generating operators $T(a)$ and $G(b)$ ($a \in$
$PC_{p,\alpha}$, $b \in \mathcal{N}$) of the algebra alg $(T(PC_{p,\alpha}), G(\mathcal{N}))$ in place of A. For $A = G(b)$
this is evident, and for $A = T(a)$ we have

$$\text{smb}\,\Phi(T(a)) = a(1+0)(I - S_{R^+})/2 \ + \ a(1-0)(I + S_{R^+})/2$$

and

$$\text{smb}'\,\text{smb}\Phi(T(a)) \ = \ \Phi(T(a(1+0)(1-\sigma)/2 \ + \ a(1-0)(1+\sigma)/2)) \ .$$

The coset $\pi(T(a - (a(1+0)(1-\sigma)/2 + a(1-0)(1+\sigma)/2)))$ lies in the ideal $\mathcal{J}$ and
we arrive at the assertion.

4 th step. By the topological isomorphy of the algebras $\mathcal{B}/\mathcal{J}$ and $\Sigma^p(\alpha)$, Costabel's decomposition of $\Sigma^p(\alpha)$ applies to the algebra $\mathcal{B}/\mathcal{J}$, i.e. each coset $\Phi(A) \in$
$\mathcal{B}/\mathcal{J}$ can be uniquely written as

$$\Phi(A) \ = \ \alpha\Phi(I) \ + \ \beta\Phi(T(\sigma)) \ + \ \Phi(G(b)) \ , \tag{8}$$

and the collection $\mathcal{K}$ of all cosets $\Phi(G(b))$ forms a closed two-sided ideal in $\mathcal{B}/\mathcal{J}$
which is isomorphic to the ideal $N^p(\alpha)$. By 2.1.2(c)(iii), the ideal $N^p(\alpha)$ can be
viewed as the smallest closed subalgebra of $\Sigma^p(\alpha)$ which contains the Mellin convolutions $N^2 = M^0(n_\pi^2)$ and $S_{R^+}N^2 = M^0(sn_\pi^2)$. Thus, $\mathcal{K}$ is the smallest closed
subalgebra of $\mathcal{B}/\mathcal{J}$ which contains the cosets $\Phi(G(n_\pi^2))$ and $\Phi(G(sn_\pi^2))$. But from

$$\Phi(G(n_\pi^2)) = \Phi(G(n_\pi)^2) = \Phi(H(f)^2) = \Phi(T(f\tilde{f}) - T(f)T(\tilde{f}))$$

by Proposition 2.11(b) and (7) and

$$\Phi(G(sn_\pi^2)) = \Phi(T(\sigma)G(n_\pi^2)) = \Phi(T(\sigma)(T(f\tilde{f}) - T(f)T(\tilde{f})))$$

by Proposition 2.11(d) we conclude that $\Phi(G(n_\pi^2)) \in \mathcal{A}/\mathcal{J}$ and $\Phi(G(sn_\pi^2)) \in \mathcal{A}/\mathcal{J}$.
Thus, $\mathcal{K} \subseteq \mathcal{A}/\mathcal{J}$ and, invoking the representation (8), we have $\mathcal{B}/\mathcal{J} \subseteq \mathcal{A}/\mathcal{J}$ which
proves our claim. ∎

2.5 Algebras of Toeplitz operators (Fredholmness)

2.5.1 Rotation invariance

Before establishing a Fredholm theory for operators in the algebras alg $T(PC_{p,\alpha})$ or alg $(T^0(PC_{p,\alpha}), P)$ let us turn our attention to one more (rather obvious) property of these algebras: their *rotation invariance*. More precisely: Given a point $t \in \mathbf{T}$ we denote by $Y_t : l_Z^p(\alpha) \to l_Z^p(\alpha)$ the operator

$$Y_t : (x_n)_{n \in Z} \mapsto (t^{-n} x_n)_{n \in Z} .$$

Then one has

$$Y_t^{-1} \text{alg} (T^0(PC_{p,\alpha}), P) Y_t = \text{alg} (T^0(PC_{p,\alpha}), P) \tag{1}$$

and

$$Y_t^{-1} \text{alg} T(PC_{p,\alpha}) Y_t = \text{alg} T(PC_{p,\alpha}) .$$

Let us check relation (1) for example. Clearly, it suffices to prove that $Y_t^{-1} A Y_t \in$ alg $(T^0(PC_{p,\alpha}), P)$ for each generating operator A of this algebra. But since both operators P and Y_t are of diagonal form, it is $Y_t^{-1} P Y_t = P$. The proof of the identity

$$Y_t^{-1} T^0(a) Y_t = T^0(a_t) \quad \text{with} \quad a_t(s) = a(ts) \tag{2}$$

is left as an exercise to the reader.

For $t \in \mathbf{T}$ we define a mapping

$$\text{smb}_t : \text{alg} (T^0(PC_{p,\alpha}), P) \to \Sigma_R^p(\alpha)$$

by

$$\text{smb}_t : A \mapsto \text{smb} (Y_t^{-1} A Y_t) = \underset{n \to \infty}{\text{s-lim}} E_n Y_t^{-1} A Y_t E_{-n} L_n .$$

This definition is correct by (1) and by Proposition 2.10, and smb_t is a continuous algebra homomorphism. The restriction of this homomorphism onto alg $T(PC_{p,\alpha})$ will also be denoted by smb_t.

2.5.2 Fredholmness

Theorem 2.2 *(a) An operator $A \in \text{alg} T(PC_{p,\alpha})$ is Fredholm if and only if each operator $\text{smb}_t A (\in \Sigma^p(\alpha))$ with $t \in \mathbf{T}$ is invertible. In particular,*

$$\text{smb}_t T(a) = a(t+0)(I - S_{R^+})/2 + a(t-0)(I + S_{R^+})/2 , \tag{1}$$

$$\text{smb}_t Y_s G(b) Y_s^{-1} = \begin{cases} M^0(b) & \text{if} \quad t = s \\ 0 & \text{if} \quad t \neq s , \end{cases} \tag{2}$$

$$\text{smb}_t K = 0 \quad \text{for compact } K . \tag{3}$$

(b) The algebra alg $^\pi T(PC_{p,\alpha})$ is inverse closed in the Calkin algebra $L(l_{Z^+}^p(\alpha))/K(l_{Z^+}^p(\alpha))$, and the algebra alg$T(PC_{p,\alpha})$ is inverse closed in $L(l_{Z^+}^p(\alpha))$.

Proof   The algebra $\mathrm{alg}\,^{\pi}T(PC_{p,\alpha})$ possesses a rich and manageable center: Indeed, by 2.3.2 Proposition 2.8(b), each coset $\pi(T(f))$ with $f \in C_{p,\alpha}$ is in the center of this algebra. This suggests to apply Allan's local principle 1.4.4 with $\mathrm{alg}\,^{\pi}T(PC_{p,\alpha})$ and $\{\pi(T(f)), f \in C_{p,\alpha}\}$ in place of $\mathcal{A}$ and $\mathcal{B}$, respectively. For this purpose, we need the maximal ideal space of the algebra $\mathcal{B} := \{\pi(T(f)), f \in C_{p,\alpha}\}$. The identity (3) (which follows immediately from Proposition 2.10) involves that $\mathrm{smb}_t\, A$ only depends on the coset $\pi(A)$. Thus, the identity (1) (which is an obvious consequence of 2.5.1(2) and of 2.3.3 (1), (2)) implies

$$\mathrm{smb}_t\, \pi(T(f)) \; = \; f(t)I \; ,$$

and we can identify the restriction of the mapping smb_t onto $\mathcal{B}$ with a linear multiplicative functional on $\mathcal{B}$ which corresponds to the maximal ideal

$$\{\pi(T(f)) \; : \; f(t) = 0 \; , \; f \in C_{p,\alpha} \} \tag{4}$$

of $\mathcal{B}$. Conversely, each maximal ideal of $\mathcal{B}$ is of this form (4). Indeed, let $\pi(T(f))$ belong to a certain maximal ideal of $\mathcal{B}$. Then there is a point $t \in \mathbf{T}$ where f vanishes. (Assume for contrary $f(t) \neq 0$ for all $t \in \mathbf{T}$. Then f is invertible in $C_{p,\alpha}$ by 2.3.1(d) and 1.4.2(i). But now Proposition 2.8 (c) leads to $\pi(T(f))\,\pi(T(f^{-1})) = \pi(I)$ and thus to the invertibility of $\pi(T(f))$ which is impossible.) Hence, $\pi(T(f))$ is contained in a maximal ideal of the form (4).

Following Allan, we let $\mathcal{J}_t$ refer to the smallest closed two-sided ideal of the algebra $\mathcal{A} = \mathrm{alg}\,^{\pi}T(PC_{p,\alpha})$ which contains the maximal ideal (4) of $\mathcal{B}$, and we write Φ_t for the canonical homomorphism from $\mathrm{alg}\,T(PC_{p,\alpha})$ onto $\mathcal{A}/\mathcal{J}_t$. Then the mapping

$$\Phi_t(A) \; \mapsto \; \Phi_1(Y_t^{-1}AY_t) \tag{5}$$

is correctly defined, and it yields an isomorphism between the local algebras $\mathcal{A}/\mathcal{J}_t$ and $\mathcal{A}/\mathcal{J}_1$. For the correctness of (5) notice that $Y_t^{-1}KY_t$ is compact whenever K is compact, whence the correctness of the mapping

$$\mathrm{alg}\,^{\pi}T(PC_{p,\alpha}) \to \mathrm{alg}\,^{\pi}T(PC_{p,\alpha}) \; , \; \pi(A) \mapsto \pi(Y_t^{-1}AY_t) \tag{6}$$

follows. It is immediate from 2.5.1 (2) that (6) sends the maximal ideal $\{\pi(T(f)), f \in C_{p,\alpha}, f(t) = 0\}$ into the maximal ideal $\{\pi(T(f)), f \in C_{p,\alpha}, f(1) = 0\}$ and hence, the ideal $\mathcal{J}_t$ into $\mathcal{J}_1$. Thus, the definition of (5) is correct, and the invertibility of the operators Y_t leads to the isomorphy between the local algebras $\mathcal{A}/\mathcal{J}_t$ and $\mathcal{A}/\mathcal{J}_1$.

So it remains to consider the local algebra $\mathcal{A}/\mathcal{J}_1$. But this algebra has already been studied: remember that the ideal $\mathcal{J}_1$ coincides with the ideal $\mathcal{J}$ defined in the proof of the Theorem 2.1, and that smb_1 just coincides with the homomorphism 2.4.3 (1). Thus, the properties of the mapping smb verified in the proof of Theorem 2.1 can be restated as follows: the homomorphism smb_1 is an isomorphism from the local algebra $\mathcal{A}/\mathcal{J}_1$ onto $\Sigma^p(\alpha)$. Moreover, since smb_t is nothing else but the composition of the mappings (6) and smb_1 we get: smb_t is an isomorphism from

$A/\mathcal{J}_t$ onto $\Sigma^p(\alpha)$. In particular, the coset $\Phi_t(A)$ is invertible if and only if the operator $\mathrm{smb}_t A$ is invertible. Allan's local principle gives: If $A \in \mathrm{alg}\, T(PC_{p,\alpha})$ then the coset $\pi(A)$ is invertible in $\mathrm{alg}\,{}^\pi T(PC_{p,\alpha})$ if and only if all operators $\mathrm{smb}_t A$ are invertible.

Next we show the inverse closedness of $\mathrm{alg}\,{}^\pi T(PC_{p,\alpha})$ in the Calkin algebra. Consider the collection of all cosets $C = \pi\left(\sum_i \prod_j T(a_{ij})\right)$ where the functions a_{ij} are piecewise constant. This collection is dense in $\mathrm{alg}\,{}^\pi T(PC_{p,\alpha})$ by 2.3.1(f), and the already proved part shows that the spectrum of C in $\mathrm{alg}\,{}^\pi T(PC_{p,\alpha})$ has no interior points. So Corollary 1.1 applies to give the inverse closedness, that is, the coset $\pi(A)$ is invertible in $\mathrm{alg}\,{}^\pi T(PC_{p,\alpha})$ if and only if A is a Fredholm operator. Further, using that all compact operators belong to $\mathrm{alg}\, T(PC_{p,\alpha})$, one easily can show that the algebra $\mathrm{alg}\, T(PC_{p,\alpha})$ is inverse closed in $L(l^p_{Z+})$ (compare also Exercise 1.11).

To complete the proof it remains to verify identity (2). By the definitions,

$$\mathrm{smb}_t\, Y_s G(b) Y_s^{-1} = \mathrm{smb}_1\, Y_{s/t} G(b) Y_{t/s} .$$

In case $s = t$ the assertion follows from the definition of $G(b)$. So let $s \neq t$, and choose a continuous multiplier f with $f(1) = 1$ and $f(s/t) = 0$. Then $\mathrm{smb}_1 T(f) = I$ and, thus,

$$\mathrm{smb}_1\, Y_{s/t} G(b) Y_{t/s} = \mathrm{smb}_1\, T(f) Y_{s/t} G(b) Y_{t/s} .$$

But, by 2.5.1(2),

$$T(f) Y_{s/t} G(b) Y_{t/s} = Y_{s/t} T(f_{s/t}) G(b) Y_{t/s} ,$$

and this operator is compact since, by Proposition 2.11(d),

$$\pi(T(f_{s/t}) G(b)) = \pi(f_{s/t}(1) G(b)) = \pi(f(s/t) G(b)) = 0 . \quad \blacksquare$$

Example 2.2 *Let $a \in PC_{p,\alpha}$, $t_1,\ldots,t_n \in \mathbf{T}$, and let $b_1,\ldots,b_n$ be continuous functions on $\bar{\mathbf{R}}$ such that $M^0(b_i) \in N^p(\alpha)$. Then the operator*

$$A = T(a) + Y_{t_1} G(b_1) Y_{t_1}^{-1} + \ldots + Y_{t_n} G(b_n) Y_{t_n}^{-1} \tag{7}$$

is Fredholm if and only if the function $A(t,z) : \mathbf{T} \times \bar{\mathbf{R}} \to \mathbf{C}$ given by

$$A(t,z) = (a(t-0) + a(t+0))/2 + (a(t-0) - a(t+0))s(z)/2$$

at the points $t \in \mathbf{T} \setminus \{t_1,\ldots,t_n\}$ and by

$$A(t_i,z) = b_i(z) + (a(t_i-0) + a(t_i+0))/2 + (a(t_i-0) - a(t_i+0))s(z)/2$$

*at $t = t_i$ $(i = 1,\ldots,n)$, does not vanish.
(The function s was introduced in 2.1.2(4).) If the operator A is Fredholm then A possesses a regularizer in $\mathrm{alg}\, T(PC_{p,\alpha})$.*

More visually: Let a be a multiplier with only finitely many jumps, say, at $x_1, \ldots, x_m \in \mathbf{T}$. Then $a(\mathbf{T})$ is a curve in the complex plane consisting of m pieces which join $a(x_i + 0)$ to $a(x_{i+1} - 0)$ $(i = 1, \ldots, m, \ x_{m+1} := x_1)$. We complete this curve to a closed curve Γ by adding the curves

$$\{a(x_i - 0)(1 + s(z))/2 + a(x_i + 0)(1 - s(z))/2 \ , \ z \in \bar{\mathbf{R}} \ \}$$

for all $x_i \notin \{t_1, \ldots, t_n\}$ (these curves join $a(x_i - 0)$ to $a(x_i + 0)$) and

$$\{b_j(z) + a(t_j - 0)(1 + s(z))/2 + a(t_j + 0)(1 - s(z))/2 \ , \ z \in \bar{\mathbf{R}} \ \}$$

for all t_j $(j = 1, \ldots, n)$ (these curves join $a(t_j - 0)$ to $a(t_j + 0)$ and produce a 'loop' at $a(t_j)$ if the function a is continuous at t_j). The operator A is Fredholm if and only if the point 0 does not lie on the curve Γ.

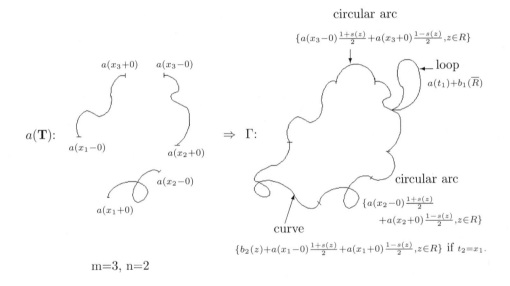

circular arc
$\{a(x_3-0)\frac{1+s(z)}{2}+a(x_3+0)\frac{1-s(z)}{2},z{\in}R\}$

$a(x_3+0)$ $a(x_3-0)$

loop
$a(t_1)+b_1(\overline{R})$

$a(\mathbf{T})$: $\Rightarrow \ \Gamma$:

$a(x_1-0)$ $a(x_2+0)$

$a(x_2-0)$

$a(x_1+0)$

circular arc
$\{a(x_2-0)\frac{1+s(z)}{2}$
$+a(x_2+0)\frac{1-s(z)}{2},z{\in}R\}$

curve
$\{b_2(z)+a(x_1-0)\frac{1+s(z)}{2}+a(x_1+0)\frac{1-s(z)}{2},z{\in}R\}$ if $t_2{=}x_1$.

m=3, n=2

For a proof notice that

$$\mathrm{smb}_t A \ = \ (a(t - 0) + a(t + 0))I/2 \ + \ (a(t - 0) - a(t + 0))S_{R^+}/2$$

if $t \in \mathbf{T} \setminus \{t_1, \ldots, t_n\}$ and

$$\mathrm{smb}_{t_i} A \ = \ M^0(b_i) + (a(t_i - 0) + a(t_i + 0))I/2 + (a(t_i - 0) - a(t_i + 0))S_{R^+}/2 \ ,$$

and take into account that $S_{R^+} = M^0(s)$ on $L_{R^+}^p(\alpha)$. ∎

Let us turn to the greater algebra alg $(T^0(PC_{p,\alpha}), P)$ now.

Theorem 2.3 *(a) An operator $A \in \mathrm{alg}\,(T^0(PC_{p,\alpha}), P)$ is Fredholm if and only if each operator $\mathrm{smb}_t A(\in \Sigma_R^p(\alpha))$ with $t \in \mathbf{T}$ is invertible. In particular,*

$$
\begin{aligned}
\mathrm{smb}\,T^0(a) &= a(t+0)(I - S_R)/2 + a(t-0)(I + S_R)/2\,, \\
\mathrm{smb}_t\,P &= \chi_{R^+} I\,, \\
\mathrm{smb}_t\,K &= 0 \quad for \quad compact \quad K\,.
\end{aligned}
$$

(b) The algebra $\mathrm{alg}\,^\pi(T^0(PC_{p,\alpha}), P)$ is inverse closed in the Calkin algebra $L(l_Z^p(\alpha))/K(l_Z^p(\alpha))$, and the algebra $\mathrm{alg}\,(T^0(PC_{p,\alpha}), P)$ is inverse closed in the algebra $L(l_Z^p(\alpha))$.

Proof   The proof proceeds as that one of Theorem 2.2. We only mark the essential steps. The algebra $\{\pi(T^0(f)), f \in C_{p,\alpha}\}$ lies in the center of the algebra $\mathcal{A} := \mathrm{alg}\,^\pi(T^0(PC_{p,\alpha}), P)$, and its maximal ideal space is homeomorphic to the unit circle $\mathbf{T}$. For each $t \in \mathbf{T}$ we form the local algebras $\mathcal{A}/\mathcal{J}_t$ and denote the canonical homomorphism from $\mathrm{alg}\,(T^0(PC_{p,\alpha}), P)$ onto $\mathcal{A}/\mathcal{J}_t$ by Φ_t. Let $A \in \mathrm{alg}\,(T^0(PC_{p,\alpha}), P)$. Then the operator $\mathrm{smb}_t A$ depends on the coset $\Phi_t(A)$ only. Hence, smb_t can be viewed as an algebra homomorphism from $\mathcal{A}/\mathcal{J}_t$ into $\Sigma_R^p(\alpha)$.

We are going to show that smb_t is the desired isomorphism. Due to the rotation invariance of the algebra $\mathrm{alg}\,(T^0(PC_{p,\alpha}), P)$ we limit ourselves to the case $t = 1$. Let $\rho : L(L_R^p(\alpha)) \to L(L_{R^+}^p(\alpha)_2)$ stand for the isomorphism $H \mapsto \eta H \eta^{-1}$ with η being defined as in Section 2.1.3. By Proposition 2.1, $\rho(\Sigma_R^p(\alpha)) \subseteq \Sigma^p(\alpha)_{2\times2}$. To each matrix $\begin{pmatrix} A & B \\ C & D \end{pmatrix} \in \rho(\Sigma_R^p(\alpha))$ we associate the operator

$$
\begin{pmatrix} E_{-n}L_n A E_n & E_{-n}L_n B E_n \\ E_{-n}L_n C E_n & E_{-n}L_n D E_n \end{pmatrix} \in (\mathrm{alg}\,T(PC_{p,\alpha}))_{2\times2}
$$

which will be denoted by $\mu \begin{pmatrix} A & B \\ C & D \end{pmatrix}$. Finally, we let $J : l_Z^p(\alpha) \to l_Z^p(\alpha)$ refer to the flip operator $(x_n) \mapsto (x_{-n-1})$, and we define the operator $\tau : (\mu \circ \rho)(\Sigma_R^p(\alpha)) \to \mathcal{A}/\mathcal{J}_1$ by

$$
\tau \begin{pmatrix} X & Y \\ Z & W \end{pmatrix} = \Phi_1(X + YJ + JZ + JWJ) \in \mathcal{A}/\mathcal{J}_1\,. \tag{8}
$$

Then $\tau \circ \mu \circ \rho$ is the inverse of the homomorphism smb_1. Let us first explain that the definition (8) is correct, i.e. that $X + YJ + JZ + JWJ$ is actually an operator in $\mathrm{alg}\,(T^0(PC_{p,\alpha}), P)$. Evidently, $JT^0(a)J = T^0(\tilde{a})$ with $\tilde{a}(t) = a(1/t)$, and $JPJ = I - P$. Thus, $JWJ \in \mathrm{alg}\,(T^0(PC_{p,\alpha}), P)$ whenever $W \in \mathrm{alg}\,T(PC_{p,\alpha})$. Further, the Proposition 2.1 tells us that the operators Y and Z are necessarily discretized Mellin convolutions. So it suffices to prove the following implication:

$$
M^0(b) \in N^p(\alpha) \;\Rightarrow\; JG(b)\,, \; G(b)J \in \mathrm{alg}\,(T^0(PC_{p,\alpha}), P)\,. \tag{9}
$$

To this end we approximate b by a function $b' n_\pi$ with $M^0(b') \in N^p(\alpha)$ as closely as desired (this is possible due to 2.1.2(c)). By Proposition 2.11(b) we have

$$\pi(G(b' n_\pi)) = \pi(G(b') G(n_\pi)) = \pi(G(b') H(f))$$

with f referring to the function 2.4.2(2). Set $Q = I - P$. A little thought shows that the Hankel operator $H(f)$ can be identified with the operator $PT^0(f)QJ$ acting on $l_Z^p(\alpha)$. Now, clearly, $G(b')H(f)J = G(b')PT^0(f)QJ^2 = G(b')PT^0(f)Q$, and the latter operator is indeed in $\mathrm{alg}\,(T^0(PC_{p,\alpha}), P)$. Analogously, the second implication of (9) can be verified.

Now it is not too hard to prove that $\tau \circ \mu \circ \rho$ is a continuous algebra homomorphism (this can be done just in the same manner as we have shown that smb' is a homomorphism, see 2.4.3). So we are left on verifying that $\tau \circ \mu \circ \rho \circ \mathrm{smb}_1 = I$ and $\mathrm{smb}_1 \circ \tau \circ \mu \circ \rho = I$, and it is sufficient to do this for the generating operators of $\mathrm{alg}\,(T^0(PC_{p,\alpha}), P)$. Let us consider the first of these identities. We claim that

$$(\tau \circ \mu \circ \rho \circ \mathrm{smb}_1)(\Phi_1(T^0(a))) = \Phi_1(T^0(a)) .$$

Indeed,

$$\mathrm{smb}_1\,\Phi_1(T^0(a)) = a(1-0)(I + S_R)/2 + a(1+0)(I - S_R)/2$$

whence, by Proposition 2.1 and formula 2.1.3(1), we obtain (using the notations $a_+ := a(1+0)$ and $a_- := a(1-0)$)

$$(\rho \circ \mathrm{smb}_1)(\Phi_1(T^0(a))) =$$
$$= \frac{1}{2} \begin{pmatrix} a_-(I + S_{R^+}) + a_+(I - S_{R^+}) & (a_+ - a_-)N \\ (a_- - a_+)N & a_-(I - S_{R^+}) + a_+(I + S_{R^+}) \end{pmatrix} .$$

Because of $E_{-n} L_n S_{R^+} E_n = T(\sigma)$ (see Section 2.2.4) and $E_{-n} L_n N E_n = G(n_\pi) = H(f) +$ compact, we have

$$(\tau \circ \mu \circ \rho \circ \mathrm{smb}_1)(\Phi_1(T^0(a))) =$$

$$= \tfrac{1}{2}\Phi_1 \big(a_- T(1+\sigma) + a_+ T(1-\sigma) + (a_+ - a_-)H(f)J \\ + (a_- - a_+)JH(f) + a_- JT(1-\sigma)J + a_+ JT(1+\sigma)J \big)$$

$$= \tfrac{1}{2}\Phi_1 \big(a_- PT^0(1+\sigma)P + a_+ PT^0(1-\sigma)P + \\ + (a_+ - a_-)PT^0(f)QJ^2 + (a_- - a_+)JPT^0(f)QJ \\ + a_- JPT^0(1-\sigma)PJ + a_+ JPT^0(1+\sigma)PJ \big) ,$$

and since $JP = QJ$ and $JT^0(a) = T^0(\tilde{a})J$, and since the functions $f + \sigma$, $\tilde{f} - \sigma$, $\tilde{\sigma} + \sigma$ are continuous on $\mathbf{T}$ and vanish at 1, this is equal to

$$\tfrac{1}{2}\Phi_1(a_-(PT^0(1+\sigma)P + PT^0(1+\sigma)Q + QT^0(1+\sigma)P + QT^0(1+\sigma)Q) + \\ + a_+(PT^0(1-\sigma)P + PT^0(1-\sigma)Q + QT^0(1-\sigma)P + QT^0(1-\sigma)Q))$$
$$= \Phi_1 \big(a_- T^0 \big(\tfrac{1+\sigma}{2} \big) + a_+ T^0 \big(\tfrac{1-\sigma}{2} \big) \big)$$
$$= \Phi_1(T^0(a))$$

as desired. Analogously, all other identities can be verified. ∎

2.5.3   Invertibility

In contrast to Fredholmness, invertibility of operators in Toeplitz algebras is of global nature and so its investigation requires methods and techniques which are quite different from that ones used above. So we restrict ourselves to stating some crucial results which are needed in what follows, but we refer to the relevant textbooks for their proofs.

Theorem 2.4 *(Coburn).* *Let a, b, c be $l^p(\alpha)$-multipliers which are not identically zero. The Toeplitz operator $T(a)$ and the operator $T^0(b)P + T^0(c)Q$ have trivial kernel or a trivial cokernel.*

Corollary 2.2 *Let a, b, c be $l^p(\alpha)$-multipliers. The operators $T(a)$ and $T^0(b)P + T^0(c)Q$ are invertible on $l^p(\alpha)$ and $l_Z^p(\alpha)$, respectively, if and only if they are Fredholm and their index is zero.*

For applying this corollary one needs some information about the index of Toeplitz operators. It can be given in terms of the curve Γ introduced in Example 2.2. The *winding number* of this curve is the increment of $\frac{1}{2\pi} \arg z$ when z runs through Γ counterclockwise.

Proposition 2.12 *(a) The index of the operator 2.5.2(7) equals the negative of the winding number of Γ.*
(b) The index of the operator $T^0(b)P + T^0(c)Q$ is equal to the index of the Toeplitz operator $T(b/c)$.

2.5.4   Matrix valued multipliers

The above considerations are not restricted to scalar valued multipliers. Indeed, both the Theorems 2.2 and 2.3 as their proofs are applicable to matrix-valued piecewise continuous multipliers in an almost obvious manner. For example, given an operator $(A_{mn})_{m,n=1}^N \in \text{alg}\,(T^0((PC_{p,\alpha})_{N \times N}), \text{diag}\,P)$ then the operators A_{mn} are in $\text{alg}\,(T^0(PC_{p,\alpha}), P)$, and so it makes sense to define the symbol of (A_{mn}) at $t \in \mathbf{T}$ by

$$\text{smb}_t(A_{mn})_{m,n=1}^N := (\text{smb}_t A_{mn})_{m,n=1}^N .$$

This new-defined mapping smb_t is again an algebra homomorphism which sends the *block Toeplitz operator* $T^0(a)$, $a \in (PC_{p,\alpha})_{N \times N}$, into the *block singular integral operator*

$$a(t+0)\,\text{diag}\,((I - S_R)/2) \; + \; a(t-0)\,\text{diag}\,((I + S_R)/2)$$

and the projection $\text{diag}\,P$ into the operator $\text{diag}(\chi_{R^+} I)$.

An operator $A \in \text{alg}\,(T^0((PC_{p,\alpha})_{N \times N}), \text{diag}\,P)$ is Fredholm if and only if $\text{smb}_t A$ is invertible for each $t \in \mathbf{T}$, and if A is Fredholm then A possesses a regularizer in $\text{alg}\,(T^0((PC_{p,\alpha})_{N \times N}), \text{diag}\,P)$.

2.6 General spline spaces

2.6.1 Definitions

The spline space S_n of all piecewise constant splines was introduced in Section 2.2.1 as follows: We took the function $\phi(t) = \chi_{[0,1)}(t)$, shifted it by $l \in \mathbf{Z}$ and streched it by $n \in \mathbf{Z}^+$ to obtain functions $\phi_{ln}(t) = \phi(nt - l)$, and then we defined S_n as the smallest closed subspace of $L^p_R(\alpha)$ containing all functions ϕ_{ln} with $l \in \mathbf{Z}$.

Now we are going to study a rather broad class of spline spaces which originates from the same procedure: we only replace the 'generating function' $\chi_{[0,1)}$ by another function, ϕ, and we start with a whole variety $\{\phi_0, \ldots, \phi_{k-1}\}$ of 'generating functions' instead of the one function ϕ. More precisely, let k be a positive integer and $F = \{\phi_0, \ldots, \phi_{k-1}\}$ be a k-tuple of *presplines*, that is, of bounded measurable non-zero functions on $\mathbf{R}$ with compact support. Given integers $r \in \{0, \ldots, k-1\}$, $l \in \mathbf{Z}$, and $n \in \mathbf{Z}^+$, we define functions ϕ_{rln} by

$$\phi_{rln}(t) = \phi_r(tn - l) \quad , t \in \mathbf{R} \, .$$

In case $k = 1$ and $F = \{\phi_0\} = \{\chi_{[0,1)}\}$, this notation is in accordance with that one of Section 2.2.1. The condition $0 < 1/p + \alpha < 1$ guarantees that all functions ϕ_{rln} belong both to the Lebesgue space $L^p_R(\alpha)$ and to its dual $L^q_R(-\alpha)$, $1/p + 1/q = 1$. So it makes sense to speak about the smallest closed subspace S^F_n of $L^p_R(\alpha)$ which contains all functions ϕ_{rln} with $r \in \{0, \ldots, k-1\}$ and $l \in \mathbf{Z}$. Clearly, this space depends on p and α. We refer to S^F_n as the *spline space generated by* F and call its elements *splines* (or F-splines). Notice that the correspondence between S^F_n and F is not one-to-one. To illustrate this, we consider an example which will accompany us throughout the following sections.

Example 2.3 *Let* $k = 1$, $F_1 = \{\chi_{[0,1]}\}$, *and* $F_2 = \{\chi_{[0,2]}/2\}$. *Then* $S^{F_1}_n = S^{F_2}_n$ *for all* n.

Indeed, first observe that we can limit ourselves to the case $n = 1$ and that the inclusion $S^{F_2}_1 \subseteq S^{F_1}_1$ is obvious. For the reverse direction it remains to show that $\chi_{[0,1]} \in S^{F_2}_1$. Given a non-negative integer m we define the function f_m by

$$f_m = \sum_{r=0}^{2^{m-1}-1} (-1)^r \frac{2^{m-1} - r}{2^m} (\chi_{[r,r+2]} + \chi_{[-r-1,-r+1]}) \, .$$

Clearly, $f_m \in S^{F_2}_1$, and rewriting f_m as

$$f_m = \chi_{[0,1]} + \frac{1}{2^m} \sum_{r=0}^{2^{m-1}-1} (-1)^r (\chi_{[r+1,r+2]} + \chi_{[-r-1,-r]})$$

we find

$$\|f_m - \chi_{[0,1]}\|^p_{L^p_R(\alpha)} =$$

$$= \frac{1}{2^{mp}} \int_R |\sum_{r=0}^{2^{m-1}-1} (-1)^r (\chi_{[r+1,r+2]}(t) + \chi_{[-r-1,-r]}(t)|^p |t|^{\alpha p} \, dt$$

$$\leq \frac{1}{2^{mp}} \int_{-2^{m-1}-2}^{2^{m-1}+2} |t|^{\alpha p} \, dt = \frac{2}{\alpha p + 1} (\frac{2^{m-1}+2}{2^{m-1}})^{\alpha p + 1} \frac{1}{2^{m(p-\alpha p - 1)+\alpha p + 1}} \, .$$

The restriction $0 < 1/p + \alpha < 1$ imposed on p and α implies that $p - \alpha p - 1 > 0$; thus, $\|f_m - \chi_{[0,1]}\|$ goes to zero as $m \to \infty$ whence $\chi_{[0,1]} \in S_n^{F_2}$. ∎

2.6.2   De Boor's estimates

As in Section 2.2.2 (where $F = \{\chi_{[0,1)}\}$), it is possible to associate with each function in S_n^F a sequence in $l_Z^p(\alpha)$ and then to derive a de Boor-type estimate. But for k-tuples of presplines it will prove to be more natural to associate with each spline a k-vector of sequences in $l_Z^p(\alpha)$. For, we abbreviate by $l_k^p(\alpha)$ the Banach space of all k-vectors $(X_0, \dots, X_{k-1})$, $X_r \in l_Z^p(\alpha)$, with norm

$$\|(X_0, \dots, X_{k-1})\|_{l_k^p(\alpha)} = \left(\sum_{r=0}^{k-1} \|X_r\|_{l_Z^p(\alpha)}^p \right)^{1/p} \, .$$

Note that $l_1^p(\alpha) = l_Z^p(\alpha)$.

If $(X_0, \dots, X_{k-1})$ is any k-vector of sequences $X_r = (x_{rl})_{l \in Z}$ (not necessarily in $l_k^p(\alpha)$) then each of the series $\sum_{l \in Z} x_{rl} \phi_{rln}$ converges on every compact interval $\mathbf{I}$ because there are only finitely many functions ϕ_{rln} living on $\mathbf{I}$. Define the operator E_n^F acting on the linear space of all k-vectors $(X_0, \dots, X_{k-1})$ by

$$E_n^F : (X_0, \dots, X_{k-1}) \mapsto \sum_{r=0}^{k-1} \sum_{l \in Z} x_{rl} \phi_{rln} \, . \tag{1}$$

The following result describes the mapping properties of the restriction of E_n^F onto the Banach space $l_k^p(\alpha)$.

Theorem 2.5 *Let* $F = \{\phi_0, \dots, \phi_{k-1}\}$ *be a* k-*tuple of presplines and* $(X_0, \dots, X_{k-1}) \in l_k^p(\alpha)$. *Then the series* $E_n^F(X_0, \dots, X_{k-1})$ *converges in* $L_R^p(\alpha)$, *i.e.* $E_n^F(X_0, \dots, X_{k-1}) \in S_n^F$ ($\subseteq L_R^p(\alpha)$), *and there is a constant* C *(depending on* F, p, α) *such that*

$$\|E_n^F(X_0, \dots, X_{k-1})\|_{L_R^p(\alpha)} \leq C \, n^{-(1/p+\alpha)} \|(X_0, \dots, X_{k-1})\|_{l_k^p(\alpha)} \, . \tag{2}$$

It is clearly sufficient to consider the case $k = 1$. Nevertheless, the proof is rather lengthy, and we defer it to Section 2.12.2.

The things become essentially more involved if one asks for the mapping properties of the "inverse" E_{-n}^F of E_n^F given by

$$E_{-n}^F : \sum_{r=0}^{k-1} \sum_{l \in Z} x_{rl} \phi_{rln} \mapsto (X_0, \dots, X_{k-1}) \tag{3}$$

with $X_r = (x_{rl})_{l \in \mathbf{Z}}$. The first reason is that (3) is not a correct definition unless each finite choice of the functions ϕ_{rln} ($r \in \{0, \ldots, k-1\}$, $l \in \mathbf{Z}$) is linearly independent. But secondly, even if these functions are linearly independent, then the coefficient sequence $E^F_{-n} f_n$ needs *not* to belong to $l^p_k(\alpha)$ for every $f_n \in S^F_n$. Let, for example, $k = 1$, $F = \{\phi_0\}$ with $\phi_0 = \chi_{[0,2]}/2$, $x_{0l} = (-1)^l (1 + |l|)^{-1/2}$, and set $f_1 = \sum_{l \in \mathbf{Z}} x_{0l} \phi_{0l1}$. Then

$$
\|f_1\|^2_{L^2_R(0)} = \left\| \sum_{l \in \mathbf{Z}} \frac{(-1)^l}{2\sqrt{|l|+1}} \chi_{[l,l+2]} \right\|^2_{L^2_R(0)}
$$

$$
= \left\| \sum_{l \in \mathbf{Z}} \left(\frac{1}{2\sqrt{|l-1|+1}} - \frac{1}{2\sqrt{|l|+1}} \right) (-1)^{l-1} \chi_{[l,l+1]} \right\|^2_{L^2_R(0)}
$$

$$
= \sum_{l \in \mathbf{Z}} \left| \frac{1}{2\sqrt{|l-1|+1}} - \frac{1}{2\sqrt{|l|+1}} \right|^2 < \infty .
$$

Thus, $f_1 \in S^F_1$, but the coefficient sequence (x_{0l}) fails to belong to $l^2_{\mathbf{Z}}(0)$. In the next section, we shall formulate a strong condition on the prebasis F which ensures that $E^F_{-n} f_n \in l^p_k(\alpha)$ whenever $f_n \in S^F_n$.

2.6.3   Canonical tuples of presplines

Let $F = \{\phi_0, \ldots, \phi_{k-1}\}$ be a k-tuple of presplines and $f \in L^p_R(\alpha)$. Our conditions imposed on p and α guarantee that $F \subseteq L^p_R(\alpha)^*$, and so it makes sense to form the integrals

$$
x_{rl}(f) = \int_R f\left(\frac{t+l}{n} \right) \overline{\phi_r(t)} \, dt
$$

where $r \in \{0, \ldots, k-1\}$ and $l \in \mathbf{Z}$. Let P^F_n stand for the linear mapping

$$
P^F_n f = (X_0, \ldots, X_{k-1}) \quad \text{with} \quad X_r = (x_{rl})_{l \in \mathbf{Z}}
$$

which associates with each $L^p_R(\alpha)$-function a k-vector of complex sequences.

Proposition 2.13 *The mapping P^F_n is a linear bounded operator from $L^p_R(\alpha)$ into $l^p_k(\alpha)$, and there is a constant $C > 0$ such that*

$$
\|P^F_n f\|_{l^p_k(\alpha)} \leq C n^{1/p+\alpha} \|f\|_{L^p_R(\alpha)} .
$$

For the proof we refer to Section 2.12.3.

The Theorem 2.5 and Proposition 2.13 involve that the operators $P^F_n E^F_n$ are bounded on $l^p_k(\alpha)$. Moreover, one has

Proposition 2.14 *Let F be a k-tuple of presplines. Then*
(a) the operators $P_n^F E_n^F : l_k^p(\alpha) \to l_k^p(\alpha)$ are independent of n.
(b) there are trigonometric polynomials $\lambda_{rs}^{F,F}$ on $\mathbf{T}$ such that $P_n^F E_n^F = T^0(\lambda^{F,F})$
with $\lambda^{F,F} = (\lambda_{rs}^{F,F})_{r,s=0}^{k-1}$.

Proof Let $F = \{\phi_0, \ldots, \phi_{k-1}\}$ and think of the operator $P_n^F E_n^F$ acting on $l_k^p(\alpha)$ as matrix $(A_{rs})_{r,s=0}^{k-1}$ of operators acting on $l_1^p(\alpha)$. The lm th entry of the infinite matrix associated with A_{rs} is

$$\int_R \phi_{smn}\left(\frac{t+l}{n}\right)\overline{\phi_r(t)}\, dt \;=\; \int_R \phi_s(t+l-m)\overline{\phi_r(t)}\, dt \;, \tag{1}$$

i.e. the operator A_{rs} is independent of n, and its lm th entry depends on the difference $l - m$ only. Consequently, $A_{rs} = T^0(\lambda_{rs}^{F,F})$, where $\lambda_{rs}^{F,F} = \lambda^{\{\phi_s\},\{\phi_r\}}$ is the $L^\infty(\mathbf{T})$-function whose $l-m$ th Fourier coefficient just coincides with (1). This function is even a trigonometric polynomial since, due to the compactness of the supports of ϕ_s, there are only finitely many differences $l - m$ such that (1) does not vanish. ∎

Let us agree to call a k-tuple F of presplines *canonical* if the matrix polynomial $\lambda^{F,F}$ is invertible. With every canonical k-tuple of presplines there is naturally associated a projection operator. Indeed, if $\lambda^{F,F}$ is invertible then the operator $T^0(\lambda^{F,F})$ is also invertible (its inverse is $T^0((\lambda^{F,F})^{-1})$), and we can define the operators $L_n^{F,F} := E_n^F T^0(\lambda^{F,F})^{-1} P_n^F$.

Proposition 2.15 *If the k-tuple F of presplines is canonical then the operator $L_n^{F,F}$ is a bounded projection of the Lebesgue space $L_R^p(\alpha)$ onto its subspace S_n^F.*

Proof The boundedness of $L_n^{F,F}$ is an immediate consequence of Theorem 2.5 and Proposition 2.13. Further

$$\begin{aligned}
L_n^{F,F} L_n^{F,F} &= E_n^F T^0(\lambda^{F,F})^{-1} P_n^F E_n^F T^0(\lambda^{F,F})^{-1} P_n^F \\
&= E_n^F T^0(\lambda^{F,F})^{-1} T^0(\lambda^{F,F}) T^0(\lambda^{F,F})^{-1} P_n^F = E_n^F T^0(\lambda^{F,F})^{-1} P_n^F \\
&= L_n^{F,F} \;,
\end{aligned}$$

that is, $L_n^{F,F}$ is a projection. It remains to show that the image of $L_n^{F,F}$ coincides with S_n^F. First observe that the invertibility of $T^0(\lambda^{F,F}) = P_n^F E_n^F$ implies that the image of P_n^F (and, thus, that of $T^0(\lambda^{F,F})^{-1} P_n^F$) equals all of $l_k^p(\alpha)$. Further, Theorem 2.5 states that the image of E_n^F is contained in S_n^F, and now it is evident that $\operatorname{Im} E_n^F$ is even dense in S_n^F (take finitely supported sequences). Thus, $\operatorname{Im} L_n^{F,F}$ is dense in S_n^F, but the image of a bounded projection is always closed. Consequently, $\operatorname{Im} L_n^{F,F}$ coincides with S_n^F. ∎

As a corollary we obtain that the operator E_n^F maps $l_k^p(\alpha)$ *onto* the spline space $S_n^F \subseteq L_R^p(\alpha)$ whenever F is canonical.

2.6.4   De Boor's estimates continued

Now we are in position to attack the reverse direction in de Boor's estimate which we left open in Section 2.6.2.

Theorem 2.6 *Let $F = \{\phi_0, \dots, \phi_{k-1}\}$ be a canonical k-tuple of presplines. Then the operators E^F_{-n} are correctly defined by 2.6.2 (3), and $E^F_{-n} = T^0(\lambda^{F,F})^{-1}P^F_n$ on S^F_n. Moreover, there is a constant $C > 0$ such that*

$$\|E^F_{-n}f\|_{l^p_k(\alpha)} \leq C\,n^{1/p+\alpha}\|f\|_{L^p_R(\alpha)} \quad for \quad all \quad f \in S^F_n\,. \tag{1}$$

Proof   Clearly, $T^0(\lambda^{F,F})^{-1}P^F_n E^F_n$ is the identity operator on $l^p_k(\alpha)$, and $E^F_n T^0(\lambda^{F,F})^{-1}P^F_n = L^{F,F}_n$ is the identity operator on $\operatorname{Im} E^F_n = S^F_n$. Thus, E^F_n is an invertible operator from $l^p_k(\alpha)$ onto all of S^F_n, and its inverse E^F_{-n} is equal to the restriction of $T^0(\lambda^{F,F})^{-1}P^F_n$ onto S^F_n. In particular, the spline functions ϕ_{rln} are linearly independent, and the norm estimate (1) is an immediate consequence of Proposition 2.13. ∎

In other words, the Banach spaces $l^p_k(\alpha)$ and $S^F_n \ (\subseteq L^p_R(\alpha))$ are topologically isomorphic if F is canonical. As a by-product of this fact we obtain:

Corollary 2.3 *If F is a canonical k-tuple of presplines then the system $\{\phi_{rln}\}$ forms a basis of S^F_n. Each spline admits a unique representation as a series $\sum_{r=0}^{k-1}\sum_{l\in Z}x_{rl}\phi_{rln}$.*

Proof   If $f \in S^F_n$ then there exists a sequence (f_m), $f_m = \sum_{r=0}^{k-1}\sum_{|l|\leq m}x^{(m)}_{rl}\phi_{rln}$, which converges to f in $L^p_R(\alpha)$. De Boor's estimate (1) implies that the sequence $(E^F_{-n}f_m)_{m\geq 0}$ is Cauchy in $l^p_k(\alpha)$. Hence, there is a vector $X = (X_0, \dots, X_{k-1}) \in l^p_k(\alpha)$ such that $E^F_{-n}f_m \to X$ as $m \to \infty$. Now employ the continuity of E^F_n (by 2.6.2(2)) to find that $E^F_n E^F_{-n}f_m = f_m \to E^F_n X$ in $L^p_R(\alpha)$ as $m \to \infty$, and this involves that $E^F_n X = f = \sum_{r=0}^{k-1}\sum_{l\in Z}x_{rl}\phi_{rln}$. The uniqueness of the coefficients x_{rl} is again a consequence of de Boor's estimates. ∎

2.7   Spline projections

2.7.1   Classes of Riemann integrable functions

Our next goal is the description of a class of projection operators which includes the Galerkin and collocation projections. Since the collocation projection cannot be defined on all of $L^p_R(\alpha)$ (the delta functional does not belong to the dual of $L^p_R(\alpha)$), we start with introducing a further Banach space which is convenient to deal with collocation operators: the weighted Riemann space $R^p_R(\alpha)$ over the real axis. The elements of $R^p_R(\alpha)$ are functions (not cosets of functions as the elements of $L_R^p(\alpha)$) which are Riemann integrable on each finite interval, and for which the norm

$$\|f\|_{R_R^p(\alpha)} := \|f\|_{L_R^p(\alpha)} + \left(\sum_{k \in \mathbb{Z}} \sup_{t \in [k, k+1]} |f(t)|^p (|k| + 1)^{\alpha p} \right)^{1/p}$$

is finite. Clearly, it would be more correct to write $\|f^\circ\|_{L_R^p(\alpha)}$ instead of $\|f\|_{L_R^p(\alpha)}$ in this definition, where f° denotes the coset in $L_R^p(\alpha)$ containing f but, as usually, the $^\circ$ will be omitted. It is not hard to see that $R_R^p(\alpha)$ is a Banach space which contains all bounded Riemann integrable functions with compact support. Further it is immediate from the definition that the embedding

$$^\circ \; : \; R_R^p(\alpha) \to L_R^p(\alpha) \, , \; f \mapsto f^\circ \, , \tag{1}$$

is continuous.

In what follows we let Y refer to either $L_R^p(\alpha)$ or $R_R^p(\alpha)$, and we write I_Y for the identity operator on $L_R^p(\alpha)$ in case $Y = L_R^p(\alpha)$ and for the operator (1) in case $Y = R_R^p(\alpha)$, and we call I_Y the embedding operator from Y into $L_R^p(\alpha)$. Let us emphasize once more that the elements of Y are common functions or cosets of functions in dependence on whether $Y = R_R^p(\alpha)$ or $L_R^p(\alpha)$, respectively.

For technical reasons we have sometimes to work in the subspace $P_R^p(\alpha)$ of $R_R^p(\alpha)$ which is the closure in $R_R^p(\alpha)$ of the linear space spanned by the characteristic functions $\chi_{[a,b)}$ with $-\infty < a < b < \infty$. This space contains all compactly supported continuous functions as well as all piecewise continuous functions with compact support which are continuous from the right and which possess one-sided finite limits from the left at each point. For brevity, we further introduce the notation Y_P to refer to either $L_R^p(\alpha)$ or $P_R^p(\alpha)$. Note that, in any case, the piecewise constant functions are dense in Y_P.

2.7.2   Riemann integrable spline functions

Let $F = \{\phi_0, \dots, \phi_{k-1}\}$ be a k-tuple of Riemann integrable presplines. One can think of F as being a subset both of the Lebesgue space $L_R^p(\alpha)$ and of the Riemann space $R_R^p(\alpha)$, and it makes sense to consider besides the spline space $S_n^F = S_n^F(L_R^p(\alpha))$ (introduced in Section 2.6.1) the spline space $S_n^F(R_R^p(\alpha))$ which is the smallest closed subspace of $R_R^p(\alpha)$ containing all functions ϕ_{rln} with $r \in \{1, \dots, k-1\}$ and $l \in \mathbb{Z}$. In the sequel, we shall write $S_n^F(Y)$ in order to indicate the dependence of the spline space on the underlying space, but we shall often use the notation S_n^F in place of $S_n^F(L_R^p(\alpha))$ for brevity.

Proposition 2.16 *Let F be a k-tuple of Riemann integrable presplines. Then*

(a) *the operator E_n^F defined by 2.6.2 (1) is bounded from $l_k^p(\alpha)$ into $R_R^p(\alpha)$ (and, thus, into $S_n^F(R_R^p(\alpha))$).*

(b) $I_Y S_n^F(Y) = S_n^F$.

Proof   For part (a) we refer to Section 2.12.2. For (b), we first observe that the inclusion $I_Y S_n^F(Y) \subseteq S_n^F$ is an immediate consequence of the boundedness of the embedding operator. Let, conversely, $f \in S_n^F$. Then $E_{-n}^F f \in l_k^p(\alpha)$ by Theorem 2.6 and $E_n^F E_{-n}^F f \in R_R^p(\alpha)$ by part (a) of the present proposition. But it is easy to see that $I_Y E_n^F E_{-n}^F f = f$ whence follows $f \in I_Y S_n^F(Y)$.

2.7.3   Premeasures

Let $Y = L_R^p(\alpha)$ or $Y = R_R^p(\alpha)$ again. The *support* of a bounded linear functional γ from Y into $\mathbf{C}$ is the complement of the largest open set $\Omega \subseteq \mathbf{C}$ such that $\gamma(f) = 0$ for all $f \in Y$ with supp $f \subseteq \Omega$. A non-zero linear bounded functional on Y with compact support is called a *premeasure on Y*. Given a k-tuple $G = \{\gamma_0, \ldots, \gamma_{k-1}\}$ of premeasures on Y we define a linear mapping P_n^G from Y into the linear space of all k-vectors of complex sequences by:

$$P_n^G f = (X_0(f), \ldots, X_{k-1}(f)) \quad \text{with} \quad X_r(f) = (x_{rl}(f))_{l \in Z} \quad \text{and}$$

$$x_{rl}(f) = \gamma_r(f(\frac{\cdot + l}{n})) = \int_R f\left(\frac{t+l}{n}\right) \overline{\gamma_r(t)} \, dt \,. \tag{1}$$

In case $Y = L_R^p(\alpha)$, we have $Y^* = L_R^q(-\alpha)$ with $1/p + 1/q = 1$, and so the integral in (1) refers to the usual Lebesgue integral in this setting.

Proposition 2.17 *Let G be a k-tuple of premeasures on Y. Then the mapping P_n^G is a linear bounded operator from Y into $l_k^p(\alpha)$, and there is a constant $C > 0$ (depending on G, p, α) such that*

$$\|P_n^G f\|_{l_k^p(\alpha)} \leq C n^{1/p+\alpha} \|f\|_Y \quad for \quad all \quad f \in Y \,.$$

For the proof we refer to Section 2.12.3.

Due to Theorem 2.5, Proposition 2.16 and Proposition 2.17, we can form the operators $P_n^G E_n^F : l_k^p(\alpha) \to l_k^p(\alpha)$. The following proposition points out that these operators have actually a very nice structure, and this structure will be employed in the subsequent step to introduce a large class of projection operators.

Proposition 2.18 *Let F be a k-tuple of presplines in Y and G a k-tuple of premeasures on Y. Then*
(a) the operators $P_n^G E_n^F : l_k^p(\alpha) \to l_k^p(\alpha)$ are independent of n.
(b) there is a matrix-valued function $\lambda^{F,G} = (\lambda_{rs}^{F,G})_{r,s=0}^{k-1}$ on $\mathbf{T}$ such that $P_n^G E_n^F = T^0(\lambda^{F,G})$, and the functions $\lambda_{rs}^{F,G}$ are trigonometric polynomials.

The proof is the same as that of Proposition 2.14.

2.7.4   Spline projections

Let again F be a k-tuple of presplines and G be a k-tuple of premeasures but now suppose moreover the matrix polynomial $\lambda^{F,G}$ to be invertible. As in Proposition 2.15 one checks that the operators $E_n^F T^0 (\lambda^{F,G})^{-1} P_n^G$ are bounded projections of the Banach space Y onto its subspace $S_n^F(Y)$. We denote this projection by $L_n^{F,G}$ and, for convenience, we introduce some more notations: The pair (F, G) is said to be *canonical* if the matrix polynomial $\lambda^{F,G}$ is invertible. Further, the k-tuple F is called *canonical* if (F, F) is a canonical pair (This definition is correct since each prespline generates a premeasure in a natural way.), and a sequence (L_n) of projection operators $L_n : Y \to S_n^F(Y)$ is *canonical* if there exists a canonical pair (F_1, G_1) such that $L_n = L_n^{F_1, G_1}$.

To make the latter situation plain: If $F_1 = \{\chi_{[0,1)}\}$ and $F_2 = \{\chi_{[0,2)}/2\}$ then $S_n^{F_1} = S_n^{F_2}$. Moreover, $\lambda^{F_1, F_1}(z) = 1$ and $\lambda^{F_2, F_2}(z) = z^{-1}/2 + 1 + z/2$ for all $z \in \mathbf{T}$ whence follows that F_1 is canonical but F_2 fails to be so since $\lambda^{F_2, F_2}(-1) = 0$. The Galerkin projections $L_n : L_R^p(\alpha) \to S_n^{F_1}$ defined in Section 2.2.1 are canonical since $L_n = L_n^{F_1, F_1}$ as one easily checks.

The following evident characterization of the projections $L_n^{F,G}$ proves often to be useful: Let $f \in Y$, then $L_n^{F,G} f$ is the uniquely determined spline function in $S_n^F(Y)$ which satisfies the equality

$$P_n^G (L_n^{F,G} f) \; = \; P_n^G f \; .$$

The proof is evident: we have $P_n^G L_n^{F,G} = P_n^G E_n^F T^0 (\lambda^{F,G})^{-1} P_n^G = P_n^G \; .$

2.7.5   The commutator property

The commutator property of canonical projections, which we are going to formulate and prove in this subsection, dominates a large part of the theory of spline projection methods for operators with variable coefficients. We begin with a technical result which concerns the commutativity of certain operator sequences modulo a sequence tending to zero in the operator norm.

Theorem 2.7 *Let F be a k-tuple of presplines, G a k-tuple of premeasures, f a continuous function on $\mathbf{R}$, and let a be a continuous $k \times k$ matrix-valued multiplier on $\mathbf{T}$. Then*

$$\| E_n^F T^0(a) P_n^G f I - f I E_n^F T^0(a) P_n^G \|_{L(Y, L_R^p(\alpha))} \to 0 \quad \text{as} \quad n \to \infty \; .$$

It is sufficient again to prove this theorem for $k = 1$. For this case, the proof will be presented in Section 2.12.4.

Theorem 2.8 *(Commutator property) If (L_n) is a canonical sequence of projections and $f \in C(\dot{\mathbf{R}})$ then*

$$\| L_n f I - f L_n \|_{L(Y, L_R^p(\alpha))} \to 0 \quad \text{as} \quad n \to \infty \; .$$

Proof   Apply Theorem 2.7 with $a = (\lambda^{F,G})^{-1}$. ∎

As a first simple but useful consequence we obtain:

Theorem 2.9 *(Asymptotic locality) If (L_n) is a canonical sequence of projections, $K \subseteq \mathbf{R}$ is a compact set, $U \subseteq \mathbf{R}$ is open, and $K \subseteq U$ then*

$$\|L_n \chi_K I - \chi_U L_n \chi_K I\|_{L(Y, L_R^p(\alpha))} \to 0 \text{ and } \|\chi_K L_n I - \chi_K L_n \chi_U I\|_{L(Y, L_R^p(\alpha))} \to 0 \,.$$

In other words: canonical projections enlarge the support of a function only unessentially.

Proof   Let f stand for a continuous function which is identical 1 on K and which vanishes outside of U. Then

$$
\begin{aligned}
\|L_n \chi_K I - \chi_U L_n \chi_K I\| &= \|(I - \chi_U) L_n f \chi_K I\| \\
&\leq \|(I - \chi_U) f L_n \chi_K I\| + \|(I - \chi_U)(L_n f I - f L_n)\chi_K I\| \\
&\leq 0 + \|I - \chi_U I\| \, \|L_n f I - f L_n\| \, \|\chi_K I\| \,,
\end{aligned}
$$

and the latter term goes to zero by Theorem 2.8. The second assertion follows analogously. ∎

2.7.6   Another representation of the function $\lambda^{F,G}$

Our next goal is criteria for a given k-tuple F of presplines to be canonical. To this end we need another representation of the polynomials $\lambda^{F,G}$. For, we suppose the functions ϕ_r in F as well as the functionals γ_r in G to be elements of the Schwartz space $\mathcal{S}(\mathbf{R})$. Then $\lambda^{F,G} = (\lambda_{rs}^{F,G})_{r,s=0}^{k-1}$, and the l th Fourier coefficient of $\lambda_{rs}^{F,G}$ is given by

$$\int_R \phi_s(t+l)\overline{\gamma_r(t)} \, dt \tag{1}$$

(compare 2.6.3(1)). Set $\tilde{\phi}_s(t) = \phi_s(-t)$. Then (1) is equal to

$$
\begin{aligned}
\int_R \tilde{\phi}_s(-l-t)\overline{\gamma_r}(t) \, dt &= (\tilde{\phi}_s * \overline{\gamma_r})(-l) \\
&= F^{-1}((F\tilde{\phi}_s)(F\overline{\gamma_r}))(-l) \\
&= \int_R e^{-2\pi i x l}(F\tilde{\phi}_s)(x)(F\overline{\gamma_r})(x) \, dx \\
&= \int_R e^{-2\pi i x l}\overline{(F\overline{\phi_s})(x)}(F\overline{\gamma_r})(x) \, dx \,.
\end{aligned}
$$

Introduce new variables $j \in \mathbf{Z}$ and $y \in [0,1)$ with $x = j + y$ to find

$$
\begin{aligned}
\int_R \phi_s(t+l)\overline{\gamma_r}(t) \, dt &= \sum_{j \in Z} \int_0^1 e^{-2\pi i y l}\overline{(F\overline{\phi_s})(y+j)}(F\overline{\gamma_r})(y+j) \, dy \\
&= \int_0^1 e^{-2\pi i y l}\left(\sum_{j \in Z} \overline{(F\overline{\phi_s})(y+j)}(F\overline{\gamma_r})(y+j)\right) dy \,. \tag{2}
\end{aligned}
$$

The exchange of summation and integration is justified by our assumption $\phi_s, \gamma_r \in \mathcal{S}(\mathbf{R})$ which involves that the functions $F\overline{\phi_s}$ and $F\overline{\gamma_r}$ decrease sufficiently fast. From (2) we conclude that (1) is just the l th Fourier coefficient of the function

$$\mathbf{T} \to \mathbf{C}, \quad e^{2\pi i y} \mapsto \sum_{j \in Z} \overline{(F\overline{\phi_s})(y+j)}(F\overline{\gamma_r})(y+j) , \tag{3}$$

thus,

$$\lambda_{rs}^{F,G}(e^{2\pi i y}) = \sum_{j \in Z} \overline{(F\overline{\phi_s})(y+j)}(F\overline{\gamma_r})(y+j) \tag{4}$$

is the desired representation. Evidently, this representation does not only hold for $F, G \subseteq \mathcal{S}(\mathbf{R})$ but also in each other situation where the series (3) converges sufficiently well. For instance, it suffices to require that, for all $y \in [0,1)$, the sequences $((F\overline{\phi_s})(y+j)(F\overline{\gamma_r})(y+j))_{j\in Z}$ belong to $l_Z^1(0)$, and that they are uniformly bounded in $l_Z^1(0)$ with respect to y. Indeed, in this case, all partial sums

$$\sum_{|j| \leq N} \overline{(F\overline{\phi_s})(y+j)}(F\overline{\gamma_r})(y+j)$$

are uniformly bounded with respect to y and N by a constant function, and Lebesgue's theorem entails the correctness of (2). Here are some special situations where this condition is satisfied:

(a) $F, G \subseteq C^1(\mathbf{R})$

(b) $F \subseteq C^2(\mathbf{R})$, G is a collection of shifted δ-functionals

(c) F, G consist of piecewise polynomial (not necessarily continuous) functions

(d) $F \subseteq L_R^p(\alpha)$, $G \subseteq C^2(\mathbf{R})$

(e) the sequences $\overline{(F\overline{\phi_s}(y+j))}_{j\in Z}$ and $(F\overline{\gamma_r}(y+j))_{j\in Z}$ belong to $l_Z^2(0)$ for all $y \in [0,1)$, and they form a bounded subset of this space

(f) the sequences $\overline{(F\overline{\phi_s}(y+j))}_{j\in Z}$ and $(F\overline{\gamma_r}(y+j))_{j\in Z}$ belong to $l_Z^1(0)$ and $l_Z^\infty(0)$ for all $y \in [0,1)$, and they form bounded subsets of these spaces, respectively.

2.7.7   Criteria for canonicity

Here we are going to derive some criteria for canonicity of k-tuples of presplines.

Theorem 2.10 *Let $F \subseteq Y$ be a k-tuple of presplines and $G \subseteq Y^*$ be a k-tuple of premeasures. If (F, G) is a canonical pair then F is canonical.*

Proof First suppose $F = \{\phi_0, \ldots, \phi_{k-1}\} \subseteq \mathcal{S}(\mathbf{R})$ and $G = \{\gamma_0, \ldots, \gamma_{k-1}\} \subseteq \mathcal{S}(\mathbf{R})$. Fix $y \in [0, 1)$, and define vectors $X_s, Y_r \in l_Z^2(0)$ by

$$X_s = ((F\overline{\phi_s})(y+j))_{j \in Z} \ , \ Y_r = ((F\overline{\gamma_r})(y+j))_{j \in Z}$$

we have, due to 2.7.6(4) $\lambda_{rs}^{F,G}(e^{2\pi i y}) = (Y_r, X_s)$ with $(.,.)$ referring to the standard Hilbert space scalar product on $l_Z^2(0)$. Further introduce vectors $X = (X_0, \ldots, X_{k-1})$ and $Y = (Y_0, \ldots, Y_{k-1})$ and associate with X and Y the matrix $G(X, Y) = ((Y_r, X_s))_{r,s=0}^{k-1}$. Clearly, $G(X, Y) = \lambda^{F,G}(e^{2\pi i y})$, and the Gram matrix $G(X, X)$ coincides with $\lambda^{F,F}(e^{2\pi i y})$. Since the Gram matrices

$$\begin{pmatrix} G(X, X) & G(X, Y) \\ G(X, Y)^* & G(Y, Y) \end{pmatrix}, \quad G(X, X) \quad \text{and} \quad G(Y, Y)$$

are non-negative, there exists a contraction $K(X, Y)$ (that is, $\|K(X, Y)\| \leq 1$) such that

$$G(X, Y) = G(X, X)^{1/2} K(X, Y) G(Y, Y)^{1/2}$$

or, equivalently,

$$\lambda^{F,G}(e^{2\pi i y}) = \lambda^{F,F}(e^{2\pi i y})^{1/2} K(X, Y) \lambda^{G,G}(e^{2\pi i y})^{1/2} . \tag{1}$$

(Schur's theorem, see e.g. ([FF], Ch. XVI, Theorem 1.1)). Thus, if $\lambda^{F,G}(e^{2\pi i y})$ is invertible then $\lambda^{F,F}(e^{2\pi i y})$ is invertible, and F (and G) is canonical.

Now we are going to treat the general case $F \subseteq Y$ and $G \subseteq Y^*$. Our first claim is to show that there is a k-tuple G' of premeasures which even belongs to $\mathcal{S}(\mathbf{R})$ and satisfies $\lambda^{F,G} = \lambda^{F,G'}$.

Choose an integer $a \geq 1$ such that $\operatorname{supp} \phi_s \subseteq [-a, a]$ and $\operatorname{supp} \gamma_r \subseteq [-a+1, a-1]$. We consider the functionals

$$\Phi_{sl} : g \mapsto \int_R \phi_s(t+l)\overline{g(t)}\, dt \, , \ l \in \mathbf{Z} \, , \tag{2}$$

on $C_0^\infty(-a, +a)$. The functionals (2) are bounded on $C_0^\infty(-a, +a)$ (since $\phi_s(t+l) \in L_R^p(\alpha)$ and $C_0^\infty(-a, +a)$ is continuously embedded into $L_R^p(\alpha)^*$).

Let us first consider those of the functionals (2) which are non-zero on $C_0^\infty(-a, +a)$. There is evidently only a finite number of such functionals, and they are linearly independent (the latter follows from the fact that each finite choice of the functions $\phi_s(t+l)$ is linearly independent, compare Section 2.6.4). Since $C_0^\infty(-a, +a)$ is infinite dimensional, there are functions $\gamma_r' \in C_0^\infty(-a, +a)$ such that

$$\Phi_{sl}(\gamma_r') = \int_R \phi_s(t+l)\overline{\gamma_r'(t)}\, dt = \int_R \phi_s(t+l)\overline{\gamma_r(t)}\, dt \tag{3}$$

for all non-zero functionals of the form (2). If, on the other hand, Φ_{sl} is the zero functional on $C_0^\infty(-a, +a)$ then $\operatorname{supp} \Phi_{sl} \cap [-a+1, a-1] = \operatorname{supp} \Phi_{sl} \cap \operatorname{supp} \gamma_r = \emptyset$ whence follows that (3) holds, too. But $\int_R \phi_s(t+l)\overline{\gamma_r(t)}\, dt$ and $\int_R \phi_s(t+l)\overline{\gamma_r'(t)}\, dt$

are the l th coefficients of the polynomials $\lambda^{F,G}$ and $\lambda^{F,G'}$, respectively, which gives our claim.

In the next step we show that $\lambda^{F,G'}$ can as closely as desired be approxiamated (in the multiplicator norm) by polynomials $\lambda^{F_n,G'}$ where the k-tuples F_n consits of functions in $\mathcal{S}(\mathbf{R})$ only. Indeed, since $G' \subseteq \mathcal{S}(\mathbf{R})$, we can now restrict ourselves to the case that $F \subseteq L_R^p(\alpha)$. But $\mathcal{S}(\mathbf{R})$ is dense in this space, and a closer look at the proof of the first part of de Boor's estimate (see Section 2.12.2) shows that the mapping

$$L_{[-a,a]}^p(\alpha) \times \ldots \times L_{[-a,a]}^p(\alpha) \quad \to \quad L(l_k^p(\alpha), L_R^p(\alpha)) \,,$$
$$F = (\phi_0, \ldots, \phi_{k-1}) \quad \mapsto \quad E_1^F \tag{4}$$

is continuous. Since $T^0(\lambda^{F,G'}) = P_1^{G'} E_1^F$, this gives what we wanted.

Now let $\lambda^{F,G} = \lambda^{F,G'}$ be invertible. Then the polynomials $\lambda^{F_n,G'}$ are invertible for n large enough, and from (1) we conclude that

$$\lambda^{F_n,G'} = (\lambda^{F_n,F_n})^{1/2} K_n (\lambda^{G',G'})^{1/2} \tag{5}$$

with certain contractions K_n. The identity (5) implies that all matrices λ^{F_n,F_n}, K_n, and $\lambda^{G',G'}$ are invertible. Write (5) as

$$K_n^{-1} = (\lambda^{G',G'})^{1/2} (\lambda^{F_n,G'})^{-1} (\lambda^{F_n,F_n})^{1/2} \,. \tag{6}$$

Taking into account that the mapping

$$L_{[-a,a]}^p(\alpha)^* \times \ldots \times L_{[-a,a]}^p(\alpha)^* \quad \to \quad L(L_R^p(\alpha), l_k^p(\alpha)) \,,$$
$$G = (\gamma_0, \ldots, \gamma_{k-1}) \quad \mapsto \quad P_1^G \tag{7}$$

is also continuous (compare Section 2.12.3), we conclude that the right hand side of (6) converges uniformly to $(\lambda^{G',G'})^{1/2} (\lambda^{F,G'})^{-1} (\lambda^{F,F})^{1/2}$ as $n \to \infty$. Thus, the limit $K := \lim_{n\to\infty} K_n^{-1}$ exists, and the matrix K is even invertible since the operators K_n^{-1} are invertible and the norms of their inverses (i.e. the norms of the K_n) are uniformly bounded by Schur's theorem. A consequence is the invertibility of the matrix $(\lambda^{G',G'})^{1/2} (\lambda^{F,G'})^{-1} (\lambda^{F,F})^{1/2}$, and this of course implies the invertibility of $\lambda^{F,F}$. ∎

For another criterion for a given k-tuple $F = \{\phi_0, \ldots, \phi_{k-1}\}$ of presplines to be canonical we suppose that supp $\phi_r \subseteq [-a,a]$ with a positive integer a, and we define functions ϕ_{rs}, $r \in \{0, \ldots, k-1\}$, $s \in \{-a, \ldots, a-1\}$, by

$$\phi_{rs}(t) = \begin{cases} \phi_r(t+s) & t \in [0,1) \\ 0 & t \notin [0,1) \,. \end{cases} \tag{8}$$

Theorem 2.11 *Let F be a k-tuple of presplines in $P_R^p(\alpha)$ and let ϕ_{rs} refer to the functions (8). Then F is canonical if and only if the functions $\sum_{s=-a}^{a-1} e^{2\pi i s y} \phi_{rs}$ $(r = 0, \ldots, k-1)$ are linearly independent for each fixed $y \in [0,1)$. If, in particular, the functions ϕ_{rs} $(r = 0, \ldots, k-1, s = -a, \ldots, a-1)$ are linearly independent then F is canonical.*

Proof The representation 2.7.6(4) makes sense under our hypotheses. Fix $y \in [0, 1)$ and denote the sequence $((F\overline{\phi_r})(y+j))_{j\in Z}$ by X_r again. Then $\lambda^{F,F}(e^{2\pi i y})$ is invertible if and only if the Gram matrix $((X_r, X_s))$ is invertible, that is, if and only if the vectors $X_0, \ldots, X_{k-1}$ are linearly independent.

Let a_r be complex numbers such that $\sum_{r=0}^{k-1} a_r X_r = 0$. Taking into account that $\phi_r = \sum_{s=-a}^{a-1} U_s \phi_{rs}$ with the shift operator $(U_s f)(t) = f(t-s)$ we find

$$0 = \sum_{r=0}^{k-1} a_r X_r = \sum_{r=0}^{k-1} a_r ((F\overline{\phi_r})(y+j))_{j\in Z} =$$

$$= \sum_{r=0}^{k-1} a_r ((F \sum_{s=-a}^{a-1} U_s \overline{\phi_{rs}})(y+j))_{j\in Z} =$$

$$= \sum_{r=0}^{k-1} \sum_{s=-a}^{a-1} a_r e^{-2\pi i s y} ((F\overline{\phi_{rs}})(y+j))_{j\in Z} . \tag{9}$$

Define a function g_y by $g_y = \sum_{r=0}^{k-1} \sum_{s=-a}^{a-1} \overline{a_r} e^{2\pi i y s} \phi_{rs}$. Then (9) can be rewritten as $0 = (F\overline{g_y})(y+j)$ for all $j \in \mathbf{Z}$ or, equivalently, as

$$0 = (F h_y)(j) , \quad j \in \mathbf{Z} , \tag{10}$$

with $h_y(z) = \overline{g_y(z)} e^{-2\pi i y z}$. Evidently, h_y is a bounded function supported on $[0, 1]$. Thinking of h_y as a function on the unit circle we derive from (10) that all Fourier coefficients of h_y vanish; thus, $h_y(z) = 0$ for all $z \in [0, 1]$ whence follows $g_y(z) = 0$ for all $z \in [0, 1]$. Conversely, if $g_y \equiv 0$ then $\sum a_r X_r = 0$. Hence, the linear independency of the vectors X_r is equivalent to the linear independency of the functions $\sum_{s=-a}^{a-1} e^{2\pi i y s} \phi_{rs}$ constituting g_y. The second assertion of the theorem is an obvious conclusion of the first one. ∎

2.8 Canonical prebases

2.8.1 Prebases

In general, spline projections do not converge strongly to the identity operator. For example, $F = \{\chi_{[0,1/2)} - \chi_{[1/2,1)}\}$ is a canonical 1-tuple of presplines, but $L_n^{F,F} \chi_{[0,1]} = 0$ for all n. In this section we consider k-tuples of presplines owning an additional property which ensures the strong convergence to I.

Again we let $Y = L_R^p(\alpha)$ or $R_R^p(\alpha)$. A k-tuple $F = \{\phi_0, \ldots, \phi_{k-1}\}$ of presplines in Y is called a Y-*prebasis* if there are complex numbers α_r^F such that

$$\sum_{r=0}^{k-1} \alpha_r^F \sum_{l\in Z} \phi_{rl1}(x) = 1 \tag{1}$$

almost everywhere in case $Y = L_R^p(\alpha)$ or for all $x \in \mathbf{R}$ in case $Y = R_R^p(\alpha)$. Clearly, if F is a Y-prebasis then $\sum_{r=0}^{k-1} \alpha_r^F \sum_{l\in Z} \phi_{rln}(t) = 1$ for all $n \geq 1$. The vector

$\alpha^F = (\alpha_0^F, \ldots, \alpha_{k-1}^F)^T$ will figure prominently in describing spline approximation methods. Finally, if F is a Y-prebasis then we call the function

$$\phi := \sum_{r=0}^{k-1} \alpha_r^F \phi_r \tag{2}$$

the *compression* of F.

Proposition 2.19 (a) *If F is an $L_R^p(\alpha)$-prebasis then $\int_R \phi(x)\, dx = 1$.*
(b) *If F is a Y-prebasis and $G = \{\gamma_0, \ldots, \gamma_{k-1}\} \subseteq Y^*$ a k-tuple of premeasures, then $\sum_{s=0}^{k-1} \alpha_s^F \lambda_{rs}^{F,G}(1) = \int_R \overline{\gamma_r(x)}\, dx$.*

Proof   From (1) we conclude that $\sum_{l \in Z} \phi(x - l) = 1$ whence

$$1 = \int_0^1 \left(\sum_l \phi(x - l) \right) dx = \sum_l \int_0^1 \phi(t - l)\, dt = \sum_l \int_l^{l+1} \phi(x)\, dx = \int_R \phi(x)\, dx$$

follows, and this is just (a). Further, if G is an arbitrary k-tuple of premeasures then, in analogy to 2.6.3(1),

$$\sum_{s=0}^{k-1} \alpha_s^F \lambda_{rs}^{F,G}(1) = \sum_{s=0}^{k-1} \alpha_s^F \sum_{l \in Z} \int_R \phi_s(x + l) \overline{\gamma_r(x)}\, dx$$

$$= \sum_{l \in Z} \int_R \phi(x + l) \overline{\gamma_r(x)}\, dx = \int_R \overline{\gamma_r(x)}\, dx \,.\ \blacksquare$$

2.8.2   Strongly convergent spline projections

Theorem 2.12 *Let (F, G) be a canonical pair with $F \subseteq Y$ being a Y-prebasis and $G \subseteq Y^*$. Then*

$$\| I_Y (L_n^{F,G} f - f) \|_{L_R^p(\alpha)} \to 0 \quad as \quad n \to \infty \tag{1}$$

for all $f \in Y_P$.

Proof   The operators $I_Y L_n^{F,G}$ are uniformly bounded with respect to n by Proposition 2.16 and Proposition 2.17. So it suffices to verify (1) for all functions belonging to a dense subset of Y_P, say, for all piecewise constant functions. Since each piecewise constant function is a linear combination of characteristic functions of intervalls, it is even sufficient to prove (1) for $\chi_{[c,d)}$ in place of f.

For $Y = L_R^p(\alpha)$ this can be readily done: Set $\chi_n := \sum_{rl} \alpha_r^F \phi_{rln}$ where the summation is taken over all (finitely many) numbers r and l for which $\mathrm{supp}\,\phi_{rln} \cap [c, d]$ is not empty. Now,

$$\| L_n^{F,G} \chi_{[c,d)} - \chi_{[c,d)} \|_{L_R^p(\alpha)}$$

$$\leq \| L_n^{F,G} (\chi_{[c,d)} - \chi_n) \|_{L_R^p(\alpha)} + \| L_n^{F,G} \chi_n - \chi_n \|_{L_R^p(\alpha)} + \| \chi_n - \chi_{[c,d)} \|_{L_R^p(\alpha)} \,.$$

The first item is less than $C \|\chi_n - \chi_{(c,d)}\|_{L_R^p(\alpha)}$ with $C = \sup_n \|L_n^{F,G}\|$, and the second one vanishes since $\chi_n \in S_n^F$ and, thus, $L_n^{F,G}\chi_n = \chi_n$. So it remains to show that $\|\chi_n - \chi_{[c,d]}\|_{L_R^p(\alpha)} \to 0$ as $n \to \infty$. But this can be easily shown if one takes into account that $\chi_n = 1$ on $[c,d]$ (which is involved by the prebasis property) and that $\|\phi_{rln}\|_{L_R^p(\alpha)} \to 0$ as $n \to \infty$.

Now let $Y = R_R^p(\alpha)$. Choose numbers c', c'', d', d'' such that $c < c' < c'' < d'' < d' < d$ and matrix polynomials λ_k of degree k such that $\lambda_k(1) = (\lambda^{F,G})^{-1}(1)$ and $\|T^0(\lambda^{F,G})^{-1} - T^0(\lambda_k)\| \to 0$ as $k \to \infty$. Then

$$\|I_Y(L_n^{F,G}\chi_{[c,d)} - \chi_{[c,d)})\|_{L_R^p(\alpha)} \le$$
$$\le \|\chi_{[c'',d'')}I_Y(L_n^{F,G}\chi_{[c,d)} - \chi_{[c,d)})\| +$$
$$+\|\chi_{R\setminus[c'',d'')}I_Y L_n^{F,G}\chi_{[c,d)}\| + \|\chi_{R\setminus[c'',d'')}\chi_{[c,d)}\| .$$

Let $\epsilon > 0$. The third item becomes smaller than ϵ if $|c - c''|$ and $|d - d''|$ are small enough. Further, the asymptotic locality (Section 2.7.5) involves, that the second item is smaller than ϵ for all n large enough if only $|c - c''|$ and $|d - d''|$ are small. For the first item we find

$$\|\chi_{[c'',d'')}I_Y(L_n^{F,G}\chi_{[c,d)} - \chi_{[c,d)})\|$$
$$\le \|\chi_{[c'',d'')}I_Y(E_n^F T^0(\lambda_k)P_n^G\chi_{[c,d)} - \chi_{[c,d)})\|$$
$$+\|\chi_{[c'',d'')}I_Y E_n^F(T^0(\lambda^{F,G})^{-1} - T^0(\lambda_k))P_n^G\chi_{[c,d)}\| . \tag{2}$$

Herein, the second item is less than

$$C \|I_Y E_n^F\|_{L(l_k^p(\alpha),L_R^p(\alpha))} \|P_n^G\|_{L(Y,l_k^p(\alpha))} \|T^0(\lambda^{F,G})^{-1} - T^0(\lambda_k)\| ,$$

and this becomes as small as desired if only k is large enough. So it remains to show that the first item in (2) goes to zero as n tends to infinity for each fixed k, c'' and d''.

Given a subset M of $\mathbf{R}$, we denote by P_M the projection

$$P_M : l_1^p(\alpha) \to l_1^p(\alpha), \ (x_r) \mapsto (y_r) \quad \text{with} \quad y_r = \begin{cases} x_r & \text{if} \quad r \in M \\ 0 & \text{if} \quad r \notin M \end{cases} ,$$

and we use the same notation for the diagonal operator $\text{diag}(P_M, \ldots, P_M)$. Further we choose a positive integer a such that $\text{supp}\,\phi_r \subseteq [-a,a]$ and $\text{supp}\,\gamma_s \subseteq [-a,a]$ for all $\phi_r \in F$ and $\gamma_s \in G$. A little thought shows that

$$\chi_{[c'',d'')}I_Y E_n^F = \chi_{[c'',d'')}I_Y E_n^F P_{[c'n,d'n)} \tag{3}$$

and

$$P_{[c'n,d'n)}T^0(\lambda_k) = P_{[c'n,d'n)}T^0(\lambda_k)P_{[cn+a,dn-a)} \tag{4}$$

for all sufficiently large n. Thus, for large n,

$$\chi_{[c'',d'')}I_Y E_n^F T^0(\lambda_k)P_n^G\chi_{[c,d)} =$$
$$= \chi_{[c'',d'')}I_Y E_n^F P_{[c'n,d'n)}T^0(\lambda_k)P_{[cn+a,dn-a)}P_n^G\chi_{[c,d)} . \tag{5}$$

Further it is easy to see that

$$P_{[cn+a,dn-a)}P_n^G\chi_{[c,d)} =$$

$$= P_{[cn+a,dn-a)}'\left(\int_R \overline{\gamma_0(x)}\,dx \begin{pmatrix} \vdots \\ 1 \\ \vdots \end{pmatrix}, \ldots, \int_R \overline{\gamma_{k-1}(x)}\,dx \begin{pmatrix} \vdots \\ 1 \\ \vdots \end{pmatrix}\right)^T,$$

and applying (3) and (4) once more we conclude that (5) coincides with

$$\chi_{[c'',d'')}I_Y E_n^F T^0(\lambda_k)\left(\int_R \overline{\gamma_0(x)}\,dx \begin{pmatrix} \vdots \\ 1 \\ \vdots \end{pmatrix}, \ldots, \int_R \overline{\gamma_{k-1}(x)}\,dx \begin{pmatrix} \vdots \\ 1 \\ \vdots \end{pmatrix}\right)^T. \qquad (6)$$

If λ is an arbitrary (scalar valued) polynomial then $T^0(\lambda)\begin{pmatrix} \vdots \\ 1 \\ \vdots \end{pmatrix} = \lambda(1)\begin{pmatrix} \vdots \\ 1 \\ \vdots \end{pmatrix}$.

Thus, (6) is the same as

$$\chi_{[c'',d'')}I_Y E_n^F \lambda_k(1)\left(\int_R \overline{\gamma_0(x)}\,dx \begin{pmatrix} \vdots \\ 1 \\ \vdots \end{pmatrix}, \ldots, \int_R \overline{\gamma_{k-1}(x)}\,dx \begin{pmatrix} \vdots \\ 1 \\ \vdots \end{pmatrix}\right)^T,$$

and Proposition 2.19(b) involves that this coincides with

$$\chi_{[c'',d'')}I_Y E_n^F\left(\alpha_0^F \begin{pmatrix} \vdots \\ 1 \\ \vdots \end{pmatrix}, \ldots, \alpha_{k-1}^F \begin{pmatrix} \vdots \\ 1 \\ \vdots \end{pmatrix}\right)^T =$$

$$= \chi_{[c'',d'')}\sum_{r=0}^{k-1}\alpha_r^F \sum_{l\in Z}\phi_{rln} = \chi_{[c'',d'')} \quad \text{as desired.} \quad \blacksquare$$

2.8.3 The reverse direction

Theorem 2.13 *Let (F,G) be a canonical pair consisting of a k-tuple $F \subseteq Y$ of presplines and a k-tuple $G \subseteq Y^*$ of premeasures. If*

$$\|I_Y(L_n^{F,G}f - f)\|_{L_R^p(\alpha)} \to 0 \qquad (1)$$

for all piecewise constant functions then F is an $L_R^p(\alpha)$-prebasis.

Proof If (1) holds then, in particular,

$$\|\chi_{[0,1)} I_Y (L_n^{F,G} \chi_{[-1,2)} - \chi_{[-1,2)})\|_{L_R^p(\alpha)} \to 0 . \tag{2}$$

Approximating $(\lambda^{F,G})^{-1}$ by a polynomial λ_k, we can argue as in the preceding section that (2) involves

$$\|\chi_{[0,1)} \sum_{r=0}^{k-1} \alpha_r^F \sum_{l \in Z} \phi_{rln} - \chi_{[0,1)}\|_{L_R^p(\alpha)} \to 0 . \tag{3}$$

Abbreviate the function $\sum_{r=0}^{k-1} \alpha_r^F \sum_{l \in Z} \phi_{rln}$ by Φ_n. Then, by (3)

$$\|\Phi_n - 1\|_{L_{[0,1]}^p(\alpha)} \to 0 \quad \text{as} \quad n \to \infty ,$$

whence follows that

$$\|\Phi_n - 1\|_{L_{[0,1]}^1(0)} \to 0 \quad \text{as} \quad n \to \infty .$$

A little thought shows that $\Phi_n(t) = \Phi_1(nt)$ $(t \in \mathbf{R})$ and that Φ_1 is a 1-periodic function. Hence,

$$0 \leftarrow \|\Phi_n - 1\|_{L_{[0,1]}^1(0)} = \int_0^1 |\Phi_n(t) - 1| \, dt$$
$$= \frac{1}{n} \int_0^n |\Phi_1(t) - 1| \, dt = \int_0^1 |\Phi_1(t) - 1| \, dt ,$$

whence follows that $\Phi_1 \equiv 1$ a.e. on $[0,1]$ and, consequently, on all of $\mathbf{R}$. ∎

As a by-product of the proofs of the preceding two theorems we obtain some other characterizations of canonical prebases.

Corollary 2.4 *Let* $F = \{\phi_0, \ldots, \phi_{k-1}\}$ *be a canonical k-tuple of presplines. The following assertions are equivalent:*

(a) F is a $L_R^p(\alpha)$-prebasis.

(b) $L_n^{F,F}$ converges strongly to the identity.

(c) there are numbers α_{rl} such that $\sum_{r=0}^{k-1} \sum_{l \in Z} \alpha_{rl} \phi_{rln}(x) = 1$ a.e. on $\mathbf{R}$.

(d) $L_n^{F,F} \chi_R = \chi_R$.

Condition (d) is meant as follows: Applying formally P_n^F to the characteristic function χ_R of the real axis, we get a k-vector of bounded sequences. The multiplication of this k-vector by $T^0(\lambda^{F,F})^{-1}$ gives a k-vector of bounded sequences again, and so $L_n^{F,F} \chi_R$ is an L_R^∞-function.

Proof The equivalence of (a) and (b) was verified in the preceding theorems.

If (c) is fulfilled then one can show as in that part of the proof of Theorem 2.12 which concerns $L_R^p(\alpha)$ that $L_n^{F,F} \to I$. Finally, (d) implies (a), as one can see by repeating the arguments in the proof of Theorem 2.12 related with $R_R^p(\alpha)$. The implications $(a) \Rightarrow (c)$, $(a) \Rightarrow (d)$ are evident. ∎

Let us consider a few examples. The 1-tuple $F_1 = \{\chi_{[0,1]}\}$ is a canonical prebasis since $\lambda^{F_1,F_1} \equiv 1$ on **T** and since $L_n^{F_1,F_1}\chi_R = \sum_{l \in \mathbb{Z}} \chi_{[l,l+1]} = 1$. The 1-tuple $F_2 = \{\chi_{[0,2]}/2\}$ is a prebasis which is not canonical as we have already seen. The 1-tuple $F_3 = \{\chi_{[0,1/2]} - \chi_{[1/2,1]}\}$ is canonical but not a prebasis. Indeed, $\lambda^{F_3,F_3} \equiv 1$ on **T**, but $L_n^{F_3,F_3}\chi_R = 0$. Finally, the 2-tuples $F_4 = \{\chi_{[0,1]}(x), x\,\chi_{[0,1]}(x)\}$ and $F_5 = \{\chi_{[0,1]}, \chi_{[0,1/2]} - \chi_{[1/2,1]}\}$ are canonical prebases since

$$\lambda^{F_4,F_4} \equiv \begin{pmatrix} 1 & 1/2 \\ 1/2 & 1/3 \end{pmatrix} \quad \text{and} \quad \lambda^{F_5,F_5} \equiv \begin{pmatrix} 1 & 0 \\ 0 & 1 \end{pmatrix}$$

are invertible matrices and $L_n^{F_4,F_4}\chi_R = L_n^{F_5,F_5}\chi_R = \sum_{l \in \mathbb{Z}} \chi_{[l,l+1]} \equiv 1$.

Thus, the projections $L_n^{F_i,F_i}$ are well defined for $i \in \{1,3,4,5\}$, and they converge strongly to the identity if and only if $i \in \{1,4,5\}$.

2.9   Concrete spline spaces

2.9.1   Piecewise constant splines

The simplest examples of spline spaces are those of piecewise constant splines, S_n^K, with $K = \{\chi_{[0,1/k)}, \dots, \chi_{[(k-1)/k,1)}\}$ for a given positive integer k. For $k = 1$ these are just the spaces S_n introduced in Section 2.2.1. One of the most pleasant properties of those spaces is that $\lambda^{K,K}$ is the $k \times k$ identity matrix. Thus, K is canonical, and it is obvious that K is also a prebasis with the constants α_r^K being equal to 1 and with compression $\phi = \chi_{[0,1)}$. Notice further that, in case $k = 1$, K is the only prebasis with supp $K \subseteq [0,1]$.

2.9.2   Prebases supported on [0,2]

Let $k = 1$ and let $F = \{\phi\}$ be a prebasis with supp $\phi \subseteq [0,2]$ which is normed in such a way that $\alpha_0^F = 1$ (otherwise replace ϕ by $\alpha_0^F\phi$). Define $\phi_{11}(x) = \phi(x)$ and $\phi_{12}(x) = \phi(x+1) = 1 - \phi_{11}(x)$ for all $x \in [0,1)$. If ϕ_{11} is *not* a non-zero multiple of the function $\chi_{[0,1)}$ then the functions ϕ_{11} and ϕ_{12} are linearly independent. In this case, F is canonical by Theorem 2.11. On the other hand, if $\phi_{11}(x) = a\chi_{[0,1)}(x)$ with a complex a then $\lambda^{F,F}(z) = (1-a)az^{-1} + (a^2 + (1-a)^2)z^0 + a(1-a)z^1$. The complex zeros of this polynomial are $a/(a-1)$ and $(a-1)/a$, hence, $\lambda^{F,F}$ has zeros on **T** if and only if $|a/(a-1)| = 1$ or, equivalently, if $\mathrm{Re}\,a = 1/2$. Summary: F is canonical if and only if $\phi_{11} \neq a\chi_{[0,1)}$ with $\mathrm{Re}\,a = 1/2$.

2.9.3 Smoothest piecewise polynomial splines

Let $k = 1$ and let d be a non-negative integer. Define recursively $\Pi^0 = \chi_{[-1/2,1/2)}$ and $\Pi^d = \Pi^0 * \Pi^{d-1}$, and set $\phi^d(x) = \Pi^d(x - 1/2)$. It is not hard to check that $F_d = \{\phi^d\}$ is a prebasis with $\alpha_0^{F_d} = 1$ (This is evident for $d = 0$ and a consequence of Exercise 2.18 if $d > 0$.). Moreover, this prebasis is even canonical. Indeed, by the convolution theorem,

$$(F\,\phi^d)(x) \;=\; \begin{cases} e^{-\pi i x}\left(\frac{\sin \pi x}{\pi x}\right)^{d+1} &,\ x \in \mathbf{R} \setminus \{0\} \\ 1 &,\ x = 0 \end{cases}$$

whence

$$\lambda^{F_d,F_d}\!\left(e^{2\pi i y}\right) \;=\; \begin{cases} \sum_{l \in Z}\left(\frac{\sin \pi(l+y)}{\pi(l+y)}\right)^{2(d+1)} &,\ y \in (0,1) \\ 1 &,\ y = 0 \end{cases} \;,$$

and this function does not vanish on $\mathbf{T}$. The functions in $S_n^{F_d}$ are called *smoothest piecewise polynomial splines of degree d*. There is another characterization of these spline spaces (compare ([PS 2], Ch.2, Part A)): $S_n^{F_d}$ is the smallest closed subspace of $L_R^p(\alpha)$ containing all $d - 1$ times continuously differentiable functions whose restrictions onto the intervals $[k/n, (k+1)/n)$ $(k \in \mathbf{Z})$ in case d is even or onto $[(k-1/2)/n, (k+1/2)/n)$ $(k \in \mathbf{Z})$ if d is odd is a polynomial of degree d.

2.9.4 Piecewise polynomial splines with defect

Let d and e be non-negative integers with $e \leq d$. A *piecewise polynomial spline of degree d with defect e* is a $d - 1 - e$ times continuously differentiable function in $L_R^p(\alpha)$ the restriction of which onto the intervals $[k/n, (k+1)/n)$ $(k \in \mathbf{Z})$ are polynomials of degree d. For $e = 0$ we just obtain the smoothest piecewise polynomial splines considered in 2.9.3 (apart from a shift by $1/2n$ for odd d). For $e = d$ the defect splines are not even continuous at the points k/n. Let us start with examining the case $e = d$. For $r = 0, \dots, e$, we define functions $\phi_r^{e,e}$ by

$$\phi_r^{e,e}(x) \;=\; \begin{cases} \binom{e}{r}(1-x)^r x^{e-r} & x \in [0,1) \\ 0 & x \in \mathbf{R} \setminus [0,1) \,. \end{cases}$$

The restriction of $\phi_r^{e,e}$ onto $[0,1)$ is just the r th *Bernstein polynomial of degree e*. The $(e+1)$-tuple $F_{e,e} = \{\phi_0^{e,e}, \dots, \phi_e^{e,e}\}$ forms a prebasis of the space of all piecewise polynomial splines of degree e with maximal defect, i.e. this space coincides with $S_n^{F_{e,e}}$. To verify this notice that the functions $\phi_r^{e,e}$ are linearly independent whence follows that each polynomial of degree e on $[0,1]$ is a linear combination of these functions. Further, the identity $\sum_{r=0}^e \phi_r^{e,e} = 1$ on $[0,1]$ is an obvious consequence of the binomial formula whence, in particular, $\alpha_r^{F_{e,e}} = 1$, $r = 0, \dots, e$. Finally, the linear independence of the functions $\phi_r^{e,e}$ involves via Theorem 2.11 that $F_{e,e}$ is canonical.

A *prebasis* of the spline spaces of degree d with defect e $(e < d)$ can be simply obtained by convolving the elements of $F_{e,e}$ $d - e$ times by the characteristic function of $[0, 1)$. But it soon turns out that the so-constructed prebases are *not* canonical in general.

A *canonical prebasis* $F_{d,e} = \{\phi_0^{d,e}, \ldots, \phi_e^{d,e}\}$ of the spline spaces with degree d and defect e can be defined by the following requirements:

1^0. $\operatorname{supp} \phi_r^{d,e} \subseteq [0, d+1-e]$.

2^0. the function $\phi_r^{d,e}$ coincides on $[0, 1)$ with the $d - e$ times convolution of $\phi_r^{e,e}$ with the characteristic function $\chi_{[0,1)}$.

3^0. the function $\phi_r^{d,e}$ is on $[1, d+1-e]$ a piecewise polynomial of degree $d - e$ and $d - e - 1$ times continuously differentiable at the knots $1, 2, \ldots, d+1-e$.

To explain these requirements: By 2^0., the functions $\phi_r^{d,e}$ are well-defined on $[0, 1)$. Further, at each of the $d + 1 - e$ knots $1, 2, \ldots, d + 1 - e$ there are $d - e$ linearly independent conditions to satisfy whence $(d + 1 - e)(d - e)$ equations for the coefficients of $\phi_r^{d,e}$ result. On the other hand, there are $(d - e)(d + 1 - e)$ freely choosable coefficients since $\phi_r^{d,e}$ is a polynomial of degree $d - e$ on each of the $d - e$ intervals $[1, 2), \ldots, [d - e, d - e + 1)$. Thus, there exist functions with properties 1^0. $- 3^0$., and these functions are uniquely determined. For example,

prebasis function		on $[0, 1)$	on $[1, 2)$	on $[2, 3)$
$\phi_0^{1,1}(x)$ $\phi_1^{1,1}(x)$	$\Big\} F_{1,1}$	x $1 - x$	0 0	0 0
$\phi_0^{2,1}(x)$ $\phi_1^{2,1}(x)$	$\Big\} F_{2,1}$	$x^2/2$ $x - x^2/2$	$1 - x/2$ $1 - x/2$	0 0
$\phi_0^{3,1}(x)$ $\phi_1^{3,1}(x)$	$\Big\} F_{3,1}$	$x^3/6$ $x^2/2 - x^3/6$	$\frac{-11x^2+34x-19}{24}$ $\frac{-13x^2+38x-17}{24}$	$\frac{5x^2-30x+45}{24}$ $\frac{7x^2-42x+63}{24}$

Proposition 2.20 *$F_{d,e}$ is a canonical prebasis of the space of all piecewise polynomial splines of degree d with defect e. Thereby, $\alpha_r^{F_{d,e}} = 1$ $(r = 0, \ldots, e)$, and the compression ϕ of $F_{d,e}$ is $\phi_0^{d-e,0}$.*

Proof Let us start with showing that

$$\phi_0^{d,e} + \ldots + \phi_e^{d,e} = \phi_0^{d-e,0}. \tag{1}$$

Indeed, the left hand side of (1) coincides on $[0, 1)$ with the $d - e$ times convolution of the function $\sum_{r=0}^e \phi_r^{e,e} = \chi_{[0,1)}$ by $\chi_{[0,1)}$, i.e. with the function $\phi_0^{d-e,0}$ (compare 2.9.3), whereas on $[1, d - e + 1)$, the left side of (1) represents a $d - e - 1$ times continuously differentiable piecewise polynomial of degree $d - e$. Thus, the left hand side of (1) is a $d - e - 1$ times continuously differentiable piecewise polynomial of degree $d - e$ on all of $[0, d - e + 1)$ which coincides with $\phi_0^{d-e,0}$ on $[0, 1)$. Now comparing the number of the coefficients with the number of the conditions at the knots yields that there is exactly one piecewise polynomial bearing these properties, namely $\phi_0^{d-e,0}$. Thus, (1) holds, and from (1) and 2.9.3 we easily conclude that

$$\sum_{r=0}^{e} \sum_{l \in Z} \phi_r^{d,e}(x - l) = \sum_{l \in Z} \phi_0^{d-e,0}(x - l) = 1 \quad \text{on} \quad \mathbf{R},$$

that is, $F_{d,e}$ is a prebasis with constants $\alpha_r^{F_{d,e}} = 1$ and compression $\phi_0^{d-e,0}$. Let us show now that this prebasis is canonical. In accordance with Theorem 2.11 we define functions ϕ_{rs} by

$$\phi_{rs}(t) = \begin{cases} \phi_r^{d,e}(t + s) & \text{if} \quad t \in [0, 1) \\ 0 & \text{if} \quad t \notin [0, 1) \end{cases}$$

and consider the equation

$$\sum_{r=0}^{e} \sum_{s=0}^{d-e} a_r b_s \phi_{rs}(x) = 0 \tag{2}$$

where we have written for brevity $b_s = e^{2\pi i s y}, y \in [0, 1)$. The functions ϕ_{rs} $(r = 0, \ldots, e)$ are polynomials of degree d for $s = 0$ and of degree $d - e$ for $s > 0$. Thus, the e powers $x^d, x^{d-1}, \ldots, x^{d-e+1}$ only appear in the items in (2) with $s = 0$. In other words: if (2) holds then $b_0 \sum_{r=0}^{e} a_r \phi_{r0}$ is a polynomial of degree $d - e$. This gives e linearly independent conditions for the $e + 1$ coefficients $a_0, \ldots, a_e$ whence follows that the linear space of all solutions $(a_0, \ldots, a_e)$ of (2) is at most one-dimensional. Further it is easy to see that $b_0 \sum_{r=0}^{e} a_r \phi_{r0}$ is actually a polynomial of degree $d - e$ if $a_0 = \ldots = a_e$, thus, the solutions $(a_0, \ldots, a_e)$ of (2) are necessarily of the form $(a_0, \ldots, a_e) = a(1, \ldots, 1)$ with some $a \in \mathbf{C}$. Replacing in (2) a_r by a we arrive at $a \sum_{r=0}^{e} \sum_{s=0}^{d-e} b_s \phi_{rs} = 0$ whence, since $\sum_{r=0}^{e} \phi_{rs} = U_{-s}(\chi_{[s,s+1)} \phi_0^{d-e,0})$ by (1), $a \sum_{s=0}^{d-e} b_s U_{-s}(\chi_{[s,s+1)} \phi_0^{d-e,0}) = 0$. Now use that $F_{d-e,0} = \{\phi_0^{d-e,0}\}$ is canonical by 2.9.3 to conclude that $a = 0$. Theorem 2.11 yields the assertion. ∎

2.9.5 Wavelets

Further important examples of spline spaces (in the sense of Section 2.6.1) can be found among spaces of wavelets. For their presentation we summarize some facts taken from Daubechies [Dau 2]; compare also [Me], [Ma].

Given a function $f \in L_R^2(0)$ and numbers $m, n \in \mathbf{Z}$, we agree upon denoting the function $x \mapsto 2^{-m/2} f(2^{-m} x - n)$ by $f_{(m,n)}$. The standard approach to wavelets is via Mallat's multiresolution analysis. A *multiresolution analysis* consists of a nested sequence

$$\ldots \subset V_2 \subset V_1 \subset V_0 \subset V_{-1} \subset V_{-2} \subset \ldots$$

of closed subspaces V_m of $L^2_R(0)$ which are subject to the following conditions:

(i) $\displaystyle\bigcap_{m \in Z} V_m = \{0\}, \quad \mathrm{clos}\,\Big(\bigcup_{m \in Z} V_m\Big) = L^2_R(0)\,.$

(ii) A function f is in V_{m-1} if and only if the function $f_{(1,0)}$ is in V_m.

(iii) There is a function ϕ in V_0 so that the set of all functions $\{\phi_{(0,n)}\,,\, n \in \mathbf{Z}\}$ forms an orthonormal basis of V_0.

One can show that then the functions $\{\phi_{(m,n)}\,,\, n \in \mathbf{Z}\}$ form an orthonormal basis of V_m, and that there exist numbers $M_1, M_2 > 0$ so that the inequality

$$M_1 \sum_{n \in Z} |c_n|^2 \le \|\sum_{n \in Z} c_n \phi_{(m,n)}\|^2_{L^2_R(0)} \le M_2 \sum_{n \in Z} |c_n|^2$$

holds for all sequences $(c_n) \in l^2_Z(0)$. The function ϕ is called the *mother wavelet*.

Let W_m denote the orthogonal complement of V_m in V_{m-1}, i.e., $V_{m-1} = V_m \oplus W_m$. Our assumptions guarantee the existence of a function ψ in W_0 having the property that the system $\{\psi_{(0,n)}\,,\, n \in \mathbf{Z}\}$ forms an orthonormal basis of W_0. Then $\{\psi_{(m,n)}\,,\, n \in \mathbf{Z}\}$ forms an orthonormal basis of W_m, and the collection of all functions $\psi_{(m,n)}$ is called the *wavelet basis* of $L^2_R(0)$. Starting with a mother wavelet ϕ one can compute the function ψ as follows: Since $V_0 \subset V_{-1}$, and since $(\phi_{(-1,n)})$ forms an orthonormal basis in V_{-1}, there are numbers c_n such that

$$\phi(x) = \sum_{n \in Z} c_n\, \phi_{(-1,n)}(x) \quad \text{for all } x \in \mathbf{R}\,.$$

Then, $\psi(x) = \sum_{n \in Z}(-1)^n c_{n+1}\, \phi_{(-1,-n)}(x)$ generates a wavelet basis. For example, starting with $\phi = \chi_{[0,1)}$ we find

$$\phi = \chi_{[0,1/2)} + \chi_{[1/2,1)} = \frac{1}{\sqrt{2}}(\phi_{(-1,0)} + \phi_{(-1,1)})$$

and, thus,

$$\psi = -\frac{1}{\sqrt{2}}\phi_{(-1,1)} + \frac{1}{\sqrt{2}}\phi_{(-1,0)} = -\chi_{[1/2,1)} + \chi_{[0,1/2)}$$

which yields the familiar Haar basis of $L^2_R(0)$.

Suppose now the mother wavelet ϕ to be a prespline (i.e. to be bounded and compactly supported). Then, obviously, $\phi_{(m,n)} = 2^{-m/2}\phi_{2-m,n}$ and, consequently, $V_m = S^{\{\phi\}}_{2-m}$ in earlier notations. Thus, in case of bounded, compactly supported mother wavelets, the theory developed in Sections 2.6-2.8 immediately applies. Let us conclude these notes with some examples and counterexamples for prespline mother wavelets.

Example 2.4 (The Haar basis) We have already seen that the prespline $\chi_{[0,1)}$ can also serve as a mother wavelet, and that the associated wavelets are just the Haar functions.

Example 2.5 (The Battle-Lemarié bases) Let ϕ^d with $d \geq 1$ denote the prespline generating spaces of smoothest piecewise polynomial splines (see 2.9.3). Orthogonalizing the system $(\phi^d_{(0,n)}, n \in \mathbf{Z})$ one obtains the Battle-Lemarié bases (compare [Bat]). The corresponding basis functions have unbounded (non-compact) supports, hence, they are *not* subject to our concept of spline spaces. But, on the other hand, these functions are rapidly decreasing, and a closer look shows that many of the results in Sections 2.6-2.8 can be transferred to the case of rapidly decreasing presplines in place of compactly supported ones.

Example 2.6 (The Daubechies bases) In [Dau 2], Daubechies proposed a method to generate systematically wavelets with compact support (which fit to our approach in 2.6-2.8). Here is her result (Theorem 3.6 in [Dau 2]).

Theorem 2.14 *(a) Let $(h(n))$ be a sequence such that*

(i) $\sum_n |h(n)| \, |n|^\epsilon < \infty$ for some $\epsilon > 0$,

(ii) $\sum_n h(n - 2k) \, h(n - 2l) = \delta_{kl}$ (the Kronecker delta),

(iii) $\sum_n h(n) = \sqrt{2}$.

Suppose further that the function $m_0(z) = \frac{1}{\sqrt{2}} \sum_n h(n)e^{inz}$ can be written as

$$m_0(z) = [\frac{1}{2}(1 + e^{iz})]^N \sum_n f(n)e^{inz}$$

where

(iv) $\sum_n |f(n)| \, |n|^\epsilon < \infty$ for some $\epsilon > 0$,

(v) $\sup_{z \in R} \left| \sum_n f(n)e^{inz} \right| < 2^{N-1}$.

Define

$$g(n) = (-1)^n \, h(1 - n) , \quad (F\phi)(z) = (2\pi)^{-1/2} \prod_{j=1}^{\infty} m_0(2^{-j}z) ,$$

$$\psi(x) = \sqrt{2} \sum_n g(n) \, \phi(2x - n) .$$

Then the functions $\phi_{(m,n)}$ generate a multiresolution analysis, and the $\psi_{(m,n)}$ are the associated wavelet basis.
(b) If the sequence $(h(n))$ is finite then the corresponding basic wavelet has compact support.

Concrete choices of finite sequences $(h(n))$ can be found in [Dau 2], page 980.

2.10   Concrete spline projections

2.10.1   Galerkin projections

Let $Y = L_R^p(\alpha)$, k a positive integer, F and G be k-tuples of presplines, and let F be moreover a prebasis.

A projection operator L_n sending $L_R^p(\alpha)$ into the spline space S_n^F is called *Galerkin projection with trial space S_n^F and test space S_n^G* if

$$(L_n f , g) = (f , g) \quad \text{for} \quad \text{all} \quad f \in L_R^p(\alpha), g \in S_n^G . \tag{1}$$

For general F and G it is by no means clear whether such an operator exists.

A sufficient condition for the existence of a Galerkin projection is that (F, G) is a canonical pair. Indeed, in this case the spline projections $L_n^{F,G}$ satisfy (1) in place of the L_n and, moreover, the $L_n^{F,G}$ are the only projections satisfying (1). To see this write (1) in the equivalent form

$$P_n^G L_n f = P_n^G f \quad , f \in L_R^p(\alpha) . \tag{2}$$

The prebasis F is canonical since (F, G) is so (Theorem 2.10), thus, the operators E_{-n}^F are well defined, and we can rewrite (2) as $P_n^G E_n^F E_{-n}^F L_n f = P_n^G f$ or

$$T^0(\lambda^{F,G}) E_{-n}^F L_n f = P_n^G f . \tag{3}$$

Multiplying both sides of (3) by $E_n^F T^0(\lambda^{F,G})^{-1}$ we find

$$L_n f = E_n^F T^0(\lambda^{F,G})^{-1} P_n^G f = L_n^{F,G} f .$$

Conversely, the invertibility of $\lambda^{F,G}$ is not necessary for the existence of the Galerkin projection with trial space S_n^F and test space S_n^G. The related counter-example is once more given by the pair (F, G) with $F = \{\chi_{[0,1]}\}$ and $G = \{\frac{1}{2}\chi_{[0,2]}\}$. In this setting we have $\lambda^{F,G}(z) = \frac{1}{2}(1 + z)$, that is, $\lambda^{F,G}$ has a zero on the unit circle, but since $S_n^G = S_n^F$ (see Example 2.3), there exists a projection with trial space S_n^F and test space S_n^G, viz. the operator $L_n^{F,F}$. Of course, the point in this counter-example is that the prebasis G is not canonical.

If both F and G are canonical, then the invertibility of $\lambda^{F,G}$ is even necessary for the existence of a corresponding Galerkin projection. Indeed, if G is canonical then $P_n^G E_n^G$ is invertible whence follows that the operators $P_n^G : L_R^p(\alpha) \to l_k^p(\alpha)$ are onto. Then (3) entails that the multiplication operator $T^0(\lambda^{F,G}) : l_k^p(\alpha) \to l_k^p(\alpha)$ is onto, too. In particular, the operator $T^0(\lambda^{F,G})$ is a Φ_--operator, and this implies the invertibility of $\lambda^{F,G}$ (see [PS 2], Theorem 4.105).

Here are some special situations where the Galerkin projections exist.

(i) Let $F = \{\phi_0, \ldots, \phi_{k-1}\}$ and $G = \{\gamma_0, \ldots, \gamma_{k-1}\}$ be prebases with supp ϕ_j, supp $\gamma_r \subseteq [0, 1]$, and let the matrix $((\phi_i, \gamma_j))_{i,j=0}^{k-1}$ be invertible. Then (F, G) is a canonical pair (and the Galerkin projection exists).

(ii) Let $F_d = \{\phi^d\}$ be the prebasis of smoothest piecewise polynomial splines of degree d (considered in 2.9.3), and define $G_{c,\epsilon} = \{\phi^{c,\epsilon}\}$ for each non-negative integer c and for each real $\epsilon \in (-1/2, 1/2)$ by $\phi^{c,\epsilon} := \phi^c(t-\epsilon)$. Then $(F_d, F_{c,\epsilon})$ is a canonical pair (see [Sc 1]).

(iii) If F is a canonical prebasis then the Galerkin projection with S_n^F both as trial and as test space exists. In particular, all concrete spline spaces considered in 2.9.1-2.9.4 give rise to their own Galerkin projections.

2.10.2 Collocation projections

Let $Y = R_R^p(\alpha)$, k a positive integer, F a k-elemental prebasis, and let $G = \{\delta_{\tau_0}, \ldots \ldots, \delta_{\tau_{k-1}}\}$ where $0 \leq \tau_0 < \tau_1 < \ldots < \tau_{k-1} < 1$ and where δ_τ refers to the shifted delta functional $\delta_\tau(f) = f(\tau)$. A projection operator L_n sending $R_R^p(\alpha)$ onto the trial space S_n^F is called *collocation (or interpolation) projection* if

$$(L_n f)((\tau_r + m)/n) = f((\tau_r + m)/n) \quad , \quad f \in R_R^p(\alpha)$$

for all $r \in \{0, \ldots, k-1\}$, $m \in \mathbf{Z}$, and $n > 0$ or, equivalently, if

$$P_n^G L_n f = P_n^G f \quad \text{for all} \quad f \in R_R^p(\alpha) .$$

As for the Galerkin method, the collocation projection exists if (F, G) is a canonical pair. Here are some special cases:

(i) Let $F = \{\phi_0, \ldots, \phi_{k-1}\}$ be a prebasis with $\text{supp}\,\phi_j \subseteq [0, 1]$, and let the matrix $(\phi_i(\tau_j))_{i,j=0}^{k-1}$ be invertible. Then (F, G) is a canonical pair.

(ii) Let F_d be the prebasis of smoothest piecewise polynomial splines of degree d, and let $\tau \in (0, 1)$. Then $(F_d, \{\delta_\tau\})$ is a canonical pair for all d, but $(F_d, \{\delta_0\})$ is canonical only if d is even (see [Sc 1]).

(iii) Let $F_{d,e}$ be the prebasis of piecewise polynomial splines of degree d with defect e (considered in 2.9.4). Then the pair $(F_{d,d}, G)$ is always canonical, and $(F_{d,d-1}, G)$ is canonical if and only if

$$\prod_{r=0}^{d-1} (1 - \tau_r) \neq \prod_{r=0}^{d-1} \tau_r .$$

Further, $(F_{d,d-2}, G)$ is canonical if and only if

$$[(-1)^d e^{\pi i t} \prod_{r=0}^{d-2} \tau_r + e^{-\pi i t} \prod_{r=0}^{d-2} (1 - \tau_r)]^2 +$$

$$+ \sum_{j=0}^{d-2} \left\{ (1 - \tau_j) \prod_{r \neq j} \frac{\tau_r^2 (1 - \tau_r)}{\tau_j - \tau_r} + (-1)^d \tau_j \prod_{r \neq j} \frac{\tau_r (1 - \tau_r)^2}{\tau_j - \tau_r} \right\} \neq 0$$

for all rational t in $[0, 1]$.

For $0 < e < d - 2$ it is in principle possible to derive similar conditions but their derivation is rather involved. The results cited in (iii) are Szyszka's ([Sz]).

2.10.3 Qualocation projections

Let $Y = R_R^p(\alpha)$ and, for simplicity, $k = 1$, and choose two prebases, $F = \{\phi\}$ and $F_1 = \{\phi_1\}$. For the Galerkin method we seeked a spline function $L_n f$ in S_n^F such that $(L_n f, f_1) = (f, f_1)$ for all $f_1 \in S_n^{F_1}$ or, equivalently,

$$(L_n f, \phi_{1ln}) = (f, \phi_{1ln}) \quad , \quad f \in R_R^p(\alpha) , \tag{1}$$

for all $l \in \mathbf{Z}$ and $n \geq 1$. In contrast to this the qualocation method results from the idea to evaluate the scalar products in (1) not exactly but by means of a certain quadrature rule, Q. Here, a quadrature rule is nothing else than a linear (not necessarily bounded) functional on Y which is interpreted as an approximation of the integral,

$$Q(g) \approx \int_R g(t)\, dt .$$

A projection operator L_n sending $R_R^p(\alpha)$ onto the spline space S_n^F is called qualocation projection with trial space S_n^F and test space $S_n^{F_1}$ if

$$Q(\overline{\phi_{1ln}}\, L_n f) = Q(\overline{\phi_{1ln}}\, f) \tag{2}$$

for all $f \in R_R^p(\alpha), l \in \mathbf{Z}, n \geq 1$. Let $\gamma : Y \to \mathbf{C}$ stand for the functional $f \mapsto Q(\overline{\phi_1} f)$, and put $G = \{\gamma\}$. Here and hereafter we suppose γ to be bounded, then γ is evidently a premeasure. In this situation, (2) can be rewritten as

$$P_n^G L_n f = P_n^G f ,$$

and as for the Galerkin or collocation method we see that the qualocation projections exist and are uniquely determined whenever (F, G) is a canonical pair, and that $L_n = L_n^{F,G}$ in this case.

Henceforth we specify the quadrature rule as follows: Choose an integer $m \geq 1$, real numbers ϵ_r with $0 \leq \epsilon_0 < \epsilon_1 < \ldots < \epsilon_{m-1} < 1$, and positive numbers w_r with $\sum_{r=0}^{m-1} w_r = 1$, and define

$$Q(g) = \sum_{k=-\infty}^{\infty} \sum_{r=0}^{m-1} w_r g(k + \epsilon_r) . \tag{3}$$

The functional γ is always bounded for this choice. For the quadrature rule (3) we can express the polynomial $\lambda^{F,G}$ more explicitly: The lm th entry of the matrix $T^0(\lambda^{F,G})$ is given by (compare 2.6.3(1)):

$$\int_R \phi(t + l - m)\, \gamma(t)\, dt =$$

$$= \sum_{k=-\infty}^{\infty} \sum_{r=0}^{m-1} w_r \overline{\phi_1(k+\epsilon_r)} \phi(k+l-m+\epsilon_r)$$

$$= \sum_{r=0}^{m-1} w_r \sum_{k=-\infty}^{\infty} \overline{\phi_1(k-l+\epsilon_r)} \phi(k-m+\epsilon_r) , \tag{4}$$

and due to

$$(\phi(k-m+\epsilon_r))_{k,m \in Z} = T^0(\lambda^{F,\{\delta_{\epsilon_r}\}})$$

and

$$(\overline{\phi_1(k-l+\epsilon_r)})_{l,k \in Z} = T^0(\lambda^{F_1,\{\delta_{\epsilon_r}\}})^* = T^0(\overline{\lambda^{F_1,\{\delta_{\epsilon_r}\}}})$$

we conclude from (4)

$$\lambda^{F,G} = \sum_{r=0}^{m-1} w_r \overline{\lambda^{F_1,\{\delta_{\epsilon_r}\}}} \lambda^{F,\{\delta_{\epsilon_r}\}} , \tag{5}$$

which is the desired representation of the qualocation lambda-function by means of collocation lambdas. This formula entails us a special situation when $\lambda^{F,G}$ is invertible. If F is a prebasis and $F = F_1$ then

$$\lambda^{F,G} = \sum_{r=0}^{m-1} w_r |\lambda^{F,\{\delta_{\epsilon_r}\}}|^2 \geq 0$$

that is, the qualocation projection exists if at least one of the pairs $(F, \{\delta_{\epsilon_r}\})$ $(r = 0, \ldots, m-1)$ is canonical.

2.11 Approximation of singular integral operators

2.11.1 Spline projection methods

In the remaining part of Chapter 2 we shall study approximation methods for singular integral operators with constant coefficients on the real axis, i.e. for the operators $aI + bS_R$ $(a, b \in \mathbf{C})$, and for elements of the algebra $\Sigma^p(\alpha)$ introduced in Section 2.1.2, i.e. for the operators $aI + bS_{R^+} + M^0(c)$ with $a, b \in \mathbf{C}$ and with $M^0(c)$ referring to a Mellin convolution belonging to the ideal $N^p(\alpha)$ of $\Sigma^p(\alpha)$ (compare Costabel's decomposition, 2.1.2(d)). Thereby, our main emphasis will be on the structure of the resulting approximation operators. We shall point out that all approximation operators under consideration can be identified with Toeplitz matrices with piecewise continuous functions, or with elements of the Toeplitz algebra $\mathrm{alg}\,(T^0(PC_{p,\alpha}), P)$ (introduced in 2.3.2), or with $k \times k$ matrices of operators in $\mathrm{alg}\,(T^0(PC_{p,\alpha}), P)$. Further we shall observe that all approximation operators have some properties in common which can be expressed in terms of the sums of the entries in each row of the corresponding matrices on $l_k^p(\alpha)$, and in terms of differences of entries in the same column.

To start with we consider spline projection methods for solving the equation

$$(aI + bS_R)\, u = f \quad , \quad f \in Y ,\tag{1}$$

where a, b are constants. Let (F, G) be a canonical pair consisting of the k-elementic prebasis $F = \{\phi_0, \ldots, \phi_{k-1}\}$ and of the k-tuple of premeasures $G = \{\gamma_0, \ldots, \gamma_{k-1}\}$. For the (F, G)-spline projection method we consider instead of (1) the approximation equation

$$L_n^{F,G}\, (aI + bS_R)\, u_n = L_n^{F,G}\, f \tag{2}$$

and seek its solutions u_n in the spline spaces $S_n^F(Y)$.

The formulation of the approximation system (2) makes sense if $Y = L_R^p(\alpha)$. In this case, the operator S_R is bounded from Y into Y, and so the application of $L_n^{F,G} = E_n^F T^0 (\lambda^{F,G})^{-1} P_n^G$ to $S_R u_n$ makes sense. In case $Y = R_R^p(\alpha)$, some more care is in order, since S_R acts not boundedly on $R_R^p(\alpha)$. (Example: the function $(S_R \chi_{[0,1)})(t) = \frac{1}{\pi i}(\ln|t-1| - \ln|t|)$ has poles at 0 and 1). So we shall always suppose in what follows that the functions $(S_R \phi_s)(. + l)$ belong to $R_R^p(\alpha)$ on a certain neighborhood of $\operatorname{supp} \gamma_r$ for all $\phi_s \in F$ and $\gamma_r \in G$ and for all integers l. This assumption guarantees that it is correct to set up approximation systems in the form (2) (compare (3) below).

To examine the structure of the approximation operators $L_n^{F,G}(aI + bS_R)|S_n^F$ it is more convenient to transform them into operators on $l_k^p(\alpha)$, that is, to deal with the operators $E_{-n}^F L_n^{F,G}(aI+bS_R)E_n^F$ (observe that the use of E_{-n}^F is correct since F is canonical for canonical (F, G) by Theorem 2.10). By the definition of $L_n^{F,G}$ and of $T^0(\lambda^{F,G})$ we have

$$
\begin{aligned}
E_{-n}^F L_n^{F,G}(aI + bS_R)E_n^F
&= T^0(\lambda^{F,G})^{-1}P_n^G(aI + bS_R)E_n^F \\
&= aT^0(\lambda^{F,G})^{-1}P_n^G E_n^F + bT^0(\lambda^{F,G})^{-1}P_n^G S_R E_n^F \\
&= aI + bT^0(\lambda^{F,G})^{-1}P_n^G S_R E_n^F .
\end{aligned}
$$

Put $P_n^G S_R E_n^F =: (A_{rs})_{r,s=0}^{k-1}$. The lm th entry of the infinite matrix associated with A_{rs} is

$$\int_R (S_R \phi_{smn})\left(\frac{t+l}{n}\right)\overline{\gamma_r}(t)\, dt = \int_R (S_R \phi_s)(t + l - m)\overline{\gamma_r}(t)\, dt .\tag{3}$$

Thus, the operator A_{rs} is independent of n, and its lm th entry depends on the difference $l - m$ only. This suggests to look for functions $\sigma^{\{\phi_s\},\{\gamma_r\}}$ whose $l - m$ th Fourier coefficient is just given by (3). If these functions are even multipliers (in fact, they are always multipliers for $Y = L_R^p(\alpha)$ due to the boundedness of the operators P_n^G, S_R, and E_n^F, and later on we shall point out that they are also multipliers for $Y = R_R^p(\alpha)$ in all concrete situations considered above) then we can introduce operators $T^0(\sigma^{F,G}) = (T^0(\sigma_{rs}^{F,G}))_{r,s=0}^{k-1}$ with $\sigma_{rs}^{F,G} = \sigma^{\{\phi_s\},\{\gamma_r\}}$, and we get

$$E_{-n}L_n^{F,G}(aI + bS_R)E_n^F = aI + bT^0(\lambda^{F,G})^{-1}T^0(\sigma^{F,G}) . \tag{4}$$

In particular, the approximation operators $L_n^{F,G}(aI + bS_R)|_{S_n^F}$ are invertible for *some* n if and only if the convolution operator $aI + bT^0(\lambda^{F,G})^{-1}T^0(\sigma^{F,G})$ is invertible, and then the approximation operators are invertible for *all* $n \geq 1$. A detailed treatise of the matrix functions $\sigma^{F,G}$ and $(\lambda^{F,G})^{-1}\sigma^{F,G}$ will be given now. For brevity, set $\rho^{F,G} := (\rho_{rs}^{F,G})_{r,s=0}^{k-1} = (\lambda^{F,G})^{-1}\sigma^{F,G}$ whenever this makes sense.

2.11.2 Another representation of the functions $\sigma^{F,G}$

Suppose the elements of F and G to be sufficiently smooth, say $F, G \subseteq \mathcal{S}(\mathbf{R})$. Taking into account that $S_R = F^{-1}\mathrm{sgn}(.)F$ on $L_R^p(\alpha) \cap L_R^p(0)$ (Section 2.1.1) and defining $\tilde{\phi}_s$ by $\tilde{\phi}_s(t) = \phi_s(-t)$ we find for the l th Fourier coefficient of $\sigma_{rs}^{F,G}$:

$$
\begin{aligned}
\int_R (S_R\phi_s)(t+l)\overline{\gamma_r}(t)\,dt &= -\int_R (S_R\tilde{\phi}_s)(-l-t)\overline{\gamma_r}(t)\,dt \\
&= -((S_R\tilde{\phi}_s) * \overline{\gamma_r})(-l) = -F^{-1}((FS_R\tilde{\phi}_s)(F\overline{\gamma_r}))(-l) \\
&= -F^{-1}(\mathrm{sgn}(.)\,(F\tilde{\phi}_s)(F\overline{\gamma_r}))(-l) \\
&= -\int_R e^{-2\pi ixl}\mathrm{sgn}x\,\overline{(F\tilde{\phi}_s)(x)}(F\overline{\gamma_r})(x)\,dx ,
\end{aligned}
$$

and, introducing new variables $j \in \mathbf{Z}$ and $y \in [0,1)$ with $x = j + y$, we get

$$
\int_R (S_R\phi_s)(t+l)\overline{\gamma_r}(t)\,dt =
$$

$$
= -\sum_{j\in Z} \int_0^1 e^{-2\pi iyl}\mathrm{sgn}(y+j)\overline{(F\tilde{\phi}_s)(y+j)}(F\overline{\gamma_r})(y+j)\,dy
$$

$$
= \int_0^1 e^{-2\pi iyl}(-\sum_{j\in Z}\mathrm{sgn}(j+1/2)\overline{(F\tilde{\phi}_s)(y+j)}(F\overline{\gamma_r})(y+j)\,dy .
$$

Hence, 2.11.1(3) is just the $l - m$ th Fourier coefficient of the function

$$\mathbf{T} \to \mathbf{C}, \quad e^{2\pi iy} \mapsto -\sum_{j\in Z}\mathrm{sgn}(j+1/2)\overline{(F\tilde{\phi}_s)(y+j)}(F\overline{\gamma_r})(y+j) \tag{1}$$

which consequently coincides with $\sigma_{rs}^{F,G}$. Formula (1) is the desired representation. The same reasoning as in Section 2.7.6 shows that the representation (1) remains valid if at least one of the conditions (a)-(e) in this section is satisfied.

2.11.3   Properties of $\sigma^{F,G}$

Let us start with examining the functions σ and ρ for Galerkin methods.

Theorem 2.15 *(a) Let $F = \{\phi_0, \ldots, \phi_{k-1}\} \subseteq L_R^p(\alpha)$ be a prebasis and $G = \{\gamma_0, \ldots, \gamma_{k-1}\} \subseteq L_R^p(\alpha)^*$ be a k-tuple of premeasures. Then the functions $\sigma_{rs}^{F,G}$ $(r, s = 0, \ldots, k-1)$ are piecewise continuous multipliers on $l_Z^p(\alpha)$ and their only possible discontinuity is at $1 \in \mathbf{T}$ where holds*

$$\sigma_{rs}^{F,G}(1+0) - \sigma_{rs}^{F,G}(1-0) = -2 \int_R \phi_s(x)\, dx \int_R \overline{\gamma_r}(x)\, dx . \tag{1}$$

Furthermore,

$$\sigma^{F,G}(1 \pm 0)\, \alpha^F = \mp \left(\int \overline{\gamma_0}(x)\, dx, \ldots, \int \overline{\gamma_{k-1}}(x)\, dx \right)^T \tag{2}$$

where α^F stands for the vector $(\alpha_0^F, \ldots, \alpha_{k-1}^F)^T$.
(b) Let (F, G) be moreover canonical. Then

$$\rho^{F,G}(1 \pm 0)\, \alpha^F = \mp \alpha^F \quad \text{and} \tag{3}$$

$$\rho_{r_1 s}^{F,G} - \rho_{r_2 s}^{F,G} \in C_{p,\alpha} . \tag{4}$$

Identity (3) and inclusion (4) are the announced column and row properties for the approximation matrix of the singular integral operator.

We prepare the proof of Theorem 2.15 by the following lemma:

Lemma 2.2 *Let ϕ be the compression of a prebasis $F \subseteq S(\mathbf{R})$ (i.e. $\phi = \sum \alpha_r^F \phi_r$). Then there is a function $h \in S(\mathbf{R})$ with compact support such that $\phi = \chi_{[0,1]} * h$.*

Proof   Let c and d be integers such that $\operatorname{supp} \phi \subseteq [c, d]$. Define a function H by

$$H(x) = \begin{cases} 0 & \text{if } x < c \\ \phi(x) & \text{if } c \leq x < c+1 \\ \phi(x) + \phi(x-1) & \text{if } c+1 \leq x < c+2 \\ \;\vdots & \qquad \vdots \\ \phi(x) + \phi(x-1) + \ldots + \phi(x-(d-c-1)) & \text{if } d-1 \leq x < d \\ 1 & \text{if } d \leq x . \end{cases}$$

The function H is in $S(\mathbf{R})$. Taking into account the identity $\sum_l \phi(x-l) \equiv 1$ (by Proposition 2.19(i)) one readily verifies that

$$\phi(x) = H(x) - H(x-1) . \tag{5}$$

Set $h = \frac{dH}{dx} \in S(\mathbf{R})$. Then (5) can be rewritten

$$\phi(x) = \int_{x-1}^x h(t)\, dt = \int_R h(x-t)\chi_{[0,1]}(t)\, dt = (h * \chi_{[0,1]})(x) , \tag{6}$$

and h has a compact support.   ∎

Proof of Theorem 2.15   To start with we suppose $F, G \subseteq \mathcal{S}(\mathbf{R})$. Then $F\overline{\phi_s}$ and $F\overline{\gamma_r}$ are smooth, and this shows via the representation 2.11.2(1) that the functions $\sigma_{rs}^{F,G}$ are smooth on $\mathbf{T} \setminus \{1\}$ and that they possess one-sided limits $\sigma_{rs}^{F,G}(1 \pm 0)$ at 1. For 2.11.3(1) observe that

$$\sigma_{rs}^{F,G}(1+0) - \sigma_{rs}^{F,G}(1-0) =$$

$$= -\sum_{j \in Z} \operatorname{sgn}(j + \tfrac{1}{2})\overline{(F\overline{\phi_s})(j)}(F\overline{\gamma_r})(j) + \sum_{j \in Z} \operatorname{sgn}(j + \tfrac{1}{2})\overline{(F\overline{\phi_s})(j+1)}(F\overline{\gamma_r})(j+1)$$

$$= -2\overline{(F\overline{\phi_s})(0)}(F\overline{\gamma_r})(0) = -2 \int_R \phi_s(x)\,dx \int_R \overline{\gamma_r}(x)\,dx \ .$$

Further, by 2.11.2(1) and the previous lemma,

$$\sum_{s=0}^{k-1} \alpha_s^F \sigma_{rs}^{F,G}(e^{2\pi i y}) =$$

$$= -\sum_{j \in Z} \operatorname{sgn}(j + 1/2)\overline{(F\overline{\phi})(y+j)}(F\overline{\gamma_r})(y+j)$$

$$= -\sum_{j \in Z} \operatorname{sgn}(j + 1/2)\overline{(F\chi_{[0,1]})(y+j)}\,\overline{(F\overline{h})(y+j)}(F\overline{\gamma_r})(y+j) \ , \qquad (7)$$

and the uniform convergence of this series entails that

$$\lim_{\left\{\substack{y \searrow 0 \\ y \nearrow 1}\right\}} \sum_{s=0}^{k-1} \alpha_s^F \sigma_{rs}^{F,G}(e^{2\pi i y}) =$$

$$= -\sum_{j \in Z} \operatorname{sgn}(j + 1/2) \lim_{\left\{\substack{y \searrow 0 \\ y \nearrow 1}\right\}} \overline{(F\chi_{[0,1]})(y+j)}\,\overline{(F\overline{h})(y+j)}(F\overline{\gamma_r})(y+j) \ .$$

Since $(F\chi_{[0,1]})(x) = e^{-\pi i x}\frac{\sin \pi x}{\pi x}$ for $x \neq 0$, and since $F\overline{h}$ and $F\overline{\gamma_r}$ are smooth and fast decreasing, we have

$$\lim_{\left\{\substack{y \searrow 0 \\ y \nearrow 1}\right\}} \overline{(F\chi_{[0,1]})(y+j)}\,\overline{(F\overline{h})(y+j)}(F\overline{\gamma_r})(y+j) =$$

$$= \begin{cases} \overline{(F\overline{h})(0)}\,(F\overline{\gamma_r})(0) & \left\{\substack{j=0 \\ j=-1}\right\} \\ 0 & \left\{\substack{j \in Z \setminus \{0\} \\ j \in Z \setminus \{-1\}}\right\} \end{cases} \ .$$

Moreover,

$$\overline{(F\overline{h})(0)} = (Fh)(0) = \int_R h(t)\,dt = \sum_{n \in Z} \int_n^{n+1} h(t)\,dt = \sum_{n \in Z} \phi(n) = 1$$

by (6) and Proposition 2.19(i), which yields the assertion 2.11.3(2) for $F, G \subseteq \mathcal{S}(\mathbf{R})$.

Concerning assertion 2.11.3(3) we recall that

$$\lambda^{F,G}(1)^{-1}\big(\int_R \overline{\gamma_0}\,dx,\dots,\int_R \overline{\gamma_{k-1}}\,dx\big)^T \;=\; \alpha^F$$

by Proposition 2.19(iii). Combining this with 2.11.3(2) we obtain

$$\lambda^{F,G}(1)^{-1}\sigma^{F,G}(1\pm0)\alpha^F \;=\; \mp\lambda^{F,G}(1)^{-1}\big(\int_R \overline{\gamma_0}\,dx,\dots,\int_R \overline{\gamma_{k-1}}\,dx\big)^T \;=\; \mp\alpha^F\,.$$

In the general case we approximate the presplines ϕ_s and the premeasures by $\mathcal{S}(\mathbf{R})$-presplines and -premeasures in the norms of $L_R^p(\alpha)$ and $L_R^p(\alpha)^*$, respectively.

The continuous dependence of the operators P_n^G and E_n^F on F and G (which was already mentioned in 2.7.7(4) and (7) involves that $T^0(\lambda^{F,G})$ and $T^0(\sigma^{F,G})$ depend continuously on F and G, too, and this yields the assertion. ∎

The results for $Y = R_R^p(\alpha)$ are less complete than those for $L_R^p(\alpha)$.

Theorem 2.16 *Let $F = \{\phi_0,\dots,\phi_{k-1}\}$ be a $R_R^p(\alpha)$-prebasis consisting of C^2-functions, and let $G = \{\gamma_0,\dots,\gamma_{k-1}\}$ be a k-tuple of premeasures possessing bounded and continuous Fourier transforms (in particular, shifted δ-functionals or the qualocation premeasure). Then the assertions of Theorem 2.15 remain valid.*

Proof If $\phi_s \in C^2$ then $\sup|x^2(F\phi_s)(x)| < \infty$. This involves the uniform convergence of the series 2.11.2(1) and shows, moreover, that $\sigma_{rs}^{F,G}$ is a piecewise continuous multiplier having its only discontinuity at 1 (note that the functions

$$e^{2\pi iy} \mapsto -\text{sign}\,(j+1/2)\overline{(F\phi_s)(y+j)}(F\overline{\gamma_r})(y+j)$$

have this property).

Further, a closer look at Lemma 2.2 shows that it holds true with $h \in \mathcal{S}(\mathbf{R})$ replaced by a function $h \in C^1(\mathbf{R})$ with compact support. Then the sequences

$$\Big(\overline{(F\chi_{[0,1)})(y+j)}\,\overline{(F\overline{h})(y+j)}\Big)_{j\in Z}$$

belong to $l_Z^1(0)$ uniformly with respect to $y \in [0,1)$, and so the remainder of the proof runs parallelly to that of Theorem 2.15. ∎

The preceding theorem covers most of the concrete projection methods for concrete spline spaces considered in 2.9.1-2.10.3. The only (but important) exceptions are the collocation and qualocation methods for spline functions in trial spaces S_n^F of piecewise polynomial splines which are not C^2. But also in these situations the analogue of Theorem 2.15 holds.

Theorem 2.17 *(a) Let F be one of the prebases $F_{d,d-1}$ $(d \geq 1)$ or $F_{d,d-2}$ $(d \geq 2)$ introduced in 2.9.4, and let G be a collection (with the same cardinality as F) of* *premeasures having bounded and continuous Fourier images. Then the assertions of Theorem 2.15 remain valid.*

(b) Let F be the prebasis $F_{d,d}$ ($d \geq 0$), and let G be a collection of shifted δ-functionals with the same cardinality as F. Then the assertions of Theorem 2.15 are valid.

Proof   (a) The compressions of $F_{d,d-1}$ and $F_{d,d-2}$ are the functions $\chi_{[0,1)} * \chi_{[0,1)}$ and $\chi_{[0,1)} * \chi_{[0,1)} * \chi_{[0,1)}$, respectively (see Proposition 2.20). The sequences

$$\left(\overline{F(\chi_{[0,1)} * \chi_{[0,1)})(y+j)} \right)_{j \in Z} = \left(\overline{(F\chi_{[0,1)})(y+j)}^2 \right)_{j \in Z}$$

and

$$\left(\overline{F(\chi_{[0,1)} * \chi_{[0,1)} * \chi_{[0,1)})(y+j)} \right)_{j \in Z} = \left(\overline{(F\chi_{[0,1)})(y+j)}^3 \right)_{j \in Z}$$

are in $l_Z^1(0)$ uniformly with respect to y; thus, the analogue of series (7) (with $h = \chi_{[0,1)}$ resp. $h = \chi_{[0,1)} * \chi_{[0,1)}$) converges uniformly, and so the arguments in the proof of Theorem 2.15 again apply.

(b) In case $F_{d,d}$ the things are more involved. In accordance with the definitions in Section 2.9.4 we set

$$\phi_s^{d,d}(x) = \begin{cases} \binom{d}{s}(1-x)^s x^{d-s} & x \in [0,1) \\ 0 & x \in \mathbf{R} \setminus [0,1), \end{cases}$$

where $d \geq 0$, and $s = 0, \ldots, d$. Straightforward computations give

$$(S_R \phi_s^{d,d})(t)$$
$$= \frac{1}{\pi i}\binom{d}{s}\sum_{k=0}^{s}(-1)^k\binom{s}{k}t^{d+k-s}\left(\ln\left|1 - \frac{1}{t}\right| + \sum_{r=0}^{d+k-s-1}\frac{1}{r+1}t^{-(r+1)} \right).$$

For $|t| > 1$, we have $|1 - 1/t| = 1 - 1/t$, and for $|x| < 1/2$,

$$\ln(1-x) = -\sum_{r=0}^{\infty}\frac{1}{r+1}x^{r+1},$$

where the series converges uniformly. Thus, for $|t| > 2$ we obtain

$$(S_R \phi_s^{d,d})(t)$$
$$= \frac{1}{\pi i}\binom{d}{s}\sum_{k=0}^{s}(-1)^k\binom{s}{k}t^{d+k-s}\left(-\sum_{r=d+k-s}^{\infty}\frac{1}{r+1}t^{-(r+1)} \right)$$
$$= \frac{1}{\pi i}\binom{d}{s}\sum_{k=0}^{s}(-1)^k\binom{s}{k}\left(-\sum_{r=1}^{\infty}\frac{1}{d+k-s+r}t^{-r} \right). \tag{8}$$

Now we use the identity (see [PBM], formula 4.2.8.4)

$$\sum_{k=0}^{n}(-1)^k\binom{n}{k}\binom{m+k}{l}^{-1}=\frac{l}{n+l}\binom{n+m}{m-l}^{-1}.$$

For the special case $l = 1$ we get

$$\sum_{k=0}^{n}(-1)^k\binom{n}{k}\frac{1}{m+k}=\frac{1}{n+1}\binom{n+m}{m-1}^{-1}=\frac{1}{n+1}\binom{n+m}{n+1}^{-1},$$

and substituting $n \mapsto s$ and $m \mapsto d+r-s$, we find

$$\sum_{k=0}^{s}(-1)^k\binom{s}{k}\frac{1}{d+r-s+k}=\frac{1}{s+1}\binom{d+r}{s+1}^{-1}.$$

The last identity and formula (8) yield

$$(S_R\phi_s^{d,d})(t)=-\frac{1}{\pi i}\binom{d}{s}\frac{1}{s+1}\sum_{r=1}^{\infty}\binom{d+r}{s+1}^{-1}t^{-r},\tag{9}$$

where the series converges absolutely and uniformly for (e.g.) $|t| > 2$. We mention the special case $d = 0$:

$$(S_R\phi_0^{0,0})(t)=-\frac{1}{\pi i}\sum_{r=1}^{\infty}\binom{r}{1}^{-1}t^{-r}=-\frac{1}{\pi i}\sum_{r=1}^{\infty}\frac{1}{r}t^{-r}=\frac{1}{\pi i}\ln\left(1-\frac{1}{t}\right).$$

Now we write

$$S_R\phi_s^{d,d}=(S_R\phi_s^{d,d}-\frac{1}{d+1}S_R\phi_0^{0,0})+\frac{1}{d+1}S_R\phi_0^{0,0}$$

and consider the term in brackets:

$$(S_R\phi_s^{d,d}-\frac{1}{d+1}S_R\phi_0^{0,0})(t)$$
$$=-\frac{1}{\pi i}\sum_{r=1}^{\infty}\left(\binom{d}{s}\frac{1}{s+1}\binom{d+r}{s+1}^{-1}-\frac{1}{d+1}\frac{1}{r}\right)t^{-r}.\tag{10}$$

Since $\binom{d}{s}\frac{1}{s+1}\binom{d+1}{s+1}^{-1}=\frac{1}{d+1}$, the term on the right-hand side of (10) is equal to

$$(S_R\phi_s^{d,d}-\frac{1}{d+1}S_R\phi_0^{0,0})(t)$$
$$=-\frac{1}{\pi i}\sum_{r=2}^{\infty}\left(\binom{d}{s}\frac{1}{s+1}\binom{d+r}{s+1}^{-1}-\frac{1}{d+1}\frac{1}{r}\right)t^{-r},$$

and this function has a decay of order t^{-2}. Let $G = \{\gamma_1, \ldots, \gamma_k\}$ with $\gamma_r = \delta_{\epsilon_r}$ (the δ-functional in $\epsilon_r \in (0,1)$). The l th Fourier coefficient of the function $\sigma_{rs}^{F,G}$ is equal to

$$
\begin{aligned}
(\sigma_{r,s}^{F,G})_l &= (S_R\phi_s^{d,d})(l + \epsilon_r) \\
&= \underbrace{(S_R\phi_s^{d,d} - \frac{1}{d+1}S_R\phi_0^{0,0})(l + \epsilon_r)}_{=b_l} + \frac{1}{d+1}\underbrace{(S_R\phi_0^{0,0})(l + \epsilon_r)}_{=s_l} .
\end{aligned}
$$

We have proved that the numbers b_l decay as l^{-2}, thus, the sequence $(b_l)_{l\in Z}$ belongs to l^1. It follows that the numbers b_l are the Fourier coefficients of a function, b_{rs} say, with absolutely convergent Fourier series. The numbers s_l are the Fourier coefficients of the function $\sigma_{0,0}^{\{\chi_{[0,1)}\},\{\delta_{\epsilon_r}\}}$, and so we have

$$
\sigma_{rs}^{F,G} = b_{rs} + \frac{1}{d+1}\sigma_{0,0}^{F',G'_r} , \tag{11}
$$

where $F' = \{\chi_{[0,1)}\} = \{\phi_0^{0,0}\}$ and $G'_r = \{\delta_{\epsilon_r}\}$. The functions b_{rs} are continuous multipliers, and the function $\sigma_{0,0}^{F',G'_r}$ is monoton (consequently, it is of bounded variation and a multiplier) and has only one discontinuity. The point of discontinuity is $1 \in \mathbf{T}$, and $\sigma_{0,0}^{F',G'_r}(1 \pm 0) = \mp 1$ (compare [Sc 1], [Sc 2]).

To prove the row property we mention that $\lambda^{F,G}$ is a constant matrix and set $(\lambda^{F,G})^{-1} = (\beta_{rs})_{r,s=0}^d$. Then we get

$$
((\lambda^{F,G})^{-1}\sigma^{F,G}\alpha^F)_r = \sum_{s=0}^d\sum_{t=0}^d \beta_{rt}\sigma_{ts}^{F,G}\alpha_s^F = \sum_{t=0}^d\beta_{rt}(\sum_{s=0}^d\sigma_{ts}^{F,G}) ,
$$

since $\alpha_s^F = 1$ for $s = 0, \ldots d$ (see Proposition 2.20). For the l th Fourier coefficient we use the identity (compare formula 2.8.1(2) and Proposition 2.20)

$$
\sum_{s=0}^d (\sigma_{ts}^{F,G})_l = \sum_{s=0}^d (S_R\phi_s^{d,d})(l + \epsilon_t) = (S_R\chi_{[0,1)})(l + \epsilon_t) = (\sigma_{0,0}^{F',G'_t})_l ,
$$

and, consequently, $\sum_{s=0}^d \sigma_{ts}^{F,G} = \sigma_{0,0}^{F',G'_t}$.
Employing once more the property $\sigma_{0,0}^{F',G'_t}(1 \pm 0) = \mp 1$ we get now

$$
(\sum_{s=0}^d\sum_{t=0}^d \beta_{rt}\sigma_{ts}^{F,G})(1 \pm 0) = \sum_{t=0}^d \beta_{rt}\sigma_{0,0}^{F',G'_t}(1 \pm 0) = \mp\sum_{t=}^d \beta_{rt} . \tag{12}
$$

In this special case ($G = \{\delta_{\epsilon_1}, \ldots \delta_{\epsilon_d}\}$) it follows that $\int_R \gamma_r(x)\,dx = 1$ for all r, and Proposition 2.19 gives

$$
\sum_{s=0}^d \alpha_s^F\lambda_{rs}^{F,G} = 1 , \quad \text{or} \quad \lambda^{F,G}\alpha^F = (1, \ldots, 1)^T .
$$

Thus, we obtain $\alpha^F = (\lambda^{F,G})^{-1}(1,\ldots,1)^T$, i.e.,

$$\sum_{t=0}^{d} \beta_{rt} = 1\,. \tag{13}$$

The identities (12) and (13) prove the validity of the row sum condition.

Now we will consider the column difference condition, i.e., we have to prove that $\sum_{t=0}^{d} \beta_{r_1 t}\sigma_{ts}^{F,G} - \sum_{t=0}^{d}\beta_{r_2 t}\sigma_{ts}^{F,G}$ is continuous for all r_1 and r_2. The only point of interest is $1 \in \mathbf{T}$. Identity (11) and the continuity of the functions b_{rs} show that it is sufficient to prove the continuity of $\sum_{t=0}^{d}(\beta_{r_1 t} - \beta_{r_2 t})\sigma_{0,0}^{F',G'_t}$ at the point 1. But from (13) we immediately get the assertion:

$$\sum_{t=0}^{d}(\beta_{r_1 t} - \beta_{r_2 t})\sigma_{0,0}^{F',G'_t}(1 \pm 0) = \mp\sum_{t=0}^{d}(\beta_{r_1 t} - \beta_{r_2 t}) = 0\,. \blacksquare$$

2.11.4 Approximation of integral operators on the semi axis

Next we explain the structure of spline approximation operators for the singular integral operator $a\chi_{R^+}I + bS_{R^+}$ with constant coefficients a, b. Let (F,G) be a canonical pair which satisfies our hypothesis from 2.11.3. Then, clearly,

$$E_{-n}^F L_n^{F,G}(a\chi_{R^+}I + bS_{R^+})E_n^F = T^0(\lambda^{F,G})^{-1}(A_{rs})_{r,s=0}^{k-1}$$

with $(A_{rs}) = P_n^G(a\chi_{R^+}I + bS_{R^+})E_n^F$. The lm th entry a_{rslm} of the infinite matrix associated with $A_{rs} \in L(l_1^p(\alpha))$ is

$$\begin{aligned}
a_{rslm} &= \int_R ((a\chi_{R^+}I + bS_{R^+})\phi_{smn})((t+l)/n)\overline{\gamma_r}(t)\,dt \\
&= \int_R ((a\chi_{R^+}I + bS_{R^+})\phi_{sm1})(t+l)\overline{\gamma_r}(t)\,dt\,.
\end{aligned}$$

Comparing this with 2.11.1(3) we find that a_{rslm} coincides with the lm th entry of the rs th component of the matrix $P_n^G(aI + bS_R)E_n^F$ for l and m being large enough (say, $l \geq l_0$ and $m \geq m_0$), and that $a_{rslm} = 0$ if l or m are small enough (say, $l \leq -l_0$ or $m \leq -m_0$). For the remaining cases (when $-l_0 < l < l_0$ or $-m_0 < m < m_0$) it is only in some concrete situations possible to identify the numbers a_{rslm}. Thus, if we denote the projection operator

$$l_1^p(\alpha) \to l_1^p(\alpha)\,, \quad (x_i) \mapsto (\ldots, 0, x_0, x_1, \ldots)$$

by P again, and if we abbreviate the diagonal operator $\mathrm{diag}(P,\ldots,P) : l_k^p(\alpha) \to l_k^p(\alpha)$ also by P (the context will always make clear which P is meant) then

$$\begin{aligned}
P_n^G(a\chi_{R^+}I + bS_{R^+})E_n^F &= PP_n^G(aI + bS_R)E_n^F P + K \\
&= PT^0(a\lambda^{F,G} + b\sigma^{F,G})P + K \\
&= T(a\lambda^{F,G} + b\sigma^{F,G}) + K \tag{1}
\end{aligned}$$

with a certain perturbation K resulting from the unidentified cases. Obviously, K is of finite rank (and compact).

2.11.5   Approximation of Mellin convolutions

Let us finally consider approximation operators for Mellin convolutions $M^0(c) \in N^p(\alpha)$ (see Section 2.1.2), where we restrict ourselves to the case $c \in \mathcal{N}$ and to real-valued functions $k = M^{-1}c$. Let (F, G) be a canonical pair. If $F \subseteq L_R^p(\alpha)$ and $G \subseteq L_R^p(\alpha)^*$ then it is always correct to form approximation operators $L_n^{F,G} M^0(c)|S_n^F$ since $M^0(c)$ is bounded on $L_R^p(\alpha)$. In case $Y = R_R^p(\alpha)$, the application of P_n^G to $M^0(c)E_n^F$ is not always possible. (Example: $(N\chi_{[0,1)})(t) = \ln|t+1| - \ln|t|$ has a pole at 0.) So we suppose in what follows that $M^0(c)\phi_{sl1}$ belongs to $R_R^p(\alpha)$ in a neighborhood of $\operatorname{supp}\gamma_r$ for all r, s and l. We claim that

$$P_n^G M^0(c) E_n^F = (A_{rs})_{r,s=0}^{k-1} \quad \text{with} \quad A_{rs} = \int_R \phi_s(x)\,dx \int_R \overline{\gamma_r}\,dx\, G(c) + K_{rs} \quad (1)$$

with certain compact operators K_{rs}. We suppose without loss of generality that $\operatorname{supp}\phi_s$, $\operatorname{supp}\gamma_r \subseteq \mathbf{R}^+$. Taking into account the integral representation 2.1.4(4) of $M^0(c)$ we find for the lm th entry of the matrix A_{rs}:

$$\int_R (M^0(c)\phi_{smn})((t+l)/n)\overline{\gamma_r}(t)\,dt =$$

$$= \int_R \int_{R^+} k((t+l)/nx)\phi_{smn}(x)\frac{dx}{x}\,\overline{\gamma_r}(t)\,dt$$

$$= \int_R \int_{R^+} k((t+l)/(y+m))\frac{1}{y+m}\phi_s(y)\,dy\,\overline{\gamma_r}(t)\,dt \,. \quad (2)$$

Now we write

$$\phi_s = \sum_{u=0}^{3} i^u \phi_{su} \quad \text{and} \quad \gamma_r = \sum_{v=0}^{3} i^v \gamma_{rv}$$

with ϕ_{su} referring to non-negative functions and γ_{rv} to non-negative functionals, and with i abbreviating the imaginary unit. Then (2) is equal to

$$\sum_{u=0}^{3}\sum_{v=0}^{3} i^{u-v} \int_R \int_{R^+} k((t+l)/(y+m))\frac{1}{y+m}\phi_{su}(y)\,dy\,\gamma_{rv}\,dt \,, \quad (3)$$

and a two-fold application of the generalized mean value theorem entails the existence of numbers $t_{lm} \in \operatorname{conv}\operatorname{supp}\gamma_r$ and $y_{ml} \in \operatorname{conv}\operatorname{supp}\phi_s$ such that (3) equals

$$\sum_{u=0}^{3}\sum_{v=0}^{3} i^{u-v} k((t_{lm}+l)/(y_{ml}+m))\frac{1}{y_{ml}+m} \int_{R^+} \phi_{su}(y)\,dy \int_R \gamma_{rv}(t)\,dt \,.$$

Hence, by Corollary 2.1 in Section 2.4.1 and Proposition 2.11(a) in Section 2.4.2, there are compact operators $K_{rsuv} \in L(l_1^p(\alpha))$ such that

$$A_{rs} = \sum_{u=0}^{3}\sum_{v=0}^{3} i^{u-v}(G(c) + K_{rsuv}) \int_{R^+} \phi_{su}(y)\,dy \int_R \gamma_{rv}(t)\,dt$$

$$= G(c) \int_R \phi_s(y)\,dy \int_R \overline{\gamma_r}(t)\,dt + K_{rs}$$

with compact operators $K_{rs} \in L(l_1^p(\alpha))$ which gives our claim (1). Let us summarize and complete the results of the Sections 2.11.3 and 2.11.4.

Proposition 2.21 *Let (F,G) with $F = \{\phi_0, \ldots, \phi_{k-1}\}$ and $G = \{\gamma_0, \ldots, \gamma_{k-1}\}$ be a canonical pair, and let $a, b \in \mathbf{C}$ and c be a continuous multiplier such that $M^0(c) \in N^p(\alpha)$ and that the general assumptions of Sections 2.11.3 and 2.11.5 are satisfied. Then*
(i) The operators $P_n^G(a\chi_{R_+}I + bS_{R^+} + M^0(c))E_n^F =: A = (A_{rs})_{r,s=0}^{k-1}$ are independent of n, and $A_{rs} \in \mathrm{alg}(T^0(PC_{p,\alpha}), P)$. In particular,

$$A_{rs} = aT(\lambda_{rs}^{F,G}) + bT(\sigma_{rs}^{F,G}) + \int_R \phi_s(x)\, dx \int_R \overline{\gamma_r}(x)\, dx\, G(c) + K_{rs}$$

with compact K_{rs}.
(ii) The operators $E_{-n}^F L_n^{F,G}(a\chi_{R^+} + bS_{R^+} + M^0(c))E_n^F =: B = (B_{rs})_{r,s=0}^{k-1}$ are independent of n, and $B_{rs} \in \mathrm{alg}(T^0(PC_{p,\alpha}), P)$. In particular,

$$B = aP + bT((\lambda^{F,G})^{-1}\sigma^{F,G}) + (\alpha_r^F \int_R \phi_s(x)\, dx\, G(c))_{r,s=0}^{k-1} + K \qquad (4)$$

with compact $K \in L(l_k^p(\alpha))$ and, moreover,

$$\sum_{s=0}^{k-1} \alpha_s^F B_{rs} = \alpha_r^F\, (T(f_r) + G(c) + K_r) \qquad (5)$$

where f_r is a piecewise continuous multiplier having its only discontinuity at $1 \in \mathbf{T}$ and satisfying

$$f_r(1 \pm 0) = a \mp b\,, \qquad (6)$$

and

$$B_{r_1 s} - B_{r_2 s} = T^0(g_{r_1 r_2}) + K_{r_1 r_2} \qquad (7)$$

where $g_{r_1 r_2}$ is continuous and $K_{r_1 r_2}$ is compact.

Identities (5)-(6) and (7) are the relevant column and row conditions, respectively.
Proof Part (i) is a summary of 2.11.4(1) and 2.11.5(1). For (ii) notice that

$$B = T^0(\lambda^{F,G})^{-1}(PT^0(a\lambda^{F,G} + b\sigma^{F,G})P +$$
$$+ ((\int \phi_s\, dx \int \overline{\gamma_r}\, dx\, G(c))_{rs})_{r,s=0}^{k-1} + (K_{rs})_{r,s=0}^{k-1}\,. \qquad (8)$$

For the first product in (8) we use Proposition 2.8(a) to find that $T^0(\lambda^{F,G})^{-1}P - PT^0(\lambda^{F,G})^{-1}$ is a compact operator and thus

$$T^0(\lambda^{F,G})^{-1}PT^0(a\lambda^{F,G} + b\sigma^{F,G})P = PT^0(a + b(\lambda^{F,G})^{-1}\sigma^{F,G})P + \text{compact}\,.$$

For the second product in (8) we observe that, by Propositions 2.8(a) and 2.11(d),

$$T^0(\lambda^{F,G})^{-1}(\int \phi_s \, dx \int \overline{\gamma_r} \, dx \, G(c)) =$$

$$= \lambda^{F,G}(1)^{-1}(\int \phi_s \, dx \int \overline{\gamma_r} \, dx \, G(c)) + \text{compact} ,$$

and the latter term equals

$$(\alpha_r^F \int \phi_s \, dx \, G(c))_{r,s=0}^{k-1} + \text{compact}$$

due to Proposition 2.19(b). This proves (4), and for (5) and (6) it remains to apply Theorem 2.15(b) and the identities (iii) and (ii) in Proposition 2.19. Finally, (7) is obvious from (4) and 2.11.3(4). ∎

2.11.6 Quadrature methods

Let Q stand for a composed quadrature rule as in 2.10.3(3), i.e. given an integer $k \geq 1$, real numbers τ_s with $0 \leq \tau_0 < \tau_1 < \ldots < \tau_{k-1} < 1$ and positive numbers w_s with $\sum_{s=0}^{k-1} w_s = 1$ we have

$$\int_R g(t) \, dt \approx Q(g) = \sum_{m=-\infty}^{\infty} \sum_{s=0}^{k-1} w_s g(m + \tau_s) .$$

Further, given $n \geq 1$ we define the n th refinement of Q by

$$Q_n(g) = \sum_{m=-\infty}^{\infty} \sum_{s=0}^{k-1} w_s g(\frac{m+\tau_s}{n})\frac{1}{n} . \tag{1}$$

As we shall see later on, an immediate application of Q_n to the singular integral S_R does sometimes not produce convergent approximation methods. So we first regularize the singular integral by

$$(S_R u)(t) = \frac{1}{\pi i} \int_R \frac{u(x)}{x-t} \, dx = \frac{1}{\pi i} \int_R \frac{u(x) - u(t)}{x-t} \, dx . \tag{2}$$

Applying now Q_n to the right side of (2) yields

$$(S_R u)(t) \approx \frac{1}{\pi i} \sum_{m=-\infty}^{\infty} \sum_{s=0}^{k-1} w_s \frac{u(\frac{m+\tau_s}{n}) - u(t)}{\frac{m+\tau_s}{n} - t} \frac{1}{n} . \tag{3}$$

For the approximate solution of the equation

$$(aI + bS_R)u = f , \quad (a, b \in \mathbf{C}, f \in R_R^p(\alpha))$$

by the quadrature method one first replaces the singular integral $S_R u$ by (3) and then evaluates both sides of the resulting equation at the points $(l + \epsilon_r)/n$ where

$l \in \mathbf{Z}$ and the ϵ_r are real numbers satisfying $0 \le \epsilon_0 < \epsilon_1 < \ldots < \epsilon_{k-1} < 1$ and $\{\epsilon_0, \ldots, \epsilon_{k-1}\} \cap \{\tau_0, \ldots, \tau_{k-1}\} = \emptyset$. Explicitly written, the approximation equations are

$$au(\frac{l+\epsilon_r}{n}) + \frac{b}{\pi i} \sum_{m=-\infty}^{\infty} \sum_{s=0}^{k-1} w_s \frac{u(\frac{m+\tau_s}{n}) - u(\frac{l+\epsilon_r}{n})}{\frac{m+\tau_s}{n} - \frac{l+\epsilon_r}{n}} \frac{1}{n} = f(\frac{l+\epsilon_r}{n}), \; l \in \mathbf{Z}. \qquad (4)$$

Now abbreviate $u(\frac{l+\epsilon_r}{n})$ by u_{rln}, substitute $u(\frac{m+\tau_s}{n})$ by $u(\frac{m+\epsilon_s}{n}) = u_{smn}$, and write

$$\sum_{m=-\infty}^{\infty} \sum_{s=0}^{k-1} w_s \frac{u_{smn} - u_{rln}}{m + \tau_s - l - \epsilon_r} =$$
$$= \sum_{m=-\infty}^{\infty} \sum_{s=0}^{k-1} w_s \frac{u_{smn}}{m - l + \tau_s - \epsilon_r} - u_{rln} \sum_{s=0}^{k-1} w_s \sum_{m=-\infty}^{\infty} \frac{1}{m - l + \tau_s - \epsilon_r}.$$

Then the well known identity

$$\cot(\pi x) = \frac{1}{\pi}(\frac{1}{x} + \sum_{j=1}^{\infty}(\frac{1}{x - j} + \frac{1}{x + j})) \qquad (5)$$

yields the following reformulation of (4):

$$(a + ib \sum_{j=0}^{k-1} w_j \cot \pi(\tau_j - \epsilon_r)) u_{rln} +$$
$$+ \frac{b}{\pi i} \sum_{j=0}^{k-1} w_j \sum_{m=-\infty}^{\infty} \frac{u_{jmn}}{m - l + \tau_j - \epsilon_r} = f(\frac{l+\epsilon_r}{n}). \qquad (6)$$

On defining sequences F_r and U_s by $F_r = (f(\frac{l+\epsilon_r}{n}))_{l \in \mathbf{Z}}$ and $U_s = (u_{smn})_{m \in \mathbf{Z}}$ we can write (6) as the system

$$(A_{rs})_{r,s=0}^{k-1}(U_s)_{s=0}^{k-1} = (F_r)_{r=0}^{k-1} \qquad (7)$$

with the system matrices

$$A_{rs} = \begin{cases} \left(\dfrac{b}{\pi i} \dfrac{w_s}{m - l + \tau_s - \epsilon_r}\right)_{l,m \in \mathbf{Z}} & \text{if } r \neq s \\[2em] \left(\dfrac{b}{\pi i} \dfrac{w_r}{m - l + \tau_r - \epsilon_r} + \delta_{ml}(a + ib \sum_{j=0}^{k-1} w_j \cot \pi(\tau_j - \epsilon_r))\right)_{l,m \in \mathbf{Z}} & \text{if } r = s. \end{cases}$$

Let $\nu \in (-1, 1) \setminus \{0\}$, and consider the functions

$$a^{(\nu)} : \mathbf{T} \to \mathbf{C}, \quad e^{2\pi i x} \mapsto 2 \frac{e^{-\pi i \nu x} \sin(\pi \nu x)}{e^{-\pi i \nu} \sin(\pi \nu)} - 1, \quad x \in [0, 1).$$

These functions are continuous on $\mathbf{T} \setminus \{1\}$ and have finite limits $a^{(\nu)}(1 \pm 0) = \mp 1$. Hence, they represent piecewise continuous multipliers on $l_Z^p(\alpha)$. Since, moreover, the k th Fourier coefficient of $a^{(\nu)}$ equals

$$\frac{1}{\pi i} \frac{1}{-k - \nu} - i\delta_{k0} \cot \nu \pi \tag{8}$$

we arrive at the following proposition.

Proposition 2.22 *The system matrices (7) of the quadrature method (4) represent bounded operators on $l_Z^p(\alpha)$, and these operators belong to the Toeplitz algebra* $\mathrm{alg}\,(T^0(PC_{p,\alpha}), P)$. *In particular,*

$$(A_{rs})_{r,s=0}^{k-1} = aI + bT^0(\sigma^{Q_1}) \quad \text{with} \quad \sigma^{Q_1} = (\sigma_{rs}^{Q_1})_{r,s=0}^{k-1} \quad \text{and}$$

$$
\sigma_{rs}^{Q_1} =
\begin{cases}
w_s(a^{(\epsilon_r - \tau_s)} + i \cot \pi(\epsilon_r - \tau_s)) & \text{if} \quad r \neq s \\[2mm]
w_r a^{(\epsilon_r - \tau_r)} - i \displaystyle\sum_{\substack{j=0 \\ j \neq r}}^{k-1} w_j \cot \pi(\epsilon_r - \tau_j) & \text{if} \quad r = s
\end{cases}
$$

Moreover,

$$\sum_{s=0}^{k-1} \sigma_{rs}^{Q_1}(1 \pm 0) = \mp 1 \,,$$

and

$$\sigma_{r_1 s} - \sigma_{r_2 s} \in C_{p,\alpha} \,.$$

Again we see, that a row sum and a column difference condition are satisfied.

Observe that the approximation operators A_{rr} take an admittedly simple form if $\sum_{j=0}^{k-1} w_j \cot \pi(\tau_j - \epsilon_r) = 0$. In case $k = 1$ this condition is equivalent to $|\tau_0 - \epsilon_0| = 1/2$, and the corresponding quadrature method is called *the method of discrete whirls*. Let us emphasize that the quadrature method (7) can also be viewed as a spline approximation method. For we choose a k-elementic prebasis K in such a way that (K, G) with $G = \{\delta_{\epsilon_0}, \ldots, \delta_{\epsilon_{k-1}}\}$ is a canonical pair (for simplicity one could take

$$K = \{\chi_{[0, \frac{\epsilon_0 + \epsilon_1}{2})}, \chi_{[\frac{\epsilon_0 + \epsilon_1}{2}, \frac{\epsilon_1 + \epsilon_2}{2})}, \cdots \chi_{[\frac{\epsilon_{k-2} + \epsilon_{k-1}}{2}, 1)}\} \text{ if } \epsilon_0 > 0$$

in which case $\lambda^{K,G} = I$), and we associate with (7) the spline approximation operators

$$E_n^K (A_{rs})_{r,s=0}^{k-1} E_{-n}^K : S_n^K \to S_n^K \,.$$

Clearly, the quadrature system (7) and the spline system

$$E_n^K T^0 (\lambda^{K,G})^{-1} (A_{rs}) E_{-n}^K u^{(n)} = L_n^{K,G} f \tag{9}$$

are simultaneously solvable or not. Moreover we shall employ this correspondence to *define* the convergence of quadrature methods: The method (6) converges if the associated spline approximation method (9) converges. One can show that the so-defined convergence is independent of the choice of K.

2.11.7   Another quadrature method

For another method we return to the system 2.11.6(4) and suppose for a moment the function u to be differentiable. If we let τ_r go to ϵ_r for $r = 0, \ldots, k-1$, then 2.11.6(4) goes over into

$$
au(\frac{l+\epsilon_r}{n})+
$$
$$
+\frac{b}{\pi i}\left(\sum_{(m,s)} \frac{u(\frac{m+\epsilon_s}{n}) - u(\frac{l+\epsilon_r}{n})}{\frac{m+\epsilon_s}{n} - \frac{l+\epsilon_r}{n}}\frac{w_s}{n} + \frac{w_r}{n}u'(\frac{l+\epsilon_r}{n}) \right) = f(\frac{l+\epsilon_r}{n}) ,
$$

where the summation is taken over all pairs $(m,s) \in \mathbf{Z} \times \{0, \ldots, k-1\}$ with $(m,s) \neq (l,r)$. Neglecting the small terms $\frac{w_r}{n}u'(\frac{l+\epsilon_r}{n})$ and abbreviating $u(\frac{l+\epsilon_r}{n})$ by u_{rln}, we arrive at

$$
au_{rln} + \frac{b}{\pi i} \sum_{\substack{(m,s)\in Z\times\{0,\ldots,k-1\}\\(m,s)\neq(l,r)}} w_s\frac{u_{smn} - u_{rln}}{m - l + \epsilon_s - \epsilon_r} = f(\frac{l+\epsilon_r}{n}) . \tag{1}
$$

Using besides 2.11.6(5) the identity $\sum_{j=1}^{\infty}(\frac{1}{j} + \frac{1}{-j}) = 0$ and writing (1) as a system

$$
(A_{rs})_{r,s=0}^{k-1} (U_s)_{s=0}^{k-1} = (F_r)_{r=0}^{k-1} \tag{2}
$$

we get in analogy to Proposition 2.22 :

Proposition 2.23 *The system matrices (2) of the quadrature method (1) represent bounded operators on $l_Z^p(\alpha)$, and these operators belong to the Toeplitz algebra* $\mathrm{alg}\,(T^0(PC_{p,\alpha}), P)$. *In particular,*

$$
(A_{rs})_{r,s=0}^{k-1} = aI + bT^0(\sigma^{Q_2}) \quad with \quad \sigma^{Q_2} = (\sigma_{rs}^{Q_2})_{r,s=0}^{k-1}
$$

and

$$
\sigma_{rs}^{Q_2} = \begin{cases} w_s(a^{(\epsilon_r - \epsilon_s)} + i\cot\pi(\epsilon_r - \epsilon_s)) & if \quad r \neq s \\ w_r a^{(0)} - i\sum_{j\neq r}^{\infty} w_j\cot\pi(\epsilon_r - \epsilon_j) & if \quad r = s , \end{cases}
$$

where $a^{(0)}(e^{2\pi i x}) = 2x - 1$, $x \in [0,1)$. *Moreover,*

$$
\sum_{s=0}^{k-1}\sigma_{rs}^{Q_2}(1\pm 0) = \mp 1 , \qquad \cdot
$$

and $\sigma_{r_1 s} - \sigma_{r_2 s} \in C_{p,\alpha}$.

2.11.8 Quadrature methods for operators in $\Sigma^p(\alpha)$

Now we turn to quadrature methods for the operators $a\chi_{R^+}I + bS_{R^+} + M^0(c)$ ($c \in \mathcal{N}$) which span the algebra $\Sigma^p(\alpha)$. Writing $M^0(c)$ in the form

$$(M^0(c)u)(t) = \int_0^\infty k(\frac{t}{s})u(s)\,\frac{ds}{s}\,, \quad k = M^{-1}c\,,$$

the analogues of the quadrature methods 2.11.6(6) and 2.11.7(1) read as follows:

$$(a+bi\sum_{s=0}^{k-1} w_s \cot \pi(\tau_s - \epsilon_r))u_{rln} \;+\; \frac{b}{\pi i}\sum_{s=0}^{k-1} w_s \sum_{m=0}^{\infty} \frac{u_{smn}}{m-l+\tau_s - \epsilon_r} +$$

$$+\sum_{s=0}^{k-1} w_s \sum_{m=0}^{\infty} k(\frac{l+\epsilon_r}{m+\tau_s})\,\frac{u_{smn}}{m+\tau_s} \;=\; f(\frac{l+\epsilon_r}{n})\,, \quad l \in \mathbf{Z}^+ \tag{1}$$

and

$$au_{rln} + \frac{b}{\pi i}\sum_{\substack{(m,s)\in Z^+\times\{0,\dots,k-1\} \\ (m,s)\neq(l,r)}} w_s \frac{u_{smn} - u_{rln}}{m-l+\epsilon_s - \epsilon_r} +$$

$$+\sum_{s=0}^{k-1} w_s \sum_{m=0}^{\infty} k(\frac{l+\epsilon_r}{m+\epsilon_s})\,\frac{u_{smn}}{m+\epsilon_s} = f(\frac{l+\epsilon_r}{n})\,, \quad l \in \mathbf{Z}^+\,. \tag{2}$$

Let $(A_{rs}^{(1)})_{r,s=0}^{k-1}$ and $(A_{rs}^{(2)})_{r,s=0}^{k-1}$ stand for the system matrices of (1) and (2), respectively. Then an immediate application of Lemma 2.1 and Proposition 2.11(a) yields

Proposition 2.24 *The system matrices $A_{rs}^{(1)}, A_{rs}^{(2)}$ represent bounded operators on $l_Z^p(\alpha)$, and these operators belong to the Toeplitz algebra $\mathrm{alg}\,(T^0(PC_{p,\alpha}), P)$ (compare Section 2.3.2). In particular,*

$$(A_{rs}^{(i)})_{r,s=0}^{k-1} = aI + bT(\sigma^{Q_i}) + (w_s G(c))_{r,s=0}^{k-1} + K \quad (i = 1, 2)\,,$$

where K is compact and the functions σ^{Q_i} are defined in the preceding propositions.

The operators A_{rs} again satisfy conditions at the point 1 which can most advantageously be expressed in terms of local cosets: There is an operator A in the Toeplitz algebra $\mathrm{alg}\,(T^0(PC_{p,\alpha}), P)$ such that

$$\sum_s \Phi_1^\pi(A_{rs})\alpha_s^F = \alpha_r^F \Phi_1^\pi(A)\,,$$

and, moreover,

$$\Phi_1^\pi(A_{r_1 s}) - \Phi_1^\pi(A_{r_2 s}) \in \Phi_1^\pi(\mathrm{alg}\,(T^0(C_{p,\alpha}), P))\,.$$

For all approximation methods considered up to now (viz. the projection methods such as Galerkin, collocation or qualocation methods as well as the quadrature methods), the corresponding system matrices belong to the Toeplitz algebra alg $(T^0(PC_{p,\alpha}), P)$, and they are generated by piecewise continuous multipliers having at $1 \in \mathbf{T}$ their only discontinuity. Next we present some methods which lead to multipliers with two or more discontinuities. For simplicity we restrict ourselves to modifications of the rectangle rule instead of the more involved quadrature method 2.11.6(1).

2.11.9   Chained quadrature methods

Prössdorf and Rathsfeld ([PR 1]) proposed the following method for solving the equation $(aI + bS_R)u = f$:

$$au_{ln} + \frac{2b}{\pi i} \sum_{\substack{m \in \mathbf{Z} \\ m \equiv l+1(2)}} \frac{u_{mn}}{m-l} = f(l/n)\,, \ l \in \mathbf{Z} \tag{1}$$

which could be viewed as a chained quadrature method. The analogue of (1) for the operator $a\chi_{R^+} I + bS_{R^+} + M^0(c)$ is

$$au_{ln} + \frac{2b}{\pi i} \sum_{\substack{m=0 \\ m \equiv l+1(2)}}^{\infty} \frac{u_{mn}}{m-l} + 2 \sum_{\substack{m=0 \\ m \equiv l+1(2)}}^{\infty} k(l/m)\frac{u_{mn}}{m} = f(l/n)\,, \ l \in \mathbf{Z}^+. \tag{2}$$

Introduce the function $\sigma^{Q_3} : \mathbf{T} \to \mathbf{C}$ by

$$\sigma^{Q_3}(e^{2\pi i x}) = \begin{cases} -1 & \text{if} \ \ 0 \le x < 1/2 \\ 1 & \text{if} \ \ 1/2 \le x < 1\,, \end{cases}$$

and let $A^{(1)}$ and $A^{(2)}$ stand for the system matrices of (1) and (2), respectively.

Proposition 2.25 *The system matrices $A^{(1)}$ and $A^{(2)}$ represent bounded operators on $l_Z^p(\alpha)$, and these operators belong to* alg $(T^0(PC_{p,\alpha}), P)$ *and* alg $T(PC_{p,\alpha})$, *respectively. In particular,*

$$A^{(1)} = T^0(a + b\sigma^{Q_3})\,,$$

$$A^{(2)} = T(a + b\sigma^{Q_3}) + G(c) - Y_{-1}G(c)Y_{-1}^{-1} + K\,,$$

where $Y_{-1} : l_Z^p(\alpha) \to l_Z^p(\alpha)\,, \quad (x_k) \mapsto ((-1)^k x_k)\,,$ *and K is compact.*

The proof is straightforward. Observe that not only the approximation of the singular integral part, bS_R, leads to multipliers with two discontinuities (the function σ^{Q_3}), but also the Mellin convolution $M^0(c)$ (the operator $G(c)$ corresponds with discontinuities at $1 \in \mathbf{T}$, and the operator $Y_{-1}G(c)Y_{-1}^{-1}$ with discontinuities at $-1 \in \mathbf{T}$, compare Sections 2.4.3 and 2.5.2).

To construct approximation methods for singular integral operators with arbitrarily many prescribed discontinuities one simply turns the tables by starting with a piecewise continuous multiplier, $\hat{\sigma}$ (but with $\hat{\sigma}(1 \pm 0) = \mp 1$ in order to guarantee convergence of the method, see 2.15), and then looking for an approximation method having $T^0(\hat{\sigma})$ as its system matrix. For example, let

$$\hat{\sigma}(e^{2\pi i x}) = \begin{cases} -1 & \text{if} \quad 0 \leq x < 1/4 \\ -1/3 & \text{if} \quad 1/4 \leq x < 1/2 \\ 1/3 & \text{if} \quad 1/2 \leq x < 3/4 \\ 1 & \text{if} \quad 3/4 \leq x < 1 \,. \end{cases}$$

The m th Fourier coefficient of $\hat{\sigma}$ is

$$\hat{\sigma}_m = \begin{cases} 0 & \text{if} \quad m \equiv 0(4) \\ \frac{4i}{3\pi m} & \text{if} \quad m \not\equiv 0(4) \,, \end{cases}$$

and this multiplier brings about the following system for the approximate solution of the singular integral equation $(aI + bS_R)u = f$:

$$a u_{ln} + \frac{4b}{3\pi i} \sum_{\substack{m \in Z \\ m \not\equiv l(4)}} \frac{u_{mn}}{m - l} = f(l/n) \quad (l \in \mathbf{Z}).$$

The formulation of the analogue for semi-axis operators with Mellin convolution is left as an exercise.

2.11.10   Quadrocation methods

Quadrocation methods appear as combinations of quadrature and projection methods, in particular of quadrature and qualocation methods. They are aimed to produce fully discretized and numerically well-behaved approximation methods. For their explanation we start with the quadrature rule 2.11.6(1) the application of which to the integral in the singular integral equation $(aI + bS_R)u = f$ yields the equation

$$a u(t) + \frac{b}{\pi i} \sum_{s=0}^{k-1} w_s \sum_{m \in Z} \frac{u_{smn}}{m + \tau_s - tn} = f(t) \,, \ t \in \mathbf{R} \,, \tag{1}$$

in case no regularization of the singular integral is made, and

$$(a + bi \sum_{s=0}^{k-1} w_s \cot \pi(\tau_s - nt)) u(t) +$$

$$+ \frac{b}{\pi i} \sum_{s=0}^{k-1} w_s \sum_{m \in Z} \frac{u_{smn}}{m + \tau_s - nt} = f(t) \,, \ t \in \mathbf{R} \,, \tag{2}$$

for the regularized singular integral (compare Section 2.11.6). In (1) and (2) we have written $u_{smn} = u(\frac{m+\tau_s}{n})$. There are two conceivable ways to introduce quadrocation methods. The first one is to choose a k-tuple G of premeasures and then to apply the operators P_n^G to both sides of (1) or (2) to get a discrete system. Just in this way we derived the quadrature method 2.11.6(7). Remember that the evaluation of (2) at the points $(l + \epsilon_r)/n$ is nothing else than application of P_n^G with $G = \{\delta_{\epsilon_0}, \dots, \delta_{\epsilon_{k-1}}\}$. For the second way we choose a canonical pair (F, G), put up the spline projection method

$$L_n^{F,G}(aI + bS_R)u_n = L_n^{F,G}f \quad , \quad u_n \in S_n^F , \tag{3}$$

and then we replace the operator $aI + bS_R$ in (3) by its quadrature based approximation (1) or (2).

Let us consider some examples. For simplicity we let $k = 1$; then (1) and (2) can be written as

$$au(t) + \frac{b}{\pi i} \sum_{m \in Z} \frac{u_{mn}}{m + \tau - tn} = f(t) \quad , \quad t \in \mathbf{R} , \tag{4}$$

and

$$(a + bi \cot \pi(\tau - nt))u(t) + \frac{b}{\pi i} \sum_{m \in Z} \frac{u_{mn}}{m + \tau - tn} = f(t) \quad , \quad t \in \mathbf{R} , \tag{5}$$

respectively. For the first example we let $F = \{\phi\}$ be a prebasis which is normed by $\int_R \phi(x)\, dx = 1$, and we apply the Galerkin test operator P_n^F to (4) (One can show that a regularization is not necessary in this setting.). In the resulting system

$$a \int_R u(\frac{p+t}{n})\bar{\phi}(t)\, dt +$$

$$+ \frac{b}{\pi i} \int_R \sum_{m \in Z} \frac{u_{mn}}{m + \tau - p - t}\bar{\phi}(t)\, dt = \int_R f(\frac{t+p}{n})\bar{\phi}(t)\, dt , \quad p \in \mathbf{Z} ,$$

we identify

$$\int_R u(\frac{p+t}{n})\bar{\phi}(t)\, dt \approx u(\frac{p+\xi}{n}) \int_R \bar{\phi}(t)\, dt \approx u_{pn}$$

to obtain

$$au_{pn} + \frac{b}{\pi i} \sum_{m \in Z} u_{mn} \int_R \frac{\bar{\phi}(t)}{m + \tau - p - t}\, dt = \int_R f(\frac{t+p}{n})\bar{\phi}(t)\, dt . \tag{6}$$

Proposition 2.26 *The system operator A of (6) is bounded on $l_1^p(\alpha)$. In particular,*

$$A = T^0(a + b\, \sigma^{F',G'})$$

with $F' = \{\tilde{\bar{\phi}}\}$, and $G' = \{\delta_{-\tau}\}$. Moreover,

$$(a + b\, \sigma^{F',G'})(1 \pm 0) = a \mp b .$$

(Recall that $\tilde{\bar{\phi}}(t) := \phi(-t)$.)

For a proof it is sufficient to compare (6) with the system obtained from the collocation method involved by (F', G').

As a second (and more important for applications) example we test (5) with the qualocation operator P_n^G where $G = \{\gamma\}$, $\gamma : f \mapsto Q(\bar{\phi}f)$, $\{\phi\}$ is a prebasis, and

$$Q(\bar{\phi}f) = \sum_{s=0}^{k-1} w_s \sum_{l \in Z} f(l + \epsilon_s) \bar{\phi}(l + \epsilon_s) \,,$$

with $\epsilon_s \neq \tau$ for all s (compare Sections 2.10.3 and 2.11.6). In the resulting system

$$\sum_{s=0}^{k-1} w_s \sum_{l \in Z} \bar{\phi}(l + \epsilon_s) \left((a + bi \cot \pi(\tau - \epsilon_s)) \, u((l + \epsilon_s + p)/n) + \right.$$

$$\left. + \frac{b}{\pi i} \sum_{m \in Z} \frac{u_{mn}}{m + \tau - p - l - \epsilon_s} \right) = \int_R f((t + p)/n)\gamma(t) \, dt$$

we identify $u(\frac{l+\epsilon_s+p}{n}) \approx u_{l+p,n}$ and substitute $l + p =: l$ to find

$$\sum_{s=0}^{k-1} w_s \sum_{l \in Z} \tilde{\phi}(p - l - \epsilon_s) \left((a + bi \cot \pi(\tau - \epsilon_s)) \, u_{ln} \right.\, -$$

$$- \frac{b}{\pi i} \sum_{m \in Z} \frac{u_{mn}}{l - m + \epsilon_s - \tau} = \int_R f(\frac{t+p}{n})\gamma(t) \, dt \,.$$

Observing that $\tilde{\phi}(p - l - \epsilon_s)$ is the $p - l$ th Fourier coefficient of $\lambda^{\{\tilde{\phi}\},\{\delta_{-\epsilon_s}\}}$ and that the k th Fourier coefficient of the function $a^{(\epsilon_s - \tau)}$ introduced in 2.11.6 is given by 2.11.6(8), and abbreviating $(u_{mn})_{m \in Z}$ to $U^{(n)}$ we get

$$\sum_{s=0}^{k-1} w_s T^0(\lambda^{\{\tilde{\phi}\},\{\delta_{-\epsilon_s}\}}) \, T^0(a + b \, a^{(\epsilon_s - \tau)}) U^{(n)} = P_n^{\{\gamma\}} f \tag{7}$$

which corresponds to our first way to combine quadrature and qualocation methods.

For the second way we choose the spline space $S_n^{\{\phi_1\}}$ with $\{\phi_1\}$ being any prebasis for which $(\{\phi_1\}, \{\gamma\})$ is a canonical pair. Now we consider the qualocation method corresponding to the projections $L_n^{\{\phi_1\},\{\gamma\}}$, and then we replace the operator $aI + bS_R$ by the quadrature based approximation (5). A straightforward computation yields the system

$$E_n^{\{\phi_1\}} T^0(\lambda^{\{\phi_1\},\{\gamma\}})^{-1} \left(\sum_{s=0}^{k-1} w_s T^0(\lambda^{\{\tilde{\phi}\},\{\delta_{-\epsilon_s}\}}) T^0(a + b \, a^{(\epsilon_s - \tau)}) \right) \times$$

$$\times \, T^0(\lambda^{\{\phi_1\},\{\delta_\tau\}}) E_{-n}^{\{\phi_1\}} u^{(n)} = L_n^{\{\phi_1\},\{\gamma\}} f \qquad (u^{(n)} \in S_n^{\{\phi_1\}}) \,, \tag{8}$$

which corresponds to the second way of combining qualocation and quadrature methods.

Write (7) and (8) as $A_n^{(1)} U^{(n)} = P_n^{\{\gamma\}} f$ and $(E_n^{\{\phi_1\}} A_n^{(2)} E_{-n}^{\{\phi_1\}}) u^{(n)} = L_n^{\{\phi_1\},\{\gamma\}} f$, respectively.

Proposition 2.27 *The operators $A_n^{(1)}$, $A_n^{(2)}$ are independent of n and bounded on $l_1^p(\alpha)$; they belong to the Toeplitz algebra $\mathrm{alg}\,(T^0(PC_{p,\alpha}), P)$, and $A_n^{(i)} = T^0(g^{(i)})$ $(i = 1, 2)$ with the piecewise continuous multipliers*

$$g^{(1)} = \sum_{s=0}^{k-1} w_s \lambda^{\{\tilde\phi\},\{\delta_{-\epsilon_s}\}} (a + b\,a^{(\epsilon_s - \tau)}) \quad and$$

$$g^{(2)} = \sum_{s=0}^{k-1} w_s \lambda^{\{\tilde\phi\},\{\delta_{-\epsilon_s}\}} (a + b\,a^{(\epsilon_s - \tau)}) \lambda^{\{\phi_1\},\{\delta_\tau\}} \Big/ \lambda^{\{\phi_1\},\{\gamma\}} .$$

Moreover,

$$g^{(i)}(1 \pm 0) = a \mp b \quad (i = 1, 2).$$

Let us emphasize that, if $(\{\phi_1\}, \{\delta_\tau\})$ is a canonical pair, then the functions $g^{(1)}$ and $g^{(2)}$ are simultaneously invertible or not, and so the methods $(A_n^{(1)})$ and $(A_n^{(2)})$ are actually equivalent. One example for this can be given by choosing $\phi = \phi_1 = \chi_{[0,1)}$. Then all pairs $(\{\tilde\phi\}, \{\delta_{-\epsilon_s}\})$, $(\{\phi_1\}, \{\delta_\tau\})$ and $(\{\phi_1\}, \{\gamma\})$ are canonical and, moreover,

$$\lambda^{\{\tilde\phi\},\{\delta_{-\epsilon_s}\}} = \lambda^{\{\phi_1\},\{\delta_\tau\}} = \lambda^{\{\phi_1\},\{\gamma\}} = 1 .$$

Thus, in this setting, $g^{(1)} = g^{(2)} = a + b \sum_{s=0}^{k-1} w_s\, a^{(\epsilon_s - \tau)}$, and the two quadrocation methods coincide.

2.12   Proofs

2.12.1   Proof of Lemma 2.1

Youngs inequality (see [HR] (20.18) for the unweighted case, from which the weighted version can be easily derived)

$$\|k_1 * k_2\|_{L^p_{R+}(\alpha)} \leq \|k_1\|_{L^r_{R+}(1/p - 1/r + \alpha)} \|k_2\|_{L^s_{R+}(1/p - 1/s + \alpha)}$$

with $1/r + 1/s = 1 + 1/p$ involves that the class of all functions satisfying 2.4.1(i) forms an algebra. Further, the elementary inequality

$$|a + b|^v \leq \max(2^{v-1}, 1)\,(|a|^v + |b|^v), \quad a, b \in \mathbf{C}, v > 0, \tag{1}$$

applies to give that the class of all functions satisfying 2.4.1(ii) is a linear space.

Our next claim is the following: If the functions k and l satisfy conditions 2.4.1(i) and (ii) then the function $k * l$ also satisfies (ii). For brevity we write $\sum_m$, $\sum_n$, and $\int$ for $\sum_{m=0}^{\infty}$, $\sum_{n=0}^{\infty}$, $\int_0^{\infty}$, respectively. Then

$$\sum_m (m+1)^{\alpha p} \left(\sum_n (n+1)^{-\beta v} \left| \int k(\frac{t_{mn}}{s_{nm}x}) l(x) \frac{dx}{s_{nm}x} - \int k(\frac{m+1}{(n+1)x}) l(x) \frac{dx}{(n+1)x} \right|^v \right)^{p/v}$$

$$= \sum_m (m+1)^{\alpha p} \left(\sum_n (n+1)^{-\beta v} \times \right.$$

$$\times \left. \left| \int k(\frac{t_{mn}}{x}) l(\frac{x}{s_{nm}}) \frac{dx}{s_{nm}x} - \int k(\frac{m+1}{x}) l(\frac{x}{n+1}) \frac{dx}{(n+1)x} \right|^v \right)^{p/v}$$

$$= \sum_m (m+1)^{\alpha p} \left(\sum_n (n+1)^{-\beta v} \left| \int (k(\frac{t_{mn}}{x})\frac{1}{x} - k(\frac{m+1}{x})\frac{1}{x}) l(\frac{x}{s_{nm}}) \frac{dx}{s_{nm}} + \right. \right.$$

$$\left. \left. + \int k(\frac{m+1}{x})\frac{1}{x} (l(\frac{x}{s_{nm}})\frac{1}{s_{nm}} - l(\frac{x}{n+1})\frac{1}{n+1}) \, dx \right|^v \right)^{p/v}$$

which is by (1) less than

$$2^{p-p/v} \sum_m (m+1)^{\alpha p} \left(\sum_n (n+1)^{-\beta v} \left| \int (k(\frac{t_{mn}}{x})\frac{1}{x} - k(\frac{m+1}{x})\frac{1}{x}) l(\frac{x}{s_{nm}}) \frac{dx}{s_{nm}} \right|^v + \right.$$

$$\left. + \sum_n (n+1)^{-\beta v} \left| \int k(\frac{m+1}{x})\frac{1}{x} (l(\frac{x}{s_{nm}})\frac{1}{s_{nm}} - l(\frac{x}{n+1})\frac{1}{n+1}) \, dx \right|^v \right)^{p/v}$$

and, again by (1), less than

$$C \sum_m (m+1)^{\alpha p} \left(\sum_n (n+1)^{-\beta v} \left| \int (k(\frac{t_{mn}}{x})\frac{1}{x} - k(\frac{m+1}{x})\frac{1}{x}) \, l(\frac{x}{s_{nm}}) \frac{dx}{s_{nm}} \right|^v \right)^{p/v}$$

$$+ \quad C \sum_m (m+1)^{\alpha p} \left(\sum_n (n+1)^{-\beta v} \times \right.$$

$$\times \left. \left| \int k(\frac{m+1}{x})\frac{1}{x} (l(\frac{x}{s_{nm}})\frac{1}{s_{nm}} - l(\frac{x}{n+1})\frac{1}{n+1}) dx \right|^v \right)^{p/v}$$

with $C = 2^{p-p/v} \max\left(2^{p/v-1}, 1\right)$. Finally, by the Cauchy-Schwarz inequality, this expression is less than

$$C \sum_m (m+1)^{\alpha p} \left(\sum_n (n+1)^{-\beta v} \left(\int \left| k(\frac{t_{mn}}{x})\frac{1}{x} - k(\frac{m+1}{x})\frac{1}{x} \right|^r dx \right)^{v/r} \times \right.$$

$$\times \left. \left(\int \left| l(\frac{x}{s_{nm}})\frac{1}{s_{nm}} \right|^s dx \right)^{v/s} \right)^{p/v} +$$

$$+ \ C \sum_m (m+1)^{\alpha p} \left(\sum_n (n+1)^{-\beta v} \left(\int \left| k(\frac{m+1}{x}) \frac{1}{x} \right|^s dx \right)^{v/s} \times \right.$$

$$\left. \times \left(\int \left| l(\frac{x}{s_{nm}}) \frac{1}{s_{nm}} - l(\frac{x}{n+1}) \frac{1}{n+1} \right|^r dx \right)^{v/r} \right)^{p/v} \tag{2}$$

with *arbitrary* numbers $r, s > 1$ satisfying $1/r + 1/s = 1$. Now we consider the function $t \mapsto \int |k(\frac{t}{x}) \frac{1}{x} - k(\frac{m+1}{x}) \frac{1}{x}|^r dx$ on $[m, m+d]$. By the continuity of k, it attains its maximum, say at $t_m \in [m, m+d]$. Thus,

$$\int \left| k(\frac{t_{mn}}{x}) \frac{1}{x} - k(\frac{m+1}{x}) \frac{1}{x} \right|^r dx \le \int \left| k(\frac{t_m}{x}) \frac{1}{x} - k(\frac{m+1}{x}) \frac{1}{x} \right|^r dx \ ,$$

and, analogously, there are numbers s_n and x_n in $[n, n+d]$ such that

$$\int \left| l(\frac{x}{s_{nm}}) \frac{1}{s_{nm}} \right|^s dx \le \int \left| l(\frac{x}{s_n}) \frac{1}{s_n} \right|^s dx$$

and

$$\int \left| l(\frac{x}{s_{nm}}) \frac{1}{s_{nm}} - l(\frac{x}{n+1}) \frac{1}{n+1} \right|^r dx \le \int \left| l(\frac{x}{x_n}) \frac{1}{x_n} - l(\frac{x}{n+1}) \frac{1}{n+1} \right|^r dx \ .$$

Consequently, (2) is less than

$$C \left[\sum_m (m+1)^{\alpha p} \left(\int \left| k(\frac{t_m}{x}) \frac{1}{x} - k(\frac{m+1}{x}) \frac{1}{x} \right|^r dx \right)^{p/r} \right] \times$$

$$\times \left[\sum_n (n+1)^{-\beta v} \left(\int \left| l(\frac{x}{s_n}) \frac{1}{s_n} \right|^s dx \right)^{v/s} \right]^{p/v} +$$

$$+ \ C \left[\sum_m (m+1)^{\alpha p} \left(\int \left| k(\frac{m+1}{x}) \frac{1}{x} \right|^s dx \right)^{p/s} \right] \times$$

$$\times \left[\sum_n (n+1)^{-\beta v} \left(\int \left| l(\frac{x}{x_n}) \frac{1}{x_n} - l(\frac{x}{n+1}) \frac{1}{n+1} \right|^r dx \right)^{v/r} \right]^{p/v} . \tag{3}$$

We are going to show that each factor in (3) which is written in brackets is bounded. By the mean value theorem, the first factor is equal to

$$\sum_m (m+1)^{\alpha p} \left(\sum_n \int_n^{n+1} \left| k(\frac{t_m}{x}) \frac{1}{x} - k(\frac{m+1}{x}) \frac{1}{x} \right|^r dx \right)^{p/r} =$$

$$= \sum_m (m+1)^{\alpha p} \left(\sum_n \left| k(\frac{t_m}{x_{nm}}) \frac{1}{x_{nm}} - k(\frac{m+1}{x_{nm}}) \frac{1}{x_{nm}} \right|^r \right)^{p/r}$$

with certain numbers $x_{nm} \in (n, n+1)$, and this expression is bounded due to our assumption 2.4.1(ii) for all $p, r > 1$ and for all α with $0 < 1/p + \alpha < 1$.

Next we consider the fourth factor. Again by the mean value theorem, it equals

$$\sum_n (n+1)^{-\beta v} \left(\sum_m \left| l(\frac{x_{mn}}{x_n}) \frac{1}{x_n} - l(\frac{x_{mn}}{n+1}) \frac{1}{n+1} \right|^r \right)^{v/r} \tag{4}$$

with numbers $x_{mn} \in (m, m+1)$. In case $v/r \leq 1$ the elementary inequality

$$(a+b)^{v/r} \leq a^{v/r} + b^{v/r} \quad , \quad a, b \in \mathbf{R}^+ \tag{5}$$

yields that (4) is less than

$$\sum_n (n+1)^{-\beta v} \sum_m \left| l(\frac{x_{mn}}{x_n}) \frac{1}{x_n} - l(\frac{x_{mn}}{n+1}) \frac{1}{n+1} \right|^v$$

$$= \sum_m \sum_n (n+1)^{-\beta v} \left| l(\frac{x_{mn}}{x_n}) \frac{1}{x_n} - l(\frac{x_{mn}}{n+1}) \frac{1}{n+1} \right|^v ,$$

and this is bounded by assumption 2.4.1(ii) (replace the p, v, α, β in 2.4.1(ii) by $v, v, 0, \beta$, respectively, and take into account that $\mathcal{M}$ is a linear space). Now let $v/r > 1$. Then $r/v < 1$, and applying (5) to the r/v th power of (4) we find

$$\left(\sum_n (n+1)^{-\beta v} \left(\sum_m \left| l(\frac{x_{mn}}{x_n}) \frac{1}{x_n} - l(\frac{x_{mn}}{n+1}) \frac{1}{n+1} \right|^r \right)^{v/r} \right)^{r/v}$$

$$\leq \sum_m \sum_n (n+1)^{-\beta r} \left| l(\frac{x_{mn}}{x_n}) \frac{1}{x_n} - l(\frac{x_{mn}}{n+1}) \frac{1}{n+1} \right|^r ,$$

which is again bounded for all $r > 1$ and $0 < 1/r - \beta < 1$.

Further, substituting $x/s_n = y$ in the second factor of (3) this factor becomes equal to

$$\sum_n (n+1)^{-\beta v} (\frac{1}{s_n})^{\frac{s-1}{s} v} \left(\int |l(y)|^s dy \right)^{v/s} .$$

The integral is bounded by 2.4.1(i) for all v and s and, since $s_n \in [n, n+d]$ and $\frac{s-1}{s} = \frac{1}{r} > 0$, the series is less than

$$(\frac{1}{s_0})^{v/r} + \sum_{n=1}^{\infty} (n+1)^{-\beta v} n^{-v/r} .$$

Obviously, this series converges if and only if $v/r + \beta v > 1$ or, equally, if $1/r > 1/v - \beta$. Thus, for r small enough (remember that r is freely choosable) the second factor is also bounded.

Finally, if we substitute $(m+1)/x = y$ in the third factor of (3) this factor becomes equal to

$$\sum_m (m+1)^{\alpha p + p/s - p}\left(\int |k(y)|^s y^{s-2}\,dy\right)^{p/s}. \tag{6}$$

Since $1/s + (1 - 2/s) = 1 - 1/s \in (0,1)$ we have by assumption 2.4.1(i) that $k \in L^s_{R_+}(1 - 2/s)$; thus, the integral in (6) is bounded for every $s > 1$. The series in (6) converges whenever $\alpha p + p(1/s - 1) = \alpha p - p/r < -1$, that is, whenever $1/r > 1/p + \alpha$.

Summary: if $1 < r < \min\{(1/p+\alpha)^{-1}, (1/v-\beta)^{-1}\}$ then (3) is bounded. Hence, $\mathcal{M}$ is an algebra.

It remains to show that $M^{-1}\mathcal{N}$ is a subalgebra of $\mathcal{M}$, that is, each function k satisfying together with its Mellin transform $Mk = b$ the conditions 2.1.4(1) and (2) should be subject to conditions 2.4.1(i) and (ii).

For $p > 1$ and $0 < 1/p + \alpha < 1$ we get from 2.1.4(2):

$$\int_0^\infty |k(x)|^p x^{\alpha p}\,dx \le$$

$$\le \int_0^1 |P_1(|\ln x|)|^p \left(\frac{x^\alpha}{1+x}\right)^p dx + \int_1^\infty |P_1(|\ln x|)|^p \left(\frac{x^\alpha}{1+x}\right)^p dx.$$

Since $1/p + \alpha > 0$ one can find an $\epsilon_1 > 0$ such that $1/p + \alpha > \epsilon_1/p$ and an $c_1 > 0$ such that $|P_1(|\ln x|)| \le c_1\,x^{-\epsilon_1/p}$ for all $x \in (0,1]$, and since $1 > 1/p + \alpha$ there is an $\epsilon_2 > 0$ such that $1/p + \alpha < 1 - \epsilon_2/p$ and an $c_2 > 0$ such that $|P_1(\ln x)| \le c_2 x^{\epsilon_2/p}$ for all $x > 1$. Then,

$$\int_0^\infty |k(x)|^p x^{\alpha p}\,dx \le c_1 \int_0^1 x^{\alpha p - \epsilon_1}\,dx + c_2 \int_1^\infty x^{\alpha p - p + \epsilon_2}\,dx < \infty.$$

Hence, k is subject to 2.4.1(i). We are going to show that k satisfies 2.4.1(ii), too. Obviously, the expression in 2.4.1(ii) can be estimated by a multiple of

$$\sum_m (m+1)^{\alpha p}\left(\sum_n (n+1)^{-\beta v}\left|k(\frac{m+1}{n+1})(\frac{1}{s_{nm}} - \frac{1}{n+1})\right|^v\right)^{p/v} +$$

$$+ \sum_m (m+1)^{\alpha p}\left(\sum_n (n+1)^{-\beta v}\left|\frac{1}{s_{nm}}\left(k(\frac{t_{mn}}{s_{nm}}) - k(\frac{m+1}{n+1})\right)\right|^v\right)^{p/v}. \tag{7}$$

The inequality

$$\left|k(\frac{m+1}{n+1})(\frac{1}{s_{nm}} - \frac{1}{n+1})\right| \le c_3 \frac{P_1(|\ln\frac{m+1}{n+1}|)}{1 + \frac{m+1}{n+1}}\frac{1}{(n+1)^2}$$

$$\le c_\epsilon (m+1)^\epsilon (n+1)^\epsilon \frac{1}{(m+n+2)(n+1)}$$

for all $\epsilon > 0$ with a constant c_ϵ depending on ϵ shows that the first item in (7) is less than

$$\sum_m \left(\sum_n \frac{(m+1)^{\alpha v + \epsilon v}}{(n+1)^{v + \beta v - \epsilon v}(m+n+2)^v} \right)^{p/v}$$

which is finite if ϵ is small enough. Assuming without loss of generality that k is real valued (otherwise decompose k into its real and imaginary part) we obtain for the second term in (7)

$$\frac{1}{s_{nm}} \left(k(\frac{t_{mn}}{s_{nm}}) - k(\frac{m+1}{n+1}) \right) \le \frac{c_4}{n+1} |k'(s_n)| \left| \frac{t_{mn}}{s_{nm}} - \frac{m+1}{n+1} \right|$$

$$\le \frac{c_5}{n+1} \frac{P_2(|\ln s_n|)}{(1+s_n)^2} \frac{m+n+2}{(n+1)^2} \le c_6 \frac{(m+1)^\epsilon (n+1)^\epsilon}{(n+1)(m+n+2)} \qquad (8)$$

with a certain number s_n between t_{mn}/s_{nm} and $(m+1)/(n+1)$. The estimate (8) gives the boundedness of the second item in (7). ∎

2.12.2 Proof of Theorem 2.5 and Proposition 2.16

The case $Y = L_R^p(\alpha)$ (Theorem 2.5)

Suppose without loss that $k = 1$, i.e. $F = \{\phi_0\}$. Since $\phi_{0ln}(t) = \phi_{0l1}(nt)$ and

$$\left(\int_R |g(nt)|^p |t|^{\alpha p} dt \right)^{1/p} = n^{-(1/p+\alpha)} \left(\int_R |g(t)|^p |t|^{\alpha p} dt \right)^{1/p}$$

for each function $g \in L_R^p(\alpha)$ we can limit ourselves to the case $n = 1$. Write for brevity $f := \phi_0$ and $f_l := \phi_{0l1}$, and suppose further that $\operatorname{supp} f \subseteq [-a, a]$ with a positive integer a. Given $X = (x_l)_{l \in Z} \in l_1^p(\alpha)$ we define sequences $(x_l^{(i)})_{l \in Z}$ ($i = 0, \ldots, 2a - 1$) by

$$x_l^{(i)} = \begin{cases} x_l & \text{if} \quad l \equiv i \operatorname{modulo} 2a \\ 0 & \text{if} \quad l \not\equiv i \operatorname{modulo} 2a . \end{cases}$$

Clearly, $(x_l) = \sum_{i=0}^{2a-1}(x_l^{(i)})$, and $(x_l^{(i)}) \in l_1^p(\alpha)$. We claim that the functions $X^{(i)} := \sum_{l \in Z} x_l^{(i)} f_l$ are in $L_R^p(\alpha)$ for all i, and that

$$\|X^{(i)}\|_{L_R^p(\alpha)} \le C \|(x_l^{(i)})\|_{l_1^p(\alpha)} .$$

Indeed,

$$\|X^{(i)}\|^p = \int_R |\sum_{l \in Z} x_l^{(i)} f_l(t)|^p |t|^{\alpha p} dt = \int_R |\sum_{l \in Z} x_{i+2al} f_{i+2al}(t)|^p |t|^{\alpha p} dt$$

and, since $\operatorname{supp} f_{i+2al} \cap \operatorname{supp} f_{i+2al'} = \emptyset$ whenever $l \neq l'$,

$$\|X^{(i)}\|^p = \sum_{l \in Z} \int_R |x_{i+2al} f_{i+2al}(t)|^p |t|^{\alpha p} dt = \sum_{l \in Z} |x_{i+2al}|^p \|f_{i+2al}\|^p_{L^p_R(\alpha)} . \tag{1}$$

Now we are going to verify the existence of constants $c_1, c_2 > 0$ such that

$$c_1^p \leq \frac{\|f_{i+2al}\|^p_{L^p_R(\alpha)}}{(|i+2al|+1)^{\alpha p}} \leq c_2^p . \tag{2}$$

Since $\operatorname{supp} f_i \subseteq [-a, 3a-1]$, we have

$$\|f_{i+2al}\|^p_{L^p_R(\alpha)} = \int_R |f_{i+2al}(t)|^p |t|^{\alpha p} dt = \int_R |f_i(t-2al)|^p |t|^{\alpha p} dt$$

$$= \int_{-a}^{3a-1} |f_i(t)|^p |t+2al|^{\alpha p} dt \tag{3}$$

If $t \in [-a, 3a-1]$ then $2a(|l|-2) \leq |t+2al| \leq 2a(|l|+2)$, and, for $i \in \{0, \ldots, 2a-1\}$, it is $2a(|l|-1) \leq |i+2al|+1 \leq 2a(|l|+1)$. Thus, for all l with $|l| > 2$,

$$\frac{1}{4} \leq \frac{2a(|l|-2)}{2a(|l|+1)} \leq \frac{|t+2al|}{|i+2al|+1} \leq \frac{2a(|l|+2)}{2a(|l|-1)} \leq \frac{5}{2} .$$

This inequality yields in combination with (3) for $|l| > 2$ and $\alpha \geq 0$:

$$\frac{\|f_{i+2al}\|^p}{(|i+2al|+1)^{\alpha p}} = \int_{-a}^{3a-1} |f_i(t)|^p \frac{|t+2al|^{\alpha p}}{(|i+2al|+1)^{\alpha p}} dt \leq \left(\frac{5}{2}\right)^{\alpha p} \|f\|^p_{L^p_R(0)}$$

and, analogously,

$$\frac{\|f_{i+2al}\|^p}{(|i+2al|+1)^{\alpha p}} \geq \left(\frac{1}{4}\right)^{\alpha p} \|f\|^p_{L^p_R(0)} .$$

Because f is a prespline we have $0 < \|f\|_{L^p_R(0)} < \infty$ and, hence, (2) holds for $|l| > 2$ and $\alpha \geq 0$. The proof for $\alpha < 0$ is analogous, and in the remaining finitely many cases (where $|l| \leq 2$), the estimate (2) is obvious. Using (2) we derive from (1)

$$\|X^{(i)}\|^p_{L^p_R(\alpha)} \leq c^p \sum_{l \in Z} |x_{i+2al}|^p (|i+2al|+1)^{\alpha p} \leq c^p \|X\|^p_{l^p_1(\alpha)}$$

with $c = c_2 \|f\|_{L^p_R(0)}$. Thus,

$$\|E_1^F X\|_{L^p_R(\alpha)} = \left\| \sum_{i=0}^{2a-1} X^{(i)} \right\|_{L^p_R(\alpha)} \leq 2ca \|X\|_{l^p_1(\alpha)} .$$

The case $Y = R^p_R(\alpha)$ (Proposition 2.16)

Suppose again $k = 1$, i.e. $F = \{f\}$, let a be a positive integer such that supp $\subseteq$ $[-a, a]$, and set $M = \sup_{t \in R} |f(t)|$. If $(x_l) \in l^p_1(\alpha)$ then

$$\sup_{t \in [r,r+1]} |\sum_{l \in Z} x_l f_{ln}(t)|^p \le \sup_R \left(\sum_{l:\text{supp} f_{ln} \cap [r,r+1] \ne \emptyset} |x_l||f_{ln}(t)| \right)^p$$

$$\le M^p \left(\sum_{l=nr-a}^{n(r+1)+a} |x_l| \right)^p \le M^p (n+2a)^{p-1} \sum_{l=nr-a}^{n(r+1)+a} |x_l|^p ,$$

because of supp $f_{ln} \subseteq [(l-a)/n, (l+a)/n]$. Thus,

$$\left(\sum_{r \in Z} \sup_{t \in [r,r+1]} \left| \sum_{l \in Z} x_l f_{ln}(t) \right|^p (|r|+1)^{\alpha p} \right)^{1/p}$$

$$\le C_1 \left(\sum_{r \in Z} \sum_{l=nr-a}^{n(r+1)+a} |x_l|^p (|r|+1)^{\alpha p} \right)^{1/p}$$

$$\le C_2 \left(\sum_{r \in Z} \sum_{l=nr-a}^{n(r+1)+a} |x_l|^p (|l|+1)^{\alpha p} \left(\frac{|r|+1}{|l|+1} \right)^{\alpha p} \right)^{1/p}$$

$$\le C_3 \left(\sum_{r \in Z} \sum_{l=nr-a}^{n(r+1)+a} |x_l|^p (|l|+1)^{\alpha p} \right)^{1/p}$$

$$= C_3 (n+2a)^{1/p} \left(\sum_{r \in Z} |x_r|^p (|r|+1)^{\alpha p} \right)^{1/p}$$

(with C_1, C_2, C_3 depending on n), whence follows

$$\| \sum_{l \in Z} x_l f_{ln} \|_{R^p_R(\alpha)} \le C \| (x_l) \|_{l^p_1(\alpha)}$$

with a constant C depending on n. ∎

2.12.3 Proof of Proposition 2.17

The case $Y = L^p_R(\alpha)$
It is evidently sufficient to treat the case $k = 1$, i.e. we let $G = \{g\}$ with a premeasure g on Y. Further we choose a positive integer a such that supp $g \subseteq$ $[-a, a]$. Then, for $f \in Y$,

$$\|P^G_n f\|^p_{l^p_1(\alpha)} = \sum_{l \in Z} \left| \int_{-a}^{a} f(\frac{t+l}{n}) \bar{g}(t)\,dt \right|^p (|l|+1)^{\alpha p}$$

$$= n^p \sum_{l \in Z} \left| \int_{\frac{-a+l}{n}}^{\frac{a+l}{n}} f(s)\bar{g}(ns-l)\, ds \right|^p (|l|+1)^{\alpha p}$$

$$= n^p \sum_{l \in Z} \|f\chi_{[\frac{-a+l}{n}, \frac{a+l}{n}]}\|_{L_R^p(\alpha)}^p \|g_{ln}\|_{L_R^q(-\alpha)}^p (|l|+1)^{\alpha p} \qquad (1)$$

where we set $g(nt-l) = g_{ln}(t)$. Since

$$\|g_{ln}\|_{L_R^q(-\alpha)}^q = n^{\alpha q - 1} \|g_{l1}\|_{L_R^q(-\alpha)}^q$$

we conclude from inequality 2.12.2(2) that there is a constant C (depending on G, p, α) such that

$$\|g_{ln}\|_{L_R^q(-\alpha)}^q \leq Cn^{\alpha q - 1}(|l|+1)^{-\alpha q}. \qquad (2)$$

Combining (1) and (2) we find

$$\|P_n^G f\|_{l_1^p(\alpha)}^p \leq n^p \sum_{l \in Z} \|f\chi_{[\frac{-a+l}{n}, \frac{a+l}{n}]}\|_{L_R^p(\alpha)}^p C^{p/q} n^{\alpha p - p/q}(|l|+1)^{-\alpha p + \alpha p}$$

$$= C_1 n^{\alpha p + 1} \sum_{l \in Z} \|f\chi_{[\frac{-a+l}{n}, \frac{a+l}{n}]}\|_{L_R^p(\alpha)}^p$$

$$= 2C_1 a n^{\alpha p + 1} \|f\|_{L_R^p(\alpha)}^p$$

which yields the assertion.

The case $Y = R_R^p(\alpha)$

Let, as above, $G = \{g\}$ with $g \in Y^*$, and fix a positive integer a such that $\operatorname{supp} g \subseteq [-a, a]$. Then, for arbitrary $f \in Y$,

$$\|P_n^G f\|_{l_1^p(\alpha)}^p = \sum_{l \in Z} \left| \int_R f(\frac{t+l}{n}) \bar{g}(t)\, dt \right|^p (|l|+1)^{\alpha p}$$

$$\leq \sum_{l \in Z} \|f(\frac{\cdot + l}{n}) \chi_{[-a,a]}\|_Y^p \|g\|_{Y^*}^p (|l|+1)^{\alpha p}$$

$$\leq \|g\|_{Y^*}^p \sum_{l \in Z} \left[\left(\int_{-a}^a |f(\frac{t+l}{n})|^p |t|^{\alpha p}\, dt \right)^{1/p} + \right.$$

$$\left. + \left(\sum_{k=-a}^{a-1} \sup_{t \in [k, k+1]} |f(\frac{t+l}{n})|^p (|k|+1)^{\alpha p} \right)^{1/p} \right]^p (|l|+1)^{\alpha p}$$

$$\leq \|g\|_{Y^*}^p \sum_{l \in Z} \left[\left(\sup_{t \in [-a,a]} |f(\frac{t+l}{n})|^p \int_{-a}^a |t|^{\alpha p}\, dt \right)^{1/p} + \right.$$

$$+ \left(\sup_{t\in[-a,a]} |f(\frac{t+l}{n})|^p \sum_{k=-a}^{a-1} (|k|+1)^{\alpha p} \right)^{1/p} \Bigg]^p (|l|+1)^{\alpha p}$$

$$= C^p \sum_{l\in Z} \sup_{t\in[-a,a]} |f(\frac{t+l}{n})|^p (|l|+1)^{\alpha p} \qquad (3)$$

with

$$C = \|g\|_{Y^*} \left(\left(\int_{-a}^{a} |t|^{\alpha p} dt \right)^{1/p} + \left(\sum_{k=-a}^{a-1} (|k|+1)^{\alpha p} \right)^{1/p} \right) = \|g\|_{Y^*} \|\chi_{[-a,a]}\|_Y .$$

Substituting $l = mn + r$, $r \in \{0,\ldots,n-1\}$, we find that (3) is not greater than

$$C^p \sum_{m\in Z} \sum_{r=0}^{n-1} \sup_{t\in[-a,a]} |f(\frac{t+mn+r}{n})|^p (|mn+r|+1)^{\alpha p}$$

$$= C^p \sum_{m\in Z} \sum_{r=0}^{n-1} \sup_{s\in[m+\frac{r-a}{n},m+\frac{r+a}{n}]} |f(s)|^p (|mn+r|+1)^{\alpha p}$$

$$\leq C^p \sum_{m\in Z} \sum_{r=0}^{n-1} \sup_{s\in[m-a,m+a]} |f(s)|^p \sup_{r\in\{0,\ldots,n-1\}} (|mn+r|+1)^{\alpha p}$$

$$= C^p n \sum_{m\in Z} \sup_{s\in[m-a,m+a]} |f(s)|^p \sup_{r\in\{0,\ldots,n-1\}} (|mn+r|+1)^{\alpha p}$$

$$\leq C^p n \sum_{m\in Z} \sum_{l=m-a}^{m+a-1} \sup_{s\in[l,l+1]} |f(s)|^p \sup_{r\in\{0,\ldots,n-1\}} (|mn+r|+1)^{\alpha p}$$

$$= C^p n^{\alpha p+1} \sum_{m\in Z} \sum_{l=m-a}^{m+a-1} \sup_{s\in[l,l+1]} |f(s)|^p (|l|+1)^{\alpha p} \sup_{r\in\{0,\ldots,n-1\}} \frac{(|m+\frac{r}{n}|+\frac{1}{n})^{\alpha p}}{(|l|+1)^{\alpha p}} .$$

$$(4)$$

From

$$\frac{|m|-1}{|l|+1} \leq \frac{|m+\frac{r}{n}|+\frac{1}{n}}{|l|+1} \leq \frac{|m|+2}{|l|+1}$$

we conclude that

$$D := \sup_{r,l,m,n} \frac{(|m+\frac{r}{n}|+\frac{1}{n})^{\alpha}}{(|l|+1)^{\alpha}} < \infty ,$$

where the supremum is taken over all $n \in \mathbf{Z}^+, m \in \mathbf{Z}, r \in \{0,\ldots,n-1\}$, and $l \in \{m-a,\ldots,m+a-1\}$. Thus, (4) is less than

$$C^p D^p n^{\alpha p+1} \sum_{m\in Z} \sum_{l=m-a}^{m+a-1} \sup_{s\in[l,l+1]} |f(s)|^p (|l|+1)^{\alpha p}$$

$$= 2aC^p D^p n^{\alpha p+1} \sum_{l \in Z} \sup_{s \in [l, l+1]} |f(s)|^p (|l| + 1)^{\alpha p} ,$$

whence follows $\|P_n^G f\|_{l_1^p(\alpha)} \le c_1 n^{\alpha+1/p} \|f\|_Y$ as desired. ∎

2.12.4 Proof of Theorem 2.7

Let $k = 1, F = \{\phi\}$ and $G = \{g\}$. Further we suppose without loss (by 2.3.1(f)) that the multiplier a is a trigonometric polynomial, and since then $T^0(a)$ is a linear combination of the discrete shift operators V_r we are thus left on verifying that

$$\|E_n^F V_r P_n^G fI - f E_n^F V_r P_n^G\|_{L(Y, L_R^p(\alpha))} \to 0 \quad \text{as} \quad n \to \infty .$$

Assume f to be real valued (otherwise decompose f into its real and imaginary part), and let $y \in Y$ be a non-negative function. Then, writing for brevity

$$g_{ln}(y) = \int_R y(\frac{t+l}{n}) \bar{g}(t) \, dt \quad \text{and} \quad \phi_{ln}(t) = \phi(nt - l) ,$$

we have

$$(E_n^F V_r P_n^G fy - f E_n^F V_r P_n^G y)(t) =$$
$$= \sum_{l \in Z} (g_{l-r,n}(fy) - f(t)g_{l-r,n}(y)) \, \phi_{ln}(t) . \tag{1}$$

Now split the functional g into

$$g = g^{(0)} + i g^{(1)} - g^{(2)} - i g^{(3)} = \sum_{j=0}^{3} i^j g^{(j)}$$

with $g^{(j)} \in Y^*$ referring to non-negative functionals, and define $g_{ln}^{(j)}$ as above by

$$g_{ln}^{(j)}(y) = \int_R y(\frac{t+l}{n}) g^{(j)}(t) \, dt , \quad y \in Y .$$

(Hint: It is easy to decompose g into a sum $g^{(r)} + i g^{(i)}$ with real valued functionals $g^{(r)}$ and $g^{(i)}$. That it is further possible to write $g^{(r)}$ and $g^{(i)}$ as differences of two non-negative functionals can be seen as in ([Die], Sections 13.3.2-13.3.6), in case of functionals on spaces of continuous functions.)

Then, in particular,

$$g_{ln}(fy) = \sum_{j=0}^{3} (-i)^j g_{ln}^{(j)}(fy) ,$$

and applying the mean value theorem to each of the integrals $g_{ln}^{(j)}(fy)$ we get the existence of numbers $t_{ln}^{(j)} \in \text{conv supp } g^{(j)}$ such that

$$g_{ln}^{(j)}(fy) = f(\frac{t_{ln}^{(j)} + l}{n}) \int_R y(\frac{t+l}{n}) g^{(j)}(t) \, dt = f(\frac{t_{ln}^{(j)} + l}{n}) g_{ln}^{(j)}(y).$$

Thus, (1) is equal to

$$\sum_{l\in Z}\sum_{j=0}^{3}(-i)^{j}\left(f(\frac{t_{l-r,n}^{(j)}+l-r}{n})-f(t)\right)g_{l-r,n}^{(j)}(y)\,\phi_{ln}(t)$$

whence follows

$$\left|(E_{n}^{F}V_{r}P_{n}^{G}fy-fE_{n}^{F}V_{r}P_{n}^{G}y)(t)\right|$$
$$\leq\sum_{j=0}^{3}\sum_{l\in Z}\left|f(\frac{t_{l-r,n}^{(j)}+l-r}{n})-f(t)\right|g_{l-r,n}^{(j)}(y)\,|\phi_{ln}(t)|\,. \tag{2}$$

Let $\omega(f,c)$ denote the modul of continuity of f,

$$\omega(f,c)\ =\ \sup\{|f(t_{1})-f(t_{2})|\quad\text{with}\quad|t_{1}-t_{2}|<c\}\,,$$

and let a be a positive integer such that $\text{supp}\,g\subseteq[-a,a]$ and $\text{supp}\,\phi\subseteq[-a,a]$. Since, for $t\in\text{supp}\,\phi_{ln}$,

$$\left|\frac{t_{l-r,n}^{(j)}+l-r}{n}-t\right|\ \leq\ \frac{2a+|r|}{n}\,,$$

we find that (2) is less than

$$\omega(f,\frac{2a+|r|}{n})\sum_{j=0}^{3}\sum_{l\in Z}g_{l-r,n}^{(j)}(y)\,|\phi_{ln}(t)|\,. \tag{3}$$

Defining $G^{(j)}:=\{g^{(j)}\}$ and $|F|:=\{|\phi|\}$ we can write (3) as

$$\omega(f,\frac{2a+|r|}{n})\sum_{j=0}^{3}(E_{n}^{|F|}V_{r}P_{n}^{G^{(j)}}y)(t)\,,$$

hence,

$$\|E_{n}^{F}V_{r}P_{n}^{G}fy-fE_{n}^{F}V_{r}P_{n}^{G}y\|_{L_{R}^{p}(\alpha)}\leq$$
$$\omega(f,\frac{2a+|r|}{n})\sum_{j=0}^{3}\|E_{n}^{|F|}V_{r}P_{n}^{G^{(j)}}y\|_{L_{R}^{p}(\alpha)}$$
$$\leq\omega(f,\frac{2a+|r|}{n})\sum_{j=0}^{3}\|E_{n}^{|F|}V_{r}P_{n}^{G^{(j)}}\|_{L(Y,L_{R}^{p}(\alpha))}\,\|y\|_{Y}\,, \tag{4}$$

and now invoke Theorem 2.5 and Proposition 2.17 to find that the norms $\|E_{n}^{|F|}V_{r}P_{n}^{G^{(j)}}\|_{L(Y,L_{R}^{p}(\alpha))}$ are uniformly bounded with respect to n. Finally, observe that (4) holds for arbitrary y (split $y=\sum i^{j}y^{(j)}$ with $y^{(j)}\geq0$), thus

$$\|E_{n}^{F}V_{r}P_{n}^{G}fI-fE_{n}^{F}V_{r}P_{n}^{G}\|\ \leq\ C\,\omega(f,\frac{2a+|r|}{n})\,,$$

and the right hand side of this inequality goes to zero as $n\to\infty$ since $f\in C(\dot{\mathbf{R}})$. ∎

2.13   Exercises

E 2.1 Verify 2.1.2(5), (7) and (8).

E 2.2   1. Prove identities 2.1.2 (6) and (9) by means of the Mellin symbol calculus.

2. Let $0 < \mathrm{Re}(1/p + \alpha + \gamma) < 1$. Show that $S_{R^+,\gamma}$ is invertible on $L^p_{R^+}(\alpha)$ if and only if $\mathrm{Re}(1/p + \alpha + \gamma) \neq 1/2$. In this case,

$$S^{-1}_{R^+,\gamma} = \begin{cases} S_{R^+,\gamma+1/2} & \text{if } 0 < \mathrm{Re}(1/p + \alpha + \gamma) + 1/2 < 1 \\ S_{R^+,\gamma-1/2} & \text{if } 0 < \mathrm{Re}(1/p + \alpha + \gamma) - 1/2 < 1 . \end{cases}$$

E 2.3   1. Give a direct proof that $JN^p(\alpha) \subseteq \Sigma^p_R(\alpha)$, but neither JI nor JS_{R^+} belongs to $\Sigma^p_R(\alpha)$.
(Hint: $JN = \chi_{R^-} - S_R \chi_{R^+} \in \Sigma^p_R(\alpha)$.)

2. Show that $\Sigma^p_R(\alpha)$ is generated by two idempotents and apply Theorem 1.10 to derive a 2×2-matrix-valued symbol for invertibility in this algebra.

E 2.4 Derive assertion (e) in 2.1.2 by determining the spectrum of S_{R^+} and applying Theorem 1.9.

E 2.5 Prove that the algebra $\mathrm{alg}^\pi T(PC_{p,\alpha})$ is even commutative.
    (Hint: Start with verifying that $\pi(T(a)T(b) - T(b)T(a)) = 0$ whenever a and b have exactly one discontinuity at the same point, see ([BS 2], 6.30).)

E 2.6 Let $PC^\Omega_{p,\alpha}$ stand for the class of all piecewise continuous multipliers which are continuous on $\mathbf{T} \setminus \Omega$, and let $\mathrm{alg}\,(T^0(PC^\Omega_{p,\alpha}), P)$ stand for the smallest closed subalgebra of $L(l^p_1(\alpha))$ which contains P and all operators $T^0(a)$, $a \in PC^\Omega_{p,\alpha}$. Further let f be a continuous multiplier vanishing on Ω. Prove that for each $A \in \mathrm{alg}\,(T^0(PC^\Omega_{p,\alpha}), P)$ there are continuous multipliers g and h vanishing on Ω and a compact operator K such that

$$T^0(f)\, A \;=\; T^0(g) \;+\; T^0(h)P \;+\; K .$$

E 2.7 Verify 2.4.2 (3) and (4).
    (Hint: For (3) apply Lemma 2.1 to the difference $G(n_\pi) - H(f)$. For (4) show that the kernel function of $S_{R^+}N$ is given by $k(x) = \dfrac{1}{\pi^2}\dfrac{\ln x}{1+x}$, $x \in \mathbf{R}^+$. Thus,

$$G(sn_\pi) - \frac{1}{\pi^2}\left(\frac{\ln \frac{m+1}{n+1}}{m+n+1} \right)_{m,n} \quad \text{is a compact operator. Now verify that the operator}$$

$$\frac{1}{\pi^2}\left(\frac{\ln \frac{m+1}{n+1}}{m+n+1} \right) - \frac{1}{\pi^2}\left(\frac{\sum_{l=1}^m \frac{1}{l} - \sum_{l=1}^n \frac{1}{l}}{m+n+1} \right)$$

is also compact and that $\dfrac{1}{\pi^2}\left(\dfrac{\sum_{l=1}^m \frac{1}{l} - \sum_{l=1}^n \frac{1}{l}}{m+n+1}\right)$ is just the matrix representation of the operator $-T(f)\,H(f)$.)

E 2.8 Prove that $\operatorname{alg}(T^0(PC_{p,\alpha}),P) = \Lambda \operatorname{alg}(T^0(PC_{p,0}),P)\Lambda^{-1}$.

E 2.9 Verify the identity $Y_t^{-1} T^0(a)\,Y_t = T^0(a_t)$ with $a_t(s) = a(ts)$.

E 2.10 The mapping $\operatorname{smb}_1 : \operatorname{alg}^\pi T(PC_{p,\alpha})/J_1 \to \Sigma_R^p(\alpha)$ is a continuous isomorphism as we have seen in the proof of Theorem 2.1. Show that it is even an isometry. Prove an analogous result for the algebra $\operatorname{alg}^\pi(T^0(PC_{p,\alpha}),P)$.

(Hint: Define diagonal operators $D_n \in L(l_Z^p(\alpha))$ such that $E_n D_n$ and $D_n E_n^{-1}$ are isometries, and then use that $D_n = I+$ compact. For another proof see [R 6]).

E 2.11 Let $a \in PC_{p,\alpha}$. Show that the Hankel operator $H(a)$ with $a \in PC_{p,\alpha}$ belongs to $\operatorname{alg} T(PC_{p,\alpha})$ if and only if the function a has no discontinuities on $\mathbf{T} \setminus \{-1,1\}$. (For the Hilbert space case and the "only if" direction see ([BS 2], 4.51).)

E 2.12 Show that the local algebras $\operatorname{alg}^\pi T(PC_{p,\alpha})/\mathcal{J}_t$ (introduced in the proof of 2.2) are singly generated, and derive a Fredholm criterion for operators in $\operatorname{alg} T(PC_{p,\alpha})$ by applying Theorem 1.9 to these local algebras and by using Allan's local principle.

E 2.13 Show that the local algebras $\operatorname{alg}^\pi(T^0(PC_{p,\alpha}),P)/\mathcal{J}_t$ (introduced in the proof of 2.3) are generated by two idempotents, and derive a Fredholm criterion for operators in $\operatorname{alg}(T^0(PC_{p,\alpha}),P)$ by employing Theorem 1.10.

E 2.14 Given positive integers a, k let $F_{a,k}$ denote the collection of all k-tuples $F = \{\varphi_0,\ldots,\varphi_{k-1}\}$ of presplines with $\operatorname{supp}\varphi_r \subseteq [0,a]\,(r = 0,\ldots,k-1)$. Show that the operator E_1^F depends linearly and continuously on F varying over $F_{a,k}$.

E 2.15 Prove the analogue of Exercise 2.14 for operators P_1^G.

E 2.16 Let $F = \{\varphi_0,\ldots\varphi_{k-1}\}$ be a prebasis, and let a be a compactly supported bounded measurable function with $\int_R a(x)dx = 1$. Prove that $a * F := \{a*\varphi_0,\ldots,a*\varphi_{k-1}\}$ is also a prebasis, and that $\alpha_r^F = \alpha_r^{a*F}$ for all $r = 0,\ldots,k-1$. Is the prebasis $a * F$ canonical whenever F is so?

(Hint: Consider the prebasis $F = \{\varphi_0^{1,1}, \varphi_1^{1,1}\}$ consisting of the Bernstein polynomials $\varphi_0^{1,1}$ and $\varphi_1^{1,1}$, and let $a = \chi_{[0,1]}$.)

E 2.17 Compute the canonical prebases $F_{d,d-1}$ and $F_{d,d-2}$ of the spaces of piecewise polynomial defect splines (confer Section 2.9.3).

Solution: $F_{d,d-1} = \left\{\varphi_0^{d,d-1},\ldots,\varphi_{d-1}^{d,d-1}\right\}$ with

$$\varphi_r^{d,d-1}(x) = \begin{cases} \displaystyle\int_0^x \binom{d-1}{x}(1-t)^{d-1-r}t^r\,dt & \text{if}\quad x \in (0,1) \\[2mm] 2/d - x/d & \text{if}\quad x \in (1,2) \\ 0 & \text{if}\quad x \notin (0,2)\,. \end{cases}$$

and $F_{d,d-2} = \left\{ \varphi_0^{d,d-2}, \ldots, \varphi_{d-2}^{d,d-2} \right\}$ with

$$\varphi_r^{d,d-2}(x) = \begin{cases} \displaystyle\int_0^x \int_0^y \binom{d-2}{r}(1-z)^{d-2-r}z^r\,dz\,dy & x \in (0,1) \\[3mm] \dfrac{(2r+2-5d)x^2+(14d-4r-4)x-(2r+2+5d)}{4d(d-1)} & x \in (1,2) \\[3mm] \dfrac{(3d-2r-2)x^2+(12r+12-18d)x+(27d-18r-18)}{4d(d-1)} & x \in (2,3) \\[3mm] 0 & x \notin (0,3)\,. \end{cases}$$

E 2.18 Let $\varphi_r^{d,e}$ stand of the presplines considered in Section 2.9.4.

1. Prove that there are non-zero constants $c_r^{d,e}$ such that
$$(\varphi_r^{d,e})'(x) = \varphi_r^{d-1,e}(x) + c_r^{d,e}\varphi_0^{d-e-1,0}(x-1)\,, \quad x \in \mathbf{R}.$$

2. Use this identity to prove that $\displaystyle\int_R \varphi_r^{d,e}(x)dx \neq 0$ for all r,d,e.

3. Show that

   (i) $c_0^{d,0} = -1$

   (ii) $c_r^{d,d-2} = \dfrac{r - 3d/2 + 1}{d(d-1)}$

   (iii) $\displaystyle\sum_{r=0}^{e} c_r^{d,e} = -1$

   (iv) $c_r^{d,e} = (d-e-1)!\,\left(\varphi_r^{d,e}\right)'(d-e)$.

E 2.19 Prove that $\lambda_{rs}^{F,G}(z) = \sum_{l\in Z}(\tilde{\varphi}_s * \bar{\gamma}_r)(-l)z^l$, $|z| = 1$, (the sum being actually finite), and use this result to prove that the system $\{\varphi_{rl1}\}$ is orthogonal if and only if $\lambda^{F,F} \equiv I$.

E 2.20 Prove the following converse of the de Boor estimate: Let S be a closed subspace of $L_R^p(\alpha)$ which is continuously isomorphic to $l_1^p(\alpha)$ and suppose the isomorphism $E : l_1^p(\alpha) \to S$ to intertwine the shift operators $V_1 \in L(l_1^p(\alpha))$ and $U_1 \in L(L_R^p(\alpha))$, that is, $EV_1 = U_1E$. Then there is a function φ such that $S = S_1^{\{\varphi\}}$, i.e. S is a spline space. Formulate and prove the analogous assertion for $l_k^p(\alpha)$, $k > 1$.

E 2.21 Let F, G, Y be as in Theorem 2.12 and let $K \subseteq Y$ be a k-elementic prebasis with $\alpha_r^F = \alpha_r^K$ $(r = 1, \ldots, k)$. Show that

$$\| I_y(E_n^K E_{-n}^F L_n^{F,G} f - f) \|_{L_R^p} \to 0 \quad \text{as } n \to \infty$$

for all $f \in Y_P$. (Hint: Compare the proof of Theorem 2.12.)

E 2.22 Let the operator $A : L_R^p(\alpha) \to Y$ be homogeneous of degree 0. Verify that then the operators $P_n^G A E_n^F \in L(l_k^p(\alpha))$ are independent of n.

E 2.23 Let $F_{d,e}$ be as in Section 2.9.4. Prove

1. $(F_{2,1}, F_{1,1})$ is not canonical.

2. $(F_{2,1}, G)$ with $G = \{\chi_{[0,1)}, \delta_\varepsilon\}$ and $\varepsilon \in [0, 1)$ is canonical.

E 2.24 Let the situation be as in Section 2.10.3. Show that

$$\sigma^{F,G} = \sum_{r=0}^{m-1} w_r \overline{\lambda^{F_1,\{\delta_{\epsilon_r}\}}} \, \sigma^{F,\{\delta_{\epsilon_r}\}} .$$

E 2.25 Let F be a canonical k-tuple of presplines, and let F' be a subtuple of F. Prove that F' is canonical, too.

E 2.26 Let $F = \{\varphi_0, \ldots, \varphi_{k-1}\}$ be a canonical prebasis, and let β_r be complex numbers such that

$$\sum_{r=0}^{k-1} \beta_r \sum_{m \in Z} \varphi_{rm1} = \chi_R.$$

Prove that $\beta_r = \alpha_r^F$ (i.e. the α_r^F are uniquely determined by identity (i) in Proposition 2.19).

E 2.27 Let (F, G) be a canonical pair constituted by a prebasis F and a k-tuple $G = \{\gamma_0, \ldots, \gamma_{k-1}\}$ of premeasures. Show that not all of the numbers $\int \gamma_r(x) dx$ are zero.

E 2.28 Prove the following implications:

$\lambda^{F,F}$ invertible $\Longleftrightarrow$ $\sigma^{F,F}$ invertible ,

$\lambda^{F,F}$ invertible $\Longleftarrow$ $\lambda^{F,G}$ invertible $\Longrightarrow$ $\sigma^{F,F}$ invertible ,

$\lambda^{F,F}$ invertible $\Longleftarrow$ $\sigma^{F,G}$ invertible $\Longrightarrow$ $\sigma^{F,F}$ invertible .

E 2.29 Let $F = \{\varphi\}$ be a prebasis, let $G = \{\gamma\}$, and suppose (F, G) to be canonical. Prove that φ and γ can always be normed in such a way that $\alpha^F = 1$, $\int \varphi dx = 1$, $\int \gamma dx = 1$.

E 2.30 Let $F = \{\varphi_1, \varphi_2\}$ be a prebasis supported on $(0, 1)$. Prove that F is canonical if and only if φ_1 is not a multiple of $\chi_{[0,1]}$.

2.14 Comments and references

2.1 The most of the presented material is well known and can be found in the monographs on singular integrals and Wiener-Hopf operators theory cited in the bibliography.

In particular, see Gohberg/Krupnik [GK 1] (Section 1.1 and 1.5), Mikhlin/Prößdorf [MP] (Chapter 2, 1-3), or Duduchava [Du 1] for the definition and elementary properties of the singular integral operators, and [Du 1] or Böttcher/Silbermann [BS 2] (9.2, 9.3) for Fourier multipliers. There the inequalities 2.1.1(3) and (4) are stated for the unweighted case ($\alpha = 1$), but the proofs are evidently carry over to symmetric weights (as considered in 2.1.1: $|t|^\alpha = |-t|^\alpha$). For non-symmetric weights consult Schneider [Sch].

Property (a) in 2.1.2 goes back to Simonenko/Chin Ngok Minh [SC], and properties (b)-(d) are Costabel's results [Co 1]. Let us mention that the membership of weighted singular integral operators and certain types of Hankel operators to the algebra $\Sigma^p(\alpha)$ was verified by Nyaga and Costabel by using different (and quite technical) methods. See the report [RS 1] for further discussion of the topic.

2.2 This is a baby version of what follows in Sections 2.6.1-2.11.10. For detailed comments we refer to these sections.

2.3 For elementary facts on piecewise continuous multipliers and Toeplitz operators on $l^{p,\alpha}$-spaces we refer to [BS 1] and ([BS 2], 2.5-2.8). Proposition 2.10 is in [R 6].

2.4 There are several ways to discretize Mellin convolutions. In case of discretization by a collocation method, Prößdorf and Rathsfeld [PR 1] succeeded in showing that the corresponding system matrices always belong to the Toeplitz algebra $\mathrm{alg}\,T(PC_{p,\alpha})$. Their proof is based on the residue theorem for operator valued functions. In [R 6] one of the authors was able to carry over this result to discretizations of Mellin convolutions via Galerkin methods. The main ingredients for this were Lemma 2.1 and its corollary which imply that the difference between the Galerkin and the collocation discretization of a Mellin convolution is always compact (see Proposition 2.11(a)). Assertions (b)-(d) of Proposition 2.11 are also in [R 6]. The presented proof of the basic Theorem 2.1, which consequently employs local techniques, is new.

2.5 A famous result by Gohberg and Krupnik (see e.g. [BS 2], Proposition 6.28) concerns the determination of the maximal ideal space of the (commutative) Toeplitz algebra $\mathrm{alg}^\pi\,T(PC_{p,\alpha})$. They showed that this space is homeomorphic to the cylinder $\mathbf{T} \times \overline{\mathbf{R}}$ (provided with an exotic topology) and that the Gelfand transform of the coset $T(a) + K(l^{p,\alpha})$ is the function

$$a^\circ : \mathbf{T} \times \overline{\mathbf{R}} \to \mathbf{C}, \quad (t,z) \mapsto a(t+0)\frac{1 - \coth \pi(z + (1/p + \alpha)i)}{2} +$$
$$a(t-0)\frac{1 + \coth \pi(z + (1/p + \alpha)i)}{2} \ .$$

The image of the function a° results from that one of the function a by filling in circular arcs into each gap between $a(t-0)$ and $a(t+0)$. These results can be easily derived from our Theorem 2.2 by observing that the Mellin transform of the kernel function of the Mellin convolution operator $\mathrm{smb}_t T(a)$ is just $a^\circ(t, \cdot) : \overline{\mathbf{R}} \to \mathbf{C}$.

Moreover, Theorem 2.2 (and its proof) explains why one has to add *circular arcs* to the image of a: these arcs correspond to the spectrum of the singular integral operator over the semi axis $\mathbf{R}^+$. The assertions concerning discretized Mellin convolutions, in particular the computation of the Gelfand transform of $G(b) + K(l^{p,\alpha})$, viz. the function

$$
(t, z) \mapsto \begin{cases} b(z) & \text{if} \quad t = 1 \\ 0 & \text{if} \quad t \neq 1 \,, \end{cases}
$$

are new (see [R 6]). They show that discretized Mellin convolution act (up to compact operators) strictly local.

Other approaches to Theorems 2.2 and 2.3 (which also invoke local techniques) can be found in [BS 2] and [RS 7]. These approaches base on the observations that the local algebras of $\operatorname{alg}^\pi T(PC_{p,\alpha})$ and $\operatorname{alg}^\pi (T^0(PC_{p,\alpha}), P)$ are singly generated and generated by two idempotents, respectively. This offers a direct way of deriving symbol calculi for these algebras (compare also Exercises 2.12 and 2.13). On the other hand, these approaches give a less complete picture of Toeplitz algebras: for example, they cannot guarantee the semi-simplicity of the local algebras (which immediately follows from Theorems 2.2 and 2.3).

Coburn's theorem 2.4 is, e.g., in Böttcher/Silbermann [BS 2], 2.38, Corollary 2.2 in [BS 2], 2.40, and the cited index formula in [BS 2], 6.40.

2.6-2.8 The central results of these sections are the de Boor estimates (which show that the functions ϕ_{rln}, $r = 0, \dots, k - 1$, $l \in \mathbf{Z}$, form an unconditional basis of S_n), the commutator property, and the strong convergence of the spline projections. These results are certainly well known to specialists in spline analysis, at least in the "classical" case of piecewise polynomial splines. (See Sections 2.2.1-2.4.1 in [PS 2] and the references cited there.) Nevertheless we tried to present all results with their full proofs since many of the proofs we found in literature only concern special settings (no weights, only piecewise polynomial generating functions, and/or no spline spaces generated by k-tuples of "mother" splines).

The representation of $\lambda^{F,G}$ is due to Schmidt (see [Sc 1]).

One feature of spaces of piecewise polynomial splines, which is not in evidence in our brief summary, is the existence of finite two-scale relations $\phi(x) = \sum_{n=0}^{N} \beta_n \phi(2x - n)$, $x \in \mathbf{R}$. Relations of this type are extremely useful in order to construct fast algorithms for the practical solution of operator equations by spline projection methods and wavelet techniques (compare Section 2.9.5, the related monographs cited in the bibliography and the papers by Beylkin, Coifman, Rokhlin [BCR], Alpert, Beylkin, Coifman, Rokhlin [ABCR], Dahmen, Prößdorf, Schneider [DPS 1]-[DPS 3] and Dahmen, Kleemann, Prößdorf, Schneider [DKPS]).

2.9-2.10 Here we have summarized some known facts about concrete spline spaces and spline projections. For splines with defect we refer to Szyszka [Sz], where , in particular, the collocation projections onto these spaces were investigated. The idea of the qualocation method goes back to [Sl] who proposed this method in order to combine the advantages of Galerkin methods (high convergence rates) and collocation methods (low expense).

2.11 For the characterization of the approximation operators for spline projection methods applied to pure singular integral operators $aI + bS$ see Hagen/Silbermann [HS 1], Prößdorf/Rathsfeld [PR 1], [PR 2], Schmidt [Sc 1], Szyszka [Sz], and compare also Chapter 10 in [PS 2]. For quadrature methods the reader should consult [PR 1], [PR 4].

The quadrocation methods or fully discretized methods for singular integral operators were studied by Saranen for the first time (see [Sa]).

For the applicability of projection methods to Mellin convolution operators see [PR 1].

The most important new result in this section is the discovery of the structure of the approximation methods for operators in $\Sigma^p(\alpha)$ in case of general spline spaces (including defect splines). These results form the basis for our stability analysis in the following chapter (see, in particular, Section 3.1.2 below).

Chapter 3

Algebras of spline approximation sequences

The main goal of the present chapter is to establish stability criteria for the approximate solution of singular integral equations with piecewise continuous coefficients on the real axis. Based on our knowledge of the structure of approximation systems for the singular integral operator we shall construct a Banach algebra of approximation sequences which includes all concrete methods considered above (such as Galerkin, collocation, qualocation, quadrature, and quadrocation methods) simultaneously. Banach algebra techniques (in particular the KMS-concept) will provide us with a complete description of this algebra up to isomorphy and, thus, with a unique approach to stability investigations of approximation methods.

To point out parallels we start with treating the analytic analogue of the announced algebra, viz. the algebra $\mathrm{alg}\,(S_R, PC)$ generated by all singular integral operators with piecewise continuous coefficients. The result will be as follows: We settle a family $\{W_t\}_{t \in \dot{R}}$ of continuous Banach algebra homomorphisms from $\mathrm{alg}\,(S_R, PC)$ into algebras of Mellin convolution operators in such a way that an operator $A \in \mathrm{alg}\,(S_R, PC)$ is Fredholm if and only if all operators $W_t(A)$ are invertible.

After this, in the main part of this chapter, we shall derive an analogous result for the algebra of approximation sequences. In this setting, two families of Banach algebra homomorphisms will be needed to describe stability conditions. The homomorphisms in the first of these families take values in the algebra $\mathrm{alg}\,(S_R, PC)$, and the other ones in the Toeplitz algebra $\mathcal{T}^p_{k,\kappa,\Omega}(\alpha)$. We conclude this chapter by further examples of approximation methods which can be associated with elements of the approximation algebra, too.

3.1   Algebras of singular integral operators

3.1.1   Definitions and basic properties

Let $p > 1$ and $0 < 1/p + \alpha < 1$ again. A function a on $\mathbf{R}$ is called *piecewise continuous* if it possesses one-sided limits $a(x \pm 0)$ at each point $x \in \mathbf{R}$, if the limits $a(\pm\infty) = \lim_{x\to\pm\infty} a(x)$ exist, and if all these limits are finite. The collection PC of all piecewise continuous functions forms a Banach algebra with respect to pointwise operations and the supremum norm. If $a \in PC$ then the operator aI of multiplication by a is bounded on $L_R^p(\alpha)$, and

$$\|aI\|_{L(L_R^p(\alpha))} = \operatorname*{ess\,sup}_{x\in R} |a(x)| = \|a\|_\infty .$$

In the following sections we examine the smallest closed subalgebra $\mathcal{O}_R^p(\alpha)$ of $L(L_R^p(\alpha))$ which contains the singular integral operator S_R and all multiplication operators aI with $a \in PC$. Here are some fundamental properties of the operators in $\mathcal{O}_R^p(\alpha)$, and these facts are all we need to analyse this algebra.

(i) The algebra $\mathcal{O}_R^p(\alpha)$ contains the ideal $K(L_R^p(\alpha))$ of all compact operators on $L_R^p(\alpha)$.

(ii) If $A \in \mathcal{O}_R^p(\alpha)$ and $f \in C(\dot{R})$ then the commutator $fA - AfI$ is compact.

(iii) The operators in $\mathcal{O}_R^p(\alpha)$ can be homogenized. Precisely: For s being a positive number define the operator $Z_s : L_R^p(\alpha) \to L_R^p(\alpha)$ by $(Z_s a)(x) = a(x/s)$. Then the strong limits

$$s\text{--}\lim_{s\searrow 0} Z_s^{-1} A Z_s =: W_0(A) \tag{1}$$

and

$$s\text{--}\lim_{s\to\infty} Z_s^{-1} A Z_s =: W_\infty(A) \tag{2}$$

exist for each $A \in \mathcal{O}_R^p(\alpha)$, and the operators $W_0(A)$ and $W_\infty(A)$ are homogeneous: $Z_t^{-1} W_0(A) Z_t = W_0(A)$ and $Z_t^{-1} W_\infty(A) Z_t = W_\infty(A)$ for all $t > 0$. Moreover, the mappings $W_0, W_\infty : \mathcal{O}_R^p(\alpha) \to L(L_R^p(\alpha))$ are continuous algebra homomorphisms which act on the generating operators of $\mathcal{O}_R^p(\alpha)$ as follows:

$$W_0(S_R) = W_\infty(S_R) = S_R , \tag{3}$$
$$W_0(aI) = a(0+0)\chi_{R^+}I + a(0-0)\chi_{R^-}I , \tag{4}$$
$$W_\infty(aI) = a(+\infty)\chi_{R^+}I + a(-\infty)\chi_{R^-}I \tag{5}$$

and, for K compact,

$$W_0(K) = W_\infty(K) = 0 . \tag{6}$$

(iv) The algebra $\mathcal{O}_R^p(0)$ is translation invariant. That is: For $t \in \dot{\mathbf{R}} \backslash \{0, \infty\}$, let the shift operator $U_t : L(L_R^p(0)) \to L(L_R^p(0))$ be defined by $(U_t f)(s) = f(s-t)$. Then $U_{-t} A U_t \in \mathcal{O}_R^p(0)$ whenever $A \in \mathcal{O}_R^p(0)$.

The proofs of (i)-(iv) are left as exercises.

By (i) it makes sense to form the quotient algebra $\mathcal{O}_R^p(\alpha)/K(L_R^p(\alpha))$. We abbreviate this algebra by $\mathcal{O}_R^p(\alpha)^\pi$ and let π stand for the canonical homomorphism from $\mathcal{O}_R^p(\alpha)$ onto $\mathcal{O}_R^p(\alpha)^\pi$. Our criterion for the Fredholmness of operators in $\mathcal{O}_R^p(\alpha)$ reads as follows:

Theorem 3.1 *(a) For each $t \in \dot{\mathbf{R}}$ there is a continuous Banach algebra homomorphism W_t from $\mathcal{O}_R^p(\alpha)^\pi$ into $L(L_R^p(\alpha))$ if $t \in \{0, \infty\}$ and into $L(L_R^p(0))$ if $t \in \dot{\mathbf{R}} \setminus \{0, \infty\}$ such that*

$$W_t(\pi(S_R)) = S_R, \tag{7}$$

$$W_t(\pi(aI)) = \begin{cases} a(t-0)\chi_{R^-} I + a(t+0)\chi_{R^+} I & \text{if} \quad t \in \mathbf{R} \\ a(-\infty)\chi_{R^-} I + a(+\infty)\chi_{R^+} I & \text{if} \quad t = \infty. \end{cases} \tag{8}$$

(b) An operator $A \in \mathcal{O}_R^p(\alpha)$ is Fredholm if and only if all operators $W_t(A), t \in \dot{\mathbf{R}}$, are invertible. If A is Fredholm then it has a regularizer in $\mathcal{O}_R^p(\alpha)$; hence, the algebra $\mathcal{O}_R^p(\alpha)^\pi$ is inverse closed in the Calkin algebra $L(L_R^p(\alpha))/K(L_R^p(\alpha))$, and $\mathcal{O}_R^p(\alpha)$ is inverse closed in $L(L_R^p(\alpha))$.

In other words: the mapping which assigns to each coset $\pi(A) \in \mathcal{O}_R^p(\alpha)^\pi$ the operator-valued function

$$t \mapsto W_t(\pi(A)), \quad t \in \dot{\mathbf{R}},$$

is a *symbol mapping* for $\mathcal{O}_R^p(\alpha)^\pi$.

   An explicit definition of the operator W_t will be given in the course of the proof of this theorem. Further we emphasize that our Fredholm criterion is effective in the following sense: From (7) and (8) we conclude that $W_t(\pi(A))$ belongs to $\Sigma_R^p(\alpha)$ in case $t \in \{0, \infty\}$ and to $\Sigma_R^p(0)$ else. But the invertibility of operators in $\Sigma_R^p(\alpha)$ can be effectively checked (compare Section 2.1.3).

   The proof of Theorem 3.1 is based on the scheme presented in Chapter 1. Let us demonstrate this in detail for the unweighted case $\alpha = 0$, and then point out the essence of the weighted case.

3.1.2 Localization

The algebraization of the Fredholmness leads to invertibility in the Calkin algebra $L(L_R^p(0))/K(L_R^p(0))$, and essentializing this problem leads to invertibility in the subalgebra $\mathcal{O}_R^p(0)^\pi$ of the Calkin algebra. So we are left with localizing.

   If $f \in C(\dot{\mathbf{R}})$ then, by 3.1.1(ii), the coset $\pi(fI)$ belongs to the center of $\mathcal{O}_R^p(0)^\pi$, thus, $\pi(C(\dot{\mathbf{R}})I)$ is a central and unital subalgebra of $\mathcal{O}_R^p(0)^\pi$. Afterwards we shall verify that $\pi(C(\dot{\mathbf{R}})I)$ is isometrically isomorphic to the C^*-algebra $C(\dot{\mathbf{R}})$. Antici-pating this result we conclude via the Gelfand Naimark theorem that the maximal ideal space of $\pi(C(\dot{\mathbf{R}})I)$ is homeomorphic to the one-point compactification $\dot{\mathbf{R}}$ of the real axis, and that the maximal ideal of $\pi(C(\dot{\mathbf{R}})I)$ associated with $t \in \dot{\mathbf{R}}$ is just given by

$$\{ \pi(fI) \; : \; f \in C(\dot{\mathbf{R}}) \quad \text{and} \quad f(t) = 0 \} \, . \tag{1}$$

Let I_t stand for the smallest closed two-sided ideal of $\mathcal{O}_R^p(0)^\pi$ which contains the ideal (1), and write $\mathcal{O}_R^p(0)_t^\pi$ for the quotient algebra $\mathcal{O}_R^p(0)^\pi / I_t$ and $\Phi_t^\pi : \mathcal{O}_R^p(0) \to \mathcal{O}_R^p(0)_t^\pi$ for the canonical homomorphism.

3.1.3   Identification

Our goal is the construction of locally equivalent representations of the local alge-bras $\mathcal{O}_R^p(0)_t^\pi$. For, we define algebra homomorphisms $W_t : \mathcal{O}_R^p(0) \to L(L_R^p(0))$ by 3.1.1(1) and (2) for $t = 0$ and $t = \infty$, respectively, and by

$$W_t(A) \; = \; W_0(U_{-t}AU_t) \quad \text{for} \quad t \in \dot{\mathbf{R}} \setminus \{0, \infty\} \, . \tag{1}$$

The definition (1) is correct (compare 3.1.1(iv)). Further we infer from 3.1.1 (3)-(6) that

$$W_t(S_R) \;\; = \;\; S_R \, , \tag{2}$$
$$W_t(aI) \;\; = \;\; a(t+0)\chi_{R^+}I + a(t-0)\chi_{R^-}I \, , \tag{3}$$

and

$$W_t(K) \;\; = \;\; 0 \quad \text{for} \quad \text{compact} \quad K \, .$$

Notice that the so-defined homomorphisms W_t act on $\mathcal{O}_R^p(0)$ and not on $\mathcal{O}_R^p(0)^\pi$, and so they cannot be the homomorphisms stated in the theorem. But from 3.1.1(6) we conclude that $W_t(A)$ depends on the coset $\pi(A)$ only. So we can actually think of W_t as acting on $\mathcal{O}_R^p(0)^\pi$, and these 'quotient homomorphisms' are the W_t from the theorem.

Further, because of $W_t(bI) = b(t)I$ for all $b \in C(\dot{\mathbf{R}})$ (by (3)), we have $W_t(B) = 0$ for all cosets $B \in \mathcal{O}_R^p(0)^\pi$ belonging to the ideal I_t and, thus, the operator $W_t(A)$ depends on the local coset $\Phi_t^\pi(A)$ only. So the mappings

$$\mathcal{O}_R^p(0)_t^\pi \to L(L_R^p(0)) \, , \; \Phi_t^\pi(A) \mapsto W_t(A) \tag{4}$$

are correctly defined and continuous algebra homomorphisms. We denote the ho-momorphism (4) by W_t again (the context will always make plain which W_t is meant).

Since (4) is an algebra homomorphism we conclude that $W_t(A)$ is invertible if $\Phi_t^\pi(A)$ is invertible. Let us prove the reverse direction, i.e. suppose $W_t(A)$ to be invertible in $L(L_R^p(0))$ for some $A \in \mathcal{O}_R^p(0)$. We have already mentioned that

$W_t(A) \in \Sigma_R^p(0)$, and since $\Sigma_R^p(0)$ is an inverse closed algebra (compare Section 2.1.3), the operator $W_t(A)$ is even invertible in $\Sigma_R^p(0)$.

Now define a mapping $W_t' : \Sigma_R^p(0) \to \mathcal{O}_R^p(0)_t^\pi$ as follows:

$$W_t'(B) = \begin{cases} \Phi_t^\pi(U_t B U_{-t}) & \text{if } t \in \mathbf{R} \\ \Phi_t^\pi(B) & \text{if } t = \infty. \end{cases}$$

Clearly, $U_t B U_{-t}$ is in $\mathcal{O}_R^p(0)$ for B in $\Sigma_R^p(0)$ whence the correctness of this definition follows. Moreover, W_t' is a continuous algebra homomorphism. Consequently, if $W_t(A)$ is invertible in $\Sigma_R^p(0)$ then $W_t'(W_t(A))$ is invertible in $\mathcal{O}_R^p(0)_t^\pi$. We claim that

$$W_t'(W_t(A)) = \Phi_t^\pi(A) \qquad (5)$$

for all $A \in \mathcal{O}_R^p(0)$.

Since W_t, W_t', and Φ_t^π are continuous algebra homomorphisms, it suffices to check (5) for the generating operators S_R and $aI (a \in PC)$ of $\mathcal{O}_R^p(0)$ in place of A. If $A = S_R$ then $U_t W_t(A) U_{-t} = S_R$, and (5) is evident. If $A = aI$ then

$$W_t'(W_t(A)) = \Phi_t^\pi(a(t-0)\chi_{(-\infty,t]} I + a(t+0)\chi_{[t,\infty)} I),$$

and this coset coincides with $\Phi_t^\pi(aI)$ since the function

$$a(t-0)\chi_{(-\infty,t]} + a(t+0)\chi_{[t,\infty)} - a$$

is continuous at t and has a zero there (compare the proof of Theorem 2.3). Thus, in accordance with the general scheme, the algebras $\mathcal{O}_R^p(0)_t^\pi$ and $\Sigma_R^p(0)$ are topologically isomorphic, and the coset $\Phi_t^\pi(A)$ is invertible if and only if the operator $W_t(A)$ is invertible.

3.1.4 Inverse closedness

One readily verifies that all hypotheses of the general scheme are satisfied in our situation. This completes the proof in case $\alpha = 0$.

3.1.5 Determination of the central subalgebra

Clearly, the mapping

$$C(\dot{\mathbf{R}}) \to \pi(C(\dot{\mathbf{R}})I), \ f \mapsto \pi(fI) \qquad (1)$$

is a continuous and surjective homomorphism. We claim that the kernel of this mapping is trivial. Indeed, by 3.1.3(3),

$$W_t(\pi(fI)) = f(t)I, \quad t \in \dot{\mathbf{R}}. \qquad (2)$$

Thus, if $\pi(fI) = 0$ then $f \equiv 0$. Moreover, since $\|Z_s\| \|Z_s^{-1}\| = 1$ and $\|U_t\| = 1$ we conclude from (2) that

$$|f(t)| = \|W_t(\pi(fI))\| \leq \|\pi(fI)\|$$

whence the isometry of the homomorphism (1) follows:

$$\|f\|_{C(\dot{R})} = \sup_{t \in \dot{R}} |f(t)| \leq \|\pi(fI)\| \leq \|fI\|_{L(L^p_R(\alpha))} = \|f\|_{C(\dot{R})} . \qquad (3)$$

3.1.6   The weighted case

If $\alpha \neq 0$ then localizing yields local algebras $\mathcal{O}^p_R(\alpha)^\pi_t$. For $t \in \{0, \infty\}$ one shows in the very same way as in 3.1.3 that these algebras are topologically isomorphic to $\Sigma^p_R(\alpha)$. We are going to explain that, for $t \in \dot{\mathbf{R}} \setminus \{0, \infty\}$, the local algebras $\mathcal{O}^p_R(\alpha)^\pi_t$ and $\mathcal{O}^p_R(0)^\pi_t$ are isometrically isomorphic in a natural way. For, we define a mapping $\Delta : L^p_R(0) \to L^p_R(\alpha)$ by $(\Delta f)(s) = |s|^{-\alpha} f(s)$. Evidently, Δ is an isometry, and with regard to Sections 2.1.2 and 2.1.3 we find that $\Delta^{-1} \mathcal{O}^p_R(\alpha)\Delta = \mathcal{O}^p_R(0)$. Moreover, the mapping $A \mapsto \Delta^{-1} A\Delta$ sends compact operators to compact operators and, hence, it involves an isomorphism

$$\mathcal{O}^p_R(\alpha)^\pi \to \mathcal{O}^p_R(0)^\pi , \; \pi(A) \mapsto \pi(\Delta^{-1} A\Delta) .$$

Further, this isomorphism sends $\pi(fI) \in \mathcal{O}^p_R(\alpha)^\pi$ to $\pi(\Delta^{-1} f\Delta) = \pi(fI) \in \mathcal{O}^p_R(0)^\pi$ for each multiplication operator fI whence follows that the mapping

$$\mathcal{O}^p_R(\alpha)^\pi_t \to \mathcal{O}^p_R(0)^\pi_t , \; \Phi^\pi_t(A) \mapsto \Phi^\pi_t(\Delta^{-1} A\Delta)$$

is a natural isomorphism between the local algebras at t. Now observe that, for $t \notin \{0, \infty\}$, the function $s \mapsto |s|^\alpha$ is continuous at t and does not vanish there. Thus,

$$\Phi^\pi_t(\Delta^{-1} A\Delta) = \Phi^\pi_t(|t|^\alpha A|t|^{-\alpha}) = \Phi^\pi_t(A) \in \mathcal{O}^p_R(0)^\pi_t ,$$

and this is what we wanted to prove. To identify the local algebras we can now proceed as in Section 3.1.3, and the proof of the inverse closedness should be based on Theorem 1.12.

For another way to treat the weighted local algebras $\mathcal{O}^p_R(\alpha)^\pi_t$ one defines operators W_t by

$$W_t : A \mapsto s\text{--}\lim_{s \searrow 0} Z^{-1}_s U_{-t} \Delta^{-1} A\Delta U_t Z_s \qquad (1)$$

and then reifies the general scheme with these operators. The appearance of the operators Δ, Δ^{-1} in (1) is caused by the circumstance that the operators U_t are not defined on all $L^p_R(\alpha)$ if $\alpha \neq 0$ and $t \neq 0$.

Example 3.1 Let us illustrate how Theorem 3.1 works. We consider the operator $A = aI + bS_R + cS_{R,\gamma}$ where a, b, c are functions in $C(\dot{\mathbf{R}})$ and the operator $S_{R,\gamma}$ is given by

$$(S_{R,\gamma} f)(t) = \frac{1}{\pi i} \int_R \left|\frac{t}{s}\right|^\gamma \frac{f(s)}{s - t} ds .$$

For $0 < 1/p + \gamma < 1$, this operator is bounded on $L_R^p(0)$ and, moreover, it is contained in the algebra $\Sigma_R^p(0)$ and, thus, also in $\mathcal{O}_R^p(0)$ (compare Section 2.1.3). By Theorem 3.1, the operator A is Fredholm on $L_R^p(0)$ if and only if all operators $W_t(A)$, $t \in \dot{\mathbf{R}}$, are invertible. A simple computation gives that

$$W_0(A) = a(0)I + b(0)S_R + c(0)S_{R,\gamma} \in L_R^p(0) , \tag{2}$$

$$W_\infty(A) = a(\infty)I + b(\infty)S_R + c(\infty)S_{R,\gamma} \in L_R^p(0) . \tag{3}$$

For $t \in \dot{\mathbf{R}} \setminus \{0, \infty\}$ one can argue as in 3.1.3 to find that

$$W_t(A) = a(t)I + (b(t) + c(t)) S_R . \tag{4}$$

The latter operators are invertible if and only if

$$a(t) \pm (b(t) + c(t)) \neq 0$$

as one easily checks by writing (4) as

$$(a(t) + (b(t) + c(t))) \frac{I + S_R}{2} + (a(t) - (b(t) + c(t))) \frac{I - S_R}{2}$$

and taking into account that the operators $(I \pm S_R)/2$ are complementary idempotents (i.e. $((I \pm S_R)/2)^2 = (I \pm S_R)/2$). So we are left on the invertibility of the operators (2) and (3) which are of the form $B = uI + vS_R + wS_{R,\gamma}$. We have seen in Section 2.1.3 that B is invertible if and only if the operator

$$RBR^{-1} = \begin{pmatrix} uI + vS_{R^+} + wS_{R^+,\gamma} & -vN_\pi - wN_{\pi,\gamma} \\ vN_\pi + wN_{\pi,\gamma} & uI - vS_{R^+} - wS_{R^+,\gamma} \end{pmatrix}$$

is invertible. Now employ identity 2.1.2 (9) to compute the determinant of RBR^{-1} as

$$u^2 - (v + w)^2 + 2vw(1 - \cos \pi\gamma)N_\pi N_{\pi,\gamma} .$$

This is just the operator of Mellin convolution by the function

$$u^2 - (v + w)^2 + 2vw(1 - \cos \pi\gamma)n_\pi n_{\pi,\gamma} , \tag{5}$$

thus, the operator B is invertible if and only if the function (5) does not vanish on $\overline{\mathbf{R}}$. Since $(n_\pi n_{\pi\gamma})(\pm\infty) = 0$, this necessarily implies that $u \neq \pm(v+w)$, and if this condition is satisfied we invoke 2.1.2 (8) to find that (5) is invertible if and only if

$$u^2 - (v + w)^2 + \frac{2vw(1 - \cos \pi\gamma)}{\sinh \pi(z + i/p) \sinh \pi(z + i(1/p + \gamma))} \neq 0, \quad z \in \mathbf{R} ,$$

or, equivalently

$$\frac{(u^2 - (v - w)^2) \cos \pi\gamma - 4vw}{u^2 - (v + w)^2} \neq \cosh \pi(2z + i(2/p + \gamma)) , \quad z \in \mathbf{R} .$$

Summary *The operator $A = aI + bS_R + cS_{R,\gamma}$ is Fredholm on $L_R^p(0)$ if and only if $a(t) \neq \pm(b(t) + c(t))$, $t \in \dot{\mathbf{R}}$, and if the points*

$$\frac{(a^2(t) - (b(t) - c(t))^2)\cos\pi\gamma - 4b(t)c(t)}{a^2(t) - (b(t) + c(t))^2}, \quad t \in \{0, \infty\},$$

do not lie on the hyperbola $\{\cosh\pi(2z + i(2/p + \gamma)), z \in \mathbf{R}\}$.

3.1.7 Invertibility

Let us add some facts concerning invertibility of singular integral operators.

Theorem 3.2 *(Coburn) Let $\mathbf{I}$ be an (finite or infinite) interval and a and b be piecewise continuous functions on $\mathbf{I}$ which are not identically equal to zero. Then the singular integral operator $aI + bS_I$ has a trivial kernel or a trivial cokernel.*

Corollary 3.1 *Let a and b be piecewise continuous. The singular integral operator $aI + bS_I$ is invertible if and only if it is Fredholm and its index is zero.*

The index formula will be formulated only for the case $\mathbf{I} = \mathbf{R}$. That this case covers all others is the assertion of the following proposition.

Proposition 3.1 *Let a, b be piecewise continuous on the interval $\mathbf{I}$ and define $\tilde{a}$, $\tilde{b}$ by*

$$\tilde{a}(t) = \begin{cases} a(t) & \text{if } t \in \mathbf{I} \\ 1 & \text{if } t \in \mathbf{R} \setminus \mathbf{I} \end{cases}, \quad \tilde{b}(t) = \begin{cases} b(t) & \text{if } t \in \mathbf{I} \\ 0 & \text{if } t \in \mathbf{R} \setminus \mathbf{I} \end{cases}.$$

Then the functions $\tilde{a}, \tilde{b}$ are piecewise continuous on $\mathbf{R}$, the operators $aI + bS_I$ and $\tilde{a}I + \tilde{b}S_R$ are simultaneously Fredholm or not, and if they are Fredholm then their indices coincide.

If $aI + bS_R$ is Fredholm then, as one can see via Theorem 3.1, the function $g := (a + b)/(a - b)$ is correctly defined, piecewise continuous on $\mathbf{R}$, and $g(t \pm 0) \neq 0$ for all $t \in \mathbf{R}$. We complete the range of g to a closed curve Γ by filling in circular arcs $C(t)$ into each gap $g(t - 0), g(t + 0)$. These arcs are defined as follows: Set

$$\delta(t) = \begin{cases} 1/p & \text{if } t \in \mathbf{R} \setminus \{0\} \\ 1/p + \alpha & \text{if } t = 0 \\ 1 - 1/p - \alpha & \text{if } t = \infty \end{cases}$$

and

$$f(t, \mu) = \begin{cases} \frac{\sin 2\pi(\frac{1}{2} - \delta)\mu \exp 2\pi i(\frac{1}{2} - \delta)\mu}{\sin 2\pi(\frac{1}{2} - \delta)\exp 2\pi i(\frac{1}{2} - \delta)} & \text{if } \delta(t) \in (0, 1) \setminus \{\frac{1}{2}\} \\ \mu & \text{if } \delta(t) = \frac{1}{2} \end{cases},$$

and define

$$C(t) = \{g(t + 0)f(t, \mu) + g(t - 0)(1 - f(t, \mu)), \quad \mu \in [0, 1]\}.$$

If $aI + bS_R$ is Fredholm then the curve Γ does not contain the origin, and the winding number of Γ is defined as the increment of $\frac{1}{2\pi}\arg z$ when z runs through Γ counterclockwise.

Proposition 3.2 *Let a, b be piecewise continuous. The index of the operator $aI + bS_R$ equals the negative of the winding number of Γ.*

3.2 Algebras of approximation sequences using piecewise constant splines

3.2.1 Motivation

To study Fredholmness (or even invertibility) of the homogeneous operators in $\Sigma_R^p(\alpha)$ is a simple matter: the Mellin transform associates with each of these operators a matrix-valued function the invertibility of which is necessary and sufficient for the Fredholmness of the operator (compare Section 2.1.3). The things become essentially more involved for singular integral operators with variable coefficients (see Sections 3.1.1-3.1.5): here one has to consider a whole Banach algebra of singular integral operators in order to have a place where localization techniques work, and these local techniques reduce the Fredholm problem for arbitrary operators in $\mathcal{O}_R^p(\alpha)$ to that one of homogeneous operators again.

Concerning the stability of spline approximation sequences the situation is completely parallel. Spline approximation sequences for homogeneous operators in $\Sigma_R^p(\alpha)$ can be handled easily: one only has to translate an approximation sequence of operators on a spline space into a sequence of operators on $l_Z^p(\alpha)$; this new sequence proves to be constant, and so the stability of the original sequence becomes equivalent to the invertibility of a certain operator.

Evidently, to attack approximation methods for operators with variable coefficients, it seems to be desirable to have a whole algebra of approximation sequences to work in. The most obvious way to construct such an algebra is to take the approximation sequences one is previously interested in, and to pack them (together with some other sequences which are needed for technical reasons) into an algebra. In this way, each concrete approximation sequence gives raise to its own algebra and, of course, each of these algebras must be studied individually.

In contrast to this, our goal in this chapter will be to construct an algebra which contains *all* approximation sequences considered above at once, but whose definition is independent of concrete approximation methods. To get an idea of how such a comprehensive algebra should be built, we return to operator theory once more. In fact, there is another nice characterization of the algebra $\mathcal{O}_R^p(0)$, namely, it is the smallest closed subalgebra of $L(L_R^p(0))$ which encloses the algebra $\Sigma_R^p(0)$ of homogeneous operators, and which is translation invariant. Indeed, if $A \in \mathcal{O}_R^p(0)$ and if U_t stands for the translation operator $(U_t f)(s) = f(s - t)$ then it is easy to see that $U_{-t} A U_t \in \mathcal{O}_R^p(0)$ again. For the reverse direction observe that each piecewise continuous function can be approximated as closely as desired by a piecewise constant function, and that each piecewise constant function can be figured by a linear combination of shifted characteristic functions $\chi_{R^{\pm}}$.

To derive an analogous description of the algebra $\mathcal{O}_R^p(\alpha)$ (with $\alpha \neq 0$) one has to replace $\Sigma_R^p(0)$ by $\Sigma_R^p(\alpha)$ and the shift U_t (which is no longer bounded on $L_R^p(\alpha)$) by the weighted shift $\Delta U_t \Delta^{-1}$ where $(\Delta f)(s) = |s|^{-\alpha} f(s)$.

These facts suggest the following axioms for the construction of an algebra $\mathcal{A}$ of approximation sequences for operators in $\mathcal{O}_R^p(\alpha)$:

1. $\mathcal{A}$ should contain a subalgebra whose elements are thought of as approximation sequences for operators in $\Sigma_R^p(\alpha)$ and which, in particular, contains all approximation sequences considered in Sections 2.11.1-2.11.10.

2. $\mathcal{A}$ should be a translation invariant algebra.

Let us analyse these axioms in detail.

3.2.2 How to realize the axioms

To get a first idea we specify the set F of presplines to be $\{\chi_{[0,1]}\}$ simply, and we start with examining all things on the Hilbert space $L_R^2(0)$. Let (A_n) be one of the approximation sequences considered in 2.11.1-2.11.10. Then the operators $E_{-n} A_n E_n$ are independent of n, and each of these operators is constituted by some of the following "elementary bricks": operators $T^0(a)$ with piecewise continuous multipliers a, discretized Mellin convolutions $G(b)$, the projection P, and compact operators. The smallest closed algebra containing these operators is just the Toeplitz algebra $\text{alg}\,(T^0(PC), P)$. Sometimes it proves to be more convenient to work in a smaller algebra than $\text{alg}\,(T^0(PC), P)$ which originates from prescribing the set of discontinuities. That means, given a closed subset Ω of $\mathbf{T}$ we let PC^Ω stand for the collection of all piecewise continuous multipliers which are continuous on $\mathbf{T} \setminus \Omega$, and we consider the smallest closed subalgebra of $\text{alg}\,(T^0(PC), P)$ which contains all operators $T^0(a)$ with $a \in PC^\Omega$ as well as the projection P. Let us denote this algebra by $\mathcal{T}_{1,1,\Omega}^2(0)$ (this notation will be generalized later on). For example, all operators $E_{-n} A_n E_n$ which arise from projection methods belong to $\mathcal{T}_{1,1,\{1\}}^2(0)$, and the quadrature methods in 2.11.9 lead to operators in $\mathcal{T}_{1,1,\{1,-1\}}^2(0)$ and $\mathcal{T}_{1,1,\{1,-1,i,-i\}}^2(0)$, respectively.

Thus, the first axiom involves that

the algebra $\mathcal{A}$ should contain all sequences $(E_n^F A E_{-n}^F | S_n^F)$ with A running through the Toeplitz algebra $\text{alg}\,(T^0(PC), P)$.

The fairly simple idea of making the algebra $\mathcal{A}$ translation invariant (= Axiom 2) brings up a little notational difficulty: what is a "shifted sequence"? Since the translation operator U_t maps the spline space S_n^F onto itself only if t is an integer multiple of $1/n$, it becomes evident that, in general, $U_t A U_{-t}$ is not operating on

S_n^F even if A does. That's why we define the translation of a given sequence (A_n) of operators $A_n : S_n^F \to S_n^F$ by $(U_{\frac{\{tn\}}{n}} A_n U_{-\frac{\{tn\}}{n}})$ with $\{x\}$ referring to the smallest integer which is greater than or equal to x. So the correct formulation of Axiom 2 is:

If $(A_n) \in \mathcal{A}$ *then* $(U_{\frac{\{tn\}}{n}} A_n U_{-\frac{\{tn\}}{n}}) \in \mathcal{A}$ *for all* $t \in \mathbf{R}$.

3.2.3 Definition of the algebra

To have some frame to work in we start with defining a comprehensive algebra, $\mathcal{F}$, of operator sequences. Let $\mathcal{F}$ stand for the collection of all bounded sequences (A_n), $A_n : S_n^F \to S_n^F$, with S_n^F thought of as being a subspace of $L_\mathbf{R}^2$. Provided with the operations $(A_n) + (B_n) = (A_n + B_n)$, $\alpha(A_n) = (\alpha A_n)$, and $(A_n)(B_n) = (A_n B_n)$, and with the norm

$$\|(A_n)\| = \sup_n \|A_n L_n^{F,F}\| < \infty \,,$$

the set $\mathcal{F}$ becomes a Banach algebra. Now we define the algebra $\mathcal{A}$ as the smallest closed subalgebra of $\mathcal{F}$ which contains all sequences of the form

$$(U_{\frac{\{tn\}}{n}} E_n^F A E_{-n}^F U_{-\frac{\{tn\}}{n}} | S_n^F) \,,$$

where t runs through $\mathbf{R}$ and A through alg $(T^0(PC), P)$. This algebra satisfies Axioms 1 and 2 evidently and, moreover, it is the smallest Banach algebra which owns these properties. One can show (and we shall do this in Sections 3.8.1-3.8.6 in a more general setting) that this algebra indeed covers a bulk of concrete approximation methods. For example, if A is the singular integral operator $aI + bS_R + T$ where a and b are piecewise continuous on $\dot{\mathbf{R}}$ and continuous on $\mathbf{R} \setminus \mathbf{Z}$ and where T is compact, then the sequences of the Galerkin, collocation, and qualocation methods for A belong to $\mathcal{A}$!

3.2.4 The stability theorem

The stability criterion for sequences in $\mathcal{A}$ which we are going to formulate now takes a similar form as the Fredholm criterion in Theorem 3.1. But, in contrast to the Fredholm case, now one needs *two* families of homomorphisms, say W_s and W^t, which are labeled by the points $s \in \dot{\mathbf{R}}$ (= the common variable, as in Fredholm theory) and $t \in \mathbf{T}$ (= the co-variable). In order to introduce these homomorphisms we recall the definition of the operators Y_t, $t \in \mathbf{T}$:

$$Y_t : l_Z^2(0) \to l_Z^2(0) \,, \quad (x_i) \mapsto (t^{-i} x_i) \,,$$

and of the discrete shift operators V_n, $n \in \mathbf{Z}$:

$$V_n : l_Z^2(0) \to l_Z^2(0) \,, \quad (x_i) \mapsto (x_{i-n}) \,.$$

Proposition 3.3 *Let $(A_n) \in \mathcal{A}$. Then,*

(a) *for each $t \in \mathbf{T}$, there is an operator $W^t(A_n)$ in $L(L_R^2(0))$ such that*

$$E_n^F Y_{t^{-1}} E_{-n}^F A_n E_n^F Y_t E_{-n}^F L_n^{F,F} \to W^t(A_n) \quad and$$

$$(E_n^F Y_{t^{-1}} E_{-n}^F A_n E_n^F Y_t E_{-n}^F L_n^{F,F})^* \to (W^t(A_n))^*$$

*strongly as $n \to \infty$. The mapping $W^t : \mathcal{A} \to L(L_R^2(0))$ is a continuous *-homomorphism.*

(b) *for each $s \in \mathbf{R}$, there is an operator $W_s(A_n)$ in $L(l_Z^2(0))$ such that*

$$V_{-\{sn\}} E_{-n}^F A_n E_n^F V_{\{sn\}} \to W_s(A_n) \quad and$$

$$(V_{-\{sn\}} E_{-n}^F A_n E_n^F V_{\{sn\}})^* \to (W_s(A_n))^*$$

*strongly as $n \to \infty$. The mappings $W_s : \mathcal{A} \to L(l_Z^2(0))$ are continuous *-homomorphisms.*

(c) *the cosets $\pi(E_{-n}^F A_n E_n^F) := E_{-n}^F A_n E_n^F + K(l_Z^2(0))$ are independent of n. If $W_\infty(A_n)$ denotes one of these cosets then the mapping*

$$W_\infty : \mathcal{A} \to L(l_Z^2(0))/K(l_Z^2(0)), \ (A_n) \mapsto W_\infty(A_n)$$

*is a continuous *- homomorphism.*

We shall prove this proposition in Section 3.5.1 in a more general context: for $L_R^p(\alpha)$-spaces and for arbitrary canonical prebases F.

Theorem 3.3 *(Stability theorem) A sequence $(A_n) \in \mathcal{A}$ is stable if and only if all operators (cosets) $W^t(A_n)$ and $W_s(A_n)$ with $t \in \mathbf{T}$ and $s \in \mathbf{R}$ are invertible.*

The complete proof is lengthy, and parts of it are rather technical and not very effective. So we shall give only a sketch of the proof here and refer to subsequent sections for a detailed treatment of the general case. The proof follows the scheme presented in the first chapter. Let us explain the main steps.

3.2.5   Algebraization and inverse closedness

Let $\mathcal{K}$ stand for the collection of all sequences (K_n) in $\mathcal{F}$ such that $\|K_n\| \to 0$ as $n \to \infty$. It is elementary to see, and has been repeatedly used hitherto, that $\mathcal{K}$ is an ideal in $\mathcal{F}$, and that a sequence $(A_n) \in \mathcal{F}$ is stable if and only if the coset $(A_n) + \mathcal{K}$ is invertible in the quotient algebra $\mathcal{F}/\mathcal{K}$. But the ideal $\mathcal{K}$ is not completely contained in our algebra A; the intersection $\mathcal{G} := \mathcal{A} \cap \mathcal{K}$ is described by the following proposition.

Proposition 3.4 *The set $\mathcal{G}$ is a closed two-sided ideal of $\mathcal{A}$, and it consists of all sequences (G_n) satisfying*

(a) $\lim_{n \to \infty} \|G_n\| = 0$,

(b) *the operators $E^F_{-n} G_n E^F_n$ are compact on $l^2_Z(0)$ for every n.*

(See Section 3.5.2 for a proof.)

On defining an involution on $\mathcal{F}$ by $(A_n)^* = (A_n^*)$, this algebra becomes a C^*-algebra, and $\mathcal{A}$ proves to be a C^*-subalgebra of $\mathcal{F}$. Now we conclude from 1.1.4(b) that $\mathcal{A}+\mathcal{K}$ is a C^*-subalgebra of $\mathcal{F}$ and, thus, $(\mathcal{A}+\mathcal{K})/\mathcal{K}$ is a C^*-subalgebra of $\mathcal{F}/\mathcal{K}$. By 1.1.4(d), this algebra is inverse closed in $\mathcal{F}/\mathcal{K}$, which implies that a sequence $(A_n) \in \mathcal{A}$ is stable if and only if the coset $(A_n) + \mathcal{K}$ is invertible in $(\mathcal{A} + \mathcal{K})/\mathcal{K}$. Finally, 1.1.4(b) entails that the algebras $(\mathcal{A} + \mathcal{K})/\mathcal{K}$ and $\mathcal{A}/(\mathcal{A} \cap \mathcal{K}) = \mathcal{A}/\mathcal{G}$ are isomorphic. Consequently, a sequence (A_n) is stable if and only if the coset $(A_n)+\mathcal{G}$ is invertible in $\mathcal{A}/\mathcal{G}$. This yields both the algebraization step and the proof of the inverse closedness, and we can focus our attention on the invertibility of cosets in $\mathcal{A}/\mathcal{G}$ in what follows.

3.2.6 Essentialization and localization

The algebra $\mathcal{O}^p_R(\alpha)$ has been studied by localizing over the continuous functions on $\dot{R}$. Clearly, operators of multiplication by functions do not belong to the sequence algebra $\mathcal{A}$, but there is a nice substitute for them, as the following proposition points out.

Proposition 3.5 *Let $f \in C(\dot{R})$. Then the sequence $(L^{F,F}_n fI|S^F_n)$ is in $\mathcal{A}$. In particular,*

$$W^t(L^{F,F}_n fI|S^F_n) = fI \qquad for\ all\quad t \in \mathbf{T}\ ,$$

$$W_s(L^{F,F}_n fI|S^F_n) = f(s)I \quad for\ all\quad s \in \mathbf{R}\ ,$$

$$W_\infty(L^{F,F}_n fI|S^F_n) = f(\infty)\pi(I)\ .$$

(See Section 3.5.2 for a proof.)

The affiliation of the sequences $(L^{F,F}_n fI|S^F_n)$ with $f \in C(\dot{R})$ to the algebra $\mathcal{A}$ suggests to study this algebra by the same means as the operator algebra $\mathcal{O}^p_R(\alpha)$, viz. by a localization over continuous functions on $\dot{R}$ or, more precisely, over the sequences $(L^{F,F}_n fI|S^F_n)$ with f continuous. For, we had to know that these sequences belong to the center of $\mathcal{A}$. Let us consider some examples to check this idea.

If f and g are continuous on $\dot{R}$ then the commutator of $(L^{F,F}_n fI|S^F_n)$ and $(L^{F,F}_n gI|S^F_n)$ can be written as

$$(L^{F,F}_n fI|S^F_n)\,(L^{F,F}_n gI|S^F_n) - (L^{F,F}_n gI|S^F_n)\,(L^{F,F}_n fI|S^F_n)$$

$$
\begin{aligned}
&= \; (L_n^{F,F} f L_n^{F,F} g I | S_n^F) - (L_n^{F,F} g L_n^{F,F} f I | S_n^F) \\
&= \; (L_n^{F,F} (fg - gf) I | S_n^F) + (L_n^{F,F} f (L_n^{F,F} - I) g I | S_n^F) \\
&\quad + (L_n^{F,F} g (I - L_n^{F,F}) f I | S_n^F) \, .
\end{aligned}
\tag{1}
$$

The first sequence vanishes identically, and the other two sequences tend to zero in the operator norm by the commutator property. Thus, the commutator (1) does not vanish in general but it always belongs to the ideal $\mathcal{G}$.

Next we consider the commutator of $(L_n^{F,F} f I | S_n^F)$ and $(L_n^{F,F} S_R | S_n^F)$. As in (1) we obtain

$$
\begin{aligned}
&(L_n^{F,F} f I | S_n^F)(L_n^{F,F} S_R | S_n^F) - (L_n^{F,F} S_R I | S_n^F)(L_n^{F,F} f I | S_n^F) \\
&= \; (L_n^{F,F} (f S_R - S_R f I) | S_n^F) + (L_n^{F,F} f (L_n^{F,F} - I) S_R | S_n^F) \\
&\quad + (L_n^{F,F} S_R (I - L_n^{F,F}) f I) | S_n^F) \, .
\end{aligned}
\tag{2}
$$

The latter two sequences are in $\mathcal{G}$ again, but the first one is no longer in $\mathcal{G}$ since

$$
\operatorname*{s\text{-}lim}_{n \to \infty} L_n^{F,F} (f S_R - S_R f) L_n^{F,F} \; = \; f S_R - S_R f I \, ,
\tag{3}
$$

and this operator does not vanish in general. But $f S_R - S_R f I$ is always a compact operator, that is, the commutator (2) is of the form

$$
(L_n^{F,F} K | S_n^F) + (G_n) \quad \text{with } K \text{ compact, } (G_n) \in \mathcal{G} \, .
\tag{4}
$$

A closer look will show that sequences of the same form appear as commutators of $(L_n^{F,F} f I | S_n^F)$ with arbitrary sequences $(E_n^F A E_{-n}^F | S_n^F)$ where $A \in \mathcal{T}_{1,1,\{1\}}^2(0)$ (remember that $(L_n^{F,F} S_R | S_n^F) = (E_n^F T^0 (\lambda^{F,F})^{-1} T^0 (\sigma^{F,F}) E_{-n}^F | S_n^F)$ and that the function $(\lambda^{F,F})^{-1} \sigma^{F,F}$ has its only discontinuity at $1 \in \mathbf{T}$). For sequences with A related to discontinuities at other points $z \in \mathbf{T}$ one can think of A as "rotation" of an operator A' with discontinuities at 1. This makes it plausible that the commutator of $(L_n^{F,F} f I | S_n^F)$ with this sequence should be a "rotation" of (4), namely

$$
(E_n^F Y_z E_{-n}^F L_n^{F,F} K E_n^F Y_{z-1} E_{-n}^F | S_n^F) + (G_n)
\tag{5}
$$

with K compact, $z \in \mathbf{T}$, and $(G_n) \in \mathcal{G}$.

Our further steps are as follows: First we show that any sequence of the form (5) is in $\mathcal{A}$, then we consider the smallest closed two-sided ideal $\mathcal{J}$ of $\mathcal{A}$ which contains all sequences of this form, and finally we explain that the cosets $(L_n^{F,F} f I | S_n^F) + \mathcal{J}$ belong to the center of the quotient algebra $\mathcal{A}/\mathcal{J}$ for all $f \in C(\dot{\mathbf{R}})$. Here are the exact statements.

Proposition 3.6 *Let $z \in \mathbf{T}$ and let K be a compact operator. Then the sequence*

$$
(K_n) := (E_n^F Y_z E_{-n}^F L_n^{F,F} K E_n^F Y_{z-1} E_{-n}^F | S_n^F)
\tag{6}
$$

is in $\mathcal{A}$, and

$$
W^t(K_n) = \begin{cases} K & \text{if} \quad t = z \\ 0 & \text{if} \quad t \neq z \end{cases} \, , \quad W_s(K_n) = 0 \quad \text{for all } s \in \dot{\mathbf{R}} \, .
$$

(See Section 3.5.2 for a proof.)

As already anounced, we let $\mathcal{J}$ stand for the smallest closed two-sided ideal of the algebra $\mathcal{A}$ which includes the ideal $\mathcal{G}$ and all sequences of the form (6) with $z \in \mathbf{T}$ and K compact. Further we write $\Phi^{\mathcal{J}}$ for the canonical homomorphism from $\mathcal{A}$ onto the quotient algebra $\mathcal{A}/\mathcal{J}$. Then the analogue of property 3.1.1(ii) reads as follows.

Proposition 3.7 *Let* $(A_n) \in \mathcal{A}$ *and* $f \in C(\dot{\mathbf{R}})$. *Then the commutator* $(A_n)(L_n^{F,F} fI|S_n^F) - ((L_n^{F,F} fI|S_n^F)(A_n)$ *is in* $\mathcal{J}$ *or, in other words, the coset* $\Phi^{\mathcal{J}}(L_n^{F,F} fI|S_n^F)$ *belongs to the center of* $\mathcal{A}/\mathcal{J}$.

(See Section 3.5.3 for a proof.)

Thus, the ideal $\mathcal{J}$ is large enough to generate a quotient algebra with a rich center. It remains to verify that it is, on the other hand, small enough that it can be lifted in the sense of Section 1.6.2.

Proposition 3.8 *The ideal* $\mathcal{J}$ *is subject to the conditions in the general lifting theorem.*

Proof   We want to reify the lifting theorem with the algebras $\mathcal{A}$, $\mathcal{G}$ and $\mathcal{J}$ defined above in place of the $\mathcal{A}$, $\mathcal{G}$ and $\mathcal{J}$ from Theorem 1.8. For, we identify the index set Ω in this theorem with the unit circle $\mathbf{T}$, the homomorphisms W^t, $t \in \Omega$, with the homomorphisms W^t defined in Proposition 3.3(a), and $\mathcal{C}^t$ and $\mathcal{R}^t$ with $L(L_R^2(0))$ and $K(L_R^2(0))$, respectively. Further we define for $t \in \mathbf{T}$

$$\mathcal{J}^t = \{(E_n^F Y_t E_{-n}^F L_n^{F,F} K E_n^F Y_{t-1} E_{-n}^F|S_n^F) + \mathcal{G}, \ K \ compact\}. \tag{7}$$

Then, clearly, $\mathcal{J}$ is the smallest closed two-sided ideal of $\mathcal{A}$ containing all sequences (A_n) which belong to some $\mathcal{J}^t$, $t \in \mathbf{T}$. So we are left on verifying that the $\mathcal{J}^t$ given by (7) can be identified with the $\mathcal{J}^t$ from the lifting theorem, that is, we have to show that $\mathcal{J}^t$ is a two-sided *ideal* of $\mathcal{A}/\mathcal{G}$, and that the restriction of the quotient homomorphism

$$W^t : \mathcal{A}/\mathcal{G} \to L(L_R^2(0)), \quad (A_n) + \mathcal{G} \mapsto W^t(A_n)$$

(being correctly defined since $\mathcal{G} \subseteq \ker W^t$) onto $\mathcal{J}^t$ is actually an isomorphism between $\mathcal{J}^t$ and $K(L_R^2(0))$.

The latter is almost evident: From Proposition 3.6 we know that

$$W^t((E_n^F Y_t E_{-n}^F L_n^{F,F} K E_n^F Y_{t-1} E_{-n}^F|S_n^F) + \mathcal{G}) = K,$$

thus, W^t maps $\mathcal{J}^t$ onto $K(L_R^2(0))$, and the kernel of the restriction of W^t onto $\mathcal{J}^t$ is trivial.

For showing that $\mathcal{J}^t$ is a left-sided ideal of $\mathcal{A}/\mathcal{G}$ we pick a sequence $(A_n) \in \mathcal{A}$. Then

$$((A_n) + \mathcal{G})((E_n^F Y_t E_{-n}^F L_n^{F,F} K E_n^F Y_{t-1} E_{-n}^F|S_n^F) + \mathcal{G})$$

$$= (E_n^F Y_t E_{-n}^F \left[E_n^F Y_{t-1} E_{-n}^F A_n E_n^F Y_t E_{-n}^F L_n^{F,F} - L_n^{F,F} W^t(A_n) \right] K E_n^F Y_{t-1} E_{-n}^F|S_n^F)$$

$$+ ((E_n^F Y_t E_{-n}^F L_n^{F,F} W^t(A_n) K E_n^F Y_{t^{-1}} E_{-n}^F | S_n^F) + \mathcal{G}) \ .$$

The second item is in $\mathcal{J}^t$ since $W^t(A_n)K$ is compact. For the first one observe that the sequence in brackets goes strongly to zero by the definition of W^t, thus, multiplying this sequence by a compact operator from the right yields a sequence tending to zero in the norm. For showing that $\mathcal{J}^t$ is also a right ideal one simply takes adjoints and proceeds as above then. ∎

In other words: the lifting theorem applies in our setting, that is the coset $(A_n)+\mathcal{G}$ is invertible in $\mathcal{A}/\mathcal{G}$ (or, equivalently, the sequence (A_n) is stable) if and only if all operators $W^t(A_n)$, $t \in \mathbf{T}$, are invertible and if the coset $\Phi^J(A_n)$ is invertible.

In accordance with the general scheme from Chapter 1, we are now going to investigate the invertibility of the coset $\Phi^J(A_n)$ by Allan's local principle. We have already seen that the collection $\mathcal{B}$ of all cosets $\Phi^J(L_n^{F,F} fI|S_n^F)$ with $f \in C(\dot{\mathbf{R}})$ belongs to the center of the quotient algebra $\mathcal{A}/\mathcal{J}$ (Proposition 3.7). Moreover, we have:

Proposition 3.9 *The set $\mathcal{B}$ is a subalgebra of the center of $\mathcal{A}/\mathcal{J}$ which is topologically isomorphic to $C(\dot{\mathbf{R}})$. The maximal ideal space of $\mathcal{B}$ is homeomorphic to $\dot{\mathbf{R}}$, and the maximal ideal corresponding to $s \in \dot{\mathbf{R}}$ is*

$$\{\Phi^J(L_n^{F,F} fI|S_n^F) : f(s) = 0\} : \tag{8}$$

Proof We conclude from the commutator property that

$$(L_n^{F,F} f_1 I|S_n^F)(L_n^{F,F} f_2 I|S_n^F) - (L_n^{F,F} f_1 f_2 I|S_n^F) \in \mathcal{G} \ .$$

Thus, $\mathcal{B}$ is an algebra, and the only thing which remains to prove is that the mapping

$$\mathcal{B} \to C(\dot{\mathbf{R}}) \ , \quad \Phi^J(L_n^{F,F} fI|S_n^F) \mapsto f$$

is a continuous isomorphism. This can be done similarly as in Section 3.1.5 for the Fredholm theory. The other assertions are immediate consequences of this isomorphy. ∎

In accordance with the notations in Section 1.6.3 we write I_s for the smallest closed two-sided ideal of $\mathcal{A}/\mathcal{J}$ containing the maximal ideal (8) of $\mathcal{B}$; and we let $\mathcal{A}_s$ refer to the "local" quotient algebra $(\mathcal{A}/\mathcal{J})/I_s$ and Φ_s^J to the canonical homomorphism $\mathcal{A} \to \mathcal{A}_s$.

3.2.7   Identification of the local algebras

From Propositions 3.5 and 3.6 we infer that, for sequences $(A_n) \in \mathcal{A}$, the operators $W_s(A_n)$ depend on the coset $\Phi_s^J(A_n)$ only. Thus, the mapping

$$\mathcal{A}_s \to \begin{cases} L(l_\mathbf{Z}^2(0)) & \text{if} \quad s \in \mathbf{R} \\ L(l_\mathbf{Z}^2(0))/K(l_\mathbf{Z}^2(0)) & \text{if} \quad s = \infty \end{cases} , \quad \Phi_s^J(A_n) \mapsto W_s(A_n) , \tag{1}$$

is correctly defined, and we denote it by W_s again.

Proposition 3.10 *The homomorphisms W_s given by (1) are locally equivalent representations of the local algebras $\mathcal{A}_s$.*

Proof Recall the definition of locally equivalent homomorphisms: we have to show that the image of W_s is an inverse closed Banach algebra, and that there exists a unital homomorphism

$$W_s' \; : \; \mathrm{Im}\, W_s \to \mathcal{A}_s$$

such that

$$\Phi_s^J(A_n) \; = \; W_s'(W_s(A_n)) \quad \text{for all } (A_n) \in \mathcal{A} \,. \tag{2}$$

For our first goal we claim that

$$\mathrm{Im}\, W_s = \begin{cases} \mathrm{alg}\,(T^0(PC), P) & \text{if} \quad s \in \mathbf{R} \\ \mathrm{alg}\,(T^0(PC), P)/K(l_Z^2(0)) & \text{if} \quad s = \infty \,. \end{cases} \tag{3}$$

Indeed, the sequences

$$(E_n^F T^0(a) E_{-n}^F | S_n^F) \quad \text{and} \quad (U_{\frac{\{sn\}}{n}} E_n^F P E_{-n}^F U_{-\frac{\{sn\}}{n}} | S_n^F) \tag{4}$$

belong to $\mathcal{A}$ by definition, and it is easy to see that

$$W_s(E_n^F T^0(a) E_{-n}^F) = \underset{n\to\infty}{\text{s-lim}}\, V_{-\{sn\}} E_{-n}^F E_n^F T^0(a) E_{-n}^F E_n^F V_{\{sn\}} = T^0(a)$$

and

$$\begin{aligned} W_s(U_{\frac{\{sn\}}{n}} E_n^F P E_{-n}^F U_{-\frac{\{sn\}}{n}}) \\ = \underset{n\to\infty}{\text{s-lim}}\, V_{-\{sn\}} E_{-n}^F U_{\frac{\{sn\}}{n}} E_n^F P E_{-n}^F U_{-\frac{\{sn\}}{n}} E_n^F V_{\{sn\}} \\ = \underset{n\to\infty}{\text{s-lim}}\, V_{-\{sn\}} V_{\{sn\}} P V_{-\{sn\}} V_{\{sn\}} = P \,. \end{aligned}$$

The sequences of the form (4) generate the whole algebra $\mathcal{A}$, and the operators of the form $T^0(a)$ and P generate the whole algebra $\mathrm{alg}\,(T^0(PC), P)$. So we infer that (3) holds in case $s \in \mathbf{R}$. For $s = \infty$ observe that

$$W_\infty(E_n^F T^0(a) E_{-n}^F) \; = \; \pi(T^0(a))$$

and

$$W_\infty(U_{\frac{\{sn\}}{n}} E_n^F P E_{-n}^F U_{-\frac{\{sn\}}{n}}) = \pi(V_{\{sn\}} P V_{-\{sn\}}) = \pi(P)$$

whence (3) analogously follows.

For the construction of the operators W_s' we first let $s = 0$ and define in this case

$$W_0' : \mathrm{alg}\,(T^0(PC), P) \to \mathcal{A}_0 \,, \quad A \mapsto \Phi_0^J(E_n^F A E_{-n}^F | S_n^F) \,.$$

The mappings W_0, W_0', and Φ_0^J are obviously continuous algebra homomorphisms; so it suffices to verify (2) for a collection of generating sequences (A_n) of the algebra $\mathcal{A}$. In case $(A_n) = (E_n^F T^0(a) E_{-n}^F)$ one has

$$W_0'(W_0(A_n)) = W_0'(T^0(a)) = \Phi_0^J(E_n^F T^0(a) E_{-n}^F) = \Phi_0^J(A_n) \,,$$

that is, (2) holds. If $(A_n) = (U_{\frac{\{yn\}}{n}} E_n^F P E_{-n}^F U_{-\frac{\{yn\}}{n}})$ then

$$
\begin{aligned}
W_0(A_n) &= \operatorname*{s\text{-}lim}_{n\to\infty} E_{-n}^F U_{\frac{\{yn\}}{n}} E_n^F P E_{-n}^F U_{-\frac{\{yn\}}{n}} E_n^F \\
&= \operatorname*{s\text{-}lim}_{n\to\infty} V_{\{yn\}} P V_{-\{yn\}} \\
&= \begin{cases} 0 & \text{if } y > 0 \\ I & \text{if } y < 0 \\ P & \text{if } y = 0 \,. \end{cases}
\end{aligned}
$$

For $y = 0$, the proof of (2) follows almost at once. So let $y > 0$ for example. Then (2) is equivalent to

$$\Phi_0^J(A_n) = \Phi_0^J(U_{\frac{\{yn\}}{n}} E_n^F P E_{-n}^F U_{-\frac{\{yn\}}{n}}) = 0 \,. \tag{5}$$

Taking into account Section 2.11.4 one easily verifies that

$$
\begin{aligned}
\Phi_0^J(U_{\frac{\{yn\}}{n}} E_n^F P E_{-n}^F U_{-\frac{\{yn\}}{n}} | S_n^F) &= \Phi_0^J(U_{\frac{\{yn\}}{n}} L_n^{F,F} \chi_{R^+} U_{-\frac{\{yn\}}{n}} | S_n^F) \\
&= \Phi_0^J(L_n^{F,F} \chi_{[\frac{\{yn\}}{n},\infty)} I | S_n^F) \,. \tag{6}
\end{aligned}
$$

Now let f be a continuous function with $f(0) = 1$ and compact support. Then $\Phi_0^J(L_n^{F,F} f I | S_n^F)$ is the identity element in the local algebra $\mathcal{A}_0$; thus, the right-hand side of (6) is equal to $\Phi_0^J(L_n^{F,F} f L_n^{F,F} \chi_{[\frac{yn}{n},\infty)} I | S_n^F)$, and this is, by the commutator property , nothing else than $\Phi_0^J(L_n^{F,F} f \chi_{[\frac{yn}{n},\infty)} I | S_n^F)$. If the support of f is sufficiently small and n is sufficiently large then $f \chi_{[\frac{yn}{n},\infty)} = 0$ (recall that $y > 0$), hence (5) holds.

The sequences (A_n) considered here generate the whole algebra $\mathcal{A}$; so the proof of (2) is complete in case $s = 0$.

For $s \in \mathbf{R} \setminus \{0\}$ one defines

$$W_s' : \operatorname{alg}(T^0(PC), P) \to \mathcal{A}_s \,, \quad A \mapsto \Phi_s^J(U_{\frac{\{sn\}}{n}} E_n^F A E_{-n}^F U_{-\frac{\{sn\}}{n}} S_n^F) \,,$$

and for $s = \infty$,

$$W_\infty' : \operatorname{alg}(T^0(PC), P)/K(l_Z^2(0)) \to \mathcal{A}_\infty \,, \quad \pi(A) \mapsto \Phi_\infty^J(E_n^F A E_{-n}^F | S_n^F) \,.$$

The proof of (2) proceeds in a similar way as for $s = 0$ (for details see Section 3.6.4). This finishes the proof of Proposition 3.10. ∎

Now we are going to turn to the general case, that is, we drop the restrictions for the prebasis F and for the underlying space $L_R^p(\alpha)$.

3.3 Approximation sequences for homogeneous operators

3.3.1 The set of approximation sequences for operators in $\Sigma_R^p(\alpha)$

For the sake of defining a comprehensive algebra of approximation methods for singular integral operators we start with analysing the approximation sequences for the homogeneous operators in $\Sigma_R^p(\alpha)$. In its consequence, this will lead us to a precise formulation of Axiom 1 from Section 3.2.1 in the general case.

We let F be a k-elementic canonical prebasis. Here and hereafter we suppose the elements of F to be rearranged and normed in such a way that

$$\alpha^F = (\underbrace{1, 1, \ldots, 1}_{\kappa}, 0, \ldots, 0)^T \tag{1}$$

with $1 \leq \kappa \leq k$. Clearly this is always possible (if necessary, replace ϕ_r by $\alpha_r^F \phi_r$), and it brings no restriction of generality. Further, given a closed subset Ω of $\mathbf{T}$ we let $PC_{p,\alpha}^\Omega$ stand for the collection of all piecewise continuous multipliers which are continuous on $\mathbf{T} \setminus \Omega$, and we consider the smallest closed subalgebra of $\mathrm{alg}\,(T^0(PC_{p,\alpha}), P)$ which contains all operators $T^0(a)$ with $a \in PC_{p,\alpha}^\Omega$ as well as the projection P. Let us denote this algebra by $\mathcal{T}_{1,1,\Omega}^p(\alpha)$.

As in case $k = 1$, all the above considered approximation sequences $(A_n), A_n : S_n^F \to S_n^F$ for operators in $\Sigma_R^p(\alpha)$ have some points in common. Firstly, the operators $E_{-n}^F A_n E_n^F : l_k^p(\alpha) \to l_k^p(\alpha)$ are independent of n and, denoting (one of) these operators by $R(A_n)$, we have $R(A_n) \in (\mathcal{T}_{1,1,\Omega}^p(\alpha))_{k \times k}$ with $\Omega = \mathbf{T}$, or with a suitably chosen set Ω. But, varying from the case $k = 1$, the operators $R(A_n)$ do not generate or even exhaust the whole Toeplitz algebra $(\mathcal{T}_{1,1,\Omega}^p(\alpha))_{k \times k}$! In fact, a closer look shows that the entries A_{rs} of *each* $k \times k$ matrix $R(A_n)$ arising from the methods in 2.11.1-2.11.10 are subject to some additional restrictions which are:

(a) the entries related to discontinuities are located in the upper part of the matrix (A_{rs}); precisely,

$$A_{rs} \in \left\{ \begin{array}{ll} \mathcal{T}_{1,1,\Omega}^p(\alpha) & (= \mathrm{alg}\,(T^0(PC_{p,\alpha}^\Omega), P)) \quad \text{if} \quad 0 \leq r \leq \kappa - 1 \\ \mathcal{T}_{1,1,\emptyset}^p(\alpha) & (= \mathrm{alg}\,(T^0(C_{p,\alpha}), P)) \quad \text{if} \quad \kappa \leq r \leq k - 1, \end{array} \right. \tag{2}$$

(b) the difference of elements in the same column of (A_{rs}) is free of discontinuities; precisely,

$$A_{r_1 s} - A_{r_2 s} \in \mathcal{T}_{1,1,\emptyset}^p(\alpha) \quad \text{if} \quad 0 \leq r_1 < r_2 \leq \kappa - 1, \tag{3}$$

(c) the (weighted and local) row sums of (A_{rs}) are independent of the number of the row; precisely, if $t \in \Omega$ then the sums

$$\sum_{s=0}^{k-1} \alpha_s^F \mathrm{smb}_t(A_{rs}) = \sum_{s=0}^{\kappa-1} \mathrm{smb}_t(A_{rs}) \tag{4}$$

are independent of r for $r = 0, \ldots, \kappa-1$ and equal to zero for $r = \kappa, \ldots, k-1$. Herein, smb_t is defined as in 2.5.2.

Condition (c) can simply be written in the form

$$(\mathrm{smb}_t(A_{rs}))_{r,s=1}^{k} \, \alpha^F \; = \; \alpha^F \sum_{s=0}^{\kappa-1} \mathrm{smb}_t(A_{1s}) \, .$$

Thus, we have a clear distinction between upper and lower part of the matrices $R(A_n)$, and we have conditions for the row sums and the column differences.

Definition: *Given a closed subset Ω of $\mathbf{T}$ and positive integers k and κ with $\kappa \leq k$ we let $\mathcal{T}_{k,\kappa,\Omega}^{p}(\alpha)$ stand for the collection of all operators $(A_{rs})_{r,s=0}^{k-1} \in (\mathcal{T}_{1,1,\Omega}^{p}(\alpha))_{k \times k}$ which are subject to the conditions (a),(b) and (c).*

In case $k = \kappa = 1$, the conditions (a)-(c) are needless, hence, this definition is consistent with that one for $\mathcal{T}_{1,1,\Omega}^{p}(\alpha)$ given above.

3.3.2   The algebra of approximation sequences for operators in $\Sigma_R^p(\alpha)$

The operators in $\mathcal{T}_{k,\kappa,\Omega}^{p}(\alpha)$ form even an algebra as the following theorem indicates.

Theorem 3.4 *The set $\mathcal{T}_{k,\kappa,\Omega}^{p}(\alpha)$ is a closed subalgebra of the Toeplitz algebra $(\mathcal{T}_{1,1,\Omega}^{p}(\alpha))_{k \times k}$, which contains all compact operators. This algebra is inverse closed in $(\mathcal{T}_{1,1,\Omega}^{p}(\alpha))_{k \times k}$, and the quotient algebra $\mathcal{T}_{k,\kappa,\Omega}^{p}(\alpha)/K(l_k^p(\alpha))$ is inverse closed in $(\mathcal{T}_{1,1,\Omega}^{p}(\alpha))_{k \times k}/K(l_k^p(\alpha))$.*

We shall give two proofs of this result. The first one works in case $p = 2$ and $\alpha = 0$ only but it is intended to explain how KMS-techniques simplify the things (provided they are applicable). The second one is for the general p and α. For the first proof we need the following lemma.

Lemma 3.1 *Let $\mathcal{R}$ be an algebra with identity and with a unital subalgebra $\mathcal{G}$, write $\mathcal{R}_{k \times k}$ for the algebra of all $k \times k$ matrices with entries in $\mathcal{R}$, and let $\mathcal{R}_{k,\kappa}$ (with a $\kappa \leq k$) denote the set of all matrices $a = (a_{rs})_{r,s=0}^{k-1} \in \mathcal{R}_{k \times k}$ satisfying the conditions*

(a) $a_{rs} \in \mathcal{G}$ whenever $\kappa \leq r \leq k - 1$,

(b) $a_{r_1 s} - a_{r_2 s} \in \mathcal{G}$ whenever $0 \leq r_1, r_2 \leq \kappa - 1$,

(c) there is an element $f(a) \in \mathcal{R}$ independent of $r = 0, \ldots, k - 1$, such that $\sum_{s=0}^{k-1} \alpha_s^F a_{rs} = \alpha_r^F f(a)$.

Then $\mathcal{R}_{k,\kappa}$ is a subalgebra of $\mathcal{R}_{k \times k}$, and the mapping $f : \mathcal{R}_{k,\kappa} \to \mathcal{R}$ is an algebra homomorphism. Moreover, if $\mathcal{G}_{k \times k}$ is inverse closed in $\mathcal{R}_{k \times k}$ and if each one-sided invertible element of $\mathcal{R}$ is two-sided invertible then $\mathcal{R}_{k,\kappa}$ is inverse closed in $\mathcal{R}_{k \times k}$.

Proof $\mathcal{R}_{k\kappa}$ is clearly a linear space, and f is a linear mapping. Define matrices $p, q \in \mathcal{R}_{k\times k}$ by

$$
p = \begin{pmatrix}
0 & & & \cdots & 0 & & \cdots & 0 \\
-e & e & 0 & \cdots & 0 & & \cdots & 0 \\
-e & 0 & e & 0 & 0 & & \cdots & 0 \\
\vdots & \vdots & \ddots & \ddots & \vdots & & & \vdots \\
-e & 0 & \cdots & 0 & e & & & \\
0 & & \cdots & & & e & & \\
\vdots & & & & & & \ddots & 0 \\
0 & & \cdots & & & & & e
\end{pmatrix}, q = \begin{pmatrix}
e & 0 & \cdots & 0 & 0 & & \cdots & 0 \\
e & 0 & \cdots & 0 & 0 & & \cdots & 0 \\
e & 0 & \cdots & 0 & 0 & & \cdots & 0 \\
\vdots & \vdots & \vdots & \vdots & \vdots & & & \vdots \\
e & 0 & \cdots & 0 & 0 & & & \\
0 & & & & & 0 & & \\
\vdots & & & & & & \ddots & \\
0 & & & & & & & 0
\end{pmatrix}
$$

(The upper left minors of p and q are $\kappa \times \kappa$ matrices, all other elements are equal to zero.).

One easily checks that $p, q \in \mathcal{G}_{k\times k}$ and that p and q are complementary idempotents, that is,

$$p^2 = p, \; q^2 = q, \; p + q = \mathrm{diag}\,(e, \ldots, e).$$

Observe further that the conditions (a) and (b) can be summarized to $pa \in \mathcal{G}_{k\times k}$ and that (c) is equivalent to

$$aq = q\,\mathrm{diag}\,(f(a), \ldots, f(a)). \tag{1}$$

Finally we remark that for a satisfying (1)

$$paq = p\,\mathrm{diag}\,(f(a), \ldots, f(a)) = 0,$$

hence, $pa = pap$.

Now it is easy to see that $\mathcal{R}_{k,\kappa}$ is actually an algebra: Let $a, b \in \mathcal{R}_{k,\kappa}$. Then

$$abq = aq\,\mathrm{diag}\,(f(b), \ldots, f(b)) = q\,\mathrm{diag}\,(f(a)f(b), \ldots, f(a)f(b)),$$

and $pab = papb \in \mathcal{G}_{k\times k}$. Further, if $a \in \mathcal{R}_{k,\kappa}$ is invertible in $\mathcal{R}_{k\times k}$ then

$$q = a^{-1}q\,\mathrm{diag}\,(f(a), \ldots, f(a)).$$

Evaluating the 00 th entries of these matrices we find that $f(a)$ is left-invertible and, thus, two-sided invertible by our assumption. So,

$$a^{-1}q = q\,\mathrm{diag}\,(f(a)^{-1}, \ldots, f(a)^{-1})$$

whence follows that a^{-1} satisfies (1) in place of a. Then, as we have seen above, $pa^{-1} = pa^{-1}p$, which implies the identities

$$p = paa^{-1}p = papa^{-1}p \quad \text{and} \quad p = pa^{-1}ap = pa^{-1}pap.$$

Thus, pap is invertible in the algebra $p\mathcal{R}_{k\times k}p$, and $(pap)^{-1}p = pa^{-1}p$. Since $\mathcal{G}_{k\times k}$ is inverse closed and $p \in \mathcal{G}_{k\times k}$ we conclude via Exercise 1.12 that pap is even invertible in $p\mathcal{G}_{k\times k}p$ whence follows $pa^{-1}p = pa^{-1} \in \mathcal{G}_{k\times k}$. ∎

First **proof** of Theorem 3.4. If $p = 2$ and $\alpha = 0$ then $T^2_{1,1,\Omega}(0)$ is a C^*-algebra. Thus, by the local inclusion theorem 1.14, the global conditions 3.3.1(2) and (3) are *equivalent* to the local conditions

$$\mathrm{smb}_t(A_{rs}) \in \mathrm{smb}_t(T^2_{1,1,\emptyset}(0)) \quad \text{if} \quad t \in \Omega \text{ and } \kappa \leq r \leq k - 1 \tag{2}$$

and

$$\mathrm{smb}_t(A_{r_1 s}) - \mathrm{smb}_t(A_{r_2 s}) \in \mathrm{smb}_t(T^2_{1,1,\emptyset}(0)) \quad \text{if} \quad t \in \Omega \text{ and } 0 \leq r_1, r_2 \leq \kappa - 1 . \tag{3}$$

It remains to apply the lemma with $\mathcal{R} = \mathrm{smb}_t(T^2_{1,1,\Omega}(0))$ and $\mathcal{G} = \mathrm{smb}_t(T^2_{1,1,\emptyset}(0))$. ∎

Second **proof** of Theorem 3.4. In general, we do not know whether the conditions 3.3.1(2), (3) and (2) (3) are equivalent (we conjecture that they are). So we are going to show straightforwardly that $\mathcal{T}^p_{k,\kappa,\Omega}(\alpha)$ is an algebra. Obviously, it is a linear space. Define p and q as in the proof of Lemma 3.1 but now with e being the identity operator on $l^p_1(\alpha)$. Each operator $A = (A_{rs}) \in \mathcal{T}^p_{k,\kappa,\Omega}(\alpha)$ can be written as

$$A = pA + qA . \tag{4}$$

Obviously,

$$qA = \begin{pmatrix} \alpha^F_0 A_{00} & \alpha^F_0 A_{01} & \cdots & \alpha^F_0 A_{0k-1} \\ \alpha^F_1 A_{00} & \alpha^F_1 A_{01} & \cdots & \alpha^F_1 A_{0k-1} \\ \vdots & \vdots & & \vdots \\ \alpha^F_{k-1} A_{00} & \alpha^F_{k-1} A_{01} & \cdots & \alpha^F_{k-1} A_{0k-1} \end{pmatrix} ; \tag{5}$$

thus, qA is in $\mathcal{T}^p_{k,\kappa,\Omega}(\alpha)$ again.

The conditions (2) and (3) tell us that $pA \in \mathcal{T}^p_{k,\kappa,\Omega}(\alpha)$. Taking into account Proposition 2.8 we find that necessarily

$$pA = T^0(c_1) + T^0(c_2) \operatorname{diag}(P, \dots, P) + K \tag{6}$$

with matrix valued continuous multipliers $c_1 = (c^{(1)}_{rs})$ and $c_2 = (c^{(2)}_{rs})$ and with a compact operator K. Since A and qA belong to $\mathcal{T}^p_{k,\kappa,\Omega}(\alpha)$, the operator (6) must also lie in $\mathcal{T}^p_{k,\kappa,\Omega}(\alpha)$ and, in particular, it has to be subject to the row sum condition (c).

Since the first rows of c_1 and c_2 vanish identically, the condition (c) implies that

$$\sum_{s=0}^{k-1} \alpha_s c^{(1)}_{rs}(t) = \sum_{s=0}^{k-1} \alpha_s c^{(2)}_{rs}(t) = 0 \tag{7}$$

for all $r = 0, \dots, k - 1$ and for all $t \in \Omega$. Thus, every operator in $\mathcal{T}^p_{k,\kappa,\Omega}(\alpha)$ can be written as the sum of the operators (5) and (6) with c_1 and c_2 specified by (7), and conversely: given operators $A_{00}, \dots, A_{0k-1} \in \mathcal{T}^p_{1,1,\Omega}(\alpha)$ and continuous multipliers c_1 and c_2 satisfying (7) then the sum of (5) and (6) lies in $\mathcal{T}^p_{k,\kappa,\Omega}(\alpha)$.

We claim that the product of two operators of this form is again of this form. It is evident that the product of two matrices (5) or of two operators (6) is again a matrix as in (5) or an operator as in (6). So it remains to examine products of the form

$$
\begin{pmatrix} \alpha_0^F A_{00} & \cdots & \alpha_0^F A_{0k-1} \\ \vdots & & \vdots \\ \alpha_{k-1}^F A_{00} & \cdots & \alpha_{k-1}^F A_{0k-1} \end{pmatrix} T^0(c_1) \text{ and } T^0(c_1) \begin{pmatrix} \alpha_0^F A_{00} & \cdots & \alpha_0^F A_{0k-1} \\ \vdots & & \vdots \\ \alpha_{k-1}^F A_{00} & \cdots & \alpha_{k-1}^F A_{0k-1} \end{pmatrix}
$$

These can be written as

$$
\begin{pmatrix} \alpha_0^F B_{00} & \cdots & \alpha_0^F B_{0k-1} \\ \vdots & & \vdots \\ \alpha_{k-1}^F B_{00} & \cdots & \alpha_{k-1}^F B_{0k-1} \end{pmatrix} \tag{8}
$$

and

$$
\begin{pmatrix} \alpha_0^F T^0(g_0) A_{00} & \cdots & \alpha_0^F T^0(g_0) A_{0k-1} \\ \vdots & & \vdots \\ \alpha_{k-1}^F T^0(g_{k-1}) A_{00} & \cdots & \alpha_{k-1}^F T^0(g_{k-1}) A_{0k-1} \end{pmatrix} \tag{9}
$$

with

$$
B_{0s} = \sum_{l=0}^{k-1} A_{0l} T^0(c_{ls}^{(1)}) \quad \text{and} \quad g_r = \sum_{l=0}^{k-1} \alpha_l^F c_{rs}^{(1)} \, .
$$

So it is immediate that the matrix in (8) is of the form (5), and that one in (9) is of the form (6): indeed, from (7) we know that the functions g_r vanish on Ω, and since $A_{0s} \in T_{1,1,\Omega}^p(\alpha)$ it remains to apply Exercise 2.6 to obtain that $\alpha_r^F T^0(g_r) A_{0s} = T^0(c_{rs}) + T^0(d_{rs}) P + K_{rs}$ with continuous multipliers c_{rs} and d_{rs} and a compact operator K_{rs}.

We renounce to give a proof of the inverse closedness here. Such a proof can be given by repeating all arguments of the proof of Theorem 2.2. ∎

Corollary 3.2 *The algebra $T_{k,\kappa,\Omega}^p(\alpha)$ is inverse closed in $L(l_k^p(\alpha))$, and the quotient algebra $T_{k,\kappa,\Omega}^p(\alpha)/K(l_k^p(\alpha))$ is inverse closed in the Calkin algebra $L(l_k^p(\alpha))/K(l_k^p(\alpha))$.*

Proof We need the inverse closedness of $T_{1,1,\Omega}^p(\alpha)/K(l_1^p(\alpha))$ in $L(l_1^p(\alpha))/K(l_1^p(\alpha))$ (see Theorems 2.2, 2.3 and Exercise 1.11. ∎

Now we can make our Axiom 1 in Section 3.2.1 more precise by requiring

1'. *A should contain all sequences $(E_n^F A E_{-n}^F)$ with $A \in T_{k,\kappa,\Omega}^p(\alpha)$.*

Our next goal is an alternative description of the algebra $T_{k,\kappa,\Omega}^p(\alpha)$ via its generators.

3.3.3   Alternative description of $T^p_{k,\kappa,\Omega}(\alpha)$

Let L stand for the mapping

$$L \;:\; L(l^p_1(\alpha)) \to L(l^p_k(\alpha))\,,\; A \mapsto \big(\int_R \phi_s(x)\,dx\,\alpha_r^F\,A\big)^{k-1}_{r,s=0}\,. \tag{1}$$

It is easy to see that L is a continuous and injective Banach algebra homomorphism. Further we pick an arbitrary piecewise continuous multiplier ρ which is smooth over $\mathbf{T}\setminus\{1\}$ and which satisfies $\rho(1\pm0)=\mp1$, and we associate with ρ the matrix-valued functions ρ^t, $t\in\mathbf{T}$, defined by

$$T^0(\rho^1) = L(T^0(\rho)) \quad\text{and}\quad T^0(\rho^t)=L(Y_tT^0(\rho)Y_t^{-1})=L(T^0(\rho_{t^{-1}}))$$

(compare 2.5.1(2)). Clearly, $\rho^1=(\int\phi_s\,dx\,\alpha_r^F\,\rho)^{k-1}_{r,s=0}$ and $\rho^t(s)=\rho^1(s/t)$. Finally we let $PC^{k,\kappa,\Omega}_{p,\alpha}$ stand for the collection of all multipliers $a=(a_{rs})$ in $(PC_{p,\alpha})_{k\times k}$ which are continuous on $\mathbf{T}\setminus\Omega$ and which are subject to the following conditions:

(a) a_{rs} is continuous for $\kappa\le r\le k-1$.

(b) $a_{r_1s}-a_{r_2s}$ is continuous for $0\le r_1,r_2\le\kappa-1$.

(c) if $t\in\Omega$ then the sums $\sum^{k-1}_{s=0}\alpha_s^F a_{rs}(t\pm0)$ are independent of r for $0\le r\le\kappa-1$ and equal to zero for $\kappa\le r\le k-1$.

From Lemma 3.1 we conclude that $PC^{k,\kappa,\Omega}_{p,\alpha}$ is even a Banach algebra.

Proposition 3.11 $T^p_{k,\kappa,\Omega}(\alpha)$ *coincides with each of the following subalgebras of* $(\mathrm{alg}\,(T^0(PC_{p,\alpha}),P))_{k\times k}$:

(i) *the smallest closed subalgebra which contains*

- *the ideal* $K(l^p_k(\alpha))$,
- *the matrix* $\mathrm{diag}\,(P,\ldots,P)$,
- *all matrix valued continuous multipliers* $T^0(f)$ *for which there exist numbers* $g(t)$ *such that* $f(t)\alpha^F=\alpha^Fg(t)$ *for all* $t\in\Omega$,
- *the image of* $T^p_{1,1,\Omega}(\alpha)$ *under the mapping* L.

(ii) *the smallest closed subalgebra which contains*

- *the ideal* $K(l^p_k(\alpha))$,
- *the matrix* $\mathrm{diag}\,(P,\ldots,P)$,
- *all operators* $T^0(f)$ *as under (i)*,
- *all multipliers* $T^0(\rho^t)$ *with* $t\in\Omega$.

(iii) *the smallest closed subalgebra which contains*

- *the ideal* $K(l^p_k(\alpha))$,

       - *the matrix* $\text{diag}\,(P,\dots,P)$,
       - *all multipliers* $T^0(a)$ *with* $a \in PC_{p,\alpha}^{k,\kappa,\Omega}$.

Proof Denote the algebras described in (i),(ii)and (iii) by T_1, T_2 and T_3, respectively, and let $A = (A_{rs}) \in T_{k,\kappa,\Omega}^p(\alpha)$. We claim that $A \in T_1$. As we have seen in the proof of Theorem 3.4,

$$A = T^0(c_1) + T^0(c_2)\text{diag}\,(P,\dots,P) + K + \begin{pmatrix} \alpha_0^F A_{00} & \dots & \alpha_0^F A_{0k-1} \\ \vdots & & \vdots \\ \alpha_{k-1}^F A_{00} & \dots & \alpha_{k-1}^F A_{0k-1} \end{pmatrix}.$$

The first three summands are in T_1, and the fourth summand can be written as

$$\sum_{s=0}^{k-1} \begin{pmatrix} 0 & \dots & 0 & \alpha_0^F A_{0s} & 0 & \dots & 0 \\ \vdots & \vdots & & & & \vdots & \vdots \\ 0 & \dots & 0 & \alpha_{k-1}^F A_{0s} & 0 & \dots & 0 \end{pmatrix} \tag{2}$$

where the non-vanishing column stands at the sth place. But the sth summand in (2) is nothing else than

$$L(A_{0s}) \begin{pmatrix} 0 & \dots & 0 & \alpha_0^F & 0 & \dots & 0 \\ \vdots & \vdots & & & & \vdots & \vdots \\ 0 & \dots & 0 & \alpha_{k-1}^F & 0 & \dots & 0 \end{pmatrix} \tag{3}$$

(see Proposition 2.19(ii)), again with the non-vanishing column at the sth place, and (3) is in T_1 by definition (the matrix in (3) can be regarded as a continuous multiplier). Thus, $T_{k,\kappa,\Omega}^p(\alpha) \subseteq T_1$. Let us now explain why $T_1 \subseteq T_2$. It is immediate from the definition that the algebra $T_{1,1,\Omega}^p(\alpha)$ is generated by the projection P, by the continuous multipliers $T^0(f)$, and by all multipliers $T^0(\rho_t)$ with $t \in \Omega$. Thus, since L is injective, the algebra $L(T_{1,1,\Omega}^p(\alpha))$ is generated by $L(P), L(T^0(f))$ and $L(T^0(\rho^t))$. The latter two operators are in T_2 by definition, and $L(P)$ is also in T_2 because of

$$L(P) = \text{diag}\,(P,\dots,P) \begin{pmatrix} \alpha_0^F \int \phi_0\,dx & \dots & \alpha_0^F \int \phi_{k-1}\,dx \\ \vdots & & \vdots \\ \alpha_{k-1}^F \int \phi_0\,dx & \dots & \alpha_{k-1}^F \int \phi_{k-1}\,dx \end{pmatrix}$$

which can be viewed as product of $\text{diag}\,(P,\dots,P)$ with a continuous multiplier. The remaining inclusions $T_2 \subseteq T_3$ and $T_3 \subseteq T_{k,\kappa,\Omega}^p(\alpha)$ are evident. ∎

3.3.4 Fredholmness of operators in $T_{k,\kappa,\Omega}^p(\alpha)$

Later on we shall be concerned with invertibility and Fredholmness of operators in $T_{k,\kappa,\Omega}^p(\alpha)$. Whereas the invertibility is a rather delicate problem (even in case $k = \kappa = 1$) there are satisfactory criteria for the Fredholmness. For their explanation we first consider the general situation of Lemma 3.1 again, i.e. we let $\mathcal{R}$ be a unital algebra with unital subalgebra $\mathcal{G}$, and we introduce the algebra $\mathcal{R}_{k,\kappa}$, the homomorphism f, and the idempotents p and q as in 3.3.2.

Lemma 3.2 (a) The set $q\mathcal{R}_{k,\kappa}p$ is a two-sided ideal of $\mathcal{R}_{k,\kappa}$ which belongs to the radical of this algebra.
(b) An element $a \in \mathcal{R}_{k\kappa}$ is invertible if and only if pap and qaq are invertible in $p\mathcal{R}_{k,\kappa}p$ and $q\mathcal{R}_{k,\kappa}q$, respectively.
(c) The $k \times k$ matrix pap is invertible in $p\mathcal{R}_{k,\kappa}p$ if and only if its right lower $(k-1) \times (k-1)$ block is invertible in $\mathcal{G}_{k-1 \times k-1}$.
(d) The matrix qaq is invertible in $q\mathcal{R}_{k,\kappa}q$ if and only if $f(a)$ is invertible in $\mathcal{R}$.

Proof   (a) The set $q\mathcal{R}_{k,\kappa}p$ is clearly linear. Since $p\mathcal{R}_{k,\kappa}q = \{0\}$ (see 3.3.2) we have further

$$pa = pap \quad \text{and} \quad aq = qaq$$

for all $a \in \mathcal{R}_{k,\kappa}$. Thus, for a and b in $\mathcal{R}_{k,\kappa}$,

$$b\,qap = q(bqa)p \in q\mathcal{R}_{k,\kappa}p \,,$$

$$qap\,b = q(apb)p \in q\mathcal{R}_{k,\kappa}p \,,$$

whence follows that $q\mathcal{R}_{k,\kappa}p$ is an ideal. It remains to show that $b+qap$ is invertible whenever b is invertible. This is already a consequence of the following readily verifiable identity:

$$(b + qap)^{-1} = b^{-1} - qb^{-1}qapb^{-1}p \,.$$

(b) Write $a \in \mathcal{R}_{k,\kappa}$ as

$$a = pap + paq + qap + qaq \,.$$

Then $paq = 0$ and qap lies in the radical, and so a is invertible if and only if $pap + qaq$ is invertible. Since

$$pap + qaq = (pap + q)(p + qaq) = (p + qaq)(pap + q) \,,$$

the invertibility of $pap+qaq$ is equivalent to the invertibility of $pap+q$ and $p+qaq$ in $\mathcal{R}_{k,\kappa}$, respectively to the invertibility of pap in $p\mathcal{R}_{k,\kappa}p$ and of qaq in $q\mathcal{R}_{k,\kappa}q$.
(c) Obviously, pap is in $p\mathcal{R}_{k,\kappa}p = p\mathcal{G}_{k\times k}p$ invertible if and only if $pap + q$ is in $\mathcal{G}_{k\times k}$ invertible.. Further, $pap + q$ is a matrix of the form

$$\begin{pmatrix} e & 0 \\ B & A \end{pmatrix} \tag{1}$$

with $B \in \mathcal{G}_{k-1 \times 1}$ and $A \in \mathcal{G}_{k-1 \times k-1}$, and it is an easy exercise to show that the matrix (1) is invertible if and only if the matrix A is so (see Exercise 1.16).
(d) Let qaq be invertible in $q\mathcal{R}_{k,\kappa}q$, i.e. there is a $b \in \mathcal{R}_{k,\kappa}$ such that

$$qaqbq = qbqaq = q \,. \tag{2}$$

Applying the homomorphism f to both sides of (2) we get

$$f(a)\, f(b) \;=\; f(b)\, f(a) \;=\; e\,(= f(q))\,,$$

thus, $f(a)$ is invertible.

Conversely, let $c \in \mathcal{R}$ such that $f(a)c = cf(a) = e$. Define a mapping $M : \mathcal{R} \to \mathcal{R}_{k\times k}$ by

$$M \;:\; c \mapsto \frac{1}{\kappa}(\alpha_r^F c)_{r,s=0}^{k-1}$$

where, as above, $\alpha_r^F = 1$ if $r = 0, \ldots, \kappa - 1$ and $\alpha_r^F = 0$ if $r = \kappa, \ldots, k - 1$. It is easy to see that M is an algebra homomorphism from $\mathcal{R}$ into $\mathcal{R}_{k,\kappa}$, and that

$$pM(c) \;=\; 0\,, \quad M(c)q \;=\; q\operatorname{diag}(c,\ldots,c)\,.$$

In particular, $f(M(c)) = c$, and we have

$$qaq\, qM(c)q \,-\, q \;=\; q\operatorname{diag}\left(f(a)c - e, \ldots, f(a)c - e\right) \;=\; 0$$

and analogously $qM(c)q\, qaq - q = 0$. This shows the invertibility of qaq in $q\mathcal{R}_{k,\kappa}q$. ∎

Now, for the desired Fredholm criterion, we use that $\mathcal{T}_{k,\kappa,\Omega}^p(\alpha)/K(l_k^p(\alpha))$ is inverse closed in $(\mathcal{T}_{1,1,\Omega}^p(\alpha))_{k\times k}/K(l_k^p(\alpha))$. Localizing the latter algebra over the unit circle in the same way as in Sections 2.5.2-2.6.1 we arrive at local quotient algebras $((\mathcal{T}_{1,1,\Omega}^p(\alpha))_{k\times k}/K(l_k^p(\alpha)))/I_t$ with subalgebras $\mathcal{T}_{k,\kappa,\Omega}^p(\alpha)_t^\pi$ containing the cosets of all elements in $\mathcal{T}_{k,\kappa,\Omega}^p(\alpha)$. Reifying the above lemma with $\mathcal{T}_{1,1,\Omega}^p(\alpha)_t^\pi$, $\mathcal{T}_{1,1,\emptyset}^p(\alpha)_t^\pi$ and $\mathcal{T}_{k,\kappa,\Omega}^p(\alpha)_t^\pi$ in place of $\mathcal{R}$, $\mathcal{G}$ and $\mathcal{R}_{k,\kappa}$, respectively, we see that the cosets $\mathrm{smb}_t(A) := A + K(l_k^p(\alpha)) + I_t$ for $A \in \mathcal{T}_{k,\kappa,\Omega}^p(\alpha)$ are invertible if and only if the cosets $p\mathrm{smb}_t(A)p$ and $f(\mathrm{smb}_t(A))$ are invertible, where p stands here for the coset corresponding to the $'p'$ in the proof of Lemma 3.1. The first of these cosets belongs to the algebra $(\mathrm{alg}\,(T^0(C_{p,\alpha}), P)_{k\times k}/K(l_k^p(\alpha)))/I_t$ which is easy to be handled since only continuous multipliers occur, and the second coset is in $\mathcal{T}_{1,1,\Omega}^p(\alpha)_t^\pi$; an algebra which has been already examined in Section 2.5.2.

3.4 The stability theorem

3.4.1 Translation invariance

Let us start with making Axiom 2 in Section 3.2.1 more explicit. As we have already explained in Section 3.2.2 for the $L_R^2(0)$-context, the sequence $(U_{\frac{\{tn\}}{n}} A_n U_{-\frac{\{tn\}}{n}})$ can be regarded as the translation of the approximation sequence (A_n) by the real number t. It is obvious that the same definition applies to unweighted L_R^p-spaces, too. Only in case the underlying Banach space $L_R^p(\alpha)$ has a proper weight some modifications are needed. The point is that, for $\alpha \neq 0$, the shift operators $U_{\frac{\{tn\}}{n}}$

still act on S_n^F, but they are not longer *uniformly* bounded with respect to n. To manage this situation we cannot proceed as in Section 3.1.6 since Δ is not a mapping from S_n^F onto S_n^F. But there is still another way to factorize out the weights which bases on de Boor's estimates. To that end let $\Lambda : l_1^p(0) \to l_1^p(\alpha)$ denote the isometry $\Lambda : (x_l)_{l \in Z} \mapsto ((|l| + 1)^{-\alpha} x_l)_{l \in Z}$, write $V_m : l_1^p(0) \to l_1^p(0)$ for the discrete shift operator $V_m : (x_l) \mapsto (x_{l-m})$, and use the same notations Λ and V_m for the diagonal operators diag $(\Lambda, \ldots, \Lambda)$ and diag $(V_m, \ldots, V_m)$ acting on $l_k^p(0)$. Now we can substitute the continuously weighted shift $\Delta U_{\frac{\{tn\}}{n}} \Delta^{-1}$ by the discretely weighted shift

$$E_n^F \Lambda E_{-n}^F U_{\frac{\{tn\}}{n}} E_n^F \Lambda^{-1} E_{-n}^F = E_n^F \Lambda V_{\{tn\}} \Lambda^{-1} E_{-n}^F , \tag{1}$$

and restate Axiom 3.2.1 2. as follows:

    2'. If $(A_n) \in \mathcal{A}$ then $(E_n^F \Lambda V_{\{tn\}} \Lambda^{-1} E_{-n}^F A_n E_n^F \Lambda V_{-\{tn\}} \Lambda^{-1} E_{-n}^F) \in \mathcal{A}$
        for all $t \in \mathbf{R}$.

3.4.2 Definition of the algebras of approximation sequences

Here and hereafter, let $F = \{\phi_0, \ldots, \phi_{k-1}\}$ be a canonical prebasis with

$$\sum_{r=0}^{\kappa-1} \sum_{l \in Z} \phi_{rln}(x) = 1 \quad \text{for} \quad \text{all} \quad x \in \mathbf{R} ,$$

that is, we suppose that

$$\alpha^F = (\underbrace{1, 1, \ldots, 1}_{\kappa}, \underbrace{0, 0, \ldots, 0}_{k-\kappa})^T .$$

As in Section 3.2.3, we introduce the collection $\mathcal{F}^F$ of all bounded sequences (A_n) of operators $A_n : S_n^F \to S_n^F$ (but now with viewing S_n^F as a subspace of $L_R^p(\alpha)$), and we make $\mathcal{F}^F$ to a Banach algebra on defining elementwise operations $(A_n) + (B_n) = (A_n + B_n)$, $\alpha(A_n) = (\alpha A_n)$, $(A_n)(B_n) = (A_n B_n)$ and the norm

$$\|(A_n)\| = \sup_n \|A_n L_n^{F,F}\|_{L_R^p(\alpha)} .$$

Now, given a fixed closed subset Ω of the unit circle, we define the algebra $\mathcal{A} = \mathcal{A}_\Omega^F$ of all "interesting" approximation sequences as the smallest closed subalgebra of $\mathcal{F}^F$ which contains all sequences of the form

$$(E_n^F \Lambda V_{\{tn\}} \Lambda^{-1} A \Lambda V_{-\{tn\}} \Lambda^{-1} E_{-n}^F) \tag{1}$$

where $t \in \mathbf{R}$ and $A \in \mathcal{T}_{k,\kappa,\Omega}^p(\alpha)$. Indeed, this algebra contains all sequences $(E_n^F A E_{-n}^F)$ with $A \in \mathcal{T}_{k,\kappa,\Omega}^p(\alpha)$ and, thus, it satisfies the first axiom. Further it is translation invariant in the sense of Section 3.4.1 and, finally, it is the smallest closed subalgebra of $\mathcal{F}^F$ which carries these properties.

Clearly, the algebra $\mathcal{A}_\Omega^F$ depends on F by its definition, but this dependence is not very strong: A little thought shows that the algebra of all sequences $(E_{-n}^F A E_n^F)$ with $(A_n) \in \mathcal{A}_\Omega^F$ only depends on the parameters k and κ of the prebasis F, but this algebra is topologically isomorphic to $\mathcal{A}_\Omega^F$ by de Boor's estimates. Thus, if F_1 and F_2 are k-elementic canonical prebases with the same κ then the algebras $\mathcal{A}_\Omega^{F_1}$ and $\mathcal{A}_\Omega^{F_2}$ coincide up to isomorphy.

For example, if $\kappa = k$ then it would be sufficient to study the algebra $\mathcal{A}_\Omega^F$ for the concrete prebasis $F = \{\chi_{[0,1/k)}, \dots, \chi_{[(k-1)/k,1)}\}$, and everyone who wants should think of F as being specified in such a way. But, firstly, doing so will not essentially simplify any part of what follows and, secondly, we prefer to treat a sequence (A_n), $A_n : S_n^{F_1} \to S_n^{F_1}$, in the associated algebra $\mathcal{A}_\Omega^{F_1}$ rather than the translated sequence $(E_n^{F_2} E_{-n}^{F_1} A_n E_n^{F_1} E_{-n}^{F_2})$ in the algebra $\mathcal{A}_\Omega^{F_2}$ which is not immediately related to the original spline space $S_n^{F_1}$.

Finally we remark that it is sometimes desirable to include an additional free parameter, $r \in \mathbf{R}$, into the definition of the algebra $\mathcal{A}_\Omega^F$ (see Section 3.8.3 for an example). That is, we let $\mathcal{A}_{\Omega,r}^F$ stand for the smallest closed subalgebra of $\mathcal{F}^F$ which contains all sequences

$$(E_n^F \Lambda V_{\{tn+r\}} \Lambda^{-1} A \Lambda V_{-\{tn+r\}} \Lambda^{-1} E_{-n}^F) \tag{2}$$

with $t \in \mathbf{R}$, $A \in \mathcal{T}_{k,\kappa,\Omega}^p(\alpha)$. This slight modification does not involve any complications.

3.4.3 Definition of the homomorphisms W_s and W^t

As in the special case examined in Sections 3.2.1-3.2.7, the general stability criterion also establishes equivalence between stability of a sequence and invertibility of two families of operators. To define these operators we let, as above, Y_t, V_n, and Λ refer to the mappings

$$Y_t : l_1^p(\alpha) \to l_1^p(\alpha) \quad , \quad (x_i) \mapsto (t^{-i} x_i) \,,$$

$$V_n : l_1^p(0) \to l_1^p(0) \quad , \quad (x_i) \mapsto (x_{i-n}) \,,$$

and

$$\Lambda : l_1^p(0) \to l_1^p(\alpha) \quad , \quad (x_i) \mapsto ((|i|+1)^{-\alpha} x_i) \,,$$

and we abbreviate the diagonal matrices $\operatorname{diag}(Y_t, \dots, Y_t)$, $\operatorname{diag}(V_n, \dots V_n)$, and $\operatorname{diag}(\Lambda, \dots, \Lambda)$ simply by Y_t, V_n, and Λ again.

Proposition 3.12 *Let* $(A_n) \in \mathcal{A}_{\Omega,r}^F$. *Then,*

(a) for each $t \in \Omega$, *there is an operator* $W^t(A_n)$ *in* $L(L_R^p(\alpha))$ *such that*

$$E_n^F Y_{t^{-1}} E_{-n}^F A_n E_n^F Y_t E_{-n}^F L_n^{F,F} \to W^t(A_n)$$

strongly as $n \to \infty$. *The mapping* $W^t : \mathcal{A}_{\Omega,r}^F \to L(L_R^p(\alpha))$ *is a continuous algebra homomorphism.*

(b) *for each $s \in \mathbf{R}$, there is an operator $W_s(A_n)$ in $L(l_k^p(\alpha))$ if $s = 0$ and in $L(l_k^p(0))$ if $s \neq 0$ such that*

$$\Lambda V_{-\{r\}} \Lambda^{-1} E_{-n}^F A_n E_n^F \Lambda V_{\{r\}} \Lambda^{-1} \to W_0(A_n)$$

and

$$V_{-\{sn+r\}} \Lambda^{-1} E_{-n}^F A_n E_n^F \Lambda V_{\{sn+r\}} \to W_s(A_n)$$

strongly as $n \to \infty$. The mappings $W_0 : \mathcal{A}_{\Omega,r}^F \to L(l_k^p(\alpha))$ and $W_s : \mathcal{A}_{\Omega,r}^F \to L(l_k^p(0))$ for $s \neq 0$ are continuous algebra homomorphisms.

(c) *the cosets $\pi(E_{-n}^F A_n E_n^F) := E_{-n}^F A_n E_n^F + K(l_k^p(\alpha))$ are independent of n. If $W_\infty(A_n)$ denotes one of these cosets then the mapping*

$$W_\infty \ : \ \mathcal{A}_{\Omega,r}^F \to L(l_k^p(\alpha))/K(l_k^p(\alpha)), \ (A_n) \mapsto W_\infty(A_n)$$

is a continuous algebra homomorphism.

The proof will be given in Section 3.5.1.

3.4.4   The general stability theorem

Theorem 3.5 *Let $(A_n) \in \mathcal{A}_{\Omega,r}^F$. Then the sequence (A_n) is stable if and only if all operators (cosets) $W^t(A_n)$ and $W_s(A_n)$ with $t \in \Omega$ and $s \in \dot{\mathbf{R}}$ are invertible.*

This theorem has an analogue in the matrix case: Given $l \geq 1$ we let $(\mathcal{A}_{\Omega,r}^F)_{l \times l}$ stand for the algebra of all $l \times l$-matrices with entries in $\mathcal{A}_{\Omega,r}^F$, and for $(A_n) = ((A_n^{ij}))_{i,j=1}^l$ with $(A_n^{ij}) \in \mathcal{A}_{\Omega,r}^F$ we define

$$W_s(A_n) := (W_s(A_n^{ij}))_{i,j=1}^l \quad \text{for} \quad s \in \dot{\mathbf{R}}$$

and

$$W^t(A_n) := (W^t(A_n^{ij}))_{i,j=1}^l \quad \text{for} \quad t \in \Omega \ .$$

With these notations, the theorem remains literally valid for $l > 1$, too. Also, the proof of the matrix case requires only minor modifications; that is why we focus our attention on the case $l = 1$ in what follows.

We prepare the proof of this theorem in the following sections by verifying analoga of properties 3.1.1 (i)-(iv) for the algebra $\mathcal{A}_{\Omega,r}^F$. The reader is proposed to have a look at the results but to leave out their proofs in the first reading since the proofs are rather technical and not very instructive.

3.5 Basic properties of algebras of approximation sequences

3.5.1 Strong limits

Our first goal is to prove Proposition 3.12 which is the analogue of 3.1.1(iii) for algebras of approximation sequences. Suppose for example the strong limits

$$\text{s-lim}_{n\to\infty} E_n^F Y_t^{-1} E_{-n}^F A_n E_n^F Y_t E_{-n}^F L_n^{F,F} =: W^t(A_n) \tag{1}$$

and

$$\text{s-lim}_{n\to\infty} E_n^F Y_t^{-1} E_{-n}^F B_n E_n^F Y_t E_{-n}^F L_n^{F,F} =: W^t(B_n) \tag{}$$

to exist. Then it is elementary that the strong limits $W^t(A_n + B_n)$ and $W^t(A_n B_n)$ also exist, and that W^t acts homomorphically:

$$W^t(A_n + B_n) = W^t(A_n) + W^t(B_n) , \quad W^t(A_n B_n) = W^t(A_n) W^t(B_n) .$$

Further, since $\quad \sup_n \| E_n^F Y_t^{-1} E_{-n}^F L_n^{F,F} \| \, \| E_n^F Y_t E_{-n}^F L_n^{F,F} \| < \infty ,$

the strong limits $W^t(A_n)$ depend continuously on the sequence (A_n). This makes it easy to derive that the set of all sequences in $\mathcal{F}^F$ for which the strong limit (1) exists, forms a closed subalgebra of $\mathcal{F}^F$, and W^t is a continuous algebra homomorphism acting on this subalgebra. Of course, the same arguments apply to the strong limits W_s.

Hence, we shall be essentially concerned with verifying the existence of the strong limits postulated in Proposition 3.12 for the generating sequences of the algebra $\mathcal{A}_{\Omega,r}^F$. If we abbreviate the sequence $(E_n^F \Lambda V_{\{yn+r\}} \Lambda^{-1} A \Lambda V_{-\{yn+r\}} \Lambda^{-1} E_{-n}^F)$ to (A, y) then, by the definition of $\mathcal{A}_{\Omega,r}^F$ and by Proposition 3.11, a system of generating sequences of $\mathcal{A}_{\Omega,r}^F$ is given by

all sequences $\quad (K, y) \quad$ with $\quad K \in K(l_k^p(\alpha)) , \ y \in \mathbf{R} ,$ (2)

all sequences $\quad (\text{diag}\,(P, \dots, P), y) \quad$ with $\quad y \in \mathbf{R} ,$ (3)

all sequences $\quad (T^0(a), y) \quad$ with $\quad a \in PC_{p,\alpha}^{k,\kappa,\Omega} , \ y \in \mathbf{R} .$ (4)

Proposition 3.13 *The strong limits $W^t(A_n)$ exist for each sequence in $\mathcal{A}_{\Omega,r}^F$ and for each $t \in \Omega$. In particular,*

$$W^t(K, y) = 0 \tag{5}$$

$$W^t(\text{diag}\,(P, \dots, P), y) = \chi_{[y,\infty)} I \tag{6}$$

$$W^t(T^0(a), y) = \frac{\hat{a}(t+0) + \hat{a}(t-0)}{2} I -$$

$$- \frac{\hat{a}(t+0) - \hat{a}(t-0)}{2} \Delta U_y \Delta^{-1} S_R \Delta U_{-y} \Delta^{-1} \tag{7}$$

where $\hat{a}(t \pm 0) := \sum_{s=0}^{k-1} \alpha_s^F a_{0s}(t \pm 0)$.

Remember that, by the definition of the function algebra $PC_{p,\alpha}^{k,\kappa,\Omega}$ in 3.3.3, the numbers $\sum_{s=0}^{k-1} \alpha_s^F a_{rs}(t \pm 0)$ are independent of r for $0 \leq r \leq \kappa - 1$ and $t \in \Omega$.

Proof Let us start with the unweighted case (that is, $\alpha = 0$ and $\Lambda = I \in L(l_k^p(0))$ and $\Delta = I \in L(L_R^p(0))$). First we remark that

$$W^t(A, y) = W^1(Y_{t^{-1}} A Y_t, y) \tag{8}$$

whenever one of these limits exists. This is a simple consequence of the identities

$$Y_{t^{-1}} V_{\{yn+r\}} = t^{\{yn+r\}} V_{\{yn+r\}} Y_{t^{-1}} \quad \text{and} \quad V_{-\{yn+r\}} Y_t = t^{-\{yn+r\}} Y_t V_{-\{yn+r\}} .$$

Further we have

$$W^1(Y_{t^{-1}} A Y_t, y) = U_y W^1(Y_{t^{-1}} A Y_t, 0) U_{-y} \tag{9}$$

provided one of these limits exists. Indeed,

$$
\begin{aligned}
W^1(Y_{t^{-1}} A Y_t, y) &= \\
&= \operatorname*{s-lim}_{n \to \infty} E_n^F V_{\{yn+r\}} Y_{t^{-1}} A Y_t V_{-\{yn+r\}} E_{-n}^F L_n^{F,F} \\
&= \operatorname*{s-lim}_{n \to \infty} U_{\frac{\{yn+r\}}{n}} E_n^F Y_{t^{-1}} A Y_t E_{-n}^F L_n^{F,F} U_{-\frac{\{yn+r\}}{n}} ,
\end{aligned}
$$

and now use the strong convergence $U_{\pm \frac{\{yn+r\}}{n}} \to U_{\pm y}$ as $n \to \infty$.

Let us prove (5). By (8) and (9), and since $Y_{t^{-1}} K Y_t$ is compact for compact K, it remains to show that

$$\operatorname*{s-lim}_{n \to \infty} E_n^F K E_{-n}^F L_n^{F,F} = 0$$

for all compact operators K. In other terms, we have to show that

$$\|E_n^F K E_{-n}^F L_n^{F,F} f\| = \|E_n^F K T^0 (\lambda^{F,F})^{-1} P_n^F f\| \to 0 \quad \text{as } n \to \infty$$

for all functions $f \in L_R^p(0)$ or, at least, for all functions f belonging to a dense subset of $L_R^p(0)$, say, for all continuous and compactly supported functions. Since $K T^0 (\lambda^{F,F})^{-1}$ is a compact operator, it can be approximated in the operator norm by a linear combination of the rank-one operators

$$K_{rstu} : l_k^p(0) \to l_k^p(0), \ ((x_{0i}), (x_{1i}), \ldots, (x_{k-1 i})) \mapsto ((y_{0j}), (y_{1j}), \ldots, (y_{k-1 j}))$$

with

$$
y_{lj} = \begin{cases} x_{rs} & \text{if} \quad l = t \text{ and } j = u \\ 0 & \text{else} . \end{cases}
$$

Clearly,

$$\|E_n^F K_{rstu} P_n^F f\| = \|\phi_{tun} \int_R f(\frac{x+s}{n}) \overline{\phi}_r(x) \, dx\| ,$$

and from $\|\phi_{tun}\| = n^{-1/p} \|\phi_t\|$ and $\int f(\frac{x+s}{n}) \overline{\phi}_r(x) \, dx \to f(0) \int \overline{\phi}_r(x) \, dx$ we get the assertion (5).

Now consider (6). Since P and Y_t are diagonal operators, we have

$$Y_{t^{-1}} \operatorname{diag}(P, \ldots, P) Y_t = \operatorname{diag}(P, \ldots, P).$$

Let K be the compact operator $E_{-n}^F L_n^{F,F} \chi_{R^+} E_n^F - \operatorname{diag}(P, \ldots, P)$ (see 2.11.4). Then

$$\operatorname*{s-lim}_{n \to \infty} E_n^F \operatorname{diag}(P; \ldots, P) E_{-n}^F L_n^{F,F} =$$
$$= \operatorname*{s-lim}_{n \to \infty} E_n^F (\operatorname{diag}(P; \ldots, P) + K) E_{-n}^F L_n^{F,F} - \operatorname*{s-lim}_{n \to \infty} E_n^F K E_{-n}^F L_n^{F,F}$$
$$= \operatorname*{s-lim}_{n \to \infty} L_n^{F,F} \chi_{R^+} L_n^{F,F} = \chi_{R^+} I$$

since $L_n^{F,F} \to I$ and by what has already been shown. Thus,

$$W^t(\operatorname{diag}(P, \ldots, P), y) = U_y \chi_{R^+} U_{-y} = \chi_{[y,\infty)} I.$$

For (7) we first suppose a to be continuous and notice that $W^1(Y_{t^{-1}} T^0(a) Y_t, 0) = W^1(T^0(a_t), 0)$ with $a_t(z) = a(tz)$. This reduces our attention to the strong limits $W^1(T^0(a), 0)$ for all continuous multipliers a satisfying the row sum condition at $1 \in \mathbf{T}$. Further we suppose without loss that the entries of a are trigonometric polynomials. Now write $T^0(a) = T^0(a(1)) + T^0(b)$, where b is a trigonometric polynomial with $b_{rs}(1) = 0$. Hence,

$$T^0(b) = T^0(c) \operatorname{diag}(I - V, \ldots, I - V) \tag{10}$$

with another polynomial c. For (10) we find

$$W^1(T^0(c) \operatorname{diag}(I - V, \ldots, I - V), 0) =$$
$$= \operatorname*{s-lim}_{n \to \infty} (E_n^F T^0(c) E_{-n}^F) (E_n^F \operatorname{diag}(I - V, \ldots, I - V) E_{-n}^F L_n^{F,F})$$
$$= \operatorname*{s-lim}_{n \to \infty} (E_n^F T^0(c) E_{-n}^F) (I - U_{1/n}) L_n^{F,F},$$

and this is equal to zero since $(I - U_{1/n}) L_n^{F,F} \to 0$ and since the operators $(E_n^F T^0(c) E_{-n}^F)$ are uniformly bounded by de Boor's estimates. So we are left with the summand $T^0(a(1))$. Since a satisfies the row sum condition at $1 \in \mathbf{T}$, there is a constant β such that

$$\sum_s \alpha_s^F a_{rs}(1) = \alpha_r^F \beta. \tag{11}$$

We must show that

$$\lim_{n \to \infty} E_n^F T^0(a(1)) E_{-n}^F L_n^{F,F} f = \beta f \tag{12}$$

for all f belonging to a dense subset of $L_R^p(0)$ which is now comfortably specified as the set of all piecewise constant functions. Each piecewise constant function is a linear combination of characteristic functions of intervals. Thus we are asked to prove (12) for $f = \chi_{[x,y]}$. Taking into account Theorem 2.12 it is not hard to see that

$$\| \sum_{r=0}^{k-1} \sum_{l=[nx]}^{[ny]} \alpha_r^F \phi_{rln} - \chi_{[x,y]} \|_{L_R^p(0)} \to 0 \quad \text{as } n \to \infty . \tag{13}$$

Thus,

$$\lim_{n \to \infty} E_n^F T^0(a(1)) E_{-n}^F L_n^{F,F} \chi_{[x,y]}$$

$$= \lim_{n \to \infty} E_n^F T^0(a(1)) E_{-n}^F \sum_{r=0}^{k-1} \sum_{l=[nx]}^{[ny]} \alpha_r^F \phi_{rln} = \lim_{n \to \infty} \sum_{r=0}^{k-1} \sum_{l=[nx]}^{[ny]} \sum_{s=0}^{k-1} a_{rs}(1) \alpha_s^F \phi_{rln}$$

$$= \lim_{n \to \infty} \sum_{r=0}^{k-1} \sum_{l=[nx]}^{[ny]} \beta \alpha_r^F \phi_{rln} = \beta \chi_{[x,y]}$$

by (13).

Now let $a = \rho^z$ (see the definition in 3.3.3). Obviously,

$$W^1(Y_{t^{-1}} T^0(\rho^z) Y_t, 0) = W^1(T^0(\rho^{z/t}), 0) .$$

If $t \neq z$ then we pick a continuous multiplier f with $f(1) = 1$ and $z/t \notin \operatorname{supp} f$, and write

$$T^0(\rho^{z/t}) = T^0(\rho^{z/t} \operatorname{diag}(f, \ldots, f)) + T^0(\rho^{z/t} \operatorname{diag}(1-f, \ldots, 1-f)) .$$

Since $\rho^{z/t} \operatorname{diag}(f, \ldots, f)$ is a continuous multiplier which, moreover, satisfies the row sum condition at 1 we conclude from (11) and (12) that

$$W^1(T^0(\rho^{z/t} \operatorname{diag}(f, \ldots, f)), 0) = \sum_s \alpha_s^F \rho_{0s}^{z/t}(1) I = \rho(\frac{t}{z}) I.$$

The last identity is a consequence of the definition of $\rho^{t/z}$ and of Proposition 2.19. Further,

$$W^1(T^0(\rho^{z/t} \operatorname{diag}(1-f, \ldots, 1-f)), 0) =$$
$$= \underset{n \to \infty}{\text{s-lim}}\, E_n^F T^0(\rho^{z/t}) E_{-n}^F\, E_n^F T^0(\operatorname{diag}(1-f, \ldots, 1-f)) E_{-n}^F L_n^{F,F} .$$

This limit is zero due to the uniform boundedness of the operators $E_n^F T^0(\rho^{z/t}) E_{-n}^F$ and by what has already been shown. Thus, $W^t(T^0(\rho^z), y) = \rho(\frac{t}{z}) I$ for $t \neq z$.

In case $t = z$ we simply write

$$T^0(\rho^1) = T^0(\rho^1 - (\lambda^{F,F})^{-1} \sigma^{F,F}) + T^0((\lambda^{F,F})^{-1} T^0(\sigma^{F,F}) .$$

The function $\rho^1 - (\lambda^{F,F})^{-1} \sigma^{F,F}$ is a continuous multiplier which satisfies the row sum condition at $1 \in \mathbf{T}$, and all row sums are equal to zero (compare Theorem

2.15(b) and remember the definition of ρ^1). Thus, $W^1(T^0(\rho^1 - (\lambda^{F,F})^{-1}\sigma^{F,F}), 0) = 0$ and, moreover,

$$W^1(T^0(\lambda^{F,F})^{-1}T^0(\sigma^{F,F}), 0) =$$
$$= \operatorname*{s-lim}_{n\to\infty} E_n^F T^0(\lambda^{F,F})^{-1}T^0(\sigma^{F,F})E_{-n}^F L_n^{F,F} = \operatorname*{s-lim}_{n\to\infty} L_n^{F,F}S_R L_n^{F,F} = S_R$$

by 2.11.1(4) which finally gives $W^t(T^0(\rho^t), y) = S_R$. Due to Proposition 3.11 (ii) and (iii) one immediately gets the assertion (7).

To complete the proof we shall shortly explain what is the point in the weighted case. The identity (8) remains true since the operators Y_t and Λ are both diagonal, and so they commute. Hence,

$$W^t(A, y) = W^1(Y_{t^{-1}} A Y_t, y)$$
$$= \operatorname*{s-lim}_{n\to\infty} E_n^F \Lambda V_{\{yn+r\}} \Lambda^{-1} Y_{t^{-1}} A Y_t \Lambda V_{-\{yn+r\}} \Lambda^{-1} E_{-n}^F L_n^{F,F}$$
$$= \operatorname*{s-lim}_{n\to\infty} (n^\alpha E_n^F \Lambda E_{-n}^F L_n^{F,F}) U_{\frac{\{yn+r\}}{n}} (n^{-\alpha} E_n^F \Lambda^{-1} E_{-n}^F L_n^{F,F})$$
$$(E_n^F Y_{t^{-1}} A Y_t E_{-n}^F L_n^{F,F})(n^\alpha E_n^F \Lambda E_{-n}^F L_n^{F,F}) U_{-\frac{\{yn+r\}}{n}} (n^{-\alpha} E_n^F \Lambda^{-1} E_{-n}^F L_n^{F,F}).$$

$$(14)$$

We claim that

$$\operatorname*{s-lim}_{n\to\infty} n^\alpha E_n^F \Lambda E_{-n}^F L_n^{F,F} = \Delta, \quad \operatorname*{s-lim}_{n\to\infty} n^{-\alpha} E_n^F \Lambda^{-1} E_{-n}^F L_n^{F,F} = \Delta^{-1}. \tag{15}$$

Once this had been shown we would get from (14)

$$W^t(A, y) = \Delta U_y \Delta^{-1}(\operatorname*{s-lim}_{n\to\infty} E_n^F Y_{t^{-1}} A Y_t E_{-n}^F L_n^{F,F}) \Delta U_{-y} \Delta^{-1}, \tag{16}$$

and the existence of the central strong limit can be verified as in the unweighted case. In particular, (5)-(7) hold.

Let us prove the first assertion of (15). As for the limit (12) it suffices to check that

$$\|\Delta^{-1} n^\alpha E_n^F \Lambda E_{-n}^F L_n^{F,F} \chi_{[x,y]} - \chi_{[x,y]}\|_{L_R^p(0)} \to 0 \tag{17}$$

as $n \to \infty$ for each interval $[x, y]$. Using (13) once more we must show

$$\left\| \Delta^{-1} n^\alpha E_n^F \Lambda E_{-n}^F \sum_{r=0}^{\kappa-1} \sum_{l=[xn]}^{[yn]} \phi_{rln} - \sum_{r=0}^{\kappa-1} \sum_{l=[xn]}^{[yn]} \phi_{rln} \right\| \to 0$$

in place of (17) or, equivalently,

$$\left\| \sum_{r=0}^{\kappa-1} \sum_{l=[xn]}^{[yn]} \left(\frac{\Delta^{-1} n^\alpha}{(|l|+1)^\alpha} - 1 \right) \phi_{rln} \right\|_{L_R^p(0)} \to 0. \tag{18}$$

Suppose without loss that $\operatorname{supp}\phi_r \subseteq [0, a]$ with a positive integer a. Then the norm in (18) is less than or equal to

$$\sum_{r=0}^{\kappa-1}\sum_{s=0}^{a-1} \| \sum_{l\in Z}(\frac{\Delta^{-1}n^\alpha}{(|al+s|+1)^\alpha} - 1)\phi_{r,al+s,n}\|_{L_R^p(0)} \,,$$

and since $\operatorname{supp}\phi_{r,al+s,n} \cap \operatorname{supp}\phi_{r,am+s,n} = \emptyset$ whenever $l \neq m$, this is equal to

$$\sum_{r=0}^{\kappa-1}\sum_{s=0}^{a-1}(\sum_{l\in Z}\int_R |(\frac{|x|^\alpha n^\alpha}{(|al+s|+1)^\alpha} - 1)\phi_{r,al+s,n}(x)|^p dx)^{1/p}$$

$$=\sum_{r=0}^{\kappa-1}\sum_{s=0}^{a-1}(\frac{1}{n}\sum_{l\in Z}\int_R |\frac{|x+al+s|^\alpha}{(|al+s|+1)^\alpha} - 1|^p|\phi_r(x)|^p dx)^{1/p} \,. \tag{19}$$

By the mean value theorem, there are numbers $x_l \in [0, a]$ such that (19) equals

$$\sum_{r=0}^{\kappa-1}\sum_{s=0}^{a-1}(\frac{1}{n}\sum_{l\in Z}|\frac{|x_l+al+s|^\alpha}{(|al+s|+1)^\alpha} - 1|^p)^{1/p}\|\phi_r\|_{L_R^p(0)} \,. \tag{20}$$

For l sufficiently large, there is a constant $C > 0$ such that

$$\left|\frac{|x_l+al+s|^\alpha}{(|al+s|+1)^\alpha} - 1\right| < \frac{C}{|l|+1} \,,$$

thus, the series in (20) converges whence follows that (20) goes to zero as $n \to \infty$. This proves (15). $\blacksquare$

Next we consider the strong limits W_s. It turns out that these limits exist for an essentially larger algebra than $\mathcal{A}_{\Omega,r}^F$, viz. for the subalgebra of $\mathcal{F}^F$ generated by all sequences of the form

$$(E_n^F \Lambda V_{\{yn+r\}}\Lambda^{-1}A\Lambda V_{-\{yn+r\}}\Lambda^{-1}E_{-n}^F) \tag{21}$$

but now with $A \in (T_{1,1,T}^p(\alpha))_{k\times k}$. Let us abbreviate the sequence (21) by (A, y) again. A little thought shows that this large algebra is generated by the sequences $(T^0(a), y)$ with a running through $(PC_{p,\alpha})_{k\times k}$ and $y \in \mathbf{R}$ and by the sequences $(\operatorname{diag}(P, \ldots, P), y)$, $y \in \mathbf{R}$.

Proposition 3.14 (a) *The strong limits $W_s(A_n)$ exist for each of the sequences in (21) (and thus for all sequences in $\mathcal{A}_{\Omega,r}^F$). In particular,*

$$W_s(T^0(a), y) = \begin{cases} T^0(a) & (\in L(l_k^p(\alpha))) & if & s=0, y=0 \\ \Lambda T^0(a)\Lambda^{-1} & (\in L(l_k^p(\alpha))) & if & s=0, y\neq 0 \\ \Lambda^{-1}T^0(a)\Lambda & (\in L(l_k^p(0))) & if & s\neq 0, y=s \\ T^0(a) & (\in L(l_k^p(0))) & if & s\neq 0, y\neq s \end{cases} \,, \tag{22}$$

$$W_s(\text{diag}\,(P,\ldots,P),y) = \begin{cases} \text{diag}\,(0,\ldots,0) & if \quad s < y \\ \text{diag}\,(P,\ldots,P) & if \quad s = y \\ \text{diag}\,(I,\ldots,I) & if \quad s > y \end{cases}, \tag{23}$$

and, if K is compact,

$$W_s(K,y) = \begin{cases} K & if \quad s = y = 0 \\ \Lambda^{-1}K\Lambda & if \quad s = y \neq 0 \\ 0 & if \quad s \neq y \end{cases}. \tag{24}$$

(b) If (A_n) is the sequence (21) (or, in particular, a sequence in $\mathcal{A}_{\Omega,r}^F$), then the cosets $E_{-n}^F A_n E_n^F + K(l_k^p(\alpha))$ are independent of n. Moreover, if W_∞ denotes one of these cosets, then

$$W_\infty(A,y) = \pi(A) \quad \text{for all } A \in (T_{1,1,T}^p(\alpha))_{k \times k}.$$

Proof (a) Let, for example, $s \neq 0$. The proof for $s = 0$ is analogous. Then

$$W_s(A,y) = \operatorname*{s-lim}_{n \to \infty} V_{\{yn+r\}-\{sn+r\}} \Lambda^{-1} A\Lambda V_{\{sn+r\}-\{yn+r\}}. \tag{25}$$

If $y = s$ then, clearly, $W_s(A,y) = \Lambda^{-1}A\Lambda$ which shows the third assertion of (22) and the second of (23). For $s \neq y$ and $A = T^0(a)$ we refer to ([BS 2], 6.2(b)), where the existence of the limit (25) was shown, and for $A = \text{diag}\,(P,\ldots,P)$ one takes into account that $\Lambda^{-1}P\Lambda = P$, and then the proof is almost obvious. Finally, (24) can be derived from (22) and (23) by approximating K by a linear combination of the rank-one operators K_{rstu} introduced in the proof of Proposition 3.13 and by representing these operators in terms of shift operators V_n and of the projection P which is always possible.

(b) For $\alpha = 0$ the proof is evident: In this case $E_{-n}^F A_n E_n^F = V_{\{yn+r\}}AV_{-\{yn+r\}}$, and now it remains to remember that the shift operators $V_{\{yn+r\}}$ commute with each other operator A in $(T_{1,1,\Omega}^p(\alpha))_{k \times k}$ modulo a compact operator. In particular,

$$V_{\{yn+r\}}T^0(a)V_{-\{yn+r\}} = T^0(a)$$

and

$$V_{\{yn+r\}}\text{diag}\,(P,\ldots,P)V_{-\{yn+r\}} = \text{diag}\,(P,\ldots,P) + finite\ rank\ operator.$$

In case $\alpha \neq 0$ it is most advantageous to use that $T_{1,1,T}^p(\alpha) = \Lambda T_{1,1,T}^p(0)\Lambda^{-1}$ (see Exercise 2.8). Then we have

$$E_{-n}^F(A,y)E_n^F = E_{-n}^F(\Lambda B\Lambda^{-1},y)E_n^F$$

for some $B \in (T_{1,1,T}^p(0))_{k \times k}$, and thus

$$E_{-n}^F(A,y)E_n^F = \Lambda V_{\{yn+r\}}BV_{-\{yn+r\}}\Lambda^{-1} = \Lambda B\Lambda^{-1} + compact = A + compact.$$

So, Proposition 3.12 is completely proved. ∎

3.5.2   Special sequences in $\mathcal{A}_{\Omega,r}^F$

In this section we verify that certain particular sequences, which will be needed for technical reasons, lie in $\mathcal{A}_{\Omega,r}^F$.

Proposition 3.15 *(a) $\mathcal{A}_{\Omega,r}^F$ contains the set $\mathcal{G}$ of all sequences $(G_n) \in \mathcal{F}^F$ such that $\|G_n\| \to 0$ as $n \to \infty$ and $E_{-n}^F G_n E_n^F$ is compact for all n.*
(b) $\mathcal{G}$ is a closed two-sided ideal of $\mathcal{A}_{\Omega,r}^F$.
(c) $\mathcal{G}$ is the intersection of $\mathcal{A}_{\Omega,r}^F$ with the collection of all sequences in $\mathcal{F}^F$ tending to zero in the operator norm.

Proof. (a) A little thought shows that the inclusion $\mathcal{G} \subseteq \mathcal{A}_{\Omega,r}^F$ will follow as soon as we have proved that $(G_n) \in \mathcal{A}_{\Omega,r}^F$ where (G_n) is the sequence with $G_n = 0$ for $n \neq n_0$ and $E_{-n_0}^F G_{n_0} E_{n_0}^F = K_{ij}^{st}$ with $K_{ij}^{st} \in K(l_k^p(\alpha))$ standing for the matrix whose st th entry has the matrix representation $(\delta_{il}\delta_{jm})_{l,m \in Z}$ (δ_{lm} the Kronecker delta) and whose other entries vanish.

To get this inclusion let, for definiteness, $r \in (-1,0]$. Since $V_{\{r\}} = I$ and $V_{\{n+r\}} = V_n$, and by the definition of $\mathcal{A}_{\Omega,r}^F$, the sequences

$$(E_n^F K_{ij}^{st} E_{-n}^F) = (E_n^F \Lambda V_{\{r\}} \Lambda^{-1} K_{ij}^{st} \Lambda V_{-\{r\}} \Lambda^{-1} E_{-n}^F)$$

and

$$(E_n^F V_n K_{j-n_0,j-n_0}^{tt} V_{-n} E_{-n}^F) = (E_n^F \Lambda V_{\{n+r\}} \Lambda^{-1} K_{j-n_0,j-n_0}^{tt} \Lambda V_{-\{n+r\}} \Lambda^{-1} E_{-n}^F)$$

belong to $\mathcal{A}_{\Omega,r}^F$, and because of

$$E_n^F K_{ij}^{st} E_{-n}^F E_n^F V_n K_{j-n_0,j-n_0}^{tt} V_{-n} E_{-n}^F = \begin{cases} E_n^F K_{ij}^{st} E_{-n}^F & \text{if} \quad n = n_0 \\ 0 & \text{if} \quad n \neq n_0 \end{cases}$$

we get our claim.

(b) This assertion is evident.

(c) Let $(G_n) \in \mathcal{A}_{\Omega,r}^F$ and $\|G_n\| \to 0$. Then, on the one hand, $\|E_{-n}^F G_n E_n^F + K(l_k^p(\alpha))\| \to 0$ whereas, on the other hand, $\|E_{-n}^F G_n E_n^F + K(l_k^p(\alpha))\|$ is independent of n by Proposition 3.14(b). Thus, $\|E_{-n}^F G_n E_n^F + K(l_k^p(\alpha))\| = 0$ for all n, and hence, the operators $E_{-n}^F G_n E_n^F$ are compact for all n. ∎

Proposition 3.16 *Let $f \in C(\dot{\mathbf{R}})$. Then the sequence $(L_n^{F,F} fI|S_n^F)$ is in $\mathcal{A}_{\Omega,r}^F$ for all Ω and r. In particular,*

$$W^t(L_n^{F,F} fI|S_n^F) = fI \quad \text{for all} \quad t \in \mathbf{T} \,,$$

$$W_s(L_n^{F,F} fI|S_n^F) = f(s)I \quad \text{for all} \quad s \in \mathbf{R} \,,$$

and

$$W_\infty(L_n^{F,F} fI|S_n^F) = f(\infty)\pi(I) \,.$$

Proof If f is a constant function then the sequence $(L_n^{F,F} f I | S_n^F)$ belongs obviously to $\mathcal{A}_{\Omega,r}^F$. So we can suppose without loss that $f(\infty) = 0$ (otherwise subtract the constant function $t \mapsto f(\infty)$ from f). Then, given $\epsilon > 0$, we can choose numbers $f_i \in \mathbf{C}$ and $a_i \in \mathbf{R}$ with $-\infty < a_0 < a_1 < \ldots < a_l < \infty$ such that

$$\| f - \sum_{i=0}^{l-1} f_i \chi_{[a_i, a_{i+1})} \|_\infty < \epsilon . \tag{1}$$

Our first claim is that the sequence

$$(L_n^{F,F} \sum_{i=0}^{l-1} f_i \chi_{\left[\frac{\{a_i n + r\}}{n}, \frac{\{a_{i+1} n + r\}}{n} \right)} I | S_n^F) \tag{2}$$

belongs to $\mathcal{A}_{\Omega,r}^F$. Because of $\chi_{[a,b)} = \chi_{[a,\infty)} - \chi_{[b,\infty)}$ it suffices to check whether the sequence

$$(L_n^{F,F} \chi_{\left[\frac{\{a n + r\}}{n}, \infty \right)} I | S_n^F) \tag{3}$$

is in $\mathcal{A}_{\Omega,r}^F$ for all Ω, r and a. By 2.11.4(1) there is a finite rank operator K such that

$$
\begin{aligned}
(L_n^{F,F} \chi_{\left[\frac{\{a n + r\}}{n}, \infty \right)} I | S_n^F) &= \\
&= (U_{\frac{\{a n + r\}}{n}} L_n^{F,F} \chi_{\mathbf{R}^+} I \, U_{-\frac{\{a n + r\}}{n}} | S_n^F) \\
&= (E_n^F V_{\{a n + r\}} (\mathrm{diag}\,(P, \ldots, P) + K) V_{-\{a n + r\}} E_{-n}^F) .
\end{aligned} \tag{4}
$$

The sequence $(E_n^F V_{\{a n + r\}} \mathrm{diag}\,(P, \ldots, P) V_{-\{a n + r\}} E_{-n}^F)$ can be written as

$$(E_n^F \Lambda V_{\{a n + r\}} \Lambda^{-1} \mathrm{diag}\,(P, \ldots, P) \Lambda V_{-\{a n + r\}} \Lambda^{-1} E_{-n}^F) ,$$

hence, it is in $\mathcal{A}_{\Omega,r}^F$. Let us explain that the sequence $(E_n^F V_{\{a n + r\}} K V_{-\{a n + r\}} E_{-n}^F)$ is in $\mathcal{A}_{\Omega,r}^F$, too. It is clearly sufficient to prove this for K replaced by the operator K_{ij}^{st} introduced in the proof of Proposition 3.15. For these operators we have

$$V_{\{a n + r\}} K_{ij}^{st} V_{-\{a n + r\}} = K_{i+\{a n + r\}, j+\{a n + r\}}^{st}$$

and

$$\Lambda V_{\{a n + r\}} K_{ij}^{st} V_{-\{a n + r\}} \Lambda^{-1} = \frac{(|j + \{a n + r\}| + 1)^\alpha}{(|i + \{a n + r\}| + 1)^\alpha} K_{i+\{a n + r\}, j+\{a n + r\}}^{st}$$

and, hence,

$$
\begin{aligned}
\| V_{\{a n + r\}} K_{ij}^{st} V_{-\{a n + r\}} - \Lambda V_{\{a n + r\}} K_{ij}^{st} V_{-\{a n + r\}} \Lambda^{-1} \|_{L(l_k^p(\alpha))} &= \\
&= \left| \frac{(|j + \{a n + r\}| + 1)^\alpha}{(|i + \{a n + r\}| + 1)^\alpha} - 1 \right| \| K_{i+\{a n + r\}, j+\{a n + r\}}^{st} \|_{L(l_k^p(\alpha))} .
\end{aligned} \tag{5}
$$

Since the norms $\|K_{i+\{an+r\},j+\{an+r\}}^{st}\|$ are uniformly bounded with respect to n, the norms (5) tend to zero as $n \to \infty$. Hence, the sequence

$$(E_n^F V_{\{an+r\}} K_{ij}^{st} V_{-\{an+r\}} E_{-n}^F - E_n^F \Lambda V_{\{an+r\}} K_{ij}^{st} V_{-\{an+r\}} \Lambda^{-1} E_{-n}^F)$$

lies in the ideal $\mathcal{G}$ and so, by Proposition 3.15, in $\mathcal{A}_{\Omega,r}^F$. Since further

$$(E_n^F \Lambda V_{\{an+r\}} K_{ij}^{st} V_{-\{an+r\}} \Lambda^{-1} E_{-n}^F) =$$
$$= (E_n^F \Lambda V_{\{an+r\}} \Lambda^{-1} (\Lambda K_{ij}^{st} \Lambda^{-1}) \Lambda V_{-\{an+r\}} \Lambda^{-1} E_{-n}^F) \qquad (6)$$

with the compact operator $\Lambda K_{ij}^{st} \Lambda^{-1}$, the sequence (6) is in $\mathcal{A}_{\Omega,r}^F$. Consequently, the sequences (4), (3) and (2) are in $\mathcal{A}_{\Omega,r}^F$ as we have claimed.

Now consider the sequence

$$(A_n) := (L_n^{F,F} fI|S_n^F) - \left(L_n^{F,F} \sum_{i=0}^{l-1} f_i \chi_{\left[\frac{\{a_i n+r\}}{n}, \frac{\{a_{i+1} n+r\}}{n}\right)} I|S_n^F\right).$$

We shall show that (A_n) is even in $\mathcal{G}$. Indeed, by the uniform boundedness of the projections $L_n^{F,F}$, there is a constant $C > 0$ such that

$$\|A_n\| \leq C \left\| f - \sum_{i=0}^{l-1} f_i \chi_{[\{a_i n+r\}/n, \{a_{i+1} n+r\}/n)} \right\|_\infty$$

$$\leq C \left\| f - \sum_{i=0}^{l-1} f_i \chi_{[a_i, a_{i+1})} \right\|_\infty +$$

$$+ C \left\| \sum_{i=0}^{l-1} f_i (\chi_{[a_i, a_{i+1})} - \chi_{[\{a_i n+r\}/n, \{a_{i+1} n+r\}/n)}) \right\|_\infty . \qquad (7)$$

The first term is less than $C\epsilon$ by (1). Moreover, if $r \geq 0$ (the case $r < 0$ is analogous) and if n is large enough (say, $n \geq n_0$) then

$$[\{a_i n + r\}/n , \{a_{i+1} n + r\}/n) \subseteq [a_i, a_{i+2})$$

which implies that the second summand in (7) is less than

$$C \max_i |f_i - f_{i+1}| \leq 2C\epsilon$$

for $n \geq n_0$. Summarizing these facts we find that $\sup_{n \geq n_0} \|A_n\| \leq 3C\epsilon$ and thus, $\|A_n\| \to 0$ as $n \to \infty$. Finally, the compactness of each of the operators $E_{-n}^F A_n E_n^F$ is evident if f has a compact support, and in the general case we approximate f by a function with compact support which is possible since $f(\infty) = 0$. Thus, $(A_n) \in \mathcal{G} \subseteq \mathcal{A}_{\Omega,r}^F$, and this finishes the proof of the inclusion $(L_n^{F,F} fI|S_n) \in \mathcal{A}_{\Omega,r}^F$.

It remains to compute the strong limits W^t and W_s for this sequence. The above considerations reduce this problem to the computation of these limits for the sequences $(\mathrm{diag}\,(P,\ldots,P),a_i)$ and (K,a_i) with $K \in K(l_k^p(\alpha))$, and this has already been done in Propositions 3.13 and 3.14. For $s = \infty$ we remember once more that

$$W_\infty(L_n^{F,F} fI|S_n^F) = f(\infty)W_\infty(L_n^{F,F}|S_n^F) + W_\infty(L_n^{F,F}(f - f(\infty))I|S_n^F)$$

and that $W_\infty(L_n^{F,F}|S_n^F) = \pi(I)$ and all operators $E_{-n}^F L_n^{F,F}(f - f(\infty))E_n^F$ are compact. ∎

Proposition 3.17 *Let $z \in \Omega$ and let K be a compact operator. Then the sequence*

$$(K_n) := (E_n^F Y_z E_{-n}^F L_n^{F,F} K E_n^F Y_{z^{-1}} E_{-n}^F|S_n^F) \tag{8}$$

is in $\mathcal{A}_{\Omega,r}^F$ and

$$W^t(K_n) = \begin{cases} K & if \quad t = z \\ 0 & if \quad t \neq z \end{cases} ,$$

$$W_s(K_n) = 0 \quad for\,all \quad s \in \dot{\mathbf{R}} .$$

Proof   A little thought shows that we can restrict ourselves to the case $z = 1$ when proving the inclusion $(K_n) \in \mathcal{A}_{\Omega,r}^F$ (compare Exercise 3.11). Further, as in the proof of property 3.1.1(i) (see Exercise 3.3 and the hints given there) it is sufficient to verify this inclusion for K being replaced by the compact operator $S_R fI - f S_R$ where $f \in C(\dot{\mathbf{R}})$. Taking into account the commutator property (see the Theorems 2.7 and 2.8) we are so left on verifying that

$$(L_n^{F,F} S_R L_n^{F,F} fI|S_n^F - L_n^{F,F} f L_n^{F,F} S_R|S_n^F) \in \mathcal{A}_{\Omega,r}^F . \tag{9}$$

In case $\Omega = \{1\}$ this is obvious since then $(L_n^{F,F} fI|S_n^F) \in \mathcal{A}_{\Omega,r}^F$ by Proposition 3.16 and $(L_n^{F,F} S_R|S_n^F) \in \mathcal{A}_{\Omega,r}^F$ by Theorem 2.15. If 1 is properly contained in Ω the things are more involved since $(L_n^{F,F} S_R|S_n^F)$ is not in $\mathcal{A}_{\Omega,r}^F$ in general (there is no reason for the function $(\lambda^{F,F})^{-1}\sigma^{F,F}$ to satisfy a row sum condition at points $t \in \Omega \backslash \{1\}$). Nevertheless, the commutator (9) lies in each algebra $\mathcal{A}_{\Omega,r}^F$ with $1 \in \Omega$ which can be seen as follows.

Let g stand for the function $\rho^1 - (\lambda^{F,F})^{-1}\sigma^{F,F}$ (see 3.3.3). Then we have

$$\begin{aligned} L_n^{F,F} S_R|S_n^F &= E_n^F T^0(\lambda^{F,F})^{-1} T^0(\sigma^{F,F})E_{-n}^F|S_n^F \\ &= E_n^F T^0(\rho^1)E_{-n}^F|S_n^F - E_n^F T^0(g)E_{-n}^F|S_n^F , \end{aligned}$$

and thus,

$$\begin{aligned} (L_n^{F,F} S_R L_n^{F,F} fI|S_n^F &- L_n^{F,F} f L_n^{F,F} S_R|S_n^F) = \\ &= (E_n^F T^0(\rho^1)E_{-n}^F L_n^{F,F} fI|S_n^F - L_n^{F,F} f L_n^{F,F} E_n^F T^0(\rho^1)E_{-n}^F|S_n^F) \\ &\quad -(E_n^F T^0(g)E_{-n}^F L_n^{F,F} fI|S_n^F - L_n^{F,F} f L_n^{F,F} E_n^F T^0(g)E_{-n}^F|S_n^F) . \end{aligned} \tag{10}$$

The first sequence of the right hand side of (10) is in $\mathcal{A}_{\Omega,r}^F$ for all Ω containing $1 \in \mathbf{T}$ since $(E_n^F T^0(\rho^1) E_{-n}^F | S_n^F)$ is in each of these algebras. For the second sequence we find

$$\|E_n^F T^0(g) E_{-n}^F L_n^{F,F} fI - L_n^{F,F} f L_n^{F,F} E_n^F T^0(g) E_{-n}^F L_n^{F,F}\|$$

$$\leq \|L_n^{F,F}\| \, \|E_n^F T^0(g) T^0(\lambda^{F,F})^{-1} P_n^F fI - f E_n^F T^0(g) T^0(\lambda^{F,F})^{-1} P_n^F\| \, ,$$

and since g is a continuous multiplier, the commutator property yields that this sequence is even in $\mathcal{G}$ (and thus in $\mathcal{A}_{\Omega,r}^F$ by Proposition 3.15).

It remains to determine the strong limits. For $t = z$ we have

$$W^t(K_n) \;=\; \underset{n\to\infty}{\text{s-lim}} \; L_n^{F,F} K L_n^{F,F} \;=\; K \, .$$

That all other limits vanish can be seen by computing these limits for the sequence (10) (take into account Propositions 3.13-3.16). Another way to verify this is sketched in the Exercises 3.15 and 3.16. ∎

3.5.3   Ideals and commutators

Let $\mathcal{J}_{\Omega,r}^F$ stand for the smallest closed two-sided ideal of the algebra $\mathcal{A}_{\Omega,r}^F$ which includes the ideal $\mathcal{G}$ and all sequences of the form (8) with $z \in \Omega$ and K compact. Further, write Φ^J for the canonical homomorphism from $\mathcal{A}_{\Omega,r}^F$ onto the quotient algebra $\mathcal{A}_{\Omega,r}^F / \mathcal{J}_{\Omega,r}^F$. The analogue of property 3.1.1(ii) reads as follows:

Proposition 3.18 *The commutator* $(A_n)(L_n^{F,F} fI | S_n^F) - (L_n^{F,F} fI | S_n^F)(A_n)$ *is in* $\mathcal{J}_{\Omega,r}^F$ *for all sequences* $(A_n) \in \mathcal{A}_{\Omega,r}^F$ *and all functions* $f \in C(\mathbf{R})$ *or, in other words, the coset* $\Phi^J(L_n^{F,F} fI | S_n^F)$ *belongs to the center of* $\mathcal{A}_{\Omega,r}^F / \mathcal{J}_{\Omega,r}^F$.

We prepare the proof by a somewhat surprising lemma which states that the operators $L_n^{F,F} fI | S_n^F$ are almost of diagonal form.

Lemma 3.3 *Let* (F,G) *and* (H,L) *be canonical pairs with the same* k, *and let* $f \in C(\dot{\mathbf{R}})$. *Then*

$$\|E_{-n}^F L_n^{F,G} f E_n^F - E_{-n}^H L_n^{H,L} f E_n^H\|_{L(l_k^p(\alpha))} \to 0 \, .$$

Proof   One has

$$E_{-n}^F L_n^{F,G} f E_n^F \;=\; E_{-n}^H L_n^{H,L} E_n^H E_{-n}^F L_n^{F,G} f E_n^F \, .$$

$$=\; E_{-n}^H L_n^{H,L} E_n^H T^0(\lambda^{F,G})^{-1} P_n^G f E_n^F$$

$$=\; E_{-n}^H L_n^{H,L} f E_n^H T^0(\lambda^{F,G})^{-1} P_n^G E_n^F \;+\; C_n$$

$$=\; E_{-n}^H L_n^{H,L} f E_n^H \;+\; C_n$$

with operators

$$C_n \;=\; E_{-n}^H L_n^{H,L} (E_n^H T^0(\lambda^{F,G})^{-1} P_n^G fI - f E_n^H T^0(\lambda^{F,G})^{-1} P_n^G) E_n^F$$

tending to zero in the operator norm by the commutator property. ∎

For the choice $H = L = \{\chi_{[i/k,(i+1)/k)}, i = 0, \ldots, k-1\}$, the operators $E^H_{-n} L^{H,L}_n f E^H_n$ are diagonal, thus, for *each* canonical pair, $(E^F_{-n} L^{F,G}_n f E^F_n)$ is a sequence of diagonal operators up to a perturbation tending to zero in the norm.

Proof of Proposition 3.18. We have to show that $(L^{F,F}_n f I | S^F_n)$ commutes modulo $\mathcal{J}^F_{\Omega,r}$ with each of the generating sequences 3.5.1(2)-(4) of the algebra $\mathcal{A}^F_{\Omega,r}$. Let us first treat the unweighted case $(\alpha = 0)$ in detail. The commutator of $(L^{F,F}_n f I | S^F_n)$ with 3.5.1(2) is

$$
\begin{aligned}
&(L^{F,F}_n f I | S^F_n)(K,y) - (K,y)(L^{F,F}_n f I | S^F_n) \\
&= (E^F_n V_{\{yn+r\}}[V_{-\{yn+r\}} E^F_{-n} L^{F,F}_n f E^F_n V_{\{yn+r\}} K - \\
&\quad - K V_{-\{yn+r\}} E^F_{-n} L^{F,F}_n f E^F_n V_{\{yn+r\}}] V_{-\{yn+r\}} E^F_{-n} | S^F_n).
\end{aligned}
\tag{1}
$$

From Proposition 3.16 we know that

$$
V_{-\{yn+r\}} E^F_{-n} L^{F,F}_n f E^F_n V_{\{yn+r\}} \to f(y)I \; (= W_y(L^{F,F}_n f I | S^F_n))
$$

strongly as $n \to \infty$. Analogously,

$$
(V_{-\{yn+r\}} E^F_{-n} L^{F,F}_n f E^F_n V_{\{yn+r\}})^* \to \overline{f(y)}I
$$

strongly as $n \to \infty$. Thus, the sequence in brackets in (1) goes to $f(y)K - Kf(y)I = 0$ in the operator norm (see 1.2.1(b)), and since

$$
\sup_n \|E^F_n V_{\{yn+r\}}\| \, \|V_{-\{yn+r\}} E^F_{-n} | S^F_n\| < \infty
$$

we conclude that (1) is even in $\mathcal{G}$.

Further, taking into account the Lemma 3.3, we find for the commutator of $(L^{F,F}_n f I | S^F_n)$ with 3.5.1(3)

$$
\begin{aligned}
&(L^{F,F}_n f I | S^F_n)(\mathrm{diag}\,(P,\ldots,P),y) - (\mathrm{diag}\,(P,\ldots,P),y)(L^{F,F}_n f I | S^F_n) \\
&= (E^F_n [E^F_{-n} L^{F,F}_n f E^F_n (\mathrm{diag}\,(P,\ldots,P),y) - \\
&\quad - (\mathrm{diag}\,(P,\ldots,P),y) E^F_{-n} L^{F,F}_n f E^F_n] E^F_{-n} | S^F_n) \\
&= (E^F_n [E^H_{-n} L^{H,H}_n f E^H_n (\mathrm{diag}\,(P,\ldots,P),y) - \\
&\quad - (\mathrm{diag}\,(P,\ldots,P),y) E^H_{-n} L^{H,H}_n f E^H_n] E^F_{-n} | S^F_n) + (G_n)
\end{aligned}
\tag{2}
$$

with a sequence $(G_n) \in \mathcal{G}$. Now choose H as above. Then $E^H_{-n} L^{H,H}_n f E^H_n$ and $(\mathrm{diag}\,(P,\ldots,P),y)$ are diagonal operators, and the first summand of (2) vanishes.

Now consider the sequences 3.5.1(4). Due to the Proposition 3.11 we shall prove the corresponding assertion for the operators given in Proposition 3.11(ii). If $a \in PC^{k,\kappa,\Omega}_{p,\alpha}$ is continuous then we have the commutator

$$
\begin{aligned}
&(L^{F,F}_n f I | S^F_n)(E^F_n T^0(a) E^F_{-n} | S^F_n) - (E^F_n T^0(a) E^F_{-n} | S^F_n)(L^{F,F}_n f I | S^F_n) \\
&= (L^{F,F}_n [f E^F_n T^0(a) T^0(\lambda^{F,F})^{-1} P^F_n - E^F_n T^0(a) T^0(\lambda^{F,F})^{-1} P^F_n f] I | S^F_n)
\end{aligned}
$$

which is in $\mathcal{G}$ by the commutator property (the function $a\,(\lambda^{F,F})^{-1}$ is evidently continuous). Finally we deal with the commutator of $(L_n^{F,F}fI|S_n^F)$ with $(T^0(\rho^z),0)$. Using the above lemma again we get

$$
\begin{aligned}
&(L_n^{F,F}fI|S_n^F)(E_n^FT^0(\rho^z)E_{-n}^F|S_n^F) - (E_n^FT^0(\rho^z)E_{-n}^F|S_n^F)(L_n^{F,F}fI|S_n^F)\\
&= (L_n^{F,F}fE_n^FY_zT^0(\rho^1)Y_{z^{-1}}E_{-n}^F - E_n^FY_zT^0(\rho^1)Y_{z^{-1}}E_{-n}^FL_n^{F,F}fI|S_n^F)\\
&= (E_n^FY_zE_{-n}^F[L_n^{F,F}fE_n^FT^0(\rho^1)E_{-n}^F -\\
&\quad -E_n^FT^0(\rho^1)E_{-n}^FL_n^{F,F}fI]E_n^FY_{z^{-1}}E_{-n}^F|S_n^F)
\end{aligned}
$$

modulo $\mathcal{G}$. We claim that the sequence in brackets is of the form $(L_n^{F,F}K|S_n^F+G_n)$ with K compact and $(G_n) \in \mathcal{G}$. For, introduce $g = \rho^1 - (\lambda^{F,F})^{-1}\sigma^{F,F}$ to obtain

$$
\begin{aligned}
&(L_n^{F,F}fE_n^FT^0(\rho^1)E_{-n}^F - E_n^FT^0(\rho^1)E_{-n}^FL_n^{F,F}fI|S_n^F)\\
&= (L_n^{F,F}fE_n^FT^0(g)E_{-n}^F - E_n^FT^0(g)E_{-n}^FL_n^{F,F}fI|S_n^F)\\
&\quad +(L_n^{F,F}fE_n^FT^0(\lambda^{F,F})^{-1}T^0(\sigma^{F,F})E_{-n}^F -\\
&\quad -E_n^FT^0(\lambda^{F,F})^{-1}T^0(\sigma^{F,F})E_{-n}^FL_n^{F,F}fI|S_n^F)\,.
\end{aligned}
\tag{3}
$$

We have already seen that the first term in (3) is in $\mathcal{G}$ (remember that g is continuous), and the second term equals

$$
(L_n^{F,F}fL_n^{F,F}S_R - L_n^{F,F}S_RL_n^{F,F}fI|S_n^F)
$$

which is of the desired form (compare the commutator property in Section 2.7.5).

Let us give some comments to the weighted case. Now, the generating sequences of $\mathcal{A}_{\Omega,r}^F$ are

$$
(A,y) = (E_n^F\Lambda V_{\{yn+r\}}\Lambda^{-1}A\Lambda V_{-\{yn+r\}}\Lambda^{-1}E_{-n}^F|S_n^F)
$$

with $A \in \mathcal{T}_{k,\kappa,\Omega}^p(\alpha)$. In view of Exercise 2.8 it suffices to consider sequences of the following simpler form:

$$
(E_n^F\Lambda V_{\{yn+r\}}BV_{-\{yn+r\}}\Lambda^{-1}E_{-n}^F|S_n^F)
$$

with $B \in \mathcal{T}_{k,\kappa,\Omega}^p(0)$. Further, again by the above lemma, we can commute the sequence $(L_n^{F,F}fI|S_n^F)$ and the diagonal sequences $(E_n^F\Lambda^{\pm1}E_{-n}^F|S_n^F)$ modulo $\mathcal{G}$, and so we get for the commutator

$$
\begin{aligned}
&(L_n^{F,F}fI|S_n^F)(A,y) - (A,y)(L_n^{F,F}fI|S_n^F)\\
&= (E_n^F\Lambda E_{-n}^F|S_n^F)[(L_n^{F,F}fI|S_n^F)(E_n^FV_{\{yn+r\}}BV_{-\{yn+r\}}E_{-n}^F|S_n^F) -\\
&\quad -(E_n^FV_{\{yn+r\}}BV_{-\{yn+r\}}E_{-n}^F|S_n^F)(L_n^{F,F}fI|S_n^F)](E_n^F\Lambda^{-1}E_{-n}^F|S_n^F)\,.
\end{aligned}
\tag{4}
$$

The sequence in brackets is in $\mathcal{J}_{\Omega,r}^F$ as we have just checked, and it remains to show that

$$
(E_n^F\Lambda E_{-n}^FJ_nE_n^F\Lambda^{-1}E_{-n}^F|S_n^F) \in \mathcal{J}_{\Omega,r}^F = \mathcal{J}_{\Omega,r}^F(\alpha)
\tag{5}
$$

whenever $(J_n) \in \mathcal{J}^F_{\Omega,r} = \mathcal{J}^F_{\Omega,r}(0)$. If $(J_n) \in \mathcal{G}$ this is evident. If

$$(J_n) = (E^F_n Y_z E^F_{-n} L^{F,F}_n K E^F_n Y_{z^{-1}} E^F_{-n} | S^F_n)$$

we first commute the diagonal operators Λ and Y_z. Then

$$(E^F_n \Lambda E^F_{-n} J_n E^F_n \Lambda^{-1} E^F_{-n} | S^F_n) =$$
$$= (E^F_n Y_z E^F_{-n} [E^F_n \Lambda E^F_{-n} L^{F,F}_n K E^F_n \Lambda^{-1} E^F_{-n} L^{F,F}_n] E^F_n Y_{z^{-1}} E^F_{-n} | S^F_n) \,,$$

and now remember that

$$E^F_n \Lambda E^F_{-n} L^{F,F}_n K E^F_n \Lambda^{-1} E^F_{-n} L^{F,F}_n \to \Delta K \Delta^{-1}$$

in the norm by 3.5.1(15). So, (5) equals

$$(E^F_n Y_z E^F_{-n} L^{F,F}_n \Delta K \Delta^{-1} E^F_n Y_{z^{-1}} E^F_{-n} | S^F_n)$$

up to a summand in $\mathcal{G}$ whence follows that (4) is in $\mathcal{J}^F_{\Omega,r}(\alpha)$. ∎

3.6 Proof of the stability theorem

3.6.1 Algebraization

Now all things are ready to prove the stability theorem 3.5. We will follow the general scheme presented in Chapter 1. In order to algebraize the stability question we consider the algebra $\mathcal{F}^F$ of all bounded sequences (A_n), $A_n : S^F_n \to S^F_n$, and its closed two-sided ideal $\mathcal{K}$ of all sequences tending to zero in the norm. Then, as we have already remarked, a sequence $(A_n) \in \mathcal{F}^F$ is stable if and only if the coset $(A_n) + \mathcal{K}$ is invertible in the quotient algebra $\mathcal{F}^F/\mathcal{K}$.

But, for sequences (A_n) in $\mathcal{A}^F_{\Omega,r}$, it turns out that $\mathcal{K}$ can be replaced by the smaller ideal $\mathcal{G}$ (consisting of those sequences in $\mathcal{K}$ for which $E^F_{-n} A_n E^F_n$ is compact), and this will be our starting point.

Proposition 3.19 *Let $(A_n) \in \mathcal{A}^F_{\Omega,r}$. Then (A_n) is stable if and only if the coset $(A_n) + \mathcal{G}$ is invertible in $\mathcal{F}^F/\mathcal{G}$.*

Proof Clearly, $\mathcal{G}$ is also a closed two-sided ideal of $\mathcal{F}^F$, thus, it makes sense to form the quotient algebra $\mathcal{F}^F/\mathcal{G}$.

If $(A_n)+\mathcal{G}$ is invertible then standard Neumann-series arguments yields stability of (A_n). Let, conversely, (A_n) be a stable sequence. Then there is an n_0 such that the operators A_n with $n \geq n_0$ are invertible, and the norms of their inverses are uniformly bounded. Further we know from Proposition 3.14(b) that all differences $E^F_{-n} A_n E^F_n - E^F_{-m} A_m E^F_m$ are compact. Thus, the operators A_n with $n < n_0$ are Fredholm operators with zero index. Let B_n be a two-sided regularizer of (A_n) if $n < n_0$ and the inverse of A_n if $n \geq n_0$. Then $(B_n) \in \mathcal{F}^F$ and $A_n B_n = I|S^F_n + K_n$ with

$$K_n = \begin{cases} \text{compact} & \text{if } n < n_0 \\ 0 & \text{if } n \geq n_0 \end{cases}$$

which gives $(K_n) \in \mathcal{G}$. ∎

3.6.2   Essentialization

Proposition 3.18 indicates that the quotient algebra $\mathcal{A}_{\Omega,r}^F / \mathcal{J}_{\Omega,r}^F$ is an adequate "essentialization" of the algebra $\mathcal{F}^F / \mathcal{G}$, that is, $\mathcal{G} \subseteq \mathcal{J}_{\Omega,r}^F \subseteq \mathcal{A}_{\Omega,r}^F \subseteq \mathcal{F}^F$, and $\mathcal{A}_{\Omega,r}^F / \mathcal{J}_{\Omega,r}^F$ has a sufficiently rich center. It remains to show that the ideal $\mathcal{J}_{\Omega,r}^F$ can be lifted in the sense of section 1.6.2.

Proposition 3.20 $\mathcal{J}_{\Omega,r}^F$ *is subject to the conditions in the lifting theorem 1.8.*

Proof   Define for $t \in \Omega$

$$\mathcal{J}^t = \left\{ (E_n^F Y_t E_{-n}^F L_n^{F,F} K E_n^F Y_{t^{-1}} E_{-n}^F | S_n^F) + \mathcal{G}, \; K \text{ compact} \right\} . \tag{1}$$

One shows as in the proof of Proposition 3.8 that $\mathcal{J}^t$ is a left-sided ideal of $\mathcal{A}_{\Omega,r}^F / \mathcal{G}$, that the restriction of the quotient homomorphism

$$W^t : \; \mathcal{A}_{\Omega,r}^F / \mathcal{G} \to L(L_R^p(\alpha)) \; , \; (A_n) + \mathcal{G} \mapsto W^t(A_n)$$

onto $\mathcal{J}^t$ is an isomorphism between $\mathcal{J}^t$ and the ideal $K(L_R^p(\alpha))$, and that $\mathcal{J}_{\Omega,r}^F$ is the smallest closed two-sided ideal in $\mathcal{A}_{\Omega,r}^F$ which contains all sequences (A_n) which belong to some $\mathcal{J}^t$ with $t \in \Omega$.

So it remains to prove that $\mathcal{J}^t$ is a right ideal in $\mathcal{A}_{\Omega,r}^F$. Here we *cannot* refer to the proof of Proposition 3.8 since, in general, the adjoint sequences

$$(E_n^F Y_{t^{-1}} E_{-n}^F A_n E_n^F Y_t E_{-n}^F L_n^{F,F})^* \tag{2}$$

do not strongly converge to the adjoint operator $W^t(A_n)^*$ (see Exercise 3.19). The reason is that the elements of $\mathcal{T}_{k,\kappa,\Omega}^p(\alpha)$ are "unsymmetric": there are row sums and column differences! But nevertheless we shall explain that $\mathcal{J}^t$ is a right ideal by verifying straightforwardly that

$$((E_n^F Y_t E_{-n}^F L_n^{F,F} K E_n^F Y_{t^{-1}} E_{-n}^F | S_n^F) + \mathcal{G})((A_n) + \mathcal{G}) \in \mathcal{J}^t \tag{3}$$

for all generating sequences (A_n) of $\mathcal{A}_{\Omega,r}^F$ given by 3.5.1(2)-(4). If (A_n) is the sequence 3.5.1(2) or 3.5.1(3) then (2) converges strongly to $W^t(A_n)^*$, hence, (3) holds for these sequences.

Consider the sequence $(A_n) = (T^0(a), y)$ with a as in Proposition 3.11(i). For simplicity we limit ourselves to the cases $t = 1$ and $\alpha = 0$. Then $(T^0(a), y) = (E_n^F T^0(a) E_{-n}^F | S_n^F)$. Further we can suppose without loss that $K = S_R f I - f S_R$ with a function $f \in C(\dot{\mathbf{R}})$ (compare properties 3.1.1(i) and (ii)). Now (3) reduces to

$$(L_n^{F,F}(S_R f I - f S_R) E_n^F T^0(a) E_{-n}^F | S_n^F) + \mathcal{G}$$
$$= (L_n^{F,F} S_R L_n^{F,F} f E_n^F T^0(a) E_{-n}^F | S_n^F - L_n^{F,F} f L_n^{F,F} S_R E_n^F T^0(a) E_{-n}^F | S_n^F) + \mathcal{G}$$

and abbreviating $(\lambda^{F,F})^{-1} \sigma^{F,F}$ by ρ this is the same as

$$(E_n^F T^0(\rho) E_{-n}^F L_n^{F,F} f E_n^F T^0(a) E_{-n}^F | S_n^F - L_n^{F,F} f E_n^F T^0(\rho) T^0(a) E_{-n}^F | S_n^F) + \mathcal{G}$$
$$= (E_n^F T^0(\rho a) E_{-n}^F L_n^{F,F} f I | S_n^F - L_n^{F,F} f E_n^F T^0(\rho a) E_{-n}^F | S_n^F) + \mathcal{G}$$

due to the commutator property. But we have already proved in Proposition 3.18 that the commutator of $(L_n^{F,F} fI | S_n^F)$ and $E_n^F T^0(\rho a) E_{-n}^F | S_n^F)$ is in $\mathcal{J}^1$.

Finally we consider the generating sequence $(T^0(\rho^z), y)$ under the same simplifications. Now (3) goes over into

$$
\begin{aligned}
&(L_n^{F,F} (S_R fI - f S_R) E_n^F T^0(\rho^z) E_{-n}^F | S_n^F) + \mathcal{G} \\
&= \quad (E_n^F T^0(\rho) E_{-n}^F L_n^{F,F} fI - L_n^{F,F} f E_n^F T^0(\rho) E_{-n}^F)(E_n^F T^0(\rho^z) E_{-n}^F | S_n^F) + \mathcal{G} \\
&= \quad (E_n^F T^0(\rho^1) E_{-n}^F L_n^{F,F} fI - L_n^{F,F} f E_n^F T^0(\rho^1) E_{-n}^F)(E_n^F T^0(\rho^z) E_{-n}^F | S_n^F) \\
&\quad +(E_n^F T^0(\rho - \rho^1) E_{-n}^F L_n^{F,F} fI - L_n^{F,F} f E_n^F T^0(\rho - \rho^1) E_{-n}^F) \times \\
&\quad \times (E_n^F T^0(\rho^z) E_{-n}^F | S_n^F) + \mathcal{G} .
\end{aligned}
\tag{4}
$$

The latter summand is in $\mathcal{G}$ since $\rho - \rho^1$ is continuous. For the first one we distinguish two cases: If $z = 1$ then this term is

$$
(E_n^F T^0(\rho^1) E_{-n}^F L_n^{F,F} f E_n^F T^0(\rho^1) E_{-n}^F - L_n^{F,F} f E_n^F T^0(\rho^1) T^0(\rho^1) E_{-n}^F | S_n^F) + \mathcal{G} ,
$$

and since $(\rho^1)^2$ is a continuous multiplier, this equals

$$
\begin{aligned}
&(E_n^F T^0(\rho^1) E_{-n}^F L_n^{F,F} f E_n^F T^0(\rho^1) E_{-n}^F - E_n^F T^0((\rho^1)^2) E_{-n}^F L_n^{F,F} fI | S_n^F) + \mathcal{G} \\
&= (E_n^F T^0(\rho^1) E_{-n}^F)(L_n^{F,F} f E_n^F T^0(\rho^1) E_{-n}^F - E_n^F T^0(\rho^1) E_{-n}^F L_n^{F,F} fI | S_n^F) + \mathcal{G} .
\end{aligned}
$$

Herein, the second factor is in $\mathcal{J}^1$ by Proposition 3.18, and remember that $\mathcal{J}^1$ is a left ideal.

If $z \neq 1$ then we choose a continuous (scalar-valued) multiplier g with $g(z) = 1$ and $g(1) = 0$, and form the matrix operator $L(T^0(g))$. Now the first summand in (4) can be written as

$$
\begin{aligned}
&(E_n^F T^0(\rho^1) E_{-n}^F L_n^{F,F} fI - L_n^{F,F} f E_n^F T^0(\rho^1) E_{-n}^F)(E_n^F L(T^0(g)) T^0(\rho^z) E_{-n}^F | S_n^F) \\
&\quad + \quad (E_n^F T^0(\rho^1) E_{-n}^F L_n^{F,F} fI - L_n^{F,F} f E_n^F T^0(\rho^1) E_{-n}^F) \times \\
&\quad \times (E_n^F L(T^0(1 - g)) T^0(\rho^z) E_{-n}^F | S_n^F) .
\end{aligned}
\tag{5}
$$

The multiplier $L(1 - g)\rho^z$ is continuous, hence, the second sequence of (5) is in $\mathcal{J}^1$ by which has already been proved. For the first sequence we again employ the commutator property to find

$$
\begin{aligned}
&(E_n^F T^0(\rho^1) L(T^0(g)) E_{-n}^F L_n^{F,F} fI - L_n^{F,F} f E_n^F T^0(\rho^1) L(T^0(g)) E_{-n}^F) \times \\
&\quad \times (E_n^F T^0(\rho^z) E_{-n}^F | S_n^F) .
\end{aligned}
$$

But now $\rho^1 L(g)$ is a continuous multiplier, hence, this sequence is even in $\mathcal{G}$. $\blacksquare$

In accordance with the general lifting theorem, we denote the canonical homomorphism from $\mathcal{A}_{\Omega,r}^F$ onto $\mathcal{A}_{\Omega,r}^F / \mathcal{J}_{\Omega,r}^F$ by Φ^J.

3.6.3   Localization

Let $\mathcal{B}_{\Omega,r}^F$ stand for the collection of all cosets $\Phi^J(L_n^{F,F}fI|S_n^F)$ with $f \in C(\dot{\mathbf{R}})$. Proposition 3.18 states that $\mathcal{B}_{\Omega,r}^F$ is in the center of the quotient algebra $\mathcal{A}_{\Omega,r}^F/\mathcal{J}_{\Omega,r}^F$.

Proposition 3.21 *The set $\mathcal{B}_{\Omega,r}^F$ is a subalgebra of the center of $\mathcal{A}_{\Omega,r}^F/\mathcal{J}_{\Omega,r}^F$ which is topologically isomorphic to $C(\dot{\mathbf{R}})$. The maximal ideal space of $\mathcal{B}_{\Omega,r}^F$ is homeomorphic to $\dot{\mathbf{R}}$, and the maximal ideal corresponding to $s \in \dot{\mathbf{R}}$ is*

$$\{\Phi^J(L_n^{F,F}fI|S_n^F) : f(s) = 0\} . \tag{1}$$

The proof is literally the same as that of Proposition 3.9.

In analogy to the notations used in Sections 1.6.3 and 3.2.6 we write I_s for the smallest closed two-sided ideal of $\mathcal{A}_{\Omega,r}^F/\mathcal{J}_{\Omega,r}^F$ containing the maximal ideal (1) of $\mathcal{B}_{\Omega,r}^F$, and we let $\mathcal{A}_{\Omega,r,s}^F$ refer to the "local" quotient algebra $(\mathcal{A}_{\Omega,r}^F/\mathcal{J}_{\Omega,r}^F)/I_s$ and Φ_s^J to the canonical homomorphism $\mathcal{A}_{\Omega,r}^F \to \mathcal{A}_{\Omega,r,s}^F$.

3.6.4   Identification of the local algebras

If $(A_n) \in \mathcal{A}_{\Omega,r}^F$ then the operators $W_s(A_n)$ depend on the cosets $\Phi^J(A_n)$ only (compare Propositions 3.16 and 3.17). Thus, the mapping

$$\mathcal{A}_{\Omega,r,s}^F \to \begin{cases} L(l_k^p(\alpha)) & \text{if} \quad s = 0 \\ L(l_k^p(0)) & \text{if} \quad s \in \mathbf{R} \setminus \{0\} \\ L(l_k^p(\alpha))/K(l_k^p(\alpha)) & \text{if} \quad s = \infty \end{cases}, \Phi_s^J(A_n) \mapsto W_s(A_n) , \tag{1}$$

is correctly defined, and we denote it by W_s again.

Proposition 3.22 *The homomorphisms W_s given by (1) are locally equivalent representations of the local algebras $\mathcal{A}_{\Omega,r,s}^F$.*

Proof   The homomorphisms W_s are, by definition, locally equivalent representations if their images are inverse closed and if there exist unital homomorphisms

$$W_s' : \operatorname{Im} W_s \to \mathcal{A}_{\Omega,r,s}^F$$

such that

$$\Phi_s^J(A_n) = W_s'(W_s(A_n)) \quad \text{for all } (A_n) \in \mathcal{A}_{\Omega,r}^F . \tag{2}$$

For our first goal we claim that

$$\operatorname{Im} W_s = \begin{cases} \mathcal{T}_{k,\kappa,\Omega}^p(\alpha) & \text{if} \quad s = 0 \\ \mathcal{T}_{k,\kappa,\Omega}^p(0) & \text{if} \quad s \in \mathbf{R} \setminus \{0\} \\ \mathcal{T}_{k,\kappa,\Omega}^p(\alpha)/K(l^p(\alpha)) & \text{if} \quad s = \infty \end{cases} . \tag{3}$$

Indeed, from Proposition 3.14 one easily concludes that $W_s(A_n)$ belongs to the algebras in (3), and the identities (see Proposition 3.14)

$$
\begin{aligned}
W_s(T^0(a), y) &= T^0(a) \text{ if } s = 0 \text{ and } y = 0 \text{ or } s \neq 0, \, y \neq s, \\
W_s(\operatorname{diag}(P, \ldots, P), s) &= \operatorname{diag}(P, \ldots, P), \\
W_s(\Lambda K \Lambda^{-1}) &= K \text{ for } K \text{ compact}
\end{aligned}
$$

show moreover, that all generating operators of $\mathcal{T}^p_{k,\kappa,\Omega}(\alpha)$ if $s = 0$ and of $\mathcal{T}^p_{k,\kappa,\Omega}(0)$ if $s \neq 0$ lie in the image of W_s (compare Proposition 3.11(ii) for the generating operators of $\mathcal{T}^p_{k,\kappa,\Omega}(\alpha)$). For $s = \infty$, the assertion is an immediate consequence of the identity $W_\infty(A, y) = \pi(A)$ stated in Proposition 3.14(b). Now apply Theorem 3.4 and Corollary 3.2 to find that $\operatorname{Im} W_s$ is actually inverse closed.

For the construction of W'_s we first let $s = 0$. In this case we define

$$
W'_0 : \mathcal{T}^p_{k,\kappa,\Omega}(\alpha) \to \mathcal{A}^F_{\Omega,r,0}, \quad A \mapsto \Phi^J_0(A, 0),
$$

which is obviously a continuous Banach algebra homomorphism; so it suffices to verify (2) for the generating sequences of $\mathcal{A}^F_{\Omega,r}$.

The assertion (2) is evident for generating sequences of the form $(A, 0)$ since $W_0(A, 0) = A$. So we are left on considering generating sequences (A, y) with $y \neq 0$.

If $A = K$ is compact then $W_0(K, y) = 0$, and we have to show instead of (2) that

$$
\Phi^J_0(K, y) = 0. \tag{4}
$$

For, suppose without loss that $K = K^{st}_{ij}$ is the same rank-one operator as in the proof of Proposition 3.15. Further we pick a function $f \in C(\dot{\mathbf{R}})$ with $f(0) = 1$ and $\operatorname{supp} f \subseteq [-y/2, y/2]$, and we let G refer to the canonical prebasis $\{\chi_{[i/k,(i+1)/k)}, \, i = 0, \ldots, k-1\}$. Then, by Lemma 3.3,

$$
\| L^{F,F}_n f L^{F,F}_n - E^F_n E^G_{-n} L^{G,G}_n f E^G_n E^F_{-n} L^{F,F}_n \| \to 0
$$

whence follows

$$
\begin{aligned}
&\lim_{n \to \infty} \| L^{F,F}_n f \, E^F_n \Lambda V_{\{yn+r\}} \Lambda^{-1} K \Lambda V_{-\{yn+r\}} \Lambda^{-1} E^F_{-n} L^{F,F}_n \| \\
&= \lim_{n \to \infty} \| E^F_n \left[E^G_{-n} L^{G,G}_n f E^G_n \Lambda V_{\{yn+r\}} \Lambda^{-1} K \Lambda V_{-\{yn+r\}} \Lambda^{-1} \right] E^F_{-n} L^{F,F}_n \|,
\end{aligned}
$$

and this limit is zero since the expression in brackets vanishes for large n: indeed, $\Lambda V_{\{yn+r\}} \Lambda^{-1} K \Lambda V_{-\{yn+r\}} \Lambda^{-1}$ is a matrix operator whose only non-vanishing entry stands at the $i + \{yn+r\}, j + \{yn+r\}$ th place, and $E^G_{-n} L^{G,G}_n f E^G_n$ is a diagonal matrix whose non-vanishing entries stand on the places $\{-\frac{yn}{2}\} - 1, \ldots, \{\frac{yn}{2}\}$. Consequently, $(L^{F,F}_n f I | S^F_n)(K, y) \in \mathcal{G}$, and since $\mathcal{G} \subseteq \mathcal{J}^F_{\Omega,r}$ and $\Phi^J_0(L^{F,F}_n f I | S^F_n)$ is the identity element in $\mathcal{A}^F_{\Omega,r,0}$ we arrive at (4).

If (A, y) is the sequence $(\text{diag}\,(P, \ldots, P), y)$ then the assertion (2) is equivalent to

$$\Phi_0^J(\text{diag}\,(P, \ldots, P), y) = \begin{cases} \Phi_0^J(\text{diag}\,(0, \ldots, 0), 0) & \text{if} \quad y > 0 \\ \Phi_0^J(\text{diag}\,(P, \ldots, P), 0) & \text{if} \quad y = 0 \\ \Phi_0^J(\text{diag}\,(I, \ldots, I), 0) & \text{if} \quad y < 0 \,. \end{cases}$$

Let us prove this for $y > 0$ for instance. As we have seen in Section 2.11.4, there is a compact operator K such that

$$(L_n^{F,F} \chi_{R^+} I | S_n^F) \;=\; (\text{diag}\,(P, \ldots, P) + K\,,\, 0)$$

and, consequently,

$$(L_n^{F,F} \chi_{[\frac{\{yn+r\}}{n}, \infty)} I | S_n^F) \;=\; (\text{diag}\,(P, \ldots, P), y) \;+\; (K, y) \,.$$

Let f be a continuous function with $f(0) = 1$ and compact support. Then $\Phi_0^J(L_n^{F,F} f I | S_n^F)$ is the identity element in the local algebra $\mathcal{A}_{\Omega, r, 0}^F$, and because of $\Phi_0^F(K, y) = 0$ by (4) we get

$$\begin{aligned} \Phi_0^J(\text{diag}\,(P, \ldots, P), y) &= \\ &= \Phi_0^J(L_n^{F,F} f I | S_n^F)\Phi_0^J(L_n^{F,F} \chi_{[\frac{\{yn+r\}}{n}, \infty)} I | S_n^F) \\ &= \Phi_0^J(L_n^{F,F} f L_n^{F,F} \chi_{[\frac{\{yn+r\}}{n}, \infty)} I | S_n^F) \\ &= \Phi_0^J(L_n^{F,F} f \chi_{[\frac{\{yn+r\}}{n}, \infty)} I | S_n^F) \,, \end{aligned}$$

the latter identity being a consequence of the commutator property. If the support of f is sufficiently small, and if n is large enough then $f\chi_{[\frac{\{yn+r\}}{n}, \infty)} = 0$ which yields the claim.

Now let $A = T^0(a)$ with a as in Proposition 3.11(i). Then $W_s(T^0(a), y) = \Lambda T^0(a)\Lambda^{-1}$, and (2) goes over into

$$\Phi_0^J(T^0(a), y) \;=\; \Phi_0^J(\Lambda T^0(a)\Lambda^{-1}, 0) \,.$$

This can be seen as follows: The operator $T^0(a) - \Lambda T^0(a)\Lambda^{-1}$ is compact on $l_k^p(0)$, hence, by (4),

$$\Phi_0^J(T^0(a), y) = \Phi_0^J(\Lambda T^0(a)\Lambda^{-1}, y) \,,$$

and now we get

$$\begin{aligned} \Phi_0^J(\Lambda T^0(a)\Lambda^{-1}, y) &= \\ &= \Phi_0^J(E_n^F \Lambda V_{\{yn+r\}} T^0(a) V_{-\{yn+r\}}\Lambda^{-1} E_{-n}^F) \\ &= \Phi_0^J(E_n^F \Lambda V_{\{r\}} T^0(a) V_{-\{r\}}\Lambda^{-1} E_{-n}^F) = \Phi_0^J(\Lambda T^0(a)\Lambda^{-1}, 0) \end{aligned}$$

as desired.

Finally, for $A = T^0(\rho^z)$ we only consider the unweighted situation. Then $W_0(T^0(\rho^z), y) = T^0(\rho^z)$ and, moreover,

$$\Phi_0^J(T^0(\rho^z), y) = \Phi_0^J(T^0(\rho^z), 0)$$

since $T^0(\rho^z)$ is shift invariant. The proof for $\alpha \neq 0$ is somewhat more complicated: it bases on the already mentioned equality

$$\mathcal{T}_{k,\kappa,\Omega}^p(\alpha) = \Lambda \mathcal{T}_{k,\kappa,\Omega}^p(0)\Lambda^{-1}$$

and on formula ([BS 2], 6.20).

In case $s \in \dot{\mathbf{R}} \setminus \{0, \infty\}$, the mapping W_s' is given by

$$W_s' : \mathcal{T}_{k,\kappa,\Omega}^p(0) \to \mathcal{A}_{\Omega,r,s}^F , \quad A \mapsto \Phi_s^J(\Lambda A \Lambda^{-1}, s) .$$

The proof of the equality (2) proceeds similarly to the case $s = 0$ and we omit the details here.

Finally, let $s = \infty$. We start with verifying that

$$\Phi_\infty^J(E_n^F K E_{-n}^F | S_n^F) = 0 \tag{5}$$

for each compact operator $K \in K(l_k^p(\alpha))$. Again we restrict ourselves to operators K of the form K_{ij}^{st}. Let $f \in C(\dot{\mathbf{R}})$ be a function with $f(\infty) = 1$. Since the operator $E_n^F K_{ij}^{st} E_{-n}^F$ acts via

$$\sum_{r=0}^{k-1} \sum_{l \in Z} \alpha_{rln} \phi_{rln} \mapsto \alpha_{tjn} \phi_{sin}$$

we see that $f E_n^F K_{ij}^{st} E_{-n}^F = 0$ whenever $\operatorname{supp} f \cap \operatorname{supp} \phi_{sin} = \emptyset$, i.e. whenever $\operatorname{supp} f$ is small enough. Now the same arguments as for showing (4) apply to give

$$\Phi_\infty^J(E_n^F K_{ij}^{st} E_{-n}^F | S_n^F) = \Phi_\infty^J(L_n^{F,F} f E_n^F K_{ij}^{st} E_{-n}^F | S_n^F) = 0$$

which proves (5).

Consequently, if $A \in \mathcal{T}_{k,\kappa,\Omega}^p(\alpha)$, then the coset $\Phi_\infty^J(E_n^F A E_{-n}^F | S_n^F)$ depends on the coset $\pi(A) = A + K(l_k^p(\alpha))$ only, and so it is correct to define

$$W_\infty' : \mathcal{T}_{k,\kappa,\Omega}^p(\alpha)/K(l_k^p(\alpha)) \to \mathcal{A}_{\Omega,r,\infty}^F , \quad A \mapsto \Phi_\infty^J(E_n^F A E_{-n}^F | S_n^F) .$$

It remains to show (2) for $s = \infty$. This can be done as for $s = 0$ and requires no new ideas. ∎

Now Theorem 1.11 can be restated as follows.

Corollary 3.3 *(a) For each $(A_n) \in \mathcal{A}_{\Omega,r}^F$, the coset $\Phi_s^J(A_n)$ is invertible if and only if $W_s(A_n)$ is invertible.*

(b) *For each $(A_n) \in \mathcal{A}_{\Omega,r}^F$, the coset $\Phi^J(A_n)$ is invertible in $\mathcal{A}_{\Omega,r}^F/\mathcal{J}_{\Omega,r}^F$ if all operators (cosets) $W_s(A_n)$, $s \in \dot{\mathbf{R}}$, are invertible.*

(c) *The local algebra $\mathcal{A}_{\Omega,r,s}^F$ is topologically isomorphic to $\mathcal{T}_{k,\kappa,\Omega}^p(\alpha)$ if $s = 0$, to $\mathcal{T}_{k,\kappa,\Omega}^p(0)$ if $s \in \mathbf{R} \setminus \{0\}$, and to $\mathcal{T}_{k,\kappa,\Omega}^p(\alpha)/K(l_k^p(\alpha))$ if $s = \infty$.*

3.6.5   Inverse closedness

In the final step in verifying the general scheme we have to show that stability of a sequence $(A_n) \in \mathcal{A}_{\Omega,r,s}^F$ implies the invertibility of all operators (cosets) $W^t(A_n)$ and $W_s(A_n)$.

Let us first consider the family $\{W_s\}_{s \in \dot{R}}$. For $s \neq \infty$ these homomorphisms are subject to the hypotheses in 1.6.5 as one easily checks. We emphasize that the convergence of the adjoint sequences required in 1.6.5(viii) is a consequence of Proposition 3.14(a) where the existence of the limits $W_s(A_n)$ is stated for all operators $A \in (T_{1,1,T}^p(\alpha))_{k \times k}$. Since the latter algebra is symmetric (i.e. $(T_{1,1,T}^p(\alpha))_{k \times k}^* = (T_{1,1,T}^q(-\alpha))_{k \times k}$ with $1/p + 1/q = 1$) in contrast to $T_{k,\kappa,\Omega}^p(\alpha)$, this involves the existence of the strong limits of the adjoint sequences, too. Hence, Proposition 1.9(b) yields the implication

$$(A_n) \text{ stable } \Rightarrow W_s(A_n) \text{ invertible }.$$

For $s = \infty$ it is immediate that stability of (A_n) implies the invertibility of $W_\infty(A_n)$.

Let us turn to the homomorphisms

$$W^t(A_n) = \operatorname*{s-lim}_{n \to \infty} E_n^F Y_{t^{-1}} E_{-n}^F A_n E_n^F Y_{t^{-1}} E_{-n}^F L_n^{F,F}.$$

Here the things are essentially more complicated since the adjoint sequence

$$(E_n^F Y_{t^{-1}} E_{-n}^F A_n E_n^F Y_{t^{-1}} E_{-n}^F)^* L_n^{F,F}$$

is no longer strongly convergent in general (see Exercise 3.19 for an example). To get the invertibility of $W^t(A_n)$ we can argue as follows:

The stability of (A_n) guarantees the invertibility of all operators $W_s(A_n)$ and hence, via Corollary 3.3(b), the invertibility of the coset $\Phi(A_n) + \mathcal{J}$ in the quotient algebra $\mathcal{A}_{\Omega,r}^F / \mathcal{J}_{\Omega,r}^F$. In other words, there are sequences $(B_n) \in \mathcal{A}_{\Omega,r}^F$ and $(J_n^l), (J_n^r) \in \mathcal{J}_{\Omega,r}^F$ such that

$$A_n B_n = I + J_n^r, \quad B_n A_n = I + J_n^l.$$

Applying the homomorphisms W^t to these equations we obtain

$$W^t(A_n) W^t(B_n) = I + K^r, \quad W^t(B_n) W^t(A_n) = I + K^l$$

with compact operators K^r and K^l (see Proposition 3.17) whence follows that the $W^t(A_n)$ are always Fredholm.

Further we need the following observation:

If (C_n) is any stable sequence with strong limit C then C has a trivial kernel.

Indeed, we have

$$\|x - C_n^{-1} Cx\| \leq \sup_n \|C_n^{-1}\| \, \|C_n x - Cx\| \to 0,$$

that is, if x is in the kernel of C then, necessarily, $x = 0$. Employing this observation for the stable sequences $(C_n) = (E_n^F Y_{t-1} E_{-n}^F A_n E_n^F Y_{t-1} E_{-n}^F L_n^{F,F})$ we get that all operators $W^t(A_n)$ have a trivial kernel. This together with their Fredholmness implies their one-sided invertibility from the left. Now apply the "left-sided version" of the lifting theorem 1.8 to find the left-sided invertibility of the coset $\Phi(A_n) + \mathcal{G}$ in $\mathcal{A}_{\Omega,r}^F / \mathcal{G}$. Hence, there is a sequence (C_n) in $\mathcal{A}_{\Omega,r}^F$ such that

$$((C_n) + \mathcal{G})((A_n) + \mathcal{G}) = (I|S_n^F) + \mathcal{G} \,.$$

But then $(C_n) + \mathcal{G}$ is also a left-inverse for $(A_n) + \mathcal{G}$ in $\mathcal{F}^F / \mathcal{G}$, and since (A_n) was supposed to be stable (or, equivalently, to have an invertible coset $(A_n) + \mathcal{G}$ in $\mathcal{F}^F / \mathcal{G}$), the coset $(C_n) + \mathcal{G}$ must be a two-sided inverse of $(A_n) + \mathcal{G}$. Thus, there is a sequence $(G_n) \in \mathcal{G}$ such that

$$A_n C_n = I|S_n^F + G_n \,,$$

and applying now the homomorphisms W^t to both sides of this equality gives the invertibility of $W^t(A_n)$ from the right-hand side. ∎

3.7 Sequences of local type

3.7.1 Operators of local type

Let $(A_n) \in \mathcal{F}^F$ be a sequence of approximation operators. If (A_n) is even in $\mathcal{A}_{\Omega,r}^F$ for some Ω and r then Theorem 3.5 states necessary and sufficient conditions for the stability of this sequence. Thus, we are required to decide whether (A_n) is in $\mathcal{A}_{\Omega,r}^F$. In this section we are going to explain how KMS-techniques apply to this problem. For simplicity we start with showing how the local inclusion theorem (see 1.7.3) can be employed for verifying that an operator $A \in L(L_R^p(\alpha))$ belongs to the operator algebra $\mathcal{O}_R^p(\alpha)$ introduced in Section 3.1.1.

An operator $A \in L(L_R^p(\alpha))$ is called *operator of local type* if the commutator $fA - AfI$ is compact for all functions f in $C(\dot{\mathbf{R}})$. The collection of all operators of local type on $L_R^p(\alpha)$ is clearly a Banach algebra, and it will be denoted by OLT. Further it is immediate that the compact operators $K(L_R^p(\alpha))$ form a closed two-sided ideal in OLT. The quotient algebra $OLT/K(L_R^p(\alpha))$ will be denoted by OLT^π, and the canonical homomorphism from OLT onto OLT^π by π. Finally, the definition of OLT tells us that the algebra $\pi(C(\dot{\mathbf{R}})I)$ lies in the center of OLT^π.

Theorem 3.6 *The algebra OLT^π is KMS with respect to the algebra $\pi(C(\dot{\mathbf{R}})I)$.*

Proof The mapping $\pi(fI) \mapsto \pi(\overline{f}I)$ with $\overline{f}$ referring to the complex conjugate of f is evidently an involution on $\pi(C(\dot{\mathbf{R}})I)$, and from 3.1.5(3) we know that $\|\pi(fI)\| = \sup_t |f(t)|$. Thus, $\pi(C(\dot{\mathbf{R}})I)$ is even a C^*-algebra, and it remains to show that

$$\|\pi((f + g)A)\| \leq \max(\|\pi(fA)\|, \|\pi(gA)\|) \tag{1}$$

whenever A is in OLT and f, g in $C(\dot{\mathbf{R}})$ have disjoint supports M, N. Let χ_M and χ_N stand for the characteristic functions of M and N, and let B, C be in $L(L_R^p(\alpha))$. If $x \in L_R^p(\alpha)$ then

$$
\begin{aligned}
\|\chi_M B \chi_M x + \chi_N C \chi_N x\|_{L_R^p(\alpha)}^p &= \|\chi_M B \chi_M x\|_{L_R^p(\alpha)}^p + \|\chi_N C \chi_N x\|_{L_R^p(\alpha)}^p \\
&\leq \max(\|B\|^p, \|C\|^p)(\|\chi_M x\|_{L_R^p(\alpha)}^p + \|\chi_N x\|_{L_R^p(\alpha)}^p) \\
&\leq \max(\|B\|^p, \|C\|^p) \, \|x\|_{L_R^p(\alpha)}^p \,,
\end{aligned}
$$

whence

$$
\|\chi_M B \chi_M I + \chi_N C \chi_N I\| \leq \max(\|B\|, \|C\|) \,. \tag{2}
$$

Given any $\epsilon > 0$, choose operators $K, L \in K(L_R^p(\alpha))$ such that $\|B + K\| \leq \|\pi(B)\| + \epsilon$ and $\|C + L\| \leq \|\pi(C)\| + \epsilon$. Then (2) gives

$$
\begin{aligned}
\max(\|\pi(B)\|, \|\pi(C)\|) + \epsilon &\geq \max(\|B + K\|, \|C + L\|) \\
&\geq \|\chi_M B \chi_M I + \chi_M K \chi_M I + \chi_N C \chi_N I + \chi_N L \chi_N I\| \\
&\geq \|\pi(\chi_M B \chi_M I + \chi_N C \chi_N I)\|
\end{aligned}
$$

and so

$$
\|\pi(\chi_M B \chi_M I + \chi_N C \chi_N I)\| \leq \max(\|\pi(B)\|, \|\pi(C)\|) \,. \tag{3}
$$

Now put $B = fA$, $C = gA$. Then, by (3)

$$
\begin{aligned}
\max(\|\pi(fA)\|, \|\pi(gA)\|) &\geq \|\pi(\chi_M fA \chi_M I + \chi_N gA \chi_N I)\| \\
&= \|\pi(fA - fA \chi_{R \setminus M} I + gA - gA \chi_{R \setminus N} I\| \,.
\end{aligned}
$$

Because $fA \chi_{R \setminus M} I$ and $gA \chi_{R \setminus N} I$ are compact (note that $fA \chi_{R \setminus M} I - A f \chi_{R \setminus M} I$ is compact and that $f \chi_{R \setminus M} = 0$) we arrive at (1). $\blacksquare$

3.7.2   The local inclusion theorem for OLT

Localizing the algebra OLT^π over its central subalgebra $\pi(C(\dot{\mathbf{R}})I)$ via Allan's local principle we arrive at local algebras OLT_x^π with x running through the maximal ideal space $\dot{\mathbf{R}}$ of $\pi(C(\dot{\mathbf{R}})I)$. The canonical homomorphism from OLT onto OLT_x^π will be denoted by Ψ_x^π. Since $\mathcal{O}_R^p(\alpha)$ is a subalgebra of OLT which contains $C(\dot{\mathbf{R}})I$ (whence, in particular, follows that $\pi(\mathcal{O}_R^p(\alpha))$ is a $\pi(C(\dot{\mathbf{R}})I)$-modul in OLT^π), the local inclusion theorem 1.14(b) can be specified for as follows:

Theorem 3.7 *An operator A of local type belongs to the algebra $\mathcal{O}_R^p(\alpha)$ of singular integral operators if and only if*

$$
\Psi_x^\pi(A) \in \Psi_x^\pi(\mathcal{O}_R^p(\alpha)) \quad \text{for all } x \in \dot{\mathbf{R}} \,. \tag{1}
$$

For a proof notice that the inclusions (1) entail via the KMS-property and Theorem 1.14 that $\pi(A) \in \pi(\mathcal{O}_R^p(\alpha))$, and now remember that the ideal $K(L_R^p(\alpha))$ is completely contained in $\mathcal{O}_R^p(\alpha)$ (see Section 3.1.1) whence follows $A \in \mathcal{O}_R^p(\alpha)$.

Example 3.2 Let us illustrate this theorem by a simple example. We let N stand for the Hankel operator

$$(Nf)(t) = \begin{cases} \frac{1}{\pi i} \int_0^\infty \frac{f(s)}{s+t}\, ds & , t \in \mathbf{R}^+ \\ 0 & , t \in \mathbf{R}^- \end{cases}$$

and pick an arbitrary $L^\infty(\mathbf{R})$-function a which has discontinuities of first kind at 0 and ∞. Then the operator aN^2 is in $\mathcal{O}_R^p(\alpha)$.

Indeed, aN^2 is an operator of local type, and we claim that

$$\Psi_x^\pi(aN^2) \in \Psi_x^\pi(\mathcal{O}_R^p(\alpha)) \tag{2}$$

for all $x \in \dot{\mathbf{R}}$.

If $x \in \dot{\mathbf{R}} \setminus \{0, \infty\}$ then $\Psi_x^\pi(N^2) = 0$. This is obvious for $x < 0$ and a consequence of the identity

$$\Psi_x^\pi(N^2) = \Psi_x^\pi(S_{R+}^2 - I) = \Psi_x^\pi(S_R^2 - I)$$

for $x > 0$. Thus,

$$\Psi_x^\pi(aN^2) = \Psi_x^\pi(a)\,\Psi_x^\pi(N^2) = 0 \in \Psi_x^\pi(\mathcal{O}_R^p(\alpha))$$

for these x. For $x = 0$ we can choose numbers a_+, a_- such that $a - (a_+\chi_{R^+} + a_-\chi_{R^-})$ is continuous and equal to zero at 0. Hence,

$$\Psi_0^\pi(aN^2) = \Psi_0^\pi((a_+\chi_{R^+} + a_-\chi_{R^-})N^2) \in \Psi_0^\pi(\mathcal{O}_R^p(\alpha))$$

(remember that $N^2 \in \mathcal{O}_R^p(\alpha)$ by 2.1.2(6)).

Analogously, $\Psi_\infty^\pi(aN^2) \in \Psi_\infty^\pi(\mathcal{O}_R^p(\alpha))$, which gives (2), and thus $aN^2 \in \mathcal{O}_R^p(\alpha)$. Employing 2.1.2(c)(iii) one can further show that $aM^0(b) \in \mathcal{O}_R^p(\alpha)$ for all $M^0(b) \in N^p(\alpha)$ and that $aM^0(b)$ is compact whenever a is continuous at 0 and at ∞ and has zeros at these points.

3.7.3 Sequences of local type

The analogue of operators of local type is sequences of local type, i.e. sequences which commute with all sequences $(L_n^{F,F} fI | S_n^F)$, $f \in C(\dot{\mathbf{R}})$, modulo sequences in the ideal $\mathcal{J}_{\Omega,r}^F$. This formulation is still unprecise; strictly speaking it means the following two things: the collection $SLT_{\Omega,r}^F$ of all sequences of local type consists of all sequences $(A_n) : S_n^F \to S_n^F$ such that

$$(A_n)(J_n) \in \mathcal{J}_{\Omega,r}^F \quad \text{and} \quad (J_n)(A_n) \in \mathcal{J}_{\Omega,r}^F \tag{1}$$

for all $(J_n) \in \mathcal{J}_{\Omega,r}^F$ (i.e. $\mathcal{J}_{\Omega,r}^F$ is an ideal in $SLT_{\Omega,r}^F$) and

$$(A_n)(L_n^{F,F} fI | S_n^F) - (L_n^{F,F} fI | S_n^F)(A_n) \in \mathcal{J}_{\Omega,r}^F \tag{2}$$

for all functions $f \in C(\dot{\mathbf{R}})$.

One easily checks that $SLT^F_{\Omega,r}$ forms a closed subalgebra of the algebra $\mathcal{F}^F$ of all sequences, that $\mathcal{J}^F_{\Omega,r}$ is actually a closed two-sided ideal of $SLT^F_{\Omega,r}$, and that $\mathcal{A}^F_{\Omega,r} \subseteq SLT^F_{\Omega,r}$, the latter being a consequence of Proposition 3.18. Furthermore, the set

$$\mathcal{B}^F_{\Omega,r} := \{(L^{F,F}_n fI | S^F_n) + \mathcal{J}^F_{\Omega,r} : f \in C(\dot{\mathbf{R}})\}$$

is, by (2), a central subalgebra of the quotient algebra $SLT^F_{\Omega,r}/\mathcal{J}^F_{\Omega,r}$.

Theorem 3.8 $SLT^F_{\Omega,r}/\mathcal{J}^F_{\Omega,r}$ is KMS with respect to $\mathcal{B}^F_{\Omega,r}$.

Proof   Let Ψ^J stand for the canonical homomorphism from $SLT^F_{\Omega,r}$ into $SLT^F_{\Omega,r}/\mathcal{J}^F_{\Omega,r}$. As in case of Fredholm theory one shows that the mapping

$$\Psi^J(L^{F,F}_n fI | S^F_n) \mapsto \Psi^J(L^{F,F}_n \overline{f} I | S^F_n)$$

is an involution on $\mathcal{B}^F_{\Omega,r}$ and that

$$\sup_{t\in R}|f(t)| \leq \|\Psi^J(L^{F,F}_n fI|S^F_n)\| \leq C \sup_{t\in R}|f(t)|$$

with $C = \sup_n \|L^{F,F}_n\|$. Thus, $\mathcal{B}^F_{\Omega,r}$ is topologically isomorphic to the C^*-algebra $C(\dot{\mathbf{R}})$ having maximal ideal space $\dot{\mathbf{R}}$.

We claim that there is a constant C such that, given a positive integer s, functions $f_1, \ldots, f_s \in C(\dot{\mathbf{R}})$ with pairwise disjoint supports, and a sequence $(A_n) \in SLT^F_{\Omega,r}$,

$$\|\Psi^J(L^{F,F}_n \sum_{i=1}^s f_i A_n)\| \leq C \max_i \|\Psi^J(L^{F,F}_n f_i A_n)\| .$$

Choose functions $g_i \in C(\dot{\mathbf{R}})$ with $0 \leq g_i(t) \leq 1$, $g_i(t) = 1$ for $t \in \text{supp } f_i$, and possessing pairwise disjoint supports. Then we have for all $x \in L^p_R(\alpha)$ and for all sequences $(A^{(i)}_n)_{n\geq 1} \in SLT^F_{\Omega,r}$, $i = 1, \ldots, s$,

$$\|\sum_{i=1}^s g_i L^{F,F}_n A^{(i)}_n L^{F,F}_n g_i x\|^p_{L^p_R(\alpha)}$$

$$= \sum_{i=1}^s \|g_i L^{F,F}_n A^{(i)}_n L^{F,F}_n g_i x\|^p_{L^p_R(\alpha)} \leq C^p \max_i \left\{\|A^{(i)}_n\|^p\right\} \sum_{i=1}^s \|g_i x\|^p_{L^p_R(\alpha)}$$

$$= C^p \max_i \left\{\|A^{(i)}_n\|^p\right\} \|\sum_{i=1}^s g_i x\|^p_{L^p_R(\alpha)} \leq C^p \max_i \left\{\|A^{(i)}_n\|^p\right\} \|x\|^p_{L^p_R(\alpha)}$$

whence follows

$$\|\sum_{i=1}^s g_i L^{F,F}_n A^{(i)}_n L^{F,F}_n g_i I\| \leq C \max_i \left\{\|A^{(i)}_n\|\right\} \tag{3}$$

with $C^{1/2} = \sup_n \|L^{F,F}_n\|$.

Let $\epsilon > 0$. Then, by the commutator relation (see Theorem 2.8), there is an n_0 such that

$$\|\sum_{i=1}^{s} g_i L_n^{F,F} A_n^{(i)} L_n^{F,F} g_i I - \sum_{i=1}^{s} L_n^{F,F} g_i A_n^{(i)} L_n^{F,F} g_i L_n^{F,F}\| < \epsilon$$

for all $n \geq n_0$. This shows in combination with (3) that

$$\sup_{n \geq n_0} \|\sum_{i=1}^{s} L_n^{F,F} g_i A_n^{(i)} L_n^{F,F} g_i L_n^{F,F}\| \leq C \max_{i} \sup_{n \geq n_0} \|A_n^{(i)}\| + \epsilon. \tag{4}$$

Now choose sequences $(J_n^{(i)})_{n \geq 1} \in \mathcal{J}_{\Omega,r}^{F}$, $i = 1, \ldots, s$, such that

$$\|(A_n^{(i)} + J_n^{(i)})\| = \sup_{n \geq 1} \|A_n^{(i)} + J_n^{(i)}\| < \|\Psi^J(A_n^{(i)})\| + \epsilon.$$

Then (4) yields with $A_n^{(i)} + J_n^{(i)}$ in place of $A_n^{(i)}$:

$$C \max_{i} \left\{ \|\Psi^J(A_n^{(i)})\| \right\} + C\epsilon + \epsilon$$
$$\geq C \max_{i} \|(A_n^{(i)} + J_n^{(i)})\| + \epsilon \geq C \max_{i} \sup_{n \geq n_0} \|A_n^{(i)} + J_n^{(i)}\| + \epsilon$$
$$\geq \sup_{n \geq n_0} \|\sum_{i=1}^{s} L_n^{F,F} g_i A_n^{(i)} L_n^{F,F} g_i L_n^{F,F} + \sum_{i=1}^{s} L_n^{F,F} g_i J_n^{(i)} L_n^{F,F} g_i L_n^{F,F}\|$$
$$\geq \|\sum_{i=1}^{s} \Psi^J(L_n^{F,F} g_i A_n^{(i)} L_n^{F,F} g_i L_n^{F,F})\|$$

because of $(L_n^{F,F} g_i J_n^{(i)} L_n^{F,F} g_i L_n^{F,F}) \in \mathcal{J}_{\Omega,r}^{F}$. Let ϵ go to zero we arrive at

$$\|\sum_{i=1}^{s} \Psi^J(L_n^{F,F} g_i A_n^{(i)} L_n^{F,F} g_i L_n^{F,F})\| \leq C \max_{i} \left\{ \|\Psi^J(A_n^{(i)})\| \right\}.$$

Now replace $(A_n^{(i)})$ by $(L_n^{F,F} f_i A_n) \in SLT_{\Omega,r}^{F}$ to obtain

$$C \max_{i} \left\{ \|\Psi^J(L_n^{F,F} f_i A_n)\| \right\} \geq \|\sum_{i=1}^{s} \Psi^J(L_n^{F,F} g_i L_n^{F,F} f_i A_n L_n^{F,F} g_i L_n^{F,F})\|, \tag{5}$$

and observing that

$$L_n^{F,F} g_i L_n^{F,F} f_i A_n L_n^{F,F} g_i L_n^{F,F} =$$
$$= L_n^{F,F} g_i f_i A_n L_n^{F,F} g_i L_n^{F,F} + G_n = L_n^{F,F} f_i A_n L_n^{F,F} g_i L_n^{F,F} + G_n$$
$$= A_n L_n^{F,F} f_i L_n^{F,F} g_i L_n^{F,F} + J_n = A_n L_n^{F,F} f_i L_n^{F,F} + J_n'$$
$$= L_n^{F,F} f_i A_n + J_n'',$$

where $(G_n) \in \mathcal{G}$, $(J_n), (J'_n), (J''_n) \in \mathcal{J}^F_{\Omega,r}$ (Here we have used the commutator property again.), we finally get from (5)

$$C \max_i \left\{ \|\Psi^J(L_n^{F,F} f_i A_n)\| \right\} \geq \| \sum_{i=1}^{s} \Psi^J(L_n^{F,F} f_i A_n) \|$$

as claimed. ∎

3.7.4   The local inclusion theorem for SLT

Localizing the algebra $SLT^F_{\Omega,r}/\mathcal{J}^F_{\Omega,r}$ over its central subalgebra $\mathcal{B}^F_{\Omega,r}$ we get local algebras $SLT^F_{\Omega,r,x}$ with x running through the maximal ideal space $\dot{\mathbf{R}}$ of $\mathcal{B}^F_{\Omega,r}$. Let Ψ^J_x refer to the canonical homomorphism from $SLT^F_{\Omega,r}$ onto $SLT^F_{\Omega,r,x}$. Then the local inclusion theorem 1.14(b) can be restated as

Theorem 3.9 *A sequence $(A_n) \in SLT^F_{\Omega,r}$ belongs to the algebra $\mathcal{A}^F_{\Omega,r}$ if and only if $\Psi^J_x(A_n) \in \Psi^J_x(\mathcal{A}^F_{\Omega,r})$ for all $x \in \dot{\mathbf{R}}$.*

In Section 3.8.1 we will employ this theorem for showing that certain concrete approximation sequences belong to the algebra $\mathcal{A}^F_{\Omega,r}$ (and, thus, are subject to the general stability theorem). In what follows we explain another application of KMS-techniques by giving an alternative (and possibly clearer) description of the quotient algebra $\mathcal{A}^F_{\Omega,r}/\mathcal{J}^F_{\Omega,r}$. The Sections 3.7.5 and 3.7.6 will not be needed later on.

3.7.5 $\mathcal{A}^F_{\Omega,r}/\mathcal{J}^F_{\Omega,r}$ as function algebra

Let k and κ be integers with $k \geq \kappa \geq 0$ and let Ω be a closed subset of the unit circle. Given $x \in \dot{\mathbf{R}}$ we let $\mathcal{A}(x)$ designate the algebra $\mathcal{T}^p_{k,\kappa,\Omega}(\alpha)$ if $x = 0$, $\mathcal{T}^p_{k,\kappa,\Omega}(0)$ if $x \in \mathbf{R} \setminus \{0\}$, and $\mathcal{T}^p_{k,\kappa,\Omega}(\alpha)/K(l^p_k(\alpha))$ if $x = \infty$. Further, we denote by $FUN^p_{k,\kappa,\Omega}(\alpha)$ the collection of all bounded functions on $\dot{\mathbf{R}}$ which take at $x \in \dot{\mathbf{R}}$ a value in $\mathcal{A}(x)$.

Let $A, B \in FUN^p_{k,\kappa,\Omega}(\alpha)$. On defining operations by

$$(A + B)(x) = A(x) + B(x), \quad (AB)(x) = A(x)\, B(x)$$

and a norm by

$$\|A\|_{FUN} = \sup_{x \in \dot{R}} \|A(x)\|_{\mathcal{A}(x)}$$

we make $FUN^p_{k,\kappa,\Omega}(\alpha)$ to be a Banach algebra. From Section 3.6.4 we infer that the mapping

$$\dot{\mathbf{R}} \ni x \mapsto W_x(A_n)$$

belongs to $FUN^p_{k,\kappa,\Omega}(\alpha)$ for each $(A_n) \in \mathcal{A}^F_{\Omega,r}$ whenever F and k, κ are related by
3.3.1(1). Thus, the mapping

$$\Phi : \mathcal{A}^F_{\Omega,r}/\mathcal{J}^F_{\Omega,r} \to FUN^p_{k,\kappa,\Omega}(\alpha) \,, \ (A_n) + \mathcal{J}^F_{\Omega,r} \mapsto (x \mapsto W_x(A_n))$$

is correctly defined (compare Propositions 3.16 and 3.17), and Φ is actually a
unital Banach algebra homomorphism.

Proposition 3.23 *The kernel of Φ is trivial.*

Proof Let $\Phi((A_n) + \mathcal{J}^F_{\Omega,r}) = 0$, that is, $W_x(A_n) = 0$ for all $x \in \dot{\mathbf{R}}$. This implies
via the local equivalence of the mapping W_x that $\Phi^J_x(A_n) = 0$ for all $x \in \dot{\mathbf{R}}$.
   The algebra $\mathcal{A}^F_{\Omega,r}/\mathcal{J}^F_{\Omega,r}$ is a closed subalgebra of $SLT^F_{\Omega,r}/\mathcal{J}^F_{\Omega,r}$ which contains
$\mathcal{B}^F_{\Omega,r}$ as its central subalgebra. Thus, Theorem 3.8 gives that $\mathcal{A}^F_{\Omega,r}/\mathcal{J}^F_{\Omega,r}$ is KMS
with respect to $\mathcal{B}^F_{\Omega,r}$, and from Theorem 1.13 we conclude that

$$\|(A_n) + \mathcal{J}^F_{\Omega,r}\| \leq C \max_{x \in \dot{R}} \|\Phi^J_x(A_n)\|$$

with a constant C. Hence, $(A_n) \in \mathcal{J}^F_{\Omega,r}$. $\blacksquare$

In particular, the algebra $\mathcal{A}^F_{\Omega,r}/\mathcal{J}^F_{\Omega,r}$ is isomorphic to its image $\Phi(\mathcal{A}^F_{\Omega,r}/\mathcal{J}^F_{\Omega,r})$ in
$FUN^p_{k,\kappa,\Omega}(\alpha)$, and we can think of $\mathcal{A}^F_{\Omega,r}/\mathcal{J}^F_{\Omega,r}$ as an algebra of functions.

3.7.6 $\mathcal{A}^F_{\Omega,r}/\mathcal{J}^F_{\Omega,r}$ as algebra of continuous functions

Our next goals are to introduce a subalgebra of $FUN^p_{k,\kappa,\Omega}(\alpha)$ whose elements will
be referred to as *continuous* functions, and then to show that this algebra coincides
with the image of $\mathcal{A}^F_{\Omega,r}/\mathcal{J}^F_{\Omega,r}$. Of course, the common notion of continuity fails in
this setting. One reason for this is that the algebras $\mathcal{A}(x)$ depend on x. Before
defining a modified continuity we introduce some more notations.
   Let A be a function in $FUN^p_{k,\kappa,\Omega}(\alpha)$. Given $x \in \dot{\mathbf{R}}$ we define the right (resp.
left) value $V_+(A(x))$ (resp. $V_-(A(x))$) of A at x by

$$V_+(A(x)) = \begin{cases} \text{s--}\lim_{s \to +\infty} V_{-s}\Lambda^{-1}A(x)\Lambda V_s & \text{if} \quad x = 0 \\ \text{s--}\lim_{s \to +\infty} V_{-s}A(x)V_s & \text{if} \quad x \in \mathbf{R} \setminus \{0\} \\ \text{s--}\lim_{s \to -\infty} V_{-s}\Lambda^{-1}A(x)\Lambda V_s & \text{if} \quad x = \infty \end{cases}$$

(resp.

$$V_-(A(x)) = \begin{cases} \text{s--}\lim_{s \to -\infty} V_{-s}\Lambda^{-1}A(x)\Lambda V_s & \text{if} \quad x = 0 \\ \text{s--}\lim_{s \to -\infty} V_{-s}A(x)V_s & \text{if} \quad x \in \mathbf{R} \setminus \{0\} \\ \text{s--}\lim_{s \to +\infty} V_{-s}\Lambda^{-1}A(x)\Lambda V_s & \text{if} \quad x = \infty \end{cases} \ ).$$

Notice that the existence of the strong limits at $x \in \mathbf{R}$ is well known (see, e.g.,
[BS 2], 6.2(b)) and that we have already used it in Proposition 3.14. For the limits
at ∞ we mention that

$$\text{s--}\lim_{s \to \pm\infty} V_{-s}\Lambda^{-1}K\Lambda V_s = 0$$

for any compact operator $K \in K(l_k^p(\alpha))$. Thus, $V_+(A(\infty))$ (resp. $V_-(A(\infty))$) does actually depend on the coset $A(\infty)$ only whence the correctness of the definition follows.

A function $A \in FUN_{k,\kappa,\Omega}^p(\alpha)$ will be called *continuous* if

1^0. the one-sided limits $\lim_{y \to x \pm 0} A(y)$ exist at each point $x \in \dot{\mathbf{R}}$ (convention: for $x = \infty$ we set $\infty \pm 0 = \mp \infty$),

2^0. the right- and left-sided limits of A at x are related with the right and left value of A at x by

$$\lim_{y \to x \pm 0} A(y) = V_\pm(A(x)) .$$

(The notion 'continuity' is justified by the circumstance that, if the algebras $\mathcal{A}(x)$ were independent of x, and if the right and left value of A at x were defined as being $A(x)$ itself, then the above introduced continuity would coincide with the common one.)

It is not to hard to check that the set $CON_{k,\kappa,\Omega}^p(\alpha)$ of all continuous functions forms a closed subalgebra of $FUN_{k,\kappa,\Omega}^p(\alpha)$.

Theorem 3.10 *(a)* $\Phi(\mathcal{A}_{\Omega,r}^F / \mathcal{J}_{\Omega,r}^F) = CON_{k,\kappa,\Omega}^p(\alpha)$.
(b) The algebras $\mathcal{A}_{\Omega,r}^F / \mathcal{J}_{\Omega,r}^F$ and $CON_{k,\kappa,\Omega}^p(\alpha)$ are topologically isomorphic.

Proof   Assertion (b) is an immediate consequence of part (a) and of Proposition 3.23.

For the inclusion $\Phi(\mathcal{A}_{\Omega,r}^F / \mathcal{J}_{\Omega,r}^F) \subseteq CON_{k,\kappa,\Omega}^p(\alpha)$ in (a) it suffices to show that the functions $x \mapsto W_x(A_n)$ are in $CON_{k,\kappa,\Omega}^p(\alpha)$ for every generating sequence (A_n) of $\mathcal{A}_{\Omega,r}^F$. Let us do this for example for the sequences $(T^0(a), 0)$ and $(\mathrm{diag}(P, \dots, P), 0)$. By Proposition 3.14,

$$W_x(T^0(a), 0) = \begin{cases} T^0(a) \in L(l_k^p(\alpha)) & \text{if } x = 0 \\ T^0(a) \in L(l_k^p(0)) & \text{if } x \in \mathbf{R} \setminus \{0\} \\ \pi(T^0(a)) \in L(l_k^p(\alpha))/K(l_k^p(\alpha)) & \text{if } x = \infty . \end{cases}$$

Taking into account ([BS 2], 6.2(b)) once more one shows that the one-sided values of the function $A(x) = W_x(T^0(a), 0)$ at $x \in \dot{\mathbf{R}}$ are the operators $T^0(a)$ ($\in L(l_k^p(0))$ independent of x), and these values clearly coincide with the one-sided limits of $A(x)$. Similarly, again by Proposition 3.14,

$$W_x(\mathrm{diag}\,(P, \dots, P), 0) = \begin{cases} \mathrm{diag}\,(0, \dots, 0) & \text{if } x < 0 \\ \mathrm{diag}\,(P, \dots, P) & \text{if } x = 0 \\ \mathrm{diag}\,(I, \dots, I) & \text{if } x > 0 \\ \mathrm{diag}\,(\pi(P), \dots, \pi(P)) & \text{if } x = \infty \end{cases}$$

whence follows

$$V_+(W_x(\text{diag}\,(P,\ldots,P),0)) = \begin{cases} \text{diag}\,(0,\ldots,0) & \text{if} \quad x < 0 \text{ or } x = \infty \\ \text{diag}\,(I,\ldots,I) & \text{if} \quad x \geq 0 \end{cases}$$

and

$$V_-(W_x(\text{diag}\,(P,\ldots,P),0)) = \begin{cases} \text{diag}\,(0,\ldots,0) & \text{if} \quad x \leq 0 \\ \text{diag}\,(I,\ldots,I) & \text{if} \quad x > 0 \text{ or } x = \infty \,. \end{cases}$$

These one-sided values again coincide with the corresponding one-sided limits.

The proof of the reverse inclusion $CON^p_{k,\kappa,\Omega}(\alpha) \subseteq \Phi(\mathcal{A}^F_{\Omega,r}/\mathcal{J}^F_{\Omega,r})$ can be given by means of the local inclusion theorem. For, we observe that the algebra $\Phi(\mathcal{B}^F_{\Omega,r})$ is obviously a central subalgebra of $FUN^p_{k,\kappa,\Omega}(\alpha)$ (since $W_x(L^{F,F}_n f|S^F_n) = f(x)I$), that $FUN^p_{k,\kappa,\Omega}(\alpha)$ is KMS with respect to $\Phi(\mathcal{B}^F_{\Omega,r})$ (since the norm on FUN is the supremum norm), and that $\Phi(\mathcal{B}^F_{\Omega,r})$ is a subalgebra of $\Phi(\mathcal{A}^F_{\Omega,r}/\mathcal{J}^F_{\Omega,r})$. So we can localize $FUN^p_{k,\kappa,\Omega}(\alpha)$ over $\Phi(\mathcal{B}^F_{\Omega,r})$ and, if we denote by Ψ_x ($x \in \dot{\mathbf{R}}$) the canonical homomorphism from $FUN^p_{k,\kappa,\Omega}(\alpha)$ onto the local algebra associated with x, then Theorem 1.14(b) entails that the desired inclusion holds if and only if

$$\Psi_x(A) \in \Psi_x(\Phi(\mathcal{A}^F_{\Omega,r}/\mathcal{J}^F_{\Omega,r})) \tag{1}$$

for all $A \in CON^p_{k,\kappa,\Omega}(\alpha)$ and $x \in \dot{\mathbf{R}}$.

Given a function A and a point x we choose a sequence $(A_n) \in \mathcal{A}^F_{\Omega,r}$ such that $W_x(A_n) = A(x)$. Such a sequence exists by 3.6.4(3). Now consider the function $F := A - \Phi((A_n) + \mathcal{J}^F_{\Omega,r})$. This function is in $CON^p_{k,\kappa,\Omega}(\alpha)$ by what we have already shown, and it takes the value 0 at $x \in \dot{\mathbf{R}}$. Hence, the one-sided values of F at x are also 0 whence follows that

$$\lim_{y \to x \pm 0} F(y) = 0 \,.$$

Consequently, if $f \in C(\dot{\mathbf{R}})$ is a function with $0 \leq f(y) \leq 1$, $f(x) = 1$, and with sufficiently small support then

$$\|\Phi((L^{F,F}_n f|S^F_n) + \mathcal{J}^F_{\Omega,r})\,F\| = \sup_y \|f(y)F(y)\|$$

can be made as small as desired, and now Proposition 1.10 gives that $\Psi_x(fF) = \Psi_x(F) = 0$. Thus, (1) holds, and we are done. ∎

3.8 Concrete approximation methods

3.8.1 Galerkin methods

In the remaining part of this chapter we discuss several applications of the general stability theorem, and of KMS-techniques. For Galerkin methods we will do this in detail, for other methods we only sketch the results here and left their proofs to the reader (see also Exercises 3.23-3.27).

Let $F = \{\phi_0, \ldots, \phi_{k-1}\}$ and $G = \{\gamma_0, \ldots, \gamma_{k-1}\}$ be prebases, and suppose (F, G) to be a canonical pair (for concrete examples see 2.10.1(i)-(iii)).

For the approximate solution of the equation

$$Au = f , \quad u, f \in L_R^p(\alpha)$$

by the Galerkin method with trial space S_n^F and test space S_n^G one seeks a solution $u_n \in S_n^F$ of the approximating equation

$$L_n^{F,G} A u_n = L_n^{F,G} f \tag{1}$$

or, which is equivalent, a solution $x_n = (x_{0n}, \ldots, x_{k-1,n})^T \in l_k^p(\alpha)$ with $x_{sn} = (x_{smn})_{m \in Z}$ of the infinite system

$$\sum_{s=0}^{k-1} \sum_{m \in Z} \int_R (A\phi_{smn})(\frac{t+l}{n})\overline{\gamma_r(t)}\, dt \; x_{smn} = \int_R f(\frac{t+l}{n})\overline{\gamma_r(t)}\, dt \tag{2}$$

$(r = 0, \ldots, k - 1 , \; l \in \mathbf{Z})$. The solutions to (1) and (2) are connected via

$$u_n = E_n^F x_n \quad \text{and} \quad x_n = E_{-n}^F u_n .$$

As a first example we consider the Galerkin method (1) for the operator

$$A = aI + bS_R + K \tag{3}$$

where K is a compact operator and where a and b are piecewise continuous functions on $\mathbf{R}$ having one-sided limits at infinity. Further we suppose that all discontinuities of both a and b are located at points in $\mathbf{Z}$ or at ∞.

Proposition 3.24 *The sequence $(L_n^{F,G} A | S_n^F)$ belongs to $\mathcal{A}_{\{1\},0}^F$.*

Proof First we check that $(L_n^{F,G} A | S_n^F)$ is a sequence of local type. It is immediate (notice that G is a canonical prebasis !) that the strong limits

$$W^1(L_n^{F,G} A | S_n^F) = \underset{n \to \infty}{\text{s-lim}}\, L_n^{F,G} A L_n^{F,F} = A$$

and

$$W^1(L_n^{F,G} A | S_n^F)^* = \underset{n \to \infty}{\text{s-lim}}\, (L_n^{F,G} A L_n^{F,F})^* = A^*$$

exist. This implies (in the very same way as we had shown in the proof of Proposition 3.20 that $\mathcal{J}^t$ is an ideal) that condition 3.7.3 (1) is satisfied. For condition 3.7.3(2) it remains to invoke the commutator relation (Theorem 2.8): Indeed,

$$(L_n^{F,F} fI | S_n^F)(L_n^{F,G} A | S_n^F) - (L_n^{F,G} A | S_n^F)(L_n^{F,F} fI | S_n^F)$$
$$= (L_n^{F,G}(fA - AfI) | S_n^F) + (G_n) \tag{4}$$

with $(G_n) \in \mathcal{G}$, and since $fA - AfI$ is compact and $L_n^{F,G} \to I$ strongly, we have

$$(L_n^{F,G}(fA - AfI)|S_n^F) - (L_n^{F,F}(fA - AfI)|S_n^F) \in \mathcal{G} , \tag{5}$$

that is, the sequence (4) is in $\mathcal{J}_{\{1\},0}^F$.

Now the local inclusion theorem states that $(L_n^{F,G}A|S_n^F) \in \mathcal{A}_{\{1\},0}^F$ if and only if

$$\Psi_x^J(L_n^{F,G}A|S_n^F) \in \Psi_x^J(\mathcal{A}_{\{1\},0}^F) \quad \text{for all } x \in \dot{\mathbf{R}} .$$

Let $a(x \pm 0)$ and $a(\pm\infty)$ denote the one-sided limits of a at $x \in \mathbf{R}$ and at infinity. Taking into account that

$$(L_n^{F,F}fI|S_n^F)(L_n^{F,G}A|S_n^F) - (L_n^{F,G}fA|S_n^F) \in \mathcal{G}$$

and that

$$(L_n^{F,G}K|S_n^F) \in \mathcal{J}_{\{1\},0}^F$$

for all compact operators K by the same arguments as for (5), it is not hard to see that

$$\Psi_x^J(L_n^{F,G}A|S_n^F) = \Psi_x^J(L_n^{F,G}A_x|S_n^F) \tag{6}$$

where

$$A_x = \begin{cases} a(x)I + b(x)S_R & \text{if } x \in \mathbf{R} \setminus \mathbf{Z} \\ (a(x+0)\chi_{[x,\infty)} + a(x-0)\chi_{(-\infty,x]})I + \\ \quad + (b(x+0)\chi_{[x,\infty)} + b(x-0)\chi_{(-\infty,x]})S_R & \text{if } x \in \mathbf{Z} \\ (a(+\infty)\chi_{R^+} + a(-\infty)\chi_{R^-})I \\ \quad + (b(+\infty)\chi_{R^+} + b(-\infty)\chi_{R^-})S_R & \text{if } x = \infty \end{cases}$$

is the "local representative" of A at x.

Indeed, we already know from Theorem 2.15 that $(L_n^{F,G}S_R|S_n^F) \in \mathcal{A}_{\{1\},0}^F$; thus, (6) holds for $x \in \mathbf{R} \setminus \mathbf{Z}$. For points $x \in \mathbf{Z}$ the assertion (6) essentially reduces to

$$\Psi_x^J(L_n^{F,G}\chi_{[x,\infty)}I|S_n^F), \Psi_x^J(L_n^{F,G}\chi_{[x,\infty)}S_R|S_n^F) \in \Psi_x^J(\mathcal{A}_{\{1\},0}^F) .$$

We will show that even holds

$$(L_n^{F,G}\chi_{[x,\infty)}I|S_n^F), (L_n^{F,G}\chi_{[x,\infty)}S_R|S_n^F) \in \mathcal{A}_{\{1\},0}^F . \tag{7}$$

For the first sequence in (7) we observe that

$$L_n^{F,G}\chi_{[x,\infty)}I|S_n^F = E_n^F \Lambda V_{xn} \Lambda^{-1} E_{-n}^F L_n^{F,G} \chi_{[0,\infty)} E_n^F \Lambda V_{-xn} \Lambda^{-1} E_{-n}^F|S_n^F$$

and since the algebra $\mathcal{A}_{\{1\},0}^F$ is translation invariant we conclude that $(L_n^{F,G}\chi_{[x,\infty)}I|S_n^F)$ is in $\mathcal{A}_{\{1\},0}^F$ whenever $(L_n^{F,G}\chi_{[0,\infty)}I|S_n^F)$ is so. But for the latter sequence this is a consequence of Section 2.11.4. Analogously, it suffices to consider the sequence $(L_n^{F,G}\chi_{R^+}S_R|S_n^F)$ in place of the second sequence in (7), and the inclusion $(L_n^{F,G}\chi_{R^+}S_R|S_n^F) \in \mathcal{A}_{\{1\},0}^F$ can be seen again as in 2.11.4. The proof for the case $x = \infty$ is obvious. ∎

Now the general stability theorem gives

Theorem 3.11 *The Galerkin method* $(L_n^{F,G} A | S_n^F)$ *with trial space* S_n^F *and test space* S_n^G *applies to the operator* A *in (3) if and only if this operator is invertible and if the operators (resp. the coset)*

$$
W_s(L_n^{F,G} A | S_n^F) = \begin{cases}
a(s)I + b(s)T^0(\rho^{F,G}) & \text{if } s \in \mathbf{R} \setminus \mathbf{Z} \\
(a(s+0)P + a(s-0)Q) + & \\
\quad (b(s+0)P + b(s-0)Q)T^0(\rho^{F,G}) + K_s & \text{if } s \in \mathbf{Z} \\
\pi((a(+\infty)P + a(-\infty)Q) + & \\
\quad (b(+\infty)P + b(-\infty)Q)T^0(\rho^{F,G})) & \text{if } s = \infty
\end{cases}
$$

are invertible. Herein, $\rho^{F,G} = (\lambda^{F,G})^{-1}\sigma^{F,G}$, $Q = I - P$ *and the* K_s *are certain compact operators.*

Proof   We are only left on computing the strong limits $W_s(L_n^{F,G} A | S_n^F)$. Since $(L_n^{F,G} A | S_n^F)$ is in $\mathcal{A}_{\{1\},0}^F$, and since the homomorphisms W_s are locally equivalent on $\mathcal{A}_{\{1\},0}^F$, we obtain

$$
W_s(L_n^{F,G} A | S_n^F) = W_s(L_n^{F,G} A_s | S_n^F)
$$

with the local representatives A_s being as in the proof of the preceding proposition. First let $\alpha = 0$ (= the unweighted case) and, for example, $s = 1$. Then

$$
\begin{aligned}
W_1(L_n^{F,G} \chi_{[1,\infty)} S_R | S_n^F) &= \\
&= \operatorname*{s-lim}_{n\to\infty} V_{-n} E_{-n}^F L_n^{F,G} L_n^{F,G} \chi_{[1,\infty)} S_R E_n^F V_n \\
&= \operatorname*{s-lim}_{n\to\infty} E_{-n} L_n^{F,G} U_{-1} \chi_{[1,\infty)} S_R U_1 E_n^F \\
&= \operatorname*{s-lim}_{n\to\infty} E_{-n} L_n^{F,G} \chi_{R^+} S_R E_n^F \\
&= \operatorname*{s-lim}_{n\to\infty} T^0(\lambda^{F,G})^{-1}(PT^0(\sigma^{F,G}) + K) \\
&= PT^0(\rho^{F,G}) + K'
\end{aligned}
$$

with K and K' compact. Consequently,

$$
\begin{aligned}
W_1(L_n^{F,G} A | S_n^F) &= \\
&= (a(1+0)P + a(1-0)Q) + (b(1+0)P + b(1-0)Q)T^0(\rho^{F,G}) + K_1
\end{aligned}
$$

as desired. Analogously one verifies the other limits. In case $\alpha \neq 0$ one has to take into account that

$$
\operatorname*{s-\lim}_{n\to\pm\infty} V_{-n} \Lambda T^0(a) \Lambda^{-1} V_n = T^0(a)
$$

(compare [BS 2], 6.2(b)). ∎

Let us remark two obvious special cases:

1^0. If a and b are continuous on all of $\dot{\mathbf{R}}$ then the Galerkin method (1) applies to $aI + bS_R + K$ if and only if this operator is invertible and if the matrices

$$a(s) + b(s)\rho^{F,G}(t) \in \mathbf{C}^{k \times k}$$

are invertible for all $s \in \dot{\mathbf{R}}$ and $t \in \mathbf{T}$.

This is an immediate consequence of the values of the strong limits and of the fact that the operator $T^0(\rho^{F,G})$ is invertible whenever the function $\rho^{F,G}$ is so.

2^0. Let a and b be as in the theorem but suppose moreover that $a \equiv 1$ and $b \equiv 0$ on $\mathbf{R} \setminus [0,1]$. Then the Galerkin method (1) applies to $aI + bS_R + K$ if and only if this operator is invertible, if the operators

$$W_0(L_n^{F,G} A | S_n^F) = a(0+0)P + Q + b(0+0)PT^0(\rho^{F,G}) + K_0 \qquad (8)$$

and

$$W_1(L_n^{F,G} A | S_n^F) = P + a(1-0)Q + b(1-0)QT^0(\rho^{F,G}) + K_1 \qquad (9)$$

are invertible and if the matrices

$$a(s) + b(s)\rho^{F,G}(t) \in \mathbf{C}^{k \times k}$$

are invertible for all $s \in (0,1)$ and $t \in \mathbf{T}$.

The latter example is of some importance since the operator $aI + bS_R$ proves to be invertible if and only if the operator

$$a|_{[0,1]}I + b|_{[0,1]}S_{[0,1]} \in L(L^p_{[0,1]}(\alpha))$$

is invertible (see [GK 1] or ([MP], IV.1.1.1)). Thus, the method in 2^0. can be viewed as an Galerkin method for an operator on the interval $[0,1]$. In an analogous manner one constructs Galerkin methods for other operators on other intervals (including the semi-axis). Finally we mention that the operators (8) and (9) are invertible if and only if the perturbed Toeplitz operators

$$T(a(0+0) + b(0+0)\rho^{F,G}) + K_0'$$

and

$$T(a(1-0) + b(1-0)\tilde{\rho}^{F,G}) + K_1',$$

with $\tilde{\rho}^{F,G}(t) = \rho^{F,G}(1/t)$ and K_0', K_1' compact, are invertible.

Indeed, if one thinks of $l_k^p(\alpha)$ as direct sum of $\operatorname{Im} P$ and $\operatorname{Im} Q$ then the main part of the operator (8) corresponds to the matrix

$$a(0+0) \begin{pmatrix} I & 0 \\ 0 & 0 \end{pmatrix} + b(0+0) \begin{pmatrix} I & 0 \\ 0 & 0 \end{pmatrix} \begin{pmatrix} T(\rho^{F,G}) & X \\ Y & Z \end{pmatrix} + \begin{pmatrix} 0 & 0 \\ 0 & I \end{pmatrix}$$

$$= \begin{pmatrix} a(0+0)I + b(0+0)T(\rho^{F,G}) & b(0+0)X \\ 0 & I \end{pmatrix}$$

with certain operators X, Y, Z, and this operator is invertible if and only if $a(0+0)I + b(0+0)T(\rho^{F,G})$ is so. For (9) we consider the operator $J : (x_n)_{n \in Z} \mapsto (x_{-n-1})$. Clearly, J is invertible, and

$$JW_1(L_n^{F,G} A|S_n^F)J = Q + a(1-0)P + b(1-0)PT^0(\tilde{\rho}^{F,G}) + K_1' \,,$$

and now the same arguments as for (8) apply.

Let us draw the attention to some unsatisfactory and unpleasant phenomena.

The first one is that the approximation system (2) is infinite whereas for numerical computations a finite system is needed. So we are led to the problem of combining the system (2) with a *finite section method* to get systems of the form

$$\sum_{s=0}^{k-1} \sum_{m=-m(n)}^{m(n)-1} \int_R (A\phi_{smn})(\frac{t+l}{n})\overline{\gamma_r(t)}\,dt\,x_{smn} = \int_R f(\frac{t+l}{n})\overline{\gamma_r(t)}\,dt$$

$(r = 0, \dots, k-1 \,,\ l = -m(n), \dots, m(n)-1)$. The numbers $m(n)$ have to be chosen in a way which guarantees the strong convergence of the approximation operators to the operator A, that is, $m(n)$ must increase sufficiently fast.

The second problem is involved by the 'undetermined' operators K_s appearing in the conditions in the theorem. All we know about these operators in general is that they are compact (and even of finite rank). But an explicit determination of these operators seems to be very hard. The way out of this problem is less obvious than above: it rests on a modification of the spline spaces and can be interpreted as an *'infinite section method for the identity operator'*. In that sense these two problems are dual to each other, and the whole Chapter 4 will be devoted to the solution of these problems.

It is somewhat surprising that the modified spline spaces also pertain to the solution of a third problem: The 'direct' Galerkin methods for singular integral operators with coefficients having arbitrarily located discontinuities are not contained in algebras of the form $\mathcal{A}_{\Omega,r}^F$ (confer Exercise 3.26)! By a suitable modification of the approximation methods in neighbourhoods of the discontinuity points one can force that these methods belong to $\mathcal{A}_{\Omega,r}^F$ and, consequently, become subject to the stability theorem (see Section 4.3.4).

3.8.2   Collocation methods

Let $F = \{\phi_0, \ldots, \phi_{k-1}\}$ be a prebasis consisting of Riemann integrable functions and $G = \{\delta_{\tau_0}, \ldots, \delta_{\tau_{k-1}}\}$ be a k-tupel of premeasures

$$\delta_{\tau_s}(f) \;=\; f(\tau_s) \quad, \; s = 0, \ldots k - 1 \, , \; f \in R_R^p(\alpha)$$

with $0 \le \tau_0 < \tau_1 < \ldots < \tau_{k-1} < 1$, and suppose the pair (F, G) to be canonical.
For the approximate solution of the equation

$$Au \;=\; f \quad, \; u \in L_R^p(\alpha) \, , \; f \in R_R^p(\alpha)$$

by the (F, G)-collocation method we seek a solution $u_n \in S_n^F$ of the approximating equations

$$L_n^{F,G} A u_n \;=\; L_n^{F,G} f \tag{1}$$

or, equivalently, a solution $x_n = (x_{0n}, \ldots, x_{k-1,n})^T \in l_k^p(\alpha)$ with $x_{sn} = (x_{smn})_{m \in Z}$ of the infinite system

$$\sum_{s=0}^{k-1} \sum_{m \in Z} (A\phi_{smn})(\frac{\tau_r + l}{n}) \, x_{smn} \;=\; f(\frac{\tau_r + l}{n}) \tag{2}$$

$(r = 0, \ldots, k-1 \, , \; l \in \mathbf{Z})$. The solutions to (1) and (2) are connected via $u_n = E_n^F x_n$ and $x_n = E_{-n}^F u_n$. Let $A = aI + bS_R + K$ where a and b are as in Section 3.8.1, that is, the discontinuities of a and b are located at points with integer coordinates, and let K be a compact operator which sends Lebesgue integrable functions into Riemann integrable ones.

Proposition 3.25 *If A, F and G are as above then the sequence $(L_n^{F,G} A | S_n^F)$ is in $\mathcal{A}_{\{1\},0}^F$.*

The **proof** can be given by using KMS-techniques as in 3.8.1 or, which is easier in this situation, directly. Indeed,

$$(L_n^{F,G} A | S_n^F) \;=\; (L_n^{F,G}(aI + bS_R + K) | S_n^F)$$
$$= (L_n^{F,G} aI | S_n^F) + (L_n^{F,G} bI | S_n^F)\,(L_n^{F,G} S_R | S_n^F) + (L_n^{F,G} K | S_n^F) \, .$$

Theorem 2.15 involves that the sequence $(L_n^{F,G} S_R | S_n^F)$ is in $\mathcal{A}_{\{1\},0}^F$, and the inclusion $(L_n^{F,G} K | S_n^F) \in \mathcal{A}_{\{1\},0}^F$ can be seen as the analogous result 3.8.1. Further, for the inclusion $(L_n^{F,G} aI | S_n^F) \in \mathcal{A}_{\{1\},0}^F$, we approximate a by the sum of a continuous function g and of characteristic functions $\chi_{[c,d)}$ with $c, d \in \mathbf{Z}$. The sequence $(L_n^{F,G} gI | S_n^F)$ is in $\mathcal{A}_{\{1\},0}^F$ since the sequence $(L_n^{F,F} gI | S_n^F)$ is in this algebra by Proposition 3.16 and since

$$\|(L_n^{F,F} gI - L_n^{F,G} gI) | S_n^F\| \to 0$$

(set $K = L = F$ in Lemma 3.3). Finally, $\chi_{[c,d)} = \chi_{[c,\infty)} - \chi_{[d,\infty)}$ and

$$\chi_{[c,\infty)}I = U_c\chi_{R^+}U_{-c} = U_{\frac{\{cn+r\}}{n}}\chi_{R^+}U_{-\frac{\{cn+r\}}{n}}$$

if $r = 0$. Thus,

$$(L_n^{F,G}\chi_{[c,\infty)}I|S_n^F) = (U_{\frac{\{cn+r\}}{n}}L_n^{F,G}\chi_{R^+}U_{-\frac{\{cn+r\}}{n}}|S_n^F)$$

whence follows that $(L_n^{F,G}\chi_{[c,\infty)}I|S_n^F) \in \mathcal{A}_{\{1\},0}^F$. (Remember that $(L_n^{F,G}\chi_{R^+}I|S_n^F) \in \mathcal{A}_{\{1\},0}^F$ and that $\mathcal{A}_{\{1\},0}^F$ is translation invariant.) ∎

Specifying the stability theorem to this context we obtain

Theorem 3.12 *The (F,G)-collocation method $(L_n^{F,G}A|S_n^F)$ applies to the operator A if and only if this operator is invertible and if the operators (resp. the coset)*

$$W_s(L_n^{F,G}A|S_n^F) = \begin{cases} a(s)I + b(s)T^0(\rho^{F,G}) & \text{if } s \in \mathbf{R}\backslash\mathbf{Z} \\[2mm] \begin{aligned} &T^0(\lambda^{F,G})^{-1}[(a(s{+}0)P{+}a(s{-}0)Q)T^0(\lambda^{F,G}) \\ &\quad + (b(s+0)P{+}b(s-0)Q)T^0(\sigma^{F,G})] \end{aligned} & \text{if } s \in \mathbf{Z} \\[2mm] \begin{aligned} &\pi(a(+\infty)P + a(-\infty)Q + \\ &\quad (b(+\infty)P + b(-\infty)Q)T^0(\rho^{F,G})) \end{aligned} & \text{if } s = \infty \end{cases}$$

are invertible, where $\rho^{F,G} = (\lambda^{F,G})^{-1}\sigma^{F,G}$.

Observe that, because of

$$T^0(\lambda^{F,G})^{-1}P = PT^0(\lambda^{F,G})^{-1} + \text{compact},$$

the operators $W_s(L_n^{F,G}A|S_n^F)$ have exactly the same form as in the preceding theorem but, in contrast to the Galerkin method, the compact perturbations are well determined now. Moreover, these perturbations can be completely avoided if the operators $W_s(L_n^{F,G}A|S_n^F)$ are written as in the above theorem. Indeed,

$$E_{-n}^F L_n^{F,G}\chi_{R^+}E_n^F = T^0(\lambda^{F,G})^{-1}P_n^G\chi_{R^+}E_n^F,$$

and the lm th entry a_{rslm} of the rs block of the matrix $P_n^G\chi_{R^+}E_n^F$ equals

$$\int_R (\chi_{R^+}\phi_{smn})(\frac{t+l}{n})\overline{\gamma_r}(t)\, dt$$

which reduces in case $\gamma_r = \delta_{\tau_r}$ to

$$\chi_{R^+}(\tau_r + l)\,\phi_{sm1}(\tau_r + l) = \begin{cases} \phi_{sm1}(\tau_r + l) & \text{if } l \geq 0 \\ 0 & \text{if } l < 0. \end{cases}$$

Thus,

$$P_n^G\chi_{R^+}E_n^F = PT^0(\lambda^{F,G})$$

whence the special form of the operators $W_s(L_n^{F,G}A|S_n^F)$ in case of collocation follows.

3.8.3   A special case

Another feature of collocation methods (besides the avoidance of compact perturbations) is that, in case $k = 1$, *all* sequences $(L_n^{F,G}\chi_{[c,d]}I|S_n^F)$ with arbitrary real numbers c and d belong to one of the algebras $\mathcal{A}_{\{1\},r}^F$. To explain this let $G = \{\delta_\tau\}$ with $0 \le \tau < 1$. Then, evidently,

$$\frac{\tau + k}{n} \in [c,d) \quad \text{if and only if} \quad \frac{\tau + k}{n} \in \left[\frac{\{cn - \tau\}}{n}, \frac{\{dn - \tau\}}{n}\right)$$

whence follows

$$(L_n^{F,G}\chi_{[c,d)}I|S_n^F) = (L_n^{F,G}\chi_{\left[\frac{\{cn-\tau\}}{n}, \frac{\{dn-\tau\}}{n}\right)}I|S_n^F) .$$

Since

$$\chi_{\left[\frac{\{cn-\tau\}}{n}, \frac{\{dn-\tau\}}{n}\right)} = \chi_{\left[\frac{\{cn-\tau\}}{n}, \infty\right)} - \chi_{\left[\frac{\{dn-\tau\}}{n}, \infty\right)}$$

and

$$\chi_{\left[\frac{\{cn-\tau\}}{n}, \infty\right)}I = U_{\frac{\{cn-\tau\}}{n}}\chi_{R^+}U_{-\frac{\{cn-\tau\}}{n}}$$

we get that $(L_n^{F,G}\chi_{\left[\frac{\{cn-\tau\}}{n}, \infty\right)}I|S_n^F)$ is a translation of the sequence $(L_n^{F,G}\chi_{R^+}I|S_n^F)$ in sense of 3.4.1, and this proves the inclusion

$$(L_n^{F,G}\chi_{[c,d)}I|S_n^F) \in \mathcal{A}_{\{1\},-\tau}^F . \tag{1}$$

This implies that $(L_n^{F,G}aI|S_n^F) \in \mathcal{A}_{\{1\},-\tau}^F$ for all piecewise continuous functions, and specifying the stability theorem we get

Theorem 3.13 *Let $F = \{\phi\}$, $G = \{\delta_\tau\}$ such that the pair (F,G) is canonical, and let a and b be piecewise continuous functions and K be a compact operator which sends Lebesgue-integrable funtions into Riemann integrable ones. Then the (F,G)-collocation method applies to the operator $aI + bS_R + K$ if and only if the operator A is invertible and if the operators (resp. the coset)*

$$W_s(L_n^{F,G}A|S_n^F) = \begin{cases} T^0(\lambda^{F,G})^{-1}[(a(s+0)P + a(s-0)Q)T^0(\lambda^{F,G}) \\ \quad + (b(s+0)P + b(s-0)Q)T^0(\sigma^{F,G})] & \text{if } s \in \mathbf{R} \\ \pi(a(+\infty)P + a(-\infty)Q \\ \quad + (b(+\infty)P + b(-\infty)Q)T^0(\rho^{F,G})) & \text{if } s = \infty \end{cases}$$

are invertible where $\rho^{F,G} = (\lambda^{F,G})^{-1}\sigma^{F,G}$.

This idea does not work with $k > 1$ in general since the algebra $\mathcal{A}_{\{1\},-\tau}^F$ depends on the measure δ_τ explicitly. Compare, however, Exercise 3.25.

3.8.4   Qualocation methods

Let, as in Section 2.10.3, $F = \{\phi\}$ and $F_1 = \{\phi_1\}$ be one-elementic prebases and let $G = \{\gamma\}$ with the premeasure

$$\gamma \colon R_R^p(\alpha) \to \mathbf{C} \,,\; f \mapsto Q(\overline{\phi_1}f)$$

where Q stands for a quadrature rule. Again we specify Q by choosing an integer $m \geq 1$, real numbers ϵ_r with $0 \leq \epsilon_0 < \epsilon_1 < \ldots < \epsilon_{m-1} < 1$, and positive numbers w_r with $\sum_{r=0}^{m-1} w_r = 1$ and by defining

$$Q(g) = \sum_{k=-\infty}^{\infty} \sum_{r=0}^{m-1} w_r g(k + \epsilon_r) \,.$$

Let (F, G) be a canonical pair. For the approximate solution of the equation

$$Au = f \,,\; u \in L_R^p(\alpha) \,,\; f \in R_R^p(\alpha)$$

by the (F, G)-qualocation method we seek a solution $u_n \in S_n^F$ of the appproximating equations $L_n^{F,G} A u_n = L_n^{F,G} f$ or, equivalently, a solution $x_n = (x_{mn})_{m \in Z}$ of the infinite system

$$\sum_{m \in Z} x_{mn} \sum_{k=-\infty}^{\infty} \sum_{r=0}^{m-1} w_r \overline{\phi_1}(k + \epsilon_r)\, (A\phi_{mn})(\frac{k + l + \epsilon_r}{n}) =$$

$$= \sum_{k=-\infty}^{\infty} \sum_{r=0}^{m-1} w_r \overline{\phi_1}(k + \epsilon_r)\, f(\frac{k + l + \epsilon_r}{n}) \,,$$

with $l \in \mathbf{Z}$. Observe that the sums over k are actually finite since $\operatorname{supp} \phi_1$ is compact.

Now let $A = aI + bS_R + K$ where the discontinuities of a and b are located at points with integer coordinates and where K is compact again. Further we set for brevity

$$\rho^{F,G} := \frac{\sigma^{F,G}}{\lambda^{F,G}} = \frac{\sum_{r=0}^{m-1} w_r \overline{\lambda^{F_1,\{\delta_{\epsilon_r}\}}} \sigma^{F,\{\delta_{\epsilon_r}\}}}{\sum_{r=0}^{m-1} w_r \overline{\lambda^{F_1,\{\delta_{\epsilon_r}\}}} \lambda^{F,\{\delta_{\epsilon_r}\}}}$$

(compare identity 2.10.3(5) and Exercise 2.24). Then we have

Theorem 3.14 *(a) The sequence $(L_n^{F,G} A | S_n^F)$ is in $\mathcal{A}_{\{1\},0}^F$.*
(b) The (F, G)-qualocation method $(L_n^{F,G} A | S_n^F)$ applies to the operator A if and only if A is invertible and if the operators (resp. the coset)

$$W_s(L_n^{F,G} A | S_n^F) = \begin{cases} a(s)I + b(s)T^0(\rho^{F,G}) & \text{if} \quad s \in \mathbf{R} \setminus \mathbf{Z} \\ (a(s+0)P + a(s-0)Q) + \\ \quad (b(s+0)P + b(s-0)Q)T^0(\rho^{F,G}) + K_s & \text{if} \quad s \in \mathbf{Z} \\ \pi(a(+\infty)P + a(-\infty)Q + \\ \quad (b(+\infty)P + b(-\infty)Q)T^0(\rho^{F,G})) & \text{if} \quad s = \infty \end{cases}$$

are invertible where the K_s are certain compact operators.

The proof is left as an exercise.

3.8.5 Quadrature methods

To illustrate the applicability of the stability theorem to quadrature methods for solving the singular integral equation $(aI + bS_R)u = f$ with variable coefficients we consider the quadrature method 2.11.6. that is, given numbers τ_s with $0 \leq \tau_0 < \tau_1 < \ldots < \tau_{k-1} < 1$ and positive numbers w_s with $\sum_{s=0}^{k-1} w_s = 1$ we replace the singular integral by its approximation 2.11.6(3) and then we evaluate the resulting equations at points $(l + \epsilon_r)/n$ where ϵ_r are previously given numbers with $0 \leq \epsilon_0 < \epsilon_1 < \ldots < \epsilon_{k-1} < 1$ and

$$\{\epsilon_0, \ldots, \epsilon_{k-1}\} \cap \{\tau_0, \ldots, \tau_{k-1}\} = \emptyset \,.$$

Thus, explicitly written, our approximation equations are

$$\left(a(\frac{l+\epsilon_r}{n}) + ib(\frac{l+\epsilon_r}{n}) \sum_{s=0}^{k-1} w_s \cot \pi(\tau_s - \epsilon_r) \right) u_{rln} +$$

$$+ \frac{1}{\pi i} b(\frac{l+\epsilon_r}{n}) \sum_{s=0}^{k-1} w_s \sum_{m=-\infty}^{\infty} \frac{u_{smn}}{m - l + \tau_s - \epsilon_r} = f(\frac{l+\epsilon_r}{n}) \tag{1}$$

(compare 2.11.6(6)), $l \in \mathbf{Z}$, $r = 0, \ldots, k - 1$. As in 2.11.6 we shall interprete (1) as a spline approximation system. For, we set $G = \{\delta_{\epsilon_0}, \ldots, \delta_{\epsilon_{k-1}}\}$ and introduce a prebasis by

$$K = \left\{ \chi_{[0, \frac{\epsilon_0 + \epsilon_1}{2})}, \chi_{[\frac{\epsilon_0 + \epsilon_1}{2}, \frac{\epsilon_1 + \epsilon_2}{2})}, \ldots, \chi_{[\frac{\epsilon_{k-2} + \epsilon_{k-1}}{2}, 1)} \right\} \,.$$

The pair (K, G) is clearly canonical, and if $(A_{rs})_{r,s=0}^{k-1}$ abbreviates the system matrix of (1) then the quadrature system (1) and the spline system

$$E_n^K(A_{rs}) E_{-n}^K u^{(n)} = L_n^{K,G} f \tag{2}$$

are simultaneously solvable and stable or not.

Theorem 3.15 *(a) Let the coefficients a, b be piecewise continuous on $\mathbf{R}$ and continuous on $\mathbf{R} \setminus \mathbf{Z}$. Then*

(i) the sequence $(E_n^K(A_{rs}) E_{-n}^K)$ is in $\mathcal{A}_{\{1\},0}^K$.

(ii) the method (2) is stable if and only if the operator $aI + bS_R$ is invertible and if the operators (the coset)

$$W_s(E_n^K(A_{rs})E_{-n}^K) = \begin{cases} a(s)I + b(s)T^0(\sigma^{Q_1}) & \text{if } s \in \mathbf{R}\backslash\mathbf{Z} \\[2mm] (a(s+0)P + a(s-0)Q) + \\ \quad (b(s+0)P + b(s-0)Q)T^0(\sigma^{Q_1}) & \text{if } s \in \mathbf{Z} \\[2mm] \pi(a(+\infty)P + a(-\infty)Q + \\ \quad (b(+\infty)P + b(-\infty)Q)T^0(\sigma^{Q_1})) & \text{if } s = \infty \end{cases}$$

are invertible.

(b) If $k = 1$ and a, b are arbitrary piecewise continuous functions then

(i) the sequence $(E_n^K A_{00} E_{-n}^K)$ with $K = \{\chi_{[0,1]}\}$ is in $\mathcal{A}_{\{1\},-\epsilon_0}^K$.

(ii) the method (2) is stable if and only if the operator $aI + bS_R$ is invertible and if the operators (the coset)

$$W_s(E_n^K(A_{00})E_{-n}^K) = \begin{cases} (a(s+0)P + a(s-0)Q) + \\ \quad (b(s+0)P + b(s-0)Q)T^0(\sigma^{Q_1}) & \text{if } s \in \mathbf{R} \\[2mm] \pi(a(+\infty)P + a(-\infty)Q + \\ \quad (b(+\infty)P + b(-\infty)Q)T^0(\sigma^{Q_1})) & \text{if } s = \infty \end{cases}$$

are invertible.

(See Proposition 2.22 for the definition of σ^{Q_1}.)

The **proof** of (i) is a direct consequence of Section 2.11.6 and of the identity

$$(\text{diag}\,(a(\frac{l + \epsilon_r}{n})))_{n \geq 1} = (E_{-n}^K(L_n^{F,G}a)E_n^K)_{n \geq 1}$$

which yields that the sequence $(\text{diag}\,(a(\frac{l+\epsilon_r}{n})))_{n \geq 1}$ is in $\mathcal{A}_{\{1\},0}^K$ in case (a) and in $\mathcal{A}_{\{1\},-\epsilon_0}^K$ in case (b) by Proposition 3.25 and by 3.8.3(1). Now (ii) follows easily from the stability theorem.

The reader is recommended to study the quadrature methods considered in 2.11.7 and 2.11.9 for operators with variable coefficients, too.

3.8.6 Quadrocation methods

Let us explain for example the quadrocation method combining quadrature and qualocation for the singular integral operator $aI + bS_R$ with piecewise continuous coefficients. The resulting system is

$$
\sum_{r=0}^{k-1} w_r \sum_{l\in Z} \overline{\tilde{\phi}}(p-l-\epsilon_r)\times
$$

$$
\times((a((l+\epsilon_r)/n)+b((l+\epsilon_r)/n)i\cot\pi(\tau-\epsilon_r))u_{ln} - \frac{b(\frac{l+\epsilon_r}{n})}{\pi i}\sum_{m\in Z}\frac{u_{mn}}{l-m+\epsilon_r-\tau})
$$

$$
= \sum_{r=0}^{k-1} w_r \sum_{l\in Z} \overline{\tilde{\phi}}(l+\epsilon_r)f((l+p+\epsilon_r)/n) , \quad p\in Z , \tag{1}
$$

where $\tau \in [0,1)$, $0 \le \epsilon_0 < \ldots < \epsilon_{k-1} < 1$, $\{\phi\}$ and $\{\phi_1\}$ are prebases, and $\sum_{r=0}^{k-1} w_r = 1$, $w_r \ge 0$.

Theorem 3.16 *Let the discontinuities of a and b be located at points with integer coordinates. Then the method (1) is stable if and only if the operator $aI + bS_R$ is invertible and if the operators (resp. the coset)*

$$
W_s(A_n)=\begin{cases}
\sum_{r=0}^{k-1} w_r T^0(\lambda^{\{\overline{\tilde{\phi}}\},\{\delta-\epsilon_r\}})(a(s)I + b(s)T^0(a^{(\epsilon_r-\tau)})) & \text{if } s\in\mathbf{R}\setminus\mathbf{Z} \\
\sum_{r=0}^{k-1} w_r T^0(\lambda^{\{\overline{\tilde{\phi}}\},\{\delta-\epsilon_r\}})(a(s+0)P + a(s-0)Q \\
\quad +(b(s+0)P + b(s-0)Q)T^0(a^{(\epsilon_r-\tau)})) & \text{if } s \in \mathbf{Z} \\
\pi(\sum_{r=0}^{k-1} w_r T^0(\lambda^{\{\overline{\tilde{\phi}}\},\{\delta-\epsilon_r\}})(a(+\infty)P + a(-\infty)Q \\
\quad +(b(+\infty)P + b(-\infty)Q)T^0(a^{(\epsilon_r-\tau)}))) & \text{if } s = \infty
\end{cases}
$$

are invertible.

In case $\phi=\chi_{[0,1)}$ these conditions take a form which is well known from the other methods considered above:

$$
W_s(A_n) = \begin{cases}
a(s)I + b(s)T^0(\sigma) & \text{if } s \in \mathbf{R}\setminus\mathbf{Z} \\
a(s+0)P + a(s-0)Q+ \\
\quad (b(s+0)P+b(s-0)Q)T^0(\sigma) & \text{if } s\in\mathbf{Z} \\
\pi(a(+\infty)P + a(-\infty)Q+ \\
\quad (b(+\infty)P + b(-\infty)Q)T^0(\sigma)) & \text{if } s = \infty
\end{cases}
$$

with $\sigma = \sum_{r=0}^{k-1} w_r a^{(\epsilon_r-\tau)}$.

3.9   Exercises

E 3.1 Prove (i) in Section 3.1.1.

(Hint: Put $a_n(t) = \dfrac{2i}{t-i}\left(\dfrac{t+i}{t-i}\right)^n$, $t \in \mathbf{R}$, define $P_R := (I + S_R)/2$, let u_n stand for the operator of multiplication by the function $t \mapsto \left(\dfrac{t+i}{t-i}\right)^n$, and set $P_n := u_n(I - P_R)u_{-2n}P_R u_n$. Show that

$$P_R a_n = \begin{cases} a_n & \text{if} \quad n \geq 0 \\ 0 & \text{if} \quad n < 0 \,, \end{cases}$$

that P_n is a projection from $L_R^p(\alpha)$ onto span $\{a_{-n}, \ldots, a_{n-1}\}$, and that $P_n \to I$, $P_n^* \to I^*$ strongly as $n \to \infty$. Then $\|P_n K P_n - K\| \to 0$ for each compact operator K, thus it remains to prove that $P_n K P_n \in \mathcal{O}_R^p(\alpha)$. But if (k_{ij}) is the matrix representation of $P_n K P_n$ with respect to the basis $\{a_{-n}, \ldots, a_{n-1}\}$ then

$$P_n K P_n = \sum_{i,j=-n}^{n-1} k_{ij} u_i P_1 P_R u_{-j} = \frac{1}{2} \sum_{i,j=-n}^{n-1} k_{ij} u_i (S_R u_1 - u_1 S_R)u_{-j-1}, \text{ which is}$$

obviously in $\mathcal{O}_R^p(\alpha)$.)

E 3.2 Prove (ii) in Section 3.1.1.

(Hint: The operator $r S_R - S_R r I$ is compact for each rational function r having no poles on $\mathbf{R}$.)

E 3.3 Prove (iii) in Section 3.1.1.

(Hint: First show that $\|Z_s^{-1}\|\,\|Z_s\| = 1$ to conclude that W_0 and W_∞ are continuous algebra homomorphisms. Identity 3.1.1(3) is obvious, 3.1.1(4) and (5) are straightforward, and for 3.1.1(6) prove that $s^{-1/p-\alpha}Z_s \to 0$ weakly as $s \to \infty$ and $s \to 0$.)

E 3.4 Prove that the algebras $\mathcal{O}_R^p(0)_t^\pi$ and $\Sigma_R^p(0)$ in Section 3.1.3 are isometrically isomorphic. Prove the analogous result for the weighted case.

E 3.5 Represent $\mathcal{O}_R^p(\alpha)^\pi$ as an algebra of continuous functions on $\dot{\mathbf{R}}$ taking values in $\Sigma_R^p(\alpha)$ at $x = 0$ and $x = \infty$ and in $\Sigma_R^p(0)$ at $x \in \mathbf{R} \setminus \{0\}$.

(Hint: Compare Section 3.7.6 and see [RS 8] for this and analogous representations.)

E 3.6 Let alg $(S_R, C(\dot{\mathbf{R}}))$ denote the smallest closed subalgebra of $L(L_R^p(\alpha))$ containing the singular integral operator S_R and all operators of multiplication by functions in $C(\dot{\mathbf{R}})$. Localize the quotient algebra alg $(S_R, C(\dot{\mathbf{R}}))/K(L_R^p(\alpha))$ over $\dot{\mathbf{R}}$ by means of Allan's local principle, show that the resulting local algebras are singly generated, and apply Theorem 1.9 to determine the maximal ideal space of this quotient algebra, and to find the Gelfand transform of the elements of this algebra.

E 3.7 Let $\mathcal{O}_R^p(\alpha)$ refer to the algebra introduced in 3.1.1. Show that the local algebras $\mathcal{O}_R^p(\alpha)_x^\pi$ with $x \in \dot{\mathbf{R}}$ are generated by two idempotents, and use Theorem 1.10 to derive a symbol calculus for $\mathcal{O}_R^p(\alpha)^\pi$ which assigns a 2×2 matrix function to each element of this algebra.

E 3.8 Investigate in the same manner the algebra $\mathcal{O}_T^p(0)^\pi$ where $\mathcal{O}_T^p(0)$ is the smallest closed subalgebra of $L(L_T^p(0))$ containing S_T and all operators of multiplication by functions in $PC(\mathbf{T})$.

E 3.9 Let $\mathcal{O}_{[0,1]}^p(\alpha)$ denote the smallest closed subalgebra of $L(L_{[0,1]}^p(\alpha))$ containing the singular integral operator $S_{[0,1]}$ and all operators of multiplication by functions in $PC[0,1]$). Show that $\mathcal{O}_{[0,1]}^p(\alpha)/K(L_{[0,1]}^p(\alpha))$ can be localized over $[0,1]$ by means of Allan's local principle, and that the local algebras at 0 and 1 are singly generated and at $x \in (0,1)$ generated by two idempotents. Apply Theorems 1.9 and 1.10 to get a symbol calculus for the quotient algebra $\mathcal{O}_{[0,1]}^p(\alpha)/K(L_{[0,1]}^p(\alpha))$.

E 3.10 Let $L^2(\mathbf{T})$ stand for the usual Lebesgue space over the unit circle $\mathbf{T}$ and $R(\mathbf{T})$ for the space of all Riemann integrable functions on $\mathbf{T}$ provided with the norm

$$\| f \|_{R(T)} = \| f \|_{L^2(T)} + \sup_{t \in T} | f(t) | .$$

Write e_k ($\in L^2(\mathbf{T})$) for the function $e_k(t) = t^k$, introduce projections $P_n : L^2(\mathbf{T}) \to L^2(\mathbf{T})$ by

$$P_n \left(\sum_{k \in Z} f_k e_k \right) = \sum_{k=-n}^n f_k e_k$$

and operators $L_n : R(\mathbf{T}) \to \operatorname{Im} P_n$ by

$$(L_n f)(t_j) = f(t_j) \quad \text{for all } t_j = \exp \frac{2\pi i j}{2n+1}$$

with $j = 0, \pm 1, \ldots, \pm n$, and define operators $W_n : L^2(\mathbf{T}) \to \operatorname{Im} P_n$ by

$$W_n(\sum_{k \in Z} f_k e_k) = f_{-1}e_{-n} + \ldots + f_{-n}e_{-1} + f_n e_0 + \ldots + f_0 e_n .$$

Let, finally, $\mathcal{F}$ denote the collection of all bounded sequences (A_n) of operators $A_n : \operatorname{Im} P_n \to \operatorname{Im} P_n$ for which there exist operators $W(A_n)$ and $V(A_n)$ such that

$$A_n P_n \to W(A_n), \ A_n^* P_n^* \to W(A_n)^*, \ W_n A_n W_n \to V(A_n)$$

and $W_n^* A_n^* W_n^* \to V(A_n)^*$ and let $\mathcal{J}$ stand for the subset of $\mathcal{F}$ consisting of all sequences of the form $(L_n K L_n + W_n L W_n + C_n)$ where K and L are compact and $\|C_n\| \to 0$ as $n \to \infty$.

(a) Introduce operations on $\mathcal{F}$ by

$$(A_n) + (B_n) := (A_n + B_n) \,,\ (A_n)(B_n) := (A_n B_n)$$

and a norm by $\|(A_n)\| = \sup_n \|A_n P_n\|$; and show that $\mathcal{F}$ is a Banach algebra and $\mathcal{J}$ is a closed two-sided idal in $\mathcal{F}$.

(b) Show that a sequence $(A_n) \in \mathcal{F}$ is stable if and only if the operators $W(A_n)$ and $V(A_n)$ are invertible and if the coset $(A_n) + \mathcal{J}$ is invertible in $\mathcal{F}/\mathcal{J}$. (Hint: Lifting theorem 1.8.)

(c) Let $\mathcal{A}$ denote the smallest C^*-algebra of $\mathcal{F}$ containing the sequences $(L_n(aI + bS_T)P_n)$ where $a, b \in PC(\mathbf{T})$ and the ideal $\mathcal{J}$. Show that the quotient algebra $\mathcal{A}/\mathcal{J}$ is isometrically isomorphic to the algebra $\mathcal{O}_T^2(0)^\pi$ introduced in Exercise 3.8, and use the results of this exercise to establish necessary and sufficient conditions for the trigonometric collocation method for singular integral operators on $\mathbf{T}$. (Hint: Compare [PS 2], Theorem 7.25.)

E 3.11 Prove that

(a) $Y_{t^{-1}} T^p_{k,\kappa,\Omega}(\alpha) Y_t = T^p_{k,\kappa,t^{-1}\Omega}(\alpha)$ for $t \in \mathbf{T}$.

(b) $(E_n^F Y_t E_{-n}^F) \mathcal{A}^F_{\Omega,r} (E_n^F Y_{t^{-1}} E_{-n}^F) = \mathcal{A}^F_{t^{-1}\Omega,r}$ for $t \in \mathbf{T}$.

E 3.12 If $k = 1$ and $\Omega \subseteq \Omega'$ are closed subsets of $\mathbf{T}$ then, clearly,

$$T^p_{k,\kappa,\Omega}(\alpha) \subseteq T^p_{k,\kappa,\Omega'}(\alpha). \tag{1}$$

Let $F = \{\chi_{[0,\frac{1}{2})}, \chi_{[\frac{1}{2},1)}\}$. Prove that $T^0((\lambda^{F,F})^{-1}\sigma^{F,F})$ lies in $T^p_{2,2,\{1\}}(\alpha)$ but not in $T^p_{2,2,\{1,-1\}}(\alpha)$. Thus, (1) fails for $k > 1$ in general.

E 3.13 Let $F = \{\chi_{[0,\frac{1}{k})}, \chi_{[\frac{1}{k},\frac{2}{k})}, \cdots \chi_{[\frac{k-1}{k},1)}\}$. Prove that $\sigma^{F,F} = (\sigma^{F,F}_{rs})^{k-1}_{r,s=0}$ is a Toeplitz matrix (i.e. $\sigma^{F,F}_{rs}$ depends on $r - s$ only). Is $\sigma^{F,F}$ a circulant?

E 3.14 Given integers k and κ with $k \geq \kappa > 0$ construct a k-elementic canonical prebasis F with $\alpha^F = (\underbrace{1,\ldots,1}_{\kappa},0,\ldots,0)$ such that $\lambda^{F,F}$ is a multiple of the identity matrix.

E 3.15 Show that the operators $n^{1/p+\alpha} E_n^F \Lambda V_{\{sn+r\}} : l^p_k(\alpha) \to L^p_R(\alpha)$ converge weakly to 0 as $n \to \infty$ for all s and r, and use this fact to give a direct proof that $W_s(L_n^{F,F} K | S_u^F) = 0$ for all compact K.

E 3.16 Show that the operators $E_n^F Y_t E_{-n}^F L_n^{F,F} : L_R^p(\alpha) \to L_R^p(\alpha)$ converge strongly to 0 as $n \to \infty$ whenever $t \neq 1$, and use this fact to give a direct proof that

$$W^t(L_n^{F,F} K | S_n^F) = \begin{cases} K & \text{if } t = 1 \\ 0 & \text{if } t \neq 1 \end{cases} \quad \text{for all compact } K.$$

E 3.17 (a) Let $f \in L_R^\infty$ and suppose f to be continuous at 0. Prove that $P_n^G f E_n^F \to f(0) T^0(\lambda^{F,G})$ strongly as $n \to \infty$ for all F, G (not necessarily canonical).
(b) Use (a) to show that $W_0(L_n^{F,G} f | S_n^F) = f(0)I$ for all canonical pairs (F, G).

E 3.18 Let $\mathcal{R}$ and $\mathcal{G}$ be as in Section 3.3.2. Examine the structure of matrices in $\mathcal{R}_{2,1}$ and $\mathcal{R}_{2,2}$, and verify the corresponding invertibility criteria (see 3.1).

E 3.19 (a) Let $F = \{\chi_{[0,1/2)}, \chi_{[1/2,1)}\}$ and $\Omega = \{1\}$, $r = 0$. Find a sequence $(A_n) \in \mathcal{A}_{\Omega,r}^F$ which is stable, but for which the adjoint sequence (A_n^*) does not converge strongly. In particular, $(A_n^*) \notin \mathcal{A}_{\Omega,r}^F$.

(b) Let $p = 2$, $\alpha = 0$. Then the algebra $\mathcal{F}^F$ is C^* when provided with the involution $(A_n) \to (A_n^*)$, but $\mathcal{A}_{\Omega,r}^F$ is in general not a C^*-subalgebra of $\mathcal{F}^F$.

E 3.20 Let F be a canonical prebasis and $A \in OLT$. Prove that $(L_n^{F,F} A | S_n^F) \in SLT_{\{1\},0}^F$.

E 3.21 What does 'KMS' stand for?
(Hint: [BKS] has been written during a stay of Professor Krupnik in Karl-Marx-Stadt (= the name of Chemnitz between 1953 and 1990).)

E 3.22 Compute the strong limits W_s in Theorem 3.11 for $\alpha \neq 0$.

E 3.23 Let a, b, K and F and G be as in Theorem 3.11. Prove that the modified (F, G)-Galerkin method

$$L_n^{F,G} aI | S_n^F + L_n^{F,G} b L_n^{F,G} S_R | S_n^F + L_n^{F,G} K | S_n^F = L_n^{F,G} f$$

applies to the operator $A = aI + bS_R + K$ if and only if the conditions of Theorem 3.11 are satisfied.

E 3.24 Let a, b, K and F, G be as in Proposition 3.25. Verify that the sequence $(L_n^{F,G}(aI + bS_R + K) | S_n^F)$ belongs to $\mathcal{A}_{\{1\},-r}^F$ for all $r \in [0, 1)$.

E 3.25 Let a, b be piecewise continuous functions with discontinuities only at points in $\mathbf{Z}/2$ and ∞, and let F, G and K be as in Proposition 3.25 but now with $\tau_s \in [0, 1/2)$ for $s = 0, \ldots, k - 1$. Show that

$$(L_n^{F,G}(aI + bS_R) | S_n^F) \in \mathcal{A}_{\{1\},r}^F \tag{2}$$

for all $r \in [0, 1/2)$, and establish necessary and sufficient criteria for the stability of the sequence (2).

E 3.26 Let $F = \{\chi_{[0,1)}\}$. Show that $(L_n^{F,F}\chi_{[t,\infty)}I|S_n^F) \notin \mathcal{A}_{\{1\},-r}^F$ for all $r \in [0,1)$ and $t \notin \mathbf{Z}$.

(Hint: The strong limit $W_t(L_n^{F,F}\chi_{[t,\infty)}I|S_n^F)$ does not exist.)

E 3.27 Derive Theorems 3.12-3.14 from the general stability theorem.

E 3.28 Let $\mathcal{K}$ be a Banach algebra with unit element e and $x_1, \ldots, x_k$ ($k \geq 2$) be pairwisely commuting elements of $\mathcal{K}$. Then the spectrum of the matrix

$$\begin{pmatrix} x_1 & x_2 & \cdots & x_k \\ x_1 & x_2 & \cdots & x_k \\ \vdots & \vdots & \vdots & \vdots \\ x_1 & x_2 & \cdots & x_k \end{pmatrix}$$

is equal to the union of $\{0\}$ and the spectrum of $x_1 + \cdots + x_k$ in $\mathcal{K}$.

E 3.29 Let $F = \{\varphi_0, \ldots, \varphi_{k-1}\}$ be a canonical prebasis with $\alpha_0^F = \cdots = \alpha_{\kappa-1}^F = 1$ and $\alpha_\kappa^F = \cdots = \alpha_{k-1}^F = 0$, and let $\mathcal{K}$ be a unital Banach algebra. Define a mapping

$$L : \mathcal{K} \to \mathcal{K}_{k \times k}, \quad a \mapsto \left(\alpha_r^F \int_R \varphi_s(x)dx \cdot a\right)_{r,s=0}^{k-1}.$$

Then $\sigma_{\mathcal{K}_{k \times k}}(L(a)) = \{0\} \cup \sigma_{\mathcal{K}}(a)$.

3.10   Comments and references

3.1 For singular integral operators with piecewise continuous coefficients over the real axis see the monographs Böttcher, Silbermann [BS 2], Duduchava [Du 1], Gohberg, Goldberg, Kaashoek [GGK], Gohberg, Krupnik [GK 1], Michlin, Prößdorf [MP] and Simonenko, Chin Ngok Minh [SC]. The approach presented in our book can be found in Roch, Silbermann [RS 1], [RS 4]. See also [RS 5].

For the results cited in 3.1.7 without proof see Gohberg/Krupnik [GK 1], Vol. 2, Section 9.8, or Michlin/Prößdorf [MP], IV, Section 6, Satz 6.1.

3.2 This is a simplified version of the material treated in Sections 3.3-3.6. See there for detailed references.

The idea of generating new algebras by making given algebras translation invariant (or invariant with respect to another group action) seems to be very natural and fruitful. Not only the algebra $\mathcal{O}_R^p(0)$, but also the Toeplitz algebra alg $(T^0(PC), P)$ and the algebra generated by all Wiener-Hopf operators with piecewise continuous generating functions origin in this way. For the first one, take the algebra generated by the projections P and $T^0(\chi)$, where χ refers to the characteristic function of the upper semi circle, as the basic algebra, and make this algebra invariant with respect of the action of the group $\mathbf{T}$ by

$$A \mapsto Y_{t^{-1}} A Y_t, \quad t \in \mathbf{T}.$$

In the Wiener-Hopf case, the starting algebra is that one generated by the projection $P = (I + S_R)/2$ and by the operator of multiplication by the characteristic function χ_{R^+}, whereas the group action is given by

$$A \;\mapsto\; H_{-s}\, A\, H_s\,, \quad \text{with } (H_s f)(t) \;=\; e^{2\pi i s t}\, f(t) \quad \text{for all } s \in \mathbf{R}\,.$$

3.3 These results are new and are due to the authors. It was a hard piece of work to extract the general structure of spline approximation operators for operators in $\Sigma_R^p(\alpha)$. The readers who guess that the conditions (a)-(c) pointed out in 3.3.1 are too affected and mysterious, may take comfort: the authors also guessed so. We asked for a long time whether these conditions could ever be correct and adequate. Only Lemma 3.1 (together with its proof) convinced us emphatically that they are indeed very natural.

3.4-3.6 The general stability theorem is the central result in this book. It goes back to the authors and is published here for the first time. This is the history of the algebra $\mathcal{A}_{\Omega,r}^F$: For the first time, algebras of approximation sequences were (in the case of the finite section method for Toeplitz operators) introduced and studied by Böttcher and Silbermann [BS 3]. That it is also possible to include spline approximation sequences for singular integral operators into an algebra follows from a discovery by Prößdorf and Rathsfeld [PR 2]. They observed that there is a lot of approximation methods for singular integral operators which involve approximation sequences of a characteristic structure: namely having so-called paired circulants as system matrices. Prößdorf and Rathsfeld also derived a stability criterion for single sequences of paired circulants.

Two of the authors pursued these investigations by introducing a Banach algebra of sequences of paired circulants (Hagen, Silbermann [HS 2]). Invoking local principles, they succeeded in proving necessary and sufficient stability conditions for arbitrary elements of this algebra.

The investigations by Prößdorf, Rathsfeld, Hagen and Silbermann concern periodic spline spaces (or spline spaces over the unit circle). The first direct ancestor of the algebra $\mathcal{A}_R^F(\alpha)$ appears in Roch's thesis (which deals with the finite section method for operators in the Toeplitz algebra $\mathrm{alg}\,(T^0(PC), P)$). The algebra considered there is in fact something like the algebra $\mathcal{A}_R^F(\alpha)$ with $\Omega = \mathbf{T}$, $r = 0$ and $F = \{\chi_{[0,1)}\}$ but, in contrast to the true algebra $\mathcal{A}_{T,0}^{\{\chi_{[0,1)}\}}$, this algebra is only invariant with respect to shifts by integer numbers. Many of the main ingredients for proving the general stability theorem can be found in this thesis: in particular, the lifting homomorphisms W^t and the locally equivalent representations W_s were introduced and exploited.

The algebra $\mathcal{A}_{T,0}^{\{\chi_{[0,1)}\}}$ itself was introduced and examined in [RS 6]. The general case of $\mathcal{A}_{\Omega,r}^F$ is studied in the present monograph for the first time. Basically new results are Lemma 3.3, the inverse closedness of the algebra $\mathcal{A}_{\Omega,r}^F$ in $\mathcal{F}^F$, and Propositions 3.15-3.17 which state that certain special sequences are included into this algebra. The observation that sometimes the correction term $r \neq 0$ proves to be useful goes back to [PR 2] and [HS 2].

3.7 The notion of operators of local type is due to Simonenko. Theorem 3.6 and its proof were taken from Böttcher, Krupnik, Silbermann [BKS]. The introduction of sequences of local type by the authors is aimed to get local inclusion theorems for the algebra $\mathcal{A}_{\Omega,r}^F$. Theorems 3.8 and 3.9 are new. The characterization of the quotient algebra $\mathcal{A}_{\Omega,r}^F / \mathcal{J}_{\Omega,r}^F$ as an algebra of continuous functions makes use of arguments taken from Roch, Silbermann [RS 8].

3.8 In case of spline spaces with piecewise polynomial functions with maximal smoothness, these results are the outcome of the efforts by many mathematicians. Let us only mention some of the main contributers: Prößdorf, Prößdorf/Rathsfeld, Schmidt for Galerkin methods; Prößdorf/Schmidt, Prößdorf/Rathsfeld, Arnold/Wendland, Saranen/Wendland, Prößdorf, Schmidt for collocation methods; Lifanov/Polonski, Rathsfeld, Prößdorf/Rathsfeld for composed quadrature rules; Sloan, Sloan/Wendland, Hagen/Silbermann for qualocation methods; Saranen for the quadrocation method. For detailed references see the monograph Prößdorf, Silbermann [PS 2], Notes and comments to Chapter 10.

Chapter 4

Singularities

This chapter deals with singularities and their effects on stability of approximation methods. We have already come across some types of singularities in the preceding sections: viz. the analytic singularities caused by discontinuous coefficients in the operators. These singularities bring forth undetermined compact perturbations in stability conditions (compare the operators K_s in Theorems 3.9 and 3.11) or, at worst, perturbations behaving irregularly to such an extend that the corresponding approximation sequences do not belong to any of the algebras $\mathcal{A}_{\Omega,r}^F$. Moreover, the compact perturbations also destroy the nice structure of the approximation systems (without perturbation, they are Toeplitz or Block-Toeplitz in many cases).

Other types of singularities appear as *geometric* or *numeric* ones. Let us shortly explain what is meant. For comprehensive treatment we refer to the fifth chapter. We consider the integral equation

$$(aI + bS_\square)\, u \;=\; f \quad , u, f \in L_\square^2 \,, \tag{1}$$

with constant coefficients over the oriented boundary $\square$ of the unit square (see the picture below).

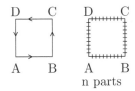

n parts

By $S_n^\square$ we denote the space of piecewise constant splines over the uniform partition of $\square$ into $4n$ parts.

For the approximate solution of (1) by the Galerkin method we look for a solution $u^{(n)} \in S_n^\square$ of the equation

$$L_n^\square (aI + bS_\square)\, u^{(n)} \;=\; L_n^\square f \,. \tag{2}$$

Herein $L_n^\square$ refers to the Galerkin projection with $S_n^\square$ both as test and as trial space. The stability of the method (2) is again accessible to Banach algebra techniques.

Localizing (2) over the boundary of the unit square we arrive at local representatives which, for example at the corner A, are of the form

$$L_n^{\measuredangle}(aI + bS_{\measuredangle})|S_n^{\measuredangle} , \tag{3}$$

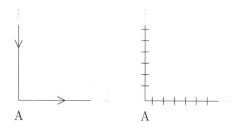

where $\measuredangle$ is an oriented infinite angle with equidistant partition, $S_n^{\measuredangle}$ is the space of piecewise constant splines over $\measuredangle$, and $L_n^{\measuredangle}$ is the corresponding Galerkin projection.

A   A

The method (3) is equivalent to a certain spline projection method for operators on the real axis. To make this clear, we map the angle $\measuredangle$ one-to-one onto the real axis by identifying the horizontal leg with the positive semi axis and the vertical leg with the negative one.

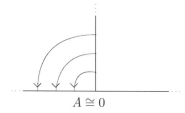

$$A \cong 0$$

Thereby, the Lebesgue space $L_{\measuredangle}^2$ goes over into L_R^2, the spline space $S_n^{\measuredangle}$ and the associated projection $L_n^{\measuredangle}$ into the spline space S_n and the projection L_n introduced in Section 2.2.1, and the singular integral operator $S_{\measuredangle}$ into the operator

$$A = \chi_{R^+} S_R \chi_{R^+} I + \chi_{R^-} S_R \chi_{R^-} I + \chi_{R^-} J N_{\pi/2} \chi_{R^+} I + \chi_{R^+} N_{3\pi/2} J \chi_{R^-} I ,$$

where $(Jf)(t) = f(-t)$ and $N_{\pi/2}, N_{3\pi/2}$ are defined as in Section 2.1.2. Thus, method (3) is stable if and only if the method

$$L_n(aI + bA)|S_n$$

is stable, but now A is an operator which involves "analytic" singularities, viz. the functions $\chi_{R^\pm}$. This explains a certain equivalence between geometric and analytic singularities; in particular, all of the unpleasant phenomena related with analytic singularities must be expected for geometric singularities again. The term "numeric singularity" will be used to paraphrase such things as changes of spline spaces or passage to non-equidistant partitions which will be briefly dealt with in the sixth chapter. For example, consider a singular integral equation on the following curve:

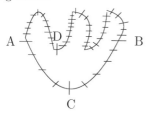

From A via C to B the curve behaves regularly, and only a few partition points are needed, whereas, from A via D to B, only a dense mesh of points can guarantee an adequate description of the geometry. The associated change of the spline spaces at A and B leads to similar effects as analytic singularities.

Finally, one often tries to compensate analytic or geometric singularities by a non-uniformly subdivided mesh in the neighbourhood of the singular point which can also be viewed as a suitable chosen singularity the effects of which are just opposite to those resulting from jumps or corners. This is actually the crucial point: whereas analytic or geometric singularities are a priori given, the numeric singularities will be initiated artificially to moderate or equalize the first ones.

The present chapter is organized as follows. We start with investigating the finite section method once again, but now for arbitrary operators in the Toeplitz algebra. The associated Banach algebra of approximation methods contains, besides the finite-section-sequences, also "infinite-section-sequences for the identity operator", and exactly these sequences will highlight the way now to evade singularities. First we discuss this for spline approximation methods for homogeneous operators (or operators with constant coefficients) which are intimately related with Toeplitz operators and their sections, and then we shall point out how to modify the approximation methods for operators with variable coefficients, i.e. the methods which belong to any algebra $\mathcal{A}_{\Omega,r}^{F}$. The crux is to introduce approximation sequences depending on two parameters, n and i, where n is the parameter counting the refinements of the mesh, whereas i controls the size of the infinite sections. We shall verify that these sequences do actually not involve undetermined terms or other perturbations if only n and i are chosen large enough.

The advantage of double indicated approximation methods is purchased by a more complicated structure of the associated symbol maps. Whereas the symbols for methods in one of the algebras $\mathcal{A}_{\Omega,r}^{F}$ are operator-valued we are now led to symbols whose values are approximation methods again. But the occuring symbol-methods are well studied and, in particular, the conditions for their stability are weaker than those for the invertibility of the operator-valued symbols.

The second main goal of this chapter is another combination of spline approximation methods for singular integral operators and finite section methods for Toeplitz operators which is intended to solve equations over unbounded domains, in particular over the real axis. In a sense, this is a "cut-off-method" for the singular point ∞.

4.1 Approximation of operators in Toeplitz algebras

4.1.1 Finite sections of operators in $T^p_{1,1,T}(\alpha)$

Recall that $T^p_{1,1,T}(\alpha)$ stands for the smallest closed subalgebra of $L(l^p_Z(\alpha))$ which contains the projection P,

$$P : (x_k) \mapsto (\dots, 0, 0, x_0, x_1, \dots) ,$$

and all operators $T^0(a)$ of multiplication by piecewise continuous multipliers $a \in PC_{p,\alpha}$, that is, $T^p_{1,1,T}(\alpha) = \mathrm{alg}\,(T^0(PC_{p,\alpha}), P)$ in earlier notations. Given $n \geq 1$ we let R_n denote the operator

$$R_n : (x_k) \mapsto (\dots, 0, 0, x_{-n}, x_{-n+1}, \dots, x_{n-2}, x_{n-1}, 0, 0, \dots) .$$

Considering $l^p_{Z+}(\alpha)$ as a closed subspace of $l^p_Z(\alpha)$ we can identify the operators PR_n with the projections P_{n-1} introduced in Section 1.5.3.

Let $A \in T^p_{1,1,T}(\alpha)$. The *finite section method* applies to the operator A if the equations

$$R_n A R_n u^{(n)} = R_n f \quad , f \in l^p_Z(\alpha) \tag{1}$$

are solvable for all $n \geq n_0$ and for all right-hand sides f and if the sequence $(u^{(n)})_{n \geq n_0}$ of their solutions converges (in the norm of $l^p_Z(\alpha)$) to a solution of the equation $Au = f$. It is evident that $R_n \to I$ and $R_n^* \to I^*$ strongly. Hence, $R_n A R_n \to A$ and $(R_n A R_n)^* \to A^*$ strongly whence follows that the finite section method (1) applies if and only if the sequence $(R_n A R_n | \mathrm{Im}\,R_n)$ of operators on $\mathrm{Im}\,R_n$ is stable (compare Proposition 1.1 and the proof of 1.9).

In what follows we would rather work with operators defined on the whole space $l^p_Z(\alpha)$ than with operators on its subspace $\mathrm{Im}\,R_n$. This can be reached by introducing operators $Q_n = I - R_n$ and adding these operators to $R_n A R_n$. Clearly, $R_n A R_n$ is invertible on $\mathrm{Im}\,R_n$ if and only if $R_n A R_n + Q_n$ is invertible on $l^p_Z(\alpha)$, and the sequence $(R_n A R_n | \mathrm{Im}\,R_n)$ is stable if and only if the sequence $(R_n A R_n + Q_n)$ is so.

To study stability of sequences of the form $(R_n A R_n + Q_n)$ we proceed contrarily to Chapter 3: now we employ de Boor's estimates to translate the stability problem (1) into an equivalent stability problem for a certain spline approximation method which, on its hand, is subject to the general stability theorem 3.5. To that end we let F refer to the canonical prebasis $\{\chi_{[0,1]}\}$.

Proposition 4.1 *If* $A \in T^p_{1,1,T}(\alpha)$ *then the sequence* $(E^F_n(R_n A R_n + Q_n)E^F_{-n})$ *belongs to* $\mathcal{A}^F_{T,0}$.

Proof It is obviously sufficient to verify that $(E^F_n A E^F_{-n})$ and $(E^F_n R_n E^F_{-n})$ belong to $\mathcal{A}^F_{T,0}$. By its definition, the algebra $\mathcal{A}^F_{T,0}$ contains all sequences of the form

$$(E^F_n \Lambda V_{\{tn\}} \Lambda^{-1} A \Lambda V_{-\{tn\}} \Lambda^{-1} E^F_{-n}) , \quad A \in T^p_{1,1,T}(\alpha) , \tag{2}$$

(see 3.4.2(1)), and setting $t = 0$ we get the desired inclusion $(E_n^F A E_{-n}^F) \in \mathcal{A}_{T,0}^F$. For the second inclusion , notice that

$$R_n = \Lambda V_{-\{n\}} \Lambda^{-1} P \Lambda V_{\{n\}} \Lambda^{-1} - \Lambda V_{\{n\}} \Lambda^{-1} P \Lambda V_{-\{n\}} \Lambda^{-1} . \tag{3}$$

Thus, it suffices to set $A = P$ and $t = +1, -1$ in (2) to obtain $(E_n^F R_n E_{-n}^F) \in \mathcal{A}_{T,0}^F$. ■

Abbreviate $E_n^F (R_n A R_n + Q_n) E_{-n}^F$ by A_n. An application of Theorem 3.5 to the sequence (A_n) gives that the finite section method (1) is stable if and only if the operators (resp. the coset) $W^t(A_n)$ $(t \in \mathbf{T})$ and $W_s(A_n)$ $(s \in \dot{\mathbf{R}})$ are invertible. But, in this particular case, an essentially finer result holds:

Theorem 4.1 *The finite section method (1) applies to the operator $A \in T_{1,1,T}^p(\alpha)$ if and only if the operators $W^t(A_n)$ $(t \in \mathbf{T})$ and $W_s(A_n)$ $(s \in \{-1,0,1\})$ are invertible. In particular, $W_0(A_n) = A$.*

Proof We claim that the invertibility of $W_s(A_n)$ for $s \in \{-1,0,1\}$ already implies invertibility of $W_s(A_n)$ for all $s \in \dot{\mathbf{R}}$. Let, for instance, $s \in (0,1)$. Taking into account the special form of the operators A_n as well as identities 3.5.1(22) and (23) in Proposition 3.14 one readily checks that the operators $W_s(A_n)$ are actually independent of $s \in (0,1)$. Hence, it suffices to verify the invertibility of $W_s(A_n)$ for at least one $s \in (0,1)$. For this one can argue as follows: If $W_0(A_n)$ is invertible then, by Corollary 3.3(a), the coset $\Phi_0^J(A_n)$ is invertible. The upper semi-continuity of the mapping $s \mapsto \|\Phi_s^J(A_n)\|$ implies the invertibility of $\Phi_s^J(A_n)$ for all s in a neighbourhood of 0 (compare Proposition 1.5). Thus, again by Corollary 3.3(a), all operators $W_s(A_n)$ with sufficiently small $s > 0$ are invertible.

For $s \in (-\infty, -1), (-1, 0)$, and $(1, \infty)$, the proof is analogous. Finally, if $s = \infty$ then, as a little thought shows,

$$W_\infty(A_n) = \pi(R_n A R_n + Q_n) = \pi(I)$$

which is always invertible. ■

4.1.2 Examples

We are going to illustrate the preceding theorem by a few examples which are partially suggested by the concrete form of certain spline approximation methods and which are partially thought to reflect some recent developments of own interest. For brevity, write Q in place of $I - P$.

Example 4.1 *Let $a, b \in PC_{p,\alpha}$ and let K be compact. The finite section method applies to the operator $A = T^0(a)P + T^0(b)Q + K$ if and only if*

(a) *the operator A is invertible in $L(l_Z^p(\alpha))$.*

(b) *the operators $PT^0(b)P + Q$ and $QT^0(a)Q + P$ are invertible in $L(l_Z^p(0))$.*

(c) the operators

$$\chi_{[-1,1]}\left(\frac{I-S_R}{2}(a(t+0)\chi_{[0,1]}+b(t+0)\chi_{[-1,0]})+\right.$$

$$\left.+\frac{I+S_R}{2}(a(t-0)\chi_{[0,1]}+b(t-0)\chi_{[-1,0]})\right)\chi_{[-1,1]}I+\chi_{R\setminus[-1,1]}I$$

are invertible in $L(L_R^p(\alpha))$ for all $t \in \mathbf{T}$.

Proof   Set $A_n := R_n A R_n + Q_n$. If (A,t) abbreviates the sequence 4.1.1(2) then we can write (A_n) as

$$((P,1)-(P,-1))(T^0(a)P+T^0(b)Q+K,0)((P,1)-(P,-1))+(P,-1)+(Q,1),$$

and now Proposition 3.13 yields immediately that $W^t(A_n)$ is just the operator under (c), whereas

$$W_{-1}(A_n) = PT^0(b)P+Q,\ W_0(A_n) = A,\ W_1(A_n) = QT^0(a)Q+P.\ \blacksquare$$

Let us emphasize that the conditions under (b) and (c) are effective since the related operators are subject to Coburn's theorem (see 2.2): The operators in (b) and (c) can be identified with the Toeplitz operators $T(b)$ and $T(\tilde{a})$ $(\tilde{a}(t) := a(1/t))$ on $l_{Z+}^p(0)$ and with the singular integral operator

$$((I-S_{[-1,1]})/2)(a(t+0)\chi_{[0,1]}+b(t+0)\chi_{[-1,0]})I+$$
$$+((I+S_{[-1,1]})/2)(a(t-0)\chi_{[0,1]}+b(t-0)\chi_{[-1,0]})I \qquad (1)$$

acting on $L_{[-1,1]}^p(\alpha)$, respectively, and these operators are invertible if and only if they are Fredholm and of index 0. For Toeplitz operators this means that the origin does not lie on the curve explained in Example 2.2 (then $T(a)$ is Fredholm) and that the negative winding number of this curve around the origin is 0 (then $T(a)$ has index 0, compare 2.5.3). The invertibility of the singular integral operator (1) is equivalent to the fact that the point 0 lies outside the curved triangle with the points $1, \hat{a} := \frac{a(t+0)}{a(t-0)}$ and $\hat{b} := \frac{b(t+0)}{b(t-0)}$ as its corner points and with the curves

$$\left\{\frac{1+\coth(z+i/p)\pi}{2}\hat{a}+\frac{1-\coth(z+i/p)\pi}{2}\ ,\ z\in\overline{\mathbf{R}}\right\},$$

$$\left\{\frac{1+\coth(z+i/p)\pi}{2}+\frac{1-\coth(z+i/p)\pi}{2}\hat{b}\ ,\ z\in\overline{\mathbf{R}}\right\},$$

and

$$\left\{\frac{1+\coth(z+i(\alpha+1/p))\pi}{2}\hat{b}+\frac{1-\coth(z+i(\alpha+1/p))\pi}{2}\hat{a}\ ,\ z\in\overline{\mathbf{R}}\right\},$$

as its sides. In case $p = 2$ and $\alpha = 0$ this criterium is usually referred to as *Verbitski's condition*:

The point 0 does not belong to the closed convex hull of the points $1, \hat{a}$ and $\hat{b}$.

Observe that this condition is automatically satisfied if both a and b are continuous at t.

Example 4.2 *Let $a_\pm, b_\pm$ be complex numbers, $c \in PC_{p,\alpha}$, and K be compact. The finite section method applies to the operator $A = (a_+ P + b_+ Q) + (a_- P + b_- Q) T^0(c) + K$ if and only if*

(a) *the operator A is invertible on $l^p_Z(\alpha)$,*

(b) *the operators $PT^0(b_+ + b_- c)P + Q$ and $QT^0(a_+ + a_- c)Q + P$ are invertible on $l^p_Z(0)$,*

(c) *the operators*

$$\chi_{R\setminus[-1,1]} I + \chi_{[-1,1]}((a_+ \chi_{R^+} + b_+ \chi_{R^-}) +$$
$$+ (a_- \chi_{R^+} + b_- \chi_{R^-})(c(t+0)\frac{I - S_R}{2} + c(t-0)\frac{I + S_R}{2}))\chi_{[-1,1]} I$$

are invertible in $L(L^p_R(\alpha))$ for all $t \in \mathbf{T}$.

The proof is immediate from Theorem 3.5 and Proposition 3.13 and 3.14.

The conditions under (b) and (c) are effective again, since the related operators can be identified with usual Toeplitz operators or with singular integral operators. For example, the operator in (c) is invertible if and only if the point 0 does not belong to the curved triangle having $1, \hat{a} := \frac{a_+ + a_- c(t-0)}{a_+ + a_- c(t+0)}$ and $\hat{b} := \frac{b_+ + b_- c(t-0)}{b_+ + b_- c(t+0)}$ as its corners and the curves

$$\left\{ \frac{1 + \coth(z + i/p)\pi}{2} + \frac{1 - \coth(z + i/p)\pi}{2}\hat{a} \, , \, z \in \overline{\mathbf{R}} \right\},$$

$$\left\{ \frac{1 + \coth(z + i/p)\pi}{2}\hat{b} + \frac{1 - \coth(z + i/p)\pi}{2} \, , \, z \in \overline{\mathbf{R}} \right\},$$

and

$$\left\{ \frac{1 + \coth(z + i(\alpha + 1/p))\pi}{2}\hat{a} + \frac{1 - \coth(z + i(\alpha + 1/p))\pi}{2}\hat{b} \, , \, z \in \overline{\mathbf{R}} \right\},$$

as its sides.

Example 4.3 *Let $a \in PC_{p,\alpha}, b \in \mathcal{N}$, and K be compact. The finite section method applies to the operator $A = PT^0(a)P + PG(b)P + Q + K$ if and only if*

(a) *the operator A is invertible in $L(l^p_Z(\alpha))$,*

(b) *the operator $P + QT^0(a)Q$ is invertible in $L(l^p_Z(0))$,*

(c) *the operators*

$$\chi_{R\backslash[0,1]}I+$$

$$+\chi_{[0,1]}\left(a(1+0)\frac{I-S_R}{2}+a(1-0)\frac{I+S_R}{2}+M^0(b)\right)\chi_{[0,1]}I$$

and

$$\chi_{R\backslash[0,1]}I+$$

$$+\chi_{[0,1]}\left(a(t+0)\frac{I-S_R}{2}+a(t-0)\frac{I+S_R}{2}\right)\chi_{[0,1]}I$$

are invertible in $L(L_R^p(\alpha))$ for all $t\in\mathbf{T}\backslash\{1\}$.

For a proof we need the strong limits $W_s(E_n^F G(b)E_{-n}^F)$ and $W^t(E_n^F G(b)E_{-n}^F)$ for $t\in\mathbf{T}$ and $s\in\{-1,0,1\}$. For completeness we compute these limits here for arbitrary prebases F.

Proposition 4.2 *Let $F=\{\phi_0,\dots,\phi_{k-1}\}$ be a canonical prebasis, $b\in\mathcal{N}$ and*

$$B:=\left(\alpha_r^F\int_R\phi_s(x)\,dx\,G(b)\right)_{r,s=0}^{k-1}.$$

Then

$$W^t(B,y)=\begin{cases}\Delta U_{-y}\Delta^{-1}M^0(b)\Delta U_y\Delta^{-1} & \text{if}\quad t=1\\ 0 & \text{if}\quad t\neq 1\end{cases}$$

and

$$W_s(B,y)=\begin{cases}B & \text{if}\quad s=y=0\\ \Lambda^{-1}B\Lambda & \text{if}\quad s=y\neq 0\\ 0 & \text{if}\quad s\neq y\end{cases}.$$

Proof   We start with the limits $W^t(B,y)$ in case $t=1$. Then, by 3.5.1(16),

$$W^1(B,y)=\Delta U_y\Delta^{-1}(\operatorname*{s-lim}_{n\to\infty}E_n^F BE_{-n}^F L_n^{F,F})\Delta U_{-y}\Delta^{-1}.$$

Let $K=B-E_{-n}^F L_n^{F,F}M^0(b)E_n^F$. This operator is compact by Proposition 2.21(ii). Thus, via Proposition 3.13(5),

$$\operatorname*{s-lim}_{n\to\infty}E_n^F BE_{-n}^F L_n^{F,F}$$

$$=\operatorname*{s-lim}_{n\to\infty}E_n^F(E_{-n}^F L_n^{F,F}M^0(b)E_n^F)E_{-n}^F L_n^{F,F}+\operatorname*{s-lim}_{n\to\infty}E_n^F KE_{-n}^F L_n^{F,F}$$

$$=\operatorname*{s-lim}_{n\to\infty}L_n^{F,F}M^0(b)L_n^{F,F}=M^0(b).$$

In case $t\neq 1$ we conclude from 3.5.1(8)

$$W^t(B,y)=W^1(Y_{t^{-1}}BY_t,y).$$

Choose a continuous function f with finite total variation which is identically 1 in a neighbourhood of 1 and which vanishes outside a sufficiently small neighbourhood U of 1. The diagonal operator $\mathrm{diag}\,(T^0(f),\dots,T^0(f))$ will be denoted by $T^0(f)$ again. Then

$$
\begin{aligned}
W^1(Y_{t-1}BY_t,y) &= W^1(Y_{t-1}BY_tT^0(f),y)+W^1(Y_{t-1}BY_tT^0(1-f),y)\\
&= W^1(Y_{t-1}BY_tT^0(f),y)+W^1(Y_{t-1}BY_t,y)\,W^1(T^0(1-f),y)
\end{aligned}
$$

and this is the zero operator since $Y_{t-1}BY_tT^0(f)$ is compact whenever $t \notin U$ by Propositions 2.11(d)) and 3.13(5), and $W^1(T^0(1-f),y)=0$ by Proposition 3.13(7).

Now consider the strong limits $W_s(B,y)$. If $s=y$ then, clearly,

$$
W_s(B,y)=\begin{cases} B & \text{if } s=y=0 \\ \Lambda^{-1}B\Lambda & \text{if } s=y\neq 0 \end{cases}.
$$

Let $s\neq y$ and suppose $s=0$ for definiteness. The proof for $s\neq 0$ is analogous. If $y>0$ then

$$
\begin{aligned}
W_0(B,y) &= W_0(B\,\mathrm{diag}\,(P,\dots,P),y)\\
&= W_0(B,y)\,W_0(\mathrm{diag}\,(P,\dots,P),y)\ =\ 0
\end{aligned}
$$

by Proposition 3.14(23).

If $y<0$ then choose a continuous function f on $\mathbf{R}$ which takes the value 1 at $-y$ and which vanishes outside a small neighbourhood of $-y$ being contained in the positive semi-axis and seperated from 0. Taking into account the compactness of the operator $B-E^F_{-n}L^{F,F}_nM^0(b)E^F_n$ in combination with Proposition 3.14(24) we conclude that

$$
W_0(B,y)\ =\ W_0(E^F_{-n}L^{F,F}_nM^0(b)E^F_n,y)
$$

which on its hand is equal to

$$
W_0(E^F_{-n}L^{F,F}_nM^0(b)fE^F_n,y)+W_0(E^F_{-n}L^{F,F}_nM^0(b)(1-f)E^F_n,y).\tag{2}
$$

The first item can be written as

$$
\Lambda V_{-\{r\}}\big(\ \underset{n\to\infty}{\text{s-lim}}\ V_{\{yn+r\}}\Lambda^{-1}E^F_{-n}L^{F,F}_nM^0(b)fE^F_n\Lambda V_{-\{yn+r\}}\big)V_{\{r\}}\Lambda^{-1}.\tag{3}
$$

Since $M^0(b)fI$ is a compact operator (see the end of 3.2), and since

$$
n^{1/p+\alpha}E^F_n\Lambda V_{\{yn+r\}}\to 0\quad\text{weakly}
$$

(Exercise 3.15) we get that (3) is the zero operator. Further, using the commutator relation (Theorem 2.8), we see that the second item of (2) equals

$$
W_0(E^F_{-n}L^{F,F}_nM^0(b)L^{F,F}_n(1-f)E^F_n,y)=
$$

$$
=\Lambda V_{-\{r\}}\big(\ \underset{n\to\infty}{\text{s-lim}}\ V_{\{yn+r\}}\Lambda^{-1}E^F_{-n}L^{F,F}_nM^0(b)L^{F,F}_n(1-f)E^F_n\Lambda V_{-\{yn+r\}}\big)V_{\{r\}}\Lambda^{-1},
$$

and now write

$$V_{\{yn+r\}}\Lambda^{-1}E_{-n}^F L_n^{F,F} M^0(b) L_n^{F,F}(1-f) E_n^F \Lambda V_{-\{yn+r\}}$$
$$= (V_{\{yn+r\}}\Lambda^{-1}E_{-n}^F L_n^{F,F} M^0(b) E_n^F \Lambda V_{-\{yn+r\}}) \times$$
$$\times (V_{\{yn+r\}}\Lambda^{-1}E_{-n}^F L_n^{F,F}(1-f) E_n^F \Lambda V_{-\{yn+r\}}) \ .$$

The first of these two factors is uniformly bounded with respect to n, whereas the second one converges strongly to $(1-f)(-y) = 0$ as we have already seen in Proposition 3.14. This shows that (2) vanishes. ■

Let us come back now to the assertion of Example 4.3. By the preceding proposition, we have with $A_n := R_n A R_n + Q_n$:

$$W_{-1}(A_n) = I\ ,\ W_0(A_n) = A\ ,\ W_1(A_n) = P + QT^0(a)Q$$

which explains conditions (a) and (b). Further, the operators under (c) are just $W^1(A_n)$ and $W^t(A_n)$ for $t \neq 1$. ■

The conditions in (b) and (c) are effective again. Indeed, for the Toeplitz operator in (b) and the singular integral operator $W^t(A_n), t \neq 1$ in (c), this has already been explained. The operator $W^1(A_n)$ can be identified with the operator

$$\frac{a(1+0) + a(1-0)}{2}\chi_{[0,1]}I - \frac{a(1+0) - a(1-0)}{2}S_{[0,1]} + \chi_{[0,1]}M^0(b)\chi_{[0,1]}I \ ,$$

thought of as acting on $L_{[0,1]}^p(\alpha)$, which is a Mellin operator with Mellin symbol

$$c(z) = \frac{a(1+0) + a(1-0)}{2} -$$
$$- \frac{a(1+0) - a(1-0)}{2}\coth\pi(z + i(1/p + \alpha)) + b(z) \ . \tag{4}$$

This operator is invertible on $L_{[0,1]}^p(\alpha)$ if and only if the operator

$$\Delta^{-1}\chi_{[0,1]}M^0(c)\chi_{[0,1]}\Delta \tag{5}$$

is invertible on $L_{[0,1]}^p(0)$. The operator (5) is again a Mellin operator, and its Mellin symbol is exactly given by (4) (this is not surprising since our definition of the Mellin transform in Section 2.1.2 depends on the weight explicitly).

Thus, $W^1(A_n)$ is invertible if and only if $\chi_{[0,1]}M^0(c)\chi_{[0,1]}I$ is invertible on $L_{[0,1]}^p(0)$. Now define $X : L_{[0,1]}^p(0) \to L_{R+}^p(0)$ by $(Xf)(t) = e^{-t/p}f(e^{-t})$. A straightforward computation shows that $X\chi_{[0,1]}M^0(c)\chi_{[0,1]}X^{-1}$ is just a Wiener-Hopf operator on $L_{R+}^p(0)$ with piecewise continuous generating function c. This operator is invertible if and only if it is Fredholm (which can be checked via a symbol calculus, see ([BS 2], Theorem 9.17)) and if its index is zero (that is, the winding number of the symbol curve is zero).

Let us finally remark that in case $b = 0$ the condition (c) is a consequence of the invertibility of the operators A and $P + QT^0(a)Q$ which can be verified by means of symbol calculus for Toeplitz and singular integral operators and of index formula.

4.1.3 Finite sections of operators in $(\mathcal{T}^p_{1,1,T}(\alpha))_{k \times k}$

Let A be an operator in $(\mathcal{T}^p_{1,1,T}(\alpha))_{k \times k}$, and abbreviate the $k \times k$ diagonal operator $\text{diag}\,(R_n, \dots, R_n)$ by R_n again. The definition of the applicability of the finite section method $R_n A R_n$ to the operator A is verbatim the same as in case $k = 1$ in Section 4.1.1. Since the general stability theorem holds in the matrix case, too (see 3.4.4), Theorem 4.1 also remains valid for matrix operators, and the assertions of Examples 4.1-4.3 in Section 4.1.2 can be maintained. The only thing that gets lost is the simplicity of the verification of some conditions since Coburn's theorem fails for matrix operators in general. Spitkovski and Tashbaev succeeded in deriving verifiable conditions for the finite section method for the operator $T^0(a)P + T^0(b)Q$, where a and b are piecewise continuous 2×2-matrix functions (see [ST]). To formulate their results, let A_t and B_t $(t \in \mathbf{T})$ refer to the 2×2-matrices

$$A_t = a(t+0)a(t-0)^{-1} \quad \text{and} \quad B_t = b(t+0)b(t-0)^{-1} , \tag{1}$$

and denote the argument of the k th eigenvalue of the matrices A_t, $A_t^{-1}B_t$, and B_t^{-1} by $\alpha^{(t)}_{0k}$, $\alpha^{(t)}_{1k}$, and $\alpha^{(t)}_{2k}$, respectively. Here we suppose that $\alpha^{(t)}_{jk} \in (-\pi, \pi]$. In case the matrices A_t and B_t have common eigenvectors, let us agree upon attaching the same subscript k to the alphas associated with the corresponding eigenvalues. If the matrices A_t and B_t share (up to linear dependence) exactly one common eigenvector, we label the corresponding alpha by the subscript $k = 2$. Finally set

$$l^{(t)}_k = \frac{1}{2\pi}(\alpha^{(t)}_{0k} + \alpha^{(t)}_{1k} + \alpha^{(t)}_{2k}) , \; k \in \{1,2\} . \tag{2}$$

Now the announced result can be restated as follows:

Theorem 4.2 *Let a and b be piecewise continuous 2×2-matrix functions on $\mathbf{T}$. Then the finite section method $(R_n C R_n + Q_n)$ is stable for the operator $C = T^0(a)P + T^0(b)Q \in L(l^2_2(0))$ if and only if*

(a) *the matrix functions a, $\tilde{b}$, and $a^{-1}b$ are Φ-factorizable in L^2_T and all their partial indices vanish.*

(b) *the matrices A_t and B_t defined in (1), and the numbers $l^{(t)}_k$ defined by (2) satisfy for all t at least one of the conditions*

(i) *A_t and B_t have no common eigenvectors, and $l^{(t)}_1 = -l^{(t)}_2$.*

(ii) A_t and B_t do not commute, they possess a common eigenvector, and $l_1^{(t)} = -l_2^{(t)} \geq 0$.

(iii) A_t and B_t commute, and $l_1^{(t)} = l_2^{(t)} = 0$.

For becoming acquainted with matrix factorization theory we recommend [CG] and [LS].

In the following section a further class of matrix operators which lead to verifiable conditions for the stability of the finite section method will be pointed out.

4.1.4   Finite sections of operators in $\mathcal{T}_{k,\kappa,\Omega}^p(\alpha)$

Let A belong to the algebra $\mathcal{T}_{k,\kappa,\Omega}^p(\alpha)$ introduced in Section 3.3.1 and let R_n be, as in the preceding section, the $k \times k$ diagonal matrix $\operatorname{diag}(R_n, \ldots, R_n)$. Further we let F denote any k-elementic canonical prebasis with

$$\alpha^F = (\underbrace{1, 1, \ldots, 1}_{\kappa}, \underbrace{0, \ldots, 0}_{k-\kappa}) \,.$$

Proposition 4.3 *The sequence $(E_n^F(R_n A R_n + Q_n)E_{-n}^F)$ belongs to $\mathcal{A}_{\Omega,r}^F$ for all $r \in \mathbf{R}$.*

Proof   If $B \in \mathcal{T}_{k,\kappa,\Omega}^p(\alpha)$ then the sequence

$$(E_n^F \Lambda V_{\{tn+r\}} \Lambda^{-1} B \Lambda V_{-\{tn+r\}} \Lambda^{-1} E_{-n}^F) \tag{1}$$

belongs to $\mathcal{A}_{\Omega,r}^F$ for all $t, r \in \mathbf{R}$ by definition. First we choose $t = 0$ and $B = \Lambda V_{-\{r\}} \Lambda^{-1} A \Lambda V_{\{r\}} \Lambda^{-1}$. This operator is in $\mathcal{T}_{k,\kappa,\Omega}^p(\alpha)$ since $\Lambda V_{\pm\{r\}} \Lambda^{-1} - V_{\pm\{r\}}$ is compact, and since each diagonal operator $\operatorname{diag}(V_{\pm\{r\}}, \ldots, V_{\pm\{r\}})$ and each compact operator belong to $\mathcal{T}_{k,\kappa,\Omega}^p(\alpha)$. This specification shows that the sequence $(E_n^F A E_{-n}^F)$ is in $\mathcal{A}_{\Omega,r}^F$.

We claim that the sequence $(E_n^F \Lambda V_{-n} \Lambda^{-1} P \Lambda V_n \Lambda^{-1} E_{-n}^F)$ also belongs to $\mathcal{A}_{\Omega,r}^F$. For, we choose $t = -1$ and $B = \Lambda V_{-\{r\}} \Lambda^{-1} P \Lambda V_{\{r\}} \Lambda^{-1}$. Then (1) goes over into

$$(E_n^F \Lambda V_{\{-n+r\}} V_{-\{r\}} \Lambda^{-1} P \Lambda V_{\{r\}} V_{-\{-n+r\}} \Lambda^{-1} E_{-n}^F) \,,$$

and since

$$V_{\{-n+r\}} V_{-\{r\}} = V_{-n+\{r\}} V_{-\{r\}} = V_{-n} \quad \text{and} \quad V_{\{r\}} V_{-\{-n+r\}} = V_{\{r\}} V_{n-\{r\}} = V_n$$

this proves our claim. Analogously, $(E_n^F \Lambda V_n \Lambda^{-1} P \Lambda V_{-n} \Lambda^{-1} E_{-n}^F) \in \mathcal{A}_{\Omega,r}^F$ and, hence, by 4.1.1(3), $(E_n^F R_n E_{-n}^F) \in \mathcal{A}_{\Omega,r}^F$ ∎.

The following theorem can be derived in the very same way as Theorem 4.1.

Theorem 4.3 *The finite section method*

$$R_n A R_n \, u^{(n)} \; = \; R_n f \tag{2}$$

applies to the operator $A \in T^p_{k,\kappa,\Omega}(\alpha)$ *if and only if the operator* A *is invertible and if the operators* $W^t(E^F_n(R_n A R_n + Q_n)E^F_{-n})$ *and* $W_s(E^F_n(R_n A R_n + Q_n)E^F_{-n})$ *are invertible for all* $t \in \Omega$ *and* $s \in \{-1,1\}$. *Herein,* W^t *and* W_s *are the strong limits associated with the algebra* $A^F_{\Omega,r}$.

Example 4.4 *Let* $a,b \in PC^{k,\kappa,\Omega}_{p,\alpha}$ *(see 3.2.6 for the definition of this subalgebra of* $(PC_{p,\alpha})_{k \times k}$*) and let* $K \in L(l^p_k(\alpha))$ *be compact. Then the finite section method* (2) *applies to the operator* $A = T^0(a)P + T^0(b)Q + K$ *if and only if*

(a) *the operator* A *is invertible in* $L(l^p_k(\alpha))$,

(b) *the operators* $PT^0(b)P + Q$ *and* $QT^0(a)Q + P$ *are invertible in* $L(l^p_k(0))$,

(c) *the operators*

$$\chi_{[-1,1]}\left(\frac{I - S_R}{2}(\alpha(t+0)\chi_{[0,1]} + \beta(t+0)\chi_{[-1,0]}) + \right.$$

$$\left. \frac{I + S_R}{2}(\alpha(t-0)\chi_{[0,1]} + \beta(t-0)\chi_{[-1,0]})\right)\chi_{[-1,1]}I + \chi_{R \setminus [-1,1]}I$$

with

$$\alpha(t \pm 0) = \sum_{s=0}^{k-1} \alpha^F_s a_{1s}(t \pm 0)\,, \;\; \beta(t \pm 0) = \sum_{s=0}^{k-1} \alpha^F_s b_{1s}(t \pm 0)\,, \; t \in \Omega\,,$$

are invertible in $L(L^p_R(\alpha))$ *for all* $t \in \Omega$.

The invertibility of the operators in (c) can be verified as in Example 4.1 in 4.1.2.

4.1.5 Kozak's identity

We let A be an arbitrary linear and bounded operator on $l^p_Z(\alpha)$ and consider the sequence $(R_n + Q_n A Q_n)$. This sequence converges strongly to the identity and can be thus thought of as an approximation method for the identity operator. This "infinite section method for the identity" is not as curious as it seems to be. Indeed, if A is invertible then there is a very close connection between this infinite section method and the usual finite section method for A^{-1} which is most strikingly expressed by *Kozak's identity*

$$(R_n A^{-1} R_n + Q_n)^{-1} = R_n A R_n + R_n A Q_n (R_n + Q_n A Q_n)^{-1} Q_n A R_n + Q_n$$

(compare Exercise 1.13). This identity can be read as follows: If the operators $R_n + Q_n A Q_n$ are invertible for all $n \geq n_0$, and if the norms of their inverses are uniformly bounded, then the operators $R_n A^{-1} R_n + Q_n$ are invertible for $n \geq n_0$, and the norms of their inverses are uniformly bounded, and conversely. In other words, we get the following constellation.

Proposition 4.4 *If A is invertible then the finite section method*

$$(R_n A R_n + Q_n)\, u^{(n)} \;=\; f$$

applies to A if and only if the infinite section method

$$(R_n + Q_n A^{-1} Q_n)\, u^{(n)} \;=\; f$$

applies to the identity operator.

4.1.6   Infinite sections for operators in Toeplitz algebras

In order to establish criteria for the infinite section method let us first note down the following analogue of Propositions 4.1 and 4.3.

Proposition 4.5 *Let* $A \in \mathcal{T}_{1,1,T}^{p}(\alpha),\, (\mathcal{T}_{1,1,T}^{p}(\alpha))_{k\times k}$, *or* $\mathcal{T}_{k,\kappa,\Omega}^{p}(\alpha)$. *Then the sequence* $(E_n^{F}(R_n + Q_n A Q_n)E_{-n}^{F})$ *is in* $\mathcal{A}_{T,0}^{\{\chi_{[0,1]}\}},\, (\mathcal{A}_{T,0}^{\{\chi_{[0,1]}\}})_{k\times k}$, *or* $\mathcal{A}_{\Omega,r}^{F}$, *respectively, where F and r are as in Section 4.1.4.*

Now the general stability theorem immediately gives

Theorem 4.4 *The infinite section method*

$$(R_n + Q_n A Q_n)\, u_n \;=\; f \tag{1}$$

with $A \in \mathcal{T}_{1,1,T}^{p}(\alpha),\, (\mathcal{T}_{1,1,T}^{p}(\alpha))_{k\times k}$, *or* $\mathcal{T}_{k,\kappa,\Omega}^{p}(\alpha)$ *applies to the identity operator if and only if the operators* $W^{t}(E_n^{F}(R_n + Q_n A Q_n)E_{-n}^{F})$ $(t \in \Omega)$ *and* $W_s(E_n^{F}(R_n + Q_n A Q_n)E_{-n}^{F})$ $(s \in \{-1,1\})$ *as well as the coset* $W_{\infty}(E_n^{F}(R_n + Q_n A Q_n)E_{-n}^{F}) = \pi(A)$ *are invertible.*

Of course, the strong limits have to be chosen in accordance with the underlying algebra.

Comparing this result with the corresponding Theorem 4.1 for the finite section method it turns out that most of the conditions (viz. the invertibility of the operators W^{t} and W_s for $t \in \Omega$ and $s \in \{-1,1\}$) prove to be of the same quality. The only significant difference is between invertibility of W_0 in Theorem 4.1 and of W_{∞} in Theorem 4.4: whereas in Theorem 4.1 *invertibility* of A is needed, now it suffices A to require to be *Fredholm*. This is just the crucial point which will be employed later on to exclude undetermined compact operators at singular points. Before doing this we add some examples.

4.1.7   Examples

Example 4.5 *Let* $a,b \in PC_{p,\alpha}$ *and K be compact. The infinite section method* 4.1.6(1) *applies with* $A = T^{0}(a)P + T^{0}(b)Q + K$ *if and only if*

 (a) *the operator* $T^{0}(a)P + T^{0}(b)Q$ *is Fredholm on* $l_{Z}^{p}(\alpha)$,

(b) the operators $QT^0(b)Q + P$ and $PT^0(a)P + Q$ are invertible on $l^p_Z(0)$,

(c) the operators

$$\chi_{R\setminus[-1,1]}\left(\frac{I - S_R}{2}(a(t+0)\chi_{[1,\infty)} + b(t+0)\chi_{(-\infty,-1]}) +\right.$$
$$\left. + \frac{I + S_R}{2}(a(t-0)\chi_{[1,\infty)} + b(t-0)\chi_{(-\infty,-1]})\right)\chi_{R\setminus[-1,1]}I + \chi_{[-1,1]}I$$

are invertible in $L(L^p_R(\alpha))$ for all $t \in \mathbf{T}$ where a or b are not continuous.

The assertion remains true in the matrix case but notice that, in scalar case, all conditions can be effectively checked. For (a) and (b) this can be done by Fredholm symbol calculus for Toeplitz operators (confer Theorems 2.2 and 2.3) whereas Coburn's theorem (see 2.2) in combination with the Fredholm symbol calculus for singular integral operators (see Theorem 3.1) states that invertibility of the operators in (c) is equivalent to the fact that the point 0 lies outside the curved triangle with the points $1, \hat{a} := \frac{a(t+0)}{a(t-0)}$ and $\hat{b} := \frac{b(t+0)}{b(t-0)}$ as its corners and with the circular arcs

$$\left\{\frac{1 + \coth(z + i/p)\pi}{2} + \frac{1 - \coth(z + i/p)\pi}{2}\hat{a} , z \in \overline{\mathbf{R}}\right\},$$

$$\left\{\frac{1 + \coth(z + i/p)\pi}{2}\hat{b} + \frac{1 - \coth(z + i/p)\pi}{2} , z \in \overline{\mathbf{R}}\right\},$$

$$\left\{\frac{1 + \coth(z + i(\alpha + 1/p))\pi}{2}\hat{b} + \frac{1 - \coth(z + i(\alpha + 1/p))\pi}{2}\hat{a} , z \in \overline{\mathbf{R}}\right\}$$

as its sides. In case $p = 2$ and $\alpha = 0$ this is just Verbitski's condition.

Example 4.6 Let $a_\pm, b_\pm$ be complex numbers, $c \in PC_{p,\alpha}$, and K be compact. The infinite section method 4.1.6(1) applies with $A = (a_+P + b_+Q) + (a_-P + b_-Q)T^0(c) + K$ if and only if

(a) the operator $(a_+P + b_+Q) + (a_-P + b_-Q)T^0(c)$ is Fredholm in $L(l^p_Z(\alpha))$,

(b) the operators $QT^0(b_+ + b_-c)Q + P$ and $PT^0(a_+ + a_-c)P + Q$ are invertible in $L(l^p_Z(0))$,

(c) the operators

$$\chi_{[-1,1]}I + \chi_{R\setminus[-1,1]}\left((a_+\chi_{R^+} + b_+\chi_{R^-})I +\right.$$
$$\left. + (a_-\chi_{R^+} + b_-\chi_{R^-})(c(t+0)\frac{I - S_R}{2} + c(t-0)\frac{I + S_R}{2})\right)\chi_{R\setminus[-1,1]}I$$

are invertible in $L(L^p_R(\alpha))$ for all $t \in \mathbf{T}$ where c is not continuous.

Here we have again brought the assertion into a form which remains valid in the matrix case. In the scalar case, all conditions are effectively verifiable: For instance, the invertibility of the operator in (c) is equivalent to the fact that the origin 0 is located outside the curved triangle with $1, \hat{a} := \frac{a_+ + a_- c(t-0)}{a_+ + a_- c(t+0)}$ and $\hat{b} := \frac{b_+ + b_- c(t-0)}{b_+ + b_- c(t+0)}$ as its corners and the curves

$$\left\{ \frac{1 + \coth(z + i/p)\pi}{2} \hat{a} + \frac{1 - \coth(z + i/p)\pi}{2} , z \in \overline{\mathbf{R}} \right\} ,$$

$$\left\{ \frac{1 + \coth(z + i/p)\pi}{2} + \frac{1 - \coth(z + i/p)\pi}{2} \hat{b} , z \in \overline{\mathbf{R}} \right\} ,$$

$$\left\{ \frac{1 + \coth(z + i(\alpha + 1/p))\pi}{2} \hat{a} + \frac{1 - \coth(z + i(\alpha + 1/p))\pi}{2} \hat{b} , z \in \overline{\mathbf{R}} \right\}$$

as its sides.

Example 4.7 *Let* $a \in PC_{p,\alpha}, b \in \mathcal{N}$, *and* K *be compact. The infinite section method 4.1.6(1) applies with* $A = PT^0(a)P + PG(b)P + Q + K$ *if and only if*

(a) *the operator* $PT^0(a)P + PG(b)P + Q$ *is Fredholm in* $L(l^p_Z(\alpha))$,

(b) *the operator* $Q + PT^0(a)P$ *is invertible in* $L(l^p_Z(0))$,

(c) *the operators*

$$\chi_{R\setminus[1,\infty)}I+$$

$$+\chi_{[1,\infty)} \left(a(1+0)\frac{I - S_R}{2} + a(1-0)\frac{I + S_R}{2} + M^0(b) \right) \chi_{[1,\infty)}I$$

and

$$\chi_{R\setminus[1,\infty)}I + \chi_{[1,\infty)} \left(a(t+0)\frac{I - S_R}{2} + a(t-0)\frac{I + S_R}{2} \right) \chi_{[1,\infty)}I$$

are invertible in $L(L^p_R(\alpha))$ *for all* $t \in \mathbf{T} \setminus \{1\}$ *where* a *is not continuous.*

These conditions are effective in the scalar case again (compare the discussion following Proposition 4.2).

4.2 Multiindiced approximation methods

4.2.1 A model situation

Let us consider the singular integral operator

$$A = (a_+ \chi_{R^+} + b_+ \chi_{R^-})I + (a_- \chi_{R^+} + b_- \chi_{R^-})S_R , \quad a_\pm, b_\pm \in \mathbf{C} ,$$

on the Hilbert space $L_R^2(0)$. If $F = \{\phi\}$ is a canonical prebasis then the Galerkin method

$$L_n^{F,F} A u_n \;=\; L_n^{F,F} f \;,\; u_n \in S_n^F \;,$$

applies to the equation $A\,u \;=\; f$ if and only if the operator

$$W_0(L_n^{F,F} A | S_n^F) = (a_+ P + b_+ Q) + (a_- P + b_- Q) T^0(\rho^{F,F}) + K \;, \tag{1}$$

with $\rho^{F,F} = (\lambda^{F,F})^{-1} \sigma^{F,F}$ and K being a certain compact operator, is invertible. Indeed, this can be derived without great effort from the general stability theorem or, which is even easier, from the observation that the operators $E_{-n}^F L_n^{F,F} A E_n^F \in L(l_Z^2(0))$ are independent of n and, therefore, coinciding with (1) (see also Exercise 2.22).

The question of invertibility of the operator (1) is a serious problem in case $K \neq 0$. On the other hand we know from Example 4.6 in 4.1.7 that the infinite section method

$$R_i + Q_i W_0(L_n^{F,F} A | S_n^F) Q_i \, u_i \;=\; f \tag{2}$$

applies if and only if

(i) the operator (1) is Fredholm on $l_Z^2(0)$,

(ii) the operators $QT^0(b_+ + b_- \rho^{F,F})Q + P$ and $PT^0(a_+ + a_- \rho^{F,F})P + Q$ are invertible on $l_Z^2(0)$,

(iii) the point 0 lies outside the (rectilinear) triangle with the points $1, \hat{a} = \frac{a_+ + a_-}{a_+ - a_-}$ and $\hat{b} = \frac{b_+ + b_-}{b_+ - b_-}$ as its corners.

Note that in (2) and in what follows , i refers to the parameter controlling the size of infinite or finite sections whereas n is reserved for counting the refinement of the spline spaces. The latter result can be reinterpreted as follows: If (i)-(iii) are satisfied then there is an i_0 such that the operators

$$R_i + Q_i W_0(L_n^{F,F} A | S_n^F) Q_i$$

are invertible for all $i \geq i_0$ and that the norms of their inverses are uniformly bounded. Using the identity

$$W_0(L_n^{F,F} A | S_n^F) \;=\; E_{-n}^F L_n^{F,F} A E_n^F$$

we can translate all things back to spline approximation sequences and find:

Proposition 4.6 *If (i)-(iii) are fulfilled then there is an i_0 such that the operators*

$$A_{n,i} := E_n^F R_i E_{-n}^F + E_n^F Q_i E_{-n}^F L_n^{F,F} A E_n^F Q_i E_{-n}^F \;:\; S_n^F \to S_n^F$$

are invertible for all $n \geq 1$ and $i \geq i_0$, and such that

$$\sup_{n \geq 1, i \geq i_0} \; \|A_{n,i}^{-1}\| < \infty \;.$$

Thus, if we fix an $i \geq i_0$, then the sequence $(A_{n,i})_{n \geq 1}$ is stable and, moreover, it can be checked that $A_{n,i} L_n^{F,F} \to A$ strongly as $n \to \infty$ for each fixed i. In other words, the modified Galerkin method

$$A_{n,i} u_{n,i} = L_n^{F,F} f$$

applies to the operator A if and only if (i)-(iii) are in force, and these conditions are well verifiable. The difficulties caused by the (analytic) singularity at 0 are cutted off by the projections Q_i. Now we are going to study this cut-off-technique for arbitrary sequences in $\mathcal{A}_{\Omega,r}^F$. Let us start with a brief introduction of multiindiced approximation methods.

4.2.2 Multiindiced operator sequences

Define a relation $\leq$ on the Cartesian product $\mathbf{Z}_+^k = \mathbf{Z}_+ \times \ldots \times \mathbf{Z}_+$ by

$$(x_1, \ldots, x_k) \leq (y_1, \ldots, y_k) \quad \text{if } x_i \leq y_i \text{ for } i = 1, \ldots, k .$$

The relation $\leq$ is a partial order on $\mathbf{Z}_+^k$, that is, it is reflexive, transitive, and anti-symmetric. If T is a subset of $\mathbf{Z}_+^k$ then T is also partially ordered by $\leq$. Here and hereafter we suppose moreover T to be a *directed* subset of $\mathbf{Z}_+^k$, i.e. given $t_1, t_2 \in T$ there exists a $t_3 \in T$ such that $t_1 \leq t_3$, $t_2 \leq t_3$, and $t_1 \neq t_3$, $t_2 \neq t_3$.

As usual, a (generalized) sequence $(x_t)_{t \in T}$ of elements of a Banach space X is said to *converge* to $x \in X$ as $t \to \infty$ if, for arbitrarily given $\epsilon > 0$, there is a $t_0 \in T$ such that $\|x_t - x\| < \epsilon$ whenever $t \geq t_0$. Further, a sequence $(A_t)_{t \in T}$ of operators $A_t \in L(X)$ is called *uniformly convergent* to $A \in L(X)$ if, given $\epsilon > 0$, there is a $t_0 \in T$ such that $\|A_t - A\| < \epsilon$ for all $t \geq t_0$. Finally, the sequence $(A_t)_{t \in T}$ *converges strongly* to A if the sequence $(A_{t(n)})_{n \in \mathbf{Z}_+}$ converges strongly to A (in the sense of 1.2.1) for *each* strongly monotonically increasing sequence $t : \mathbf{Z}_+ \to T$. In case $k = 1$ and $T = \mathbf{Z}_+$ these notations coincide with those from Section 1.2.1 (take the sequence $t(n) = n$ to see this for strong convergence). But if $k > 1$ then some care is in order. For example, let $T = \mathbf{Z}_+^2$ and consider the operator sequence $(a_{n,i} I)_{n,i \in Z_+}$ with

$$a_{n,i} = \frac{1}{\min(n+1, i+1)} .$$

This sequence converges uniformly to zero, but it does not converge strongly. Moreover, a slight modification of this example (set $a_{n,1} = n$) shows that uniformly convergent sequences are not necessarily uniformly bounded. Nevertheless, some important technical ingredients remain valid:

Proposition 4.7 *(a) Let $A_t \to A$ and $B_t^* \to B^*$ strongly as $t \to \infty$ and let K be compact. Then $A_t K B_t \to AKB$ uniformly.*
(b) If $\|A_t x - A x\| \to 0$ for all x belonging to a dense subset of X, and if $\sup_t \|A_t\| < \infty$, then $A_t \to A$ strongly.
(c) If $(A_t x)_{t \in T}$ is a bounded and convergent sequence for each $x \in X$ then

$\sup_t \|A_t\| < \infty$, *the operator defined by* $A_t x \to Ax$ *is bounded, and* $\|A\| \leq \sup_t \|A_t\|$ *(the Banach-Steinhaus theorem)*.

Proof (a) Suppose that $A_t K B_t \not\to AKB$ uniformly. Then there is an $\epsilon > 0$ and a strongly monotonically increasing sequence $t_1 < t_2 < \ldots$ in T such that

$$\|A_{t_n} K B_{t_n} - AKB\| > \epsilon \quad , \ n = 1, 2, \ldots . \tag{1}$$

On the other hand, the sequences $(A_{t_n})_{n \in Z^+}$ and $(B_{t_n}^*)_{n \in Z^+}$ converge strongly to A and B^*, respectively, and thus, by 1.2.1(b), the sequence $(A_{t_n} K B_{t_n})_{n \in Z^+}$ converges uniformly to AKB. This contradicts (1).

(b) Assume that $A_t \not\to A$ strongly. Then there is an $\epsilon > 0$, a monotonically increasing sequence $t_1 < t_2 < \ldots$ in T, and a $y \in X$ such that

$$\|A_{t_n} y - Ay\| > \epsilon \quad , \ n = 1, 2, \ldots .$$

Choose x in the dense subset figured in the formulation of (b) such that

$$\|x - y\| \leq \frac{\epsilon}{2 \sup_t \|A_t - A\|} .$$

Then

$$\begin{aligned} \|A_{t_n} x - Ax\| &\geq \|A_{t_n} y - Ay\| - \|A_{t_n}(y - x) - A(y - x)\| \\ &\geq \|A_{t_n} y - Ay\| - \sup_n \|A_{t_n} - A\| \, \|y - x\| \geq \epsilon/2 , \end{aligned}$$

which contradicts the convergence of $A_t x$ to Ax.

(c) If all sequences $(A_t x)_{t \in T}$ are bounded then the uniform boundedness principle (cp. [ReS], Vol.I, Chap. III.5, Theorem III.9) entails that $\sup_t \|A_t\| < \infty$. The operator A defined by $A_t x \to Ax$ for all $x \in X$ is clearly linear. For showing its boundedness we pick, for every given $\epsilon > 0$, a point $t_\epsilon \in T$ such that

$$\|A_t x - Ax\| < \epsilon \quad \text{if } t \geq t_\epsilon .$$

Hence, for $t \geq t_\epsilon$,

$$\|Ax\| \leq \|A_t x\| + \epsilon \leq \|A_t\| \, \|x\| + \epsilon \leq \sup_t \|A_t\| \, \|x\| + \epsilon .$$

Letting ϵ go to zero we get $\|Ax\| \leq \sup_t \|A_t\| \, \|x\|$ which yields the assertion. ■

4.2.3 Multiindiced approximation sequences

Let T be as in 4.2.2 and assume we are given a generalized sequence $(A_t)_{t \in T}$ of bounded linear operators on a Banach space X. We say that the approximation method

$$A_t x_t = y \tag{1}$$

applies to the operator $A \in L(X)$ if there is a $t_0 \in T$ such that the equations (1) possess unique solutions x_t for all $t \geq t_0$, if the sequence $(x_t)_{t \geq t_0}$ is bounded, and if the sequences $(x_{t(n)})_{n \in \mathbf{Z}^+}$ converge in the norm of X to a solution of the equation $Ax = y$ for each monotonically increasing sequence $t : \mathbf{Z}^+ \to \{t \in T : t \geq t_0\}$. The sequence (A_t) is called *stable* if there is a $t_0 \in T$ such that all operators A_t with $t \geq t_0$ are invertible and that

$$\sup_{t \geq t_0} \|A_t^{-1}\| < \infty .$$

The following result holds in analogy to Proposition 1.1.

Proposition 4.8 *Let* $A_t \to A$ *strongly. Then the approximation method* (1) *applies to the operator* A *if and only if this operator is invertible and if the sequence* $(A_t)_{t \in T}$ *is stable.*

Proof If (1) applies to A then, by definition, the operators A_t are invertible for all $t \geq t_0$, and the sequence $(A_t^{-1}y)_{t \geq t_0}$ is bounded for each $y \in X$. By the uniform boundedness principle we get

$$\sup_{t \geq t_0} \|A_t^{-1}\| < \infty ,$$

which involves stability of $(A_t)_{t \in T}$. For the invertibility of A, we choose a strongly monotonically increasing sequence $t : \mathbf{Z}^+ \to \{t \in T : t \geq t_0\}$ and consider the estimate

$$\|x - A_{t(n)}^{-1} Ax\| \leq \|A_{t(n)}^{-1}\| \, \|A_{t(n)}x - Ax\| \leq \sup_{t \geq t_0} \|A_t^{-1}\| \, \|A_{t(n)}x - Ax\| . \qquad (2)$$

Since $A_t \to A$ strongly, it shows that

$$\|x - A_{t(n)}^{-1} Ax\| \to 0 \quad \text{as } n \to \infty$$

for each $x \in X$. If, in particular, $x \in \operatorname{Ker} A$, then $x = 0$, that is, A has a trivial kernel. Since the image of A coincides with all of X by definition, the operator A is invertible.

Now let A be invertible and (A_t) be a stable sequence. Then the operators A_t are invertible for large t, say $t \geq t_0$, and the norms of their inverses are uniformly bounded. Consequently, the sequence $(x_t) = (A_t^{-1}y)$ is bounded for each y, and it remains to show the convergence of $(x_{t(n)})_{n \in \mathbf{Z}^+}$ to the solution x of $Ax = y$. But this is immediate from the estimate

$$\|A_{t(n)}^{-1}y - A^{-1}y\| = \|A_{t(n)}^{-1} Ax - x\| \leq \|A_{t(n)}^{-1}\| \, \|Ax - A_{t(n)}x\| \qquad (3)$$

in combination with the strong convergence $A_t \to A$ ∎.

The stability of a (generalized) operator sequence $(A_t)_{t \in T}$ is again equivalent to an invertibility problem in a suitably chosen Banach algebra. It is evident how this can be realized: one introduces the algebra $\mathcal{F}$ of all bounded sequences $(A_t)_{t \in T}$ and its subset $\mathcal{K}$ of all sequences $(K_t)_{t \in T}$ which converge uniformly to zero. Then $\mathcal{K}$ is a closed two-sided ideal of $\mathcal{F}$, and the sequence $(A_t)_{t \in T} \in \mathcal{F}$ is stable if and only if the coset $(A_t) + \mathcal{K}$ is invertible in $\mathcal{F}/\mathcal{K}$.

In the next sections we pursue these ideas for spline approximation methods cutting off singularities.

4.2.4 Spline projections which cut off singularities

Let F a k-elementic Y-prebasis, G a k-tupel of premeasures, and (F, G) a canonical pair. Then the projections $L_n^{F,G}$ are well-defined, and they converge strongly to the identity operator. The associated *spline projection which cuts off singularities* is the operator defined by

$$L_{n,i}^{F,G} = E_n^F Q_i E_{-n}^F L_n^{F,G} .$$

This operator is evidently a projection again and it maps Y onto a subspace of $S_n^F(Y)$. Next we investigate the strong convergence of the sequence $(L_{n,i}^{F,G})$ in sense of 4.2.2. It is easy to see that

$$I_Y E_1^F Q_i E_{-1}^F L_1^{F,G} \to 0 \quad \text{strongly} \quad \text{as} \quad i \to \infty$$

and

$$I_Y E_n^F Q_1 E_{-n}^F L_n^{F,G} \to I_Y \quad \text{strongly} \quad \text{as} \quad n \to \infty .$$

Thus, the sequence $(L_{n,i}^{F,G})_{n,i \in Z^+}$ does not converge strongly on all of $\mathbf{Z}_+^2$, but it is not too hard to find directed subsets of $\mathbf{Z}_+^2$ such that the sequence $(L_{n,i}^{F,G})_{n,i \in T}$ converges strongly (take, for example, $T = \mathbf{Z}_+ \times \{1\}$). We claim that there are even such directed subsets with the additional property that, given $(n, i) \in T$, there is a $(n', i') \in T$ with $n' > n$, $i' > i$. In order to see this we suppose that $\alpha^F = (\underbrace{1, 1, \ldots, 1}_{\kappa}, \underbrace{0, \ldots, 0}_{k-\kappa})^T$ with $\kappa \geq 1$ and introduce a new k-tupel $K = \{\psi_0, \ldots, \psi_{k-1}\}$ of presplines by

$$\psi_r = \chi_{\left[\frac{r}{\kappa}, \frac{r+1}{\kappa}\right)} \quad \text{if } r = 0, \ldots, \kappa - 1$$

and

$$\psi_r = \chi_{\left[\frac{r-\kappa}{\kappa(k-\kappa)}, \frac{r-\kappa+1/2}{\kappa(k-\kappa)}\right)} - \chi_{\left[\frac{r-\kappa+1/2}{\kappa(k-\kappa)}, \frac{r-\kappa+1}{\kappa(k-\kappa)}\right)} \quad \text{if } r = \kappa, \ldots, k - 1 .$$

The k-tupel K is even a canonical prebasis, and $\alpha^K = \alpha^F$. Now write

$$L_{n,i}^{F,G} = (E_n^F E_{-n}^K L_n^{K,K})(E_n^K Q_i E_{-n}^K L_n^{K,K})(E_n^K E_{-n}^F L_n^{F,G}) . \tag{1}$$

One can see as in Theorem 2.12 that the sequence $(E_n^F E_{-n}^K L_n^{K,K})$ and the sequence $(I_Y E_n^K E_{-n}^F L_n^{F,G})$ converge strongly to the identity operator and to the embedding operator of Y into $L_R^p(\alpha)$, respectively (compare also Exercise 2.21). Thus, the convergence of (1) depends essentially on the convergence of its "central" sequence. Since

$$E_n^K Q_i E_{-n}^K L_n^{K,K} = L_n^{K,K} \chi_{R \setminus [-\frac{i}{n}, \frac{i}{n}]} L_n^{K,K}$$

it becomes evident that the sequence $(E_n^K Q_i E_{-n}^K L_n^{K,K})_{(n,i) \in T}$ converges strongly if the sequence $(\chi_{R \setminus [-\frac{i}{n}, \frac{i}{n}]} I)_{(n,i) \in T}$ converges strongly, and this is equivalent to the convergence of the sequence $(\frac{i}{n})_{(n,i) \in T}$ to zero. Examples for index sets T which guarantee the convergence of this sequence are

$$T = \left\{ (n,i) \in \mathbf{Z}_+^2 \quad \text{with } i \leq n^{1-\delta} \text{ for some fixed } \delta > 0 \right\}$$

or

$$T = \left\{ (n,i) \in \mathbf{Z}_+^2 \quad \text{with } i \leq \frac{n}{\ln n} \right\} .$$

4.2.5 Stability of spline approximation sequences cutting off singularities

Let F be a canonical prebasis and $T \subseteq \mathbf{Z}_+^2$ an index set such that $(L_{n,i}^{F,F})_{(n,i) \in T}$ converges strongly to I as $(n,i) \to \infty$. As in Section 3.2.3 we start with defining the algebra of all (but now doubly indiced) approximation sequences and then we distingiush its subalgebra containing all "interesting" sequences. Let $\mathcal{F}^{F,T}$ denote the set of all generalized sequences $(A_{n,i})_{(n,i) \in T}$ of operators $A_{n,i} : S_n^F \to S_n^F$ such that

$$\|(A_{n,i})\| := \sup_{(n,i) \in T} \|A_{n,i} L_n^{F,F}\| < \infty . \tag{1}$$

Provided with elementwise operations and the norm (1), the class $\mathcal{F}^{F,T}$ becomes a Banach algebra. Given a closed subalgebra Ω of the unit circle $\mathbf{T}$ and a real number r we let $\mathcal{A}_{\Omega,r}^{F,T}$ stand for the smallest closed subalgebra of $\mathcal{F}^{F,T}$ which contains

(i) all sequences $(A_n)_{(n,i) \in T}$ with $(A_n)_{n \in Z^+} \in \mathcal{A}_{\Omega,r}^F$ (that is, $\mathcal{A}_{\Omega,r}^F$ can be regarded as a subalgebra of $\mathcal{A}_{\Omega,r}^{F,T}$ when identifying the sequence $(A_n)_{n \in Z^+}$ with the sequence $(A_n)_{(n,i) \in T}$ being independent of i),

(ii) all sequences

$$(E_n^F \Lambda V_{\{yn+r\}} \Lambda^{-1} Q_i \Lambda V_{-\{yn+r\}} \Lambda^{-1} E_{-n}^F | S_n^F)_{(n,i) \in T}$$

with $y \in \mathbf{R}$ (it would not suffice to include the sequence $(E_n^F Q_i E_{-n}^F)$ only since we want to work in a translation invariant algebra again).

The following proposition gives some hint how to construct the analoga of the homomorphisms W^t and W_s for the algebra $\mathcal{A}_{\Omega,r}^{F,T}$.

Proposition 4.9 Let $(A_{n,i})_{(n,i)\in T} \in \mathcal{A}_{\Omega,r}^{F,T}$, $t \in \Omega$, and $s \in \mathbf{R}$. Then

(a) the strong limits $W_s((A_{n,i})_{n\in Z^+})$ exist for each fixed i,

(b) the strong limits $W^t((A_{n,i})_{n\in Z^+})$ exist for each fixed i and are independent of i,

(c) the cosets $\pi(E_{-n}^F A_{n,i} E_n^F)$ are independent of $(n,i) \in T$,

(d) if $(Q_{n,i})$ is the sequence (ii) then, in particular,

$$W^t((Q_{n,i})_{n\in Z^+}) = I ,$$
$$W_s((Q_{n,i})_{n\in Z^+}) = \begin{cases} Q_i & \text{if } y = s \\ I & \text{if } y \neq s \end{cases} ,$$
$$W_\infty((Q_{n,i})_{n\in Z^+}) = \pi(I) .$$

Proof Since all occuring mappings are continuous homomorphisms it suffices to verify the assertion for the sequences under (i) and (ii) in place of $(A_{n,i})$.

In case $(A_{n,i}) = (A_n)_{(n,i)\in T}$ with $(A_n)_{n\in Z^+} \in \mathcal{A}_{\Omega,r}^F$ the proposition is an immediate consequence of Propositions 3.13 and 3.14. So let $A_{n,i}$ be the sequence in (ii). For $t \in \Omega$ (and even for all $t \in \mathbf{T}$) and with K as in 4.2.4 we obtain

$$W^t((A_{n,i})_{n\in Z^+}) =$$
$$= \underset{n\to\infty}{\text{s-lim}} \, E_n^F Y_{t^{-1}} \Lambda V_{\{yn+r\}} \Lambda^{-1} Q_i \Lambda V_{-\{yn+r\}} \Lambda^{-1} Y_t E_{-n}^F L_n^{F,F}$$
$$= \underset{n\to\infty}{\text{s-lim}} \, E_n^F V_{\{yn+r\}} Q_i V_{-\{yn+r\}} E_{-n}^F L_n^{F,F}$$
$$= \underset{n\to\infty}{\text{s-lim}} \, (E_n^F E_{-n}^K L_n^{K,K})(E_n^K V_{\{yn+r\}} Q_i V_{-\{yn+r\}} E_{-n}^K L_n^{K,K})(E_n^K E_{-n}^F L_n^{F,F})$$
$$= \underset{n\to\infty}{\text{s-lim}} \, (E_n^F E_{-n}^K L_n^{K,K})(L_n^{K,K} \chi_{R\setminus[\frac{-i+\{yn+r\}}{n}, \frac{i+\{yn+r\}}{n}]} L_n^{K,K})(E_n^K E_{-n}^F L_n^{F,F})$$
$$= I$$

with the same reasoning as in 4.2.4. If $s = 0$ then

$$W_0((A_{n,i})_{n\in Z^+}) =$$
$$= \underset{n\to\infty}{\text{s-lim}} \, \Lambda V_{-\{r\}} V_{\{yn+r\}} \Lambda^{-1} Q_i \Lambda V_{-\{yn+r\}} V_{\{r\}} \Lambda^{-1}$$
$$= \underset{n\to\infty}{\text{s-lim}} \, V_{\{yn+r\}-\{r\}} Q_i V_{\{r\}-\{yn+r\}}$$
$$= \begin{cases} Q_i & \text{if } y = 0 \\ I & \text{if } y \neq 0 , \end{cases}$$

and, for $s \neq 0$, one gets analogously

$$W_s((A_{n,i})_{n\in Z^+}) = \begin{cases} Q_i & \text{if } s = y \\ I & \text{if } s \neq y . \end{cases}$$

Finally,

$$\pi(E^F_{-n}A_{n,i}E^F_n) =$$
$$= \pi(\Lambda V_{\{yn+r\}}\Lambda^{-1}Q_i\Lambda V_{-\{yn+r\}}\Lambda^{-1})$$
$$= \pi(I) - \pi(\Lambda V_{\{yn+r\}}\Lambda^{-1}R_i\Lambda V_{-\{yn+r\}}\Lambda^{-1}) = \pi(I)$$

since the operators R_i are compact. ∎

Now it is correct to define the homomorphism $W^t : \mathcal{A}^{F,T}_{\Omega,r} \to L^p_R(\alpha)$ for $t \in \Omega$ by

$$W^t((A_{n,i})_{(n,i)\in T}) := W^t((A_{n,i})_{n\in Z^+}) \quad \text{for some } i$$

and the homomorphism $W_\infty : \mathcal{A}^{F,T}_{\Omega,r} \to L(l^p_k(\alpha))/K(l^p_k(\alpha))$ by

$$W_\infty((A_{n,i})_{(n,i)\in T}) := \pi(E^F_{-n}A_{n,i}E^F_n) \quad \text{for some } n \text{ and } i \;.$$

The definition of the homomorphisms W_s for $s \in \mathbf{R}$ is less obvious, and we will present it after the formulation of the following general stability theorem for spline approximation methods which cut off singularities.

Theorem 4.5 *Let* $(A_{n,i}) \in \mathcal{A}^{F,T}_{\Omega,r}$. *This sequence is stable if and only if*

(a) the operators $W^t((A_{n,i})_{(n,i)\in T})$ *are invertible for all* $t \in \Omega$,

(b) the coset $W_\infty((A_{n,i})_{(n,i)\in T})$ *is invertible,*

(c) the sequences $(W_s((A_{n,i})_{n\in Z^+}))_{i\in Z^+}$ *are stable for all* $s \in \mathbf{R}$.

Of course, the stability theorem 3.5 can be rediscovered from this theorem since, for $(A_{n,i})_{(n,i)\in T} = (A_n)_{(n,i)\in T}$ with $(A_n)_{n\in Z^+} \in \mathcal{A}^F_{\Omega,r}$,

$$W^t((A_{n,i})_{(n,i)\in T}) = W^t((A_n)_{n\in Z^+})\,,$$
$$W_\infty((A_{n,i})_{(n,i)\in T}) = W_\infty((A_n)_{n\in Z^+})$$

and

$$(W_s((A_{n,i})_{n\in Z^+}))_{i\in Z^+} = (W_s((A_n)_{n\in Z^+}))_{i\in Z^+}\,,$$

and a constant sequence is stable if and only if its generating element is invertible.

Before giving a proof of this theorem we shall define the homomorphisms W_s for $s \in \mathbf{R}$. We let $\mathcal{F}(l^p_k(\alpha))$ stand for the algebra of all sequences (A_n) with $A_n \in L(l^p_k(\alpha))$ provided, as usually, with elementwise operations and the supremum norm. The subset $\mathcal{G}(l^p_k(\alpha))$ of $\mathcal{F}(l^p_k(\alpha))$ consisting of all sequences (K_n) of compact operators with $\lim \|K_n\| = 0$ forms a closed two-sided ideal of $\mathcal{F}(l^p_k(\alpha))$, and it is easily seen that a sequence $(A_n) \in \mathcal{F}(l^p_k(\alpha))$ is stable if and only if the coset $(A_n) + \mathcal{G}(l^p_k(\alpha))$ is invertible in the quotient algebra $\mathcal{F}(l^p_k(\alpha))/\mathcal{G}(l^p_k(\alpha))$. Thus, if we define

$$W_s \;:\; \mathcal{A}^{F,T}_{\Omega,r} \to \mathcal{F}(l^p_k(\alpha(s)))/\mathcal{G}(l^p_k(\alpha(s)))\,,$$

$$(A_{n,i})_{(n,i)\in T} \mapsto (W_s((A_{n,i})_{n\in Z^+}))_{i\in Z^+} + \mathcal{G}(l^p_k(\alpha(s)))$$

with

$$\alpha(s) = \begin{cases} \alpha & \text{if} \quad s = 0 \\ 0 & \text{if} \quad s \neq 0 \end{cases}$$

then condition (c) in Theorem 4.5 can be replaced by

(c') *the cosets* $W_s((A_{n,i})_{(n,i)\in T})$ *are invertible for all* $s \in \mathbf{R}$.

4.2.6 Proof of Theorem 4.5

The proof follows the scheme in Sections 1.6.1-1.6.5 again. We only mark the essential steps.

1 st step: Algebraization. Let $\mathcal{K}^T$ stand for the subset of $\mathcal{F}^{F,T}$ consisting of all sequences $(G_{n,i})_{(n,i)\in T}$ with $\|G_{n,i}\| \to 0$ as $(n,i) \to \infty$. Then a sequence $(A_{n,i}) \in \mathcal{F}^{F,T}$ is stable if and only if the coset $(A_{n,i}) + \mathcal{K}^T$ is invertible in $\mathcal{F}^{F,T}/\mathcal{K}^T$.

2 nd step: Set $\mathcal{G}^T = \mathcal{A}_{\Omega,r}^{F,T} \cap \mathcal{K}^T$ *and let* $(G_{n,i}) \in \mathcal{G}^T$. *Then*

$$\begin{aligned} W^t((G_{n,i})_{(n,i)\in T}) &= 0, \\ W_s((G_{n,i})_{(n,i)\in T}) &= 0, \\ W_\infty((G_{n,i})_{(n,i)\in T}) &= 0. \end{aligned}$$

Indeed, given $\epsilon > 0$ there is an (n_0, i_0) such that $\|G_{n,i}\| < \epsilon$ whenever $(n,i) \geq (n_0, i_0)$. Thus, $\|W^t((G_{n,i})_{n\in Z^+})\| \leq \epsilon$ for all $i \geq i_0$, and since $W^t((G_{n,i})_{n\in Z^+})$ is independent of i for all $(G_{n,i}) \in \mathcal{A}_{\Omega,r}^{F,T}$ by Proposition 4.9(b), we find that $W^t((G_{n,i})_{n\in Z^+}) = 0$ for all i. Analogously one gets

$$\lim_{i\to\infty} \|W_s((G_{n,i})_{n\in Z^+})\| = 0.$$

The compactness of the operators $W_s((G_{n,i})_{n\in Z^+})$ for all i follows as in Proposition 3.15 by observing that

$$\pi(W_s(A_{n,i_1})_{n\in Z^+}) = \pi(W_s(A_{n,i_2})_{n\in Z^+})$$

for all sequences $(A_{n,i})_{(n,i)\in T}$ in $\mathcal{A}_{\Omega,r}^{F,T}$.

Finally, the estimate $\|\pi(G_{n,i})\| < \epsilon$ if $(n,i) \geq (n_0, i_0)$ in combination with the independence of $\pi(G_{n,i})$ of n,i (see Proposition 4.9(c)) implies that $\pi(G_{n,i}) = 0$ for all n and i.

3 rd step: The collection $\mathcal{J}_{\Omega,r}^{F,T}$ *of all sequences* $(K_{n,i})_{(n,i)\in T}$ *with*

$$K_{n,i} = K_n + G_{n,i}, \quad (K_n) \in \mathcal{J}_{\Omega,r}^F, \ (G_{n,i}) \in \mathcal{G}^T,$$

belongs to $\mathcal{A}_{\Omega,r}^{F,T}$ *and forms a closed two-sided ideal of this algebra.*

Indeed, the inclusion $\mathcal{J}_{\Omega,r}^{F,T} \subseteq \mathcal{A}_{\Omega,r}^{F,T}$ is an immediate consequence of the inclusions $\mathcal{J}_{\Omega,r}^{F} \subseteq \mathcal{A}_{\Omega,r}^{F}$ (by Proposition 3.17) and $\mathcal{G}^{T} \subseteq \mathcal{A}_{\Omega,r}^{F,T}$ (by definition), and of the definition of $\mathcal{A}_{\Omega,r}^{F,T}$. Further it is evident that the product of a sequence of the form 4.2.5(i) with a sequence in $\mathcal{J}_{\Omega,r}^{F,T}$ is again in $\mathcal{J}_{\Omega,r}^{F,T}$, and so it remains to check whether

$$(E_n^F \Lambda V_{\{yn+r\}} \Lambda^{-1} Q_i \Lambda V_{-\{yn+r\}} \Lambda^{-1} E_{-n}^F)\, \mathcal{J}_{\Omega,r}^{F,T} \subseteq \mathcal{J}_{\Omega,r}^{F,T}$$

and

$$\mathcal{J}_{\Omega,r}^{F,T}\, (E_n^F \Lambda V_{\{yn+r\}} \Lambda^{-1} Q_i \Lambda V_{-\{yn+r\}} \Lambda^{-1} E_{-n}^F) \subseteq \mathcal{J}_{\Omega,r}^{F,T}$$

for all y. Let us verify the first of these inclusions. If K is compact on $L_R^p(\alpha)$ and $\tau \in \Omega$ then

$$\begin{aligned}
&\|(E_n^F \Lambda V_{\{yn+r\}} \Lambda^{-1} Q_i \Lambda V_{-\{yn+r\}} \Lambda^{-1} E_{-n}^F)(E_n^F Y_\tau E_{-n}^F L_n^{F,F} K E_n^F Y_{\tau^{-1}} E_{-n}^F) - \\
&\qquad - (E_n^F Y_\tau E_{-n}^F L_n^{F,F} K E_n^F Y_{\tau^{-1}} E_{-n}^F)\| \\
&= \|E_n^F V_{\{yn+r\}} (Q_i - I) V_{-\{yn+r\}} Y_\tau E_{-n}^F L_n^{F,F} K E_n^F Y_{\tau^{-1}} E_{-n}^F\| \\
&= \|(E_n^F Y_\tau E_{-n}^F)(E_n^F V_{\{yn+r\}} R_i V_{-\{yn+r\}} E_{-n}^F L_n^{F,F} K)(E_n^F Y_{\tau^{-1}} E_{-n}^F)\| \\
&\leq C \, \|E_n^F V_{\{yn+r\}} R_i V_{-\{yn+r\}} E_{-n}^F L_n^{F,F} K\|
\end{aligned}$$

and this tends to zero as $(n,i) \to \infty$ since K is compact and the sequence $(E_n^F V_{\{yn+r\}} R_i V_{-\{yn+r\}} E_{-n}^F L_n^{F,F})$ converges strongly to zero as $(n,i) \to \infty$ (recall Proposition 4.7(a)). Now remember that the ideal $\mathcal{J}_{\Omega,r}^{F}$ is generated by sequences of the form $(E_n^F Y_\tau E_{-n}^F L_n^{F,F} K E_n^F Y_{\tau^{-1}} E_{-n}^F)$ to finish the proof.

4 th step: Essentialization: Denote the canonical homomorphism from $\mathcal{A}_{\Omega,r}^{F,T}$ onto $\mathcal{A}_{\Omega,r}^{F,T}/\mathcal{J}_{\Omega,r}^{F,T}$ by Φ_T^J. Specifying the lifting theorem 1.8 to the present context we obtain: A sequence $(A_{n,i}) \in \mathcal{A}_{\Omega,r}^{F,T}$ is stable if and only if the operators $W^t((A_{n,i})_{(n,i)\in T})$ are invertible for all $t \in \Omega$ and if the coset $\Phi_T^J((A_{n,i})_{(n,i)\in T})$ is invertible. The invertibility in the quotient algebra will again be studied by Allan's local principle. The basic observation for this is provided in the following step.

5 th step: If $f \in C(\dot{\mathbf{R}})$ then the coset $\Phi_T^J((L_n^{F,F} fI|S_n^F)_{(n,i)\in T})$ belongs to the center of $\mathcal{A}_{\Omega,r}^{F,T}/\mathcal{J}_{\Omega,r}^{F,T}$.
    In view of Proposition 3.18, the only thing which still requires a proof is that the commutator

$$\begin{aligned}
&(L_n^{F,F} fI|S_n^F)(E_n^F \Lambda V_{\{yn+r\}} \Lambda^{-1} Q_i \Lambda V_{-\{yn+r\}} \Lambda^{-1} E_{-n}^F) - \\
&\qquad - (E_n^F \Lambda V_{\{yn+r\}} \Lambda^{-1} Q_i \Lambda V_{-\{yn+r\}} \Lambda^{-1} E_{-n}^F)(L_n^{F,F} fI|S_n^F)
\end{aligned} \tag{1}$$

is in $\mathcal{J}_{\Omega,r}^{F,T}$. For this, remember that, by Lemma 3.3, the sequence $(E_{-n}^F L_n^{F,F} f E_n^F)$ is a sequence of diagonal operators up to a sequence in $\mathcal{G}$. Since the operators in the sequence $(E_n^F \Lambda V_{\{yn+r\}} \Lambda^{-1} Q_i \Lambda V_{-\{yn+r\}} \Lambda^{-1} E_{-n}^F)$ are also of diagonal form, the commutator (1) is even in $\mathcal{G}^T$.

6 th step: Localization. The collection $\mathcal{B}_{\Omega,r}^{F,T}$ of all cosets $\Phi_T^J(L_n^{F,F} fI|S_n^F)$ with $f \in C(\dot{\mathbf{R}})$ is a subalgebra of the center of $\mathcal{A}_{\Omega,r}^{F,T}/\mathcal{J}_{\Omega,r}^{F,T}$, and its maximal ideal space is homeomorphic to $\dot{\mathbf{R}}$ (see Proposition 3.21). Hence, we can localize over $\dot{\mathbf{R}}$ and, given $s \in \dot{\mathbf{R}}$, we write I_s^J for the smallest closed two-sided ideal of $\mathcal{A}_{\Omega,r}^{F,T}/\mathcal{J}_{\Omega,r}^{F,T}$ containing s, and we let $\mathcal{A}_{\Omega,r,s}^{F,T}$ refer to the "local" quotient algebra $(\mathcal{A}_{\Omega,r}^{F,T}/\mathcal{J}_{\Omega,r}^{F,T})/I_s^J$ and $\Phi_{s,T}^J$ to the canonical homomorphism from $\mathcal{A}_{\Omega,r}^{F,T}$ onto $\mathcal{A}_{\Omega,r,s}^{F,T}$.

7 th step: Identification. We are going to show that the mapping

$$\Phi_{s,T}^J(A_{n,i}) \mapsto W_s((A_{n,i})_{(n,i) \in T}) \in \begin{cases} \mathcal{F}(l_k^p(\alpha))/\mathcal{G}(l_k^p(\alpha)) & \text{if } s = 0 \\ \mathcal{F}(l_k^p(0))/\mathcal{G}(l_k^p(0)) & \text{if } s \in \mathbf{R} \setminus \{0\} \\ L(l_k^p(\alpha))/K(l_k^p(\alpha)) & \text{if } s = \infty \end{cases} \quad (2)$$

is a continuous and locally equivalent representation of the local algebra $\mathcal{A}_{\Omega,r,s}^{F,T}$.

Let us start with characterizing the images of W_s. By $\mathcal{F}(\mathcal{T}_{k,\kappa,\Omega}^p(\alpha))$ we denote the smallest closed subalgebra of $\mathcal{F}(l_k^p(\alpha))$ which contains all constant sequences $(A)_{n \in Z^+}$ with $A \in \mathcal{T}_{k,\kappa,\Omega}^p(\alpha)$ and the sequence $(Q_n)_{n \in Z^+}$. One can show that this algebra already encloses the ideal $\mathcal{G}(l_k^p(\alpha))$ (see ([BS 2], 7.27) or Proposition 3.15 for a similar proof), and invoking 3.6.4(3) one readily gets

$$\text{Im } W_0 = \mathcal{F}(\mathcal{T}_{k,\kappa,\Omega}^p(\alpha))/\mathcal{G}(l_k^p(\alpha)) . \quad (3)$$

Analogously, for $s \in \mathbf{R} \setminus \{0\}$,

$$\text{Im } W_s = \mathcal{F}(\mathcal{T}_{k,\kappa,\Omega}^p(0))/\mathcal{G}(l_k^p(0)) \quad (4)$$

and finally, it is immediate from 3.6.4(3) that

$$\text{Im } W_\infty = \mathcal{T}_{k,\kappa,\Omega}^p(\alpha))/K(l_k^p(\alpha)) . \quad (5)$$

The algebras (3)-(5) are inverse closed in their embedding algebras settled in (2). For (5) this has been explicitely stated in Corollary 3.2, and for (3) and (4) it can be derived as the inverse closedness of $\mathcal{A}_{\Omega,r}^F/\mathcal{G}$ in $\mathcal{F}^F/\mathcal{G}$ via Theorem 1.12.

It remains to show the existence of homomorphisms $W_s' : \text{Im } W_s \to \mathcal{A}_{\Omega,r,s}^{F,T}$ such that

$$\Phi_{s,T}^J(A_{n,i}) = W_s'(W_s(A_{n,i})) \quad (6)$$

for all $(A_{n,i}) \in \mathcal{A}_{\Omega,r}^{F,T}$.

Let $s = 0$ and $(G_i)_{i \in Z^+} \in \mathcal{G}(l_k^p(\alpha))$. The sequence $(E_n^F \Lambda V_{\{r\}} \Lambda^{-1} G_i \Lambda V_{\{r\}} \Lambda^{-1} E_{-n}^F)$ is obviously contained in $\mathcal{K}^T$. We claim that it even belongs to the smaller ideal $\mathcal{G}^T$. For this goal, it can be clearly supposed without loss that

$$G_i = \begin{cases} K_{kl}^{rs} & \text{if } i = i_0 \\ 0 & \text{if } i \neq i_0 \end{cases}$$

where the rank-one operators K_{kl}^{rs} are defined as in the proof of 3.15(a). Now the assertion can be seen as follows: The sequences $(E_n^F \Lambda V_{\{r\}} \Lambda^{-1} Q_i \Lambda V_{\{r\}} \Lambda^{-1} E_{-n}^F)$ and $(E_n^F \Lambda V_{\{r\}} \Lambda^{-1} A \Lambda V_{\{r\}} \Lambda^{-1} E_{-n}^F)$ with $A \in T_{k,\kappa,\Omega}^p(\alpha)$ belong to $\mathcal{A}_{\Omega,r}^{F,T}$ by definition; hence, the sequence

$$
\begin{aligned}
[(E_n^F \Lambda V_{\{r\}} \Lambda^{-1} V_{k-i_0} \Lambda V_{\{r\}} \Lambda^{-1} E_{-n}^F)(E_n^F \Lambda V_{\{r\}} \Lambda^{-1} P \Lambda V_{\{r\}} \Lambda^{-1} E_{-n}^F) \\
\times (E_n^F \Lambda V_{\{r\}} \Lambda^{-1} Q_i \Lambda V_{\{r\}} \Lambda^{-1} E_{-n}^F)(E_n^F \Lambda V_{\{r\}} \Lambda^{-1} V_{i_0-k} \Lambda V_{\{r\}} \Lambda^{-1} E_{-n}^F) - \\
- (E_n^F \Lambda V_{\{r\}} \Lambda^{-1} V_{k+1-i_0} \Lambda V_{\{r\}} \Lambda^{-1} E_{-n}^F)(E_n^F \Lambda V_{\{r\}} \Lambda^{-1} P \Lambda V_{\{r\}} \Lambda^{-1} E_{-n}^F) \\
\times (E_n^F \Lambda V_{\{r\}} \Lambda^{-1} Q_i \Lambda V_{\{r\}} \Lambda^{-1} E_{-n}^F)(E_n^F \Lambda V_{\{r\}} \Lambda^{-1} V_{i_0-k-1} \Lambda V_{\{r\}} \Lambda^{-1} E_{-n}^F)] \\
\times (E_n^F \Lambda V_{\{r\}} \Lambda^{-1} K_{kl}^{rs} \Lambda V_{\{r\}} \Lambda^{-1} E_{-n}^F) = \\
= (E_n^F \Lambda V_{\{r\}} \Lambda^{-1} (V_{k-i_0} P Q_i V_{i_0-k} - V_{k+1-i_0} P Q_i V_{i_0-k-1}) K_{kl}^{rs} \Lambda V_{\{r\}} \Lambda^{-1} E_{-n}^F)
\end{aligned}
$$

belongs to $\mathcal{A}_{\Omega,r}^{F,T}$, too. But

$$
(V_{k-i_0} P Q_i V_{i_0-k} - V_{k+1-i_0} P Q_i V_{i_0-k-1}) K_{kl}^{rs} = G_i
$$

as one easily checks. This gives our claim. Consequently, the mapping

$$
W_0' : \mathcal{F}(T_{k,\kappa,\Omega}^p(\alpha))/\mathcal{G}(l_k^p(\alpha)) \to \mathcal{A}_{\Omega,r,0}^{F,T} ,
$$

$$
(A_i)_{i \in Z^+} + \mathcal{G}(l_k^p(\alpha)) \mapsto \Phi_{0,T}^J((E_n^F \Lambda V_{\{r\}} \Lambda^{-1} A_i \Lambda V_{-\{r\}} \Lambda^{-1} E_{-n}^F)_{(n,i) \in T})
$$

is correctly defined. It can be verified as in Proposition 3.22 that (6) holds for all sequences $(A_n)_{(n,i) \in T}$ with $(A_n)_{n \in Z^+} \in \mathcal{A}_{\Omega,r}^F$ in place of $(A_{n,i})$. Since W_0, W_0' and $\Phi_{0,T}^J$ are continuous algebra homomorphisms it remains to verify (6) with $(A_{n,i})$ replaced by the sequence $(E_n^F V_{\{yn+r\}} Q_i V_{-\{yn+r\}} E_{-n}^F)_{(n,i) \in T}$.
If $y = 0$ then, as we have shown in the proof of Proposition 4.9,

$$
W_s(E_n^F V_{\{r\}} Q_i V_{-\{r\}} E_{-n}^F) = (Q_i)_{i \in Z^+} + \mathcal{G}(l_k^p(\alpha)) ,
$$

whence (6) follows at once. In case $y \neq 0$ we have

$$
W_s(E_n^F V_{\{yn+r\}} Q_i V_{-\{yn+r\}} E_{-n}^F) = (I)_{i \in Z^+} + \mathcal{G}(l_k^p(\alpha)) ,
$$

and so (6) reduces to the identity

$$
\Phi_{0,T}^J(E_n^F V_{\{yn+r\}} Q_i V_{-\{yn+r\}} E_{-n}^F) = \Phi_{0,T}^J(I|S_n^F) .
$$

To verify this we pick a continuous function f with $f(0) = 1$ and possessing a sufficiently small support, and we let K stand for the canonical prebasis

$$
K = \{\chi_{[0,\frac{1}{k})}, \chi_{[\frac{1}{k},\frac{2}{k})}, \ldots, \chi_{[\frac{k-1}{k},1)}\} .
$$

The following identities hold up to a summand in $\mathcal{G}^T$ by Lemma 3.3:

$$(L_n^{F,F} f E_n^F V_{\{yn+r\}} Q_i V_{-\{yn+r\}} E_{-n}^F) =$$
$$= (E_n^F E_{-n}^K L_n^{K,K} f E_n^K V_{\{yn+r\}} Q_i V_{-\{yn+r\}} E_{-n}^F)$$
$$= (E_n^F E_{-n}^K L_n^{K,K} f L_n^{K,K} \chi_{R \setminus [(-i+\{yn+r\})/n, (i-1+\{yn+r\})/n]} E_n^K E_{-n}^F)$$
$$= (E_n^F E_{-n}^K L_n^{K,K} f \chi_{R \setminus [(-i+\{yn+r\})/n, (i-1+\{yn+r\})/n]} E_n^K E_{-n}^F)$$

and this is, for large n and for all i with $(n,i) \in T$, nothing else than

$$(E_n^F E_{-n}^K L_n^{K,K} f E_n^K E_{-n}^F) = (L_n^{F,F} f I | S_n^F) \quad \text{modulo} \quad \mathcal{G}^T .$$

This completes the proof of (6) for $s = 0$; for $s \neq 0$ we define

$$W_s' : \mathcal{F}(T_{k,\kappa,\Omega}^p(0)) / \mathcal{G}(l_k^p(0)) \to \mathcal{A}_{\Omega,r,s}^{F,T} ,$$

$$(A_i)_{i \in Z^+} + \mathcal{G}(l_k^p(0)) \mapsto \Phi_{s,T}^J((E_n^F \Lambda V_{\{sn+r\}} \Lambda^{-1} A_i \Lambda V_{-\{sn+r\}} \Lambda^{-1} E_{-n}^F)_{(n,i) \in T})$$

if $s \neq \infty$ and

$$W_\infty' : T_{k,\kappa,\Omega}^p(\alpha) / K(l_k^p(\alpha)) \to \mathcal{A}_{\Omega,r,\infty}^{F,T} ,$$

$$A + K(l_k^p(\alpha)) \mapsto \Phi_{\infty,T}^J((E_n^F A E_{-n}^F)_{(n,i) \in T}) .$$

The verification of (6) runs analogously as in case $s = 0$.

8 th step: Inverse closedness. To finish the proof of Theorem 4.5 we have to show that stability of $(A_{n,i}) \in \mathcal{A}_{\Omega,r}^{F,T}$ implies invertibility of $W_s(A_{n,i})$ and $W^t(A_{n,i})$ for all $s \in \dot{\mathbf{R}}$ and $t \in \Omega$. Suppose that

$$\sup_{(n,i) \geq (n_0,i_0)} \|A_{n,i}^{-1}\| < \infty \tag{7}$$

for a certain (n_0, i_0). Then the sequences $(A_{n,i})_{n \in Z^+}$ are stable for all $i \geq i_0$, and these sequences belong to $\mathcal{A}_{\Omega,r}^F$ (recall that $Q_i = I + \text{compact}$). So we conclude from Section 3.6.2 that $W_s((A_{n,i})_{n \in Z^+})$ and $W^t((A_{n,i})_{n \in Z^+})$ are invertible for all $s \in \dot{\mathbf{R}}$, $t \in \Omega$, and $i \geq i_0$. Now Proposition 4.9 entails the assertion for $W^t((A_{n,i})_{(n,i) \in T})$ and $W_\infty((A_{n,i})_{(n,i) \in T})$, and concerning $W_s((A_{n,i}))$ we emphasize that (7) yields

$$\sup_{i \geq i_0} \|W_s((A_{n,i})_{n \in Z^+})^{-1}\| < \infty$$

whence the stability of $(W_s((A_{n,i})_{n \in Z^+}))_{i \in Z^+}$ follows. $\blacksquare$

4.2.7 Characterization of $\mathcal{A}_{\Omega,r}^{F,T}/\mathcal{J}_{\Omega,r}^{F,T}$ as function algebra

In this section we shall formulate the analogues to the results of Sections 3.7.5 and 3.7.6 pertaining to the description of $\mathcal{A}_{\Omega,r}^{F,T}/\mathcal{J}_{\Omega,r}^{F,T}$ as an algebra of continuous functions. The proofs run parallelly to those in the mentioned sections, and we omit them here.

Given $x \in \dot{\mathbf{R}}$ we let $\mathcal{A}^T(x)$ refer to the algebras $\mathcal{F}(\mathcal{T}_{k,\kappa,\Omega}^p(\alpha))/\mathcal{G}(l_k^p(\alpha))$, $\mathcal{F}(\mathcal{T}_{k,\kappa,\Omega}^p(0))/\mathcal{G}(l_k^p(0))$, and $\mathcal{T}_{k,\kappa,\Omega}^p(\alpha)/K(l_k^p(\alpha))$ in case $x = 0$, $x \in \mathbf{R} \setminus \{0\}$, and $x = \infty$, respectively. From 4.2.6 (3)-(5) and Theorem 1.11 we infer that $\mathcal{A}^T(x)$ is topologically isomorphic to the local algebra $\mathcal{A}_{\Omega,r,x}^{F,T}$.

By $FUN_{k,\kappa,\Omega}^{p,T}(\alpha)$ we denote the set of all bounded functions on $\dot{\mathbf{R}}$ which take at $x \in \dot{\mathbf{R}}$ a value in $\mathcal{A}^T(x)$. Introducing operations and a norm as in Section 3.7.5, this set becomes a Banach algebra.

Let $A \in FUN_{k,\kappa,\Omega}^{p,T}(\alpha)$ and $x \in \dot{\mathbf{R}}$, and choose sequences $(A_i^{(x)})$ and an operator A^∞ such that

$$A(x) = \begin{cases} (A_i^{(0)})_{i \in Z^+} + \mathcal{G}(l_k^p(\alpha)) & \text{if} \quad x = 0 \\ (A_i^{(x)})_{i \in Z^+} + \mathcal{G}(l_k^p(0)) & \text{if} \quad x \in \mathbf{R} \setminus \{0\} \\ A^\infty + K(l_k^p(\alpha)) & \text{if} \quad x = \infty \,. \end{cases}$$

Then we define the right (resp. left) value $V_+(A(x))$(resp. $V_-(A(x))$) of A at x by

$$V_+(A(x)) = \begin{cases} \text{s–}\lim_{s \to +\infty} V_{-s} \Lambda^{-1} A_i^{(0)} \Lambda V_s & \text{if} \quad x = 0 \\ \text{s–}\lim_{s \to +\infty} V_{-s} A_i^{(x)} V_s & \text{if} \quad x \in \mathbf{R} \setminus \{0\} \\ \text{s–}\lim_{s \to -\infty} V_{-s} \Lambda^{-1} A^\infty \Lambda V_s & \text{if} \quad x = \infty \end{cases}$$

(resp.

$$V_-(A(x)) = \begin{cases} \text{s–}\lim_{s \to -\infty} V_{-s} \Lambda^{-1} A_i^{(0)} \Lambda V_s & \text{if} \quad x = 0 \\ \text{s–}\lim_{s \to -\infty} V_{-s} A_i^{(x)} V_s & \text{if} \quad x \in \mathbf{R} \setminus \{0\} \\ \text{s–}\lim_{s \to +\infty} V_{-s} \Lambda^{-1} A^\infty \Lambda V_s & \text{if} \quad x = \infty \end{cases}).$$

This definition is correct, that is, all strong limits exist, and they are independent of the choice of the sequences $(A_i^{(x)})$, of the choice of A^∞, and of the subscript i.

A function $A \in FUN_{k,\kappa,\Omega}^{p,T}(\alpha)$ is said to be *continuous* if the one-sided limits $\lim_{y \to x \pm 0} A(y)$ exist at each point $x \in \dot{\mathbf{R}}$ (we define $\infty \pm 0 = \mp\infty$) and if

$$\lim_{y \to x \pm 0} A(y) = V_\pm(A(x)) \,.$$

The set $CON_{k,\kappa,\Omega}^{p,T}(\alpha)$ of all continuous functions forms a closed subalgebra of $FUN_{k,\kappa,\Omega}^{p,T}(\alpha)$.

Theorem 4.6 *The algebras $\mathcal{A}_{\Omega,r}^{F,T}/\mathcal{J}_{\Omega,r}^{F,T}$ and $CON_{k,\kappa,\Omega}^{p,T}(\alpha)$ are topologically isomorphic, and the isomorphism associates the coset $\Phi_T^J(A_{n,i})$ with the function $x \mapsto W_x((A_{n,i})_{(n,i) \in T})$.*

4.3 Approximation of singular integral operators

4.3.1 Galerkin methods

If $F \subseteq L_R^p(\alpha)$ and $G \subseteq L_R^p(\alpha)^*$ are k-elementic prebases and (F, G) is a canonical pair, and if a and b are piecewise continuous coefficients which are continuous on $\dot{\mathbf{R}} \setminus \{0, 1\}$ then the Galerkin approximation sequence

$$(L_n^{F,G}(aI + bS_R + K)|S_n^F)_{n \in Z^+}$$

for the singular integral operator $aI + bS_R + K$ with compact perturbation K belongs to $\mathcal{A}_{\{1\},0}^F$ (see Proposition 3.24). The discontinuities at $0, 1$ of the coefficients involve undetermined local invertibility conditions at these points (compare Theorem 3.11), and this suggests the consideration of the following *modified Galerkin method*: Given n and i in $\mathbf{Z}^+$ we define

$$Q_{n,i} := E_n^F Q_i V_n Q_i V_{-n} E_{-n}^F.$$

The sequence $(Q_{n,i})_{(n,i) \in T}$ belongs to $\mathcal{A}_{\{1\},0}^{F,T}$ since

$$Q_{n,i} = (E_n^F Q_i E_{-n}^F)(E_n^F \Lambda V_{\{n\}} \Lambda^{-1} Q_i \Lambda V_{-\{n\}} \Lambda^{-1} E_{-n}^F).$$

Fix the index set T in such a way that $\frac{i}{n} \to 0$ as $(n,i) \to \infty$.

Theorem 4.7 *(a) The modified Galerkin sequence $(A_{n,i})_{(n,i) \in T}$,*

$$A_{n,i} = Q_{n,i} L_n^{F,G}(aI + bS_R + K)Q_{n,i}|S_n^F \tag{1}$$

belongs to $\mathcal{A}_{\{1\},0}^{F,T}$.
(b) The sequence $(A_{n,i})$ is stable if and only if

(i) the operator $A = aI + bS_R + K \in L(L_R^p(\alpha))$ is invertible,

(ii) the matrix functions $a(s) + b(s)\rho^{F,G}$ with $\rho^{F,G} = (\lambda^{F,G})^{-1}\sigma^{F,G}$ are invertible for all $s \in \dot{\mathbf{R}} \setminus \{0, 1\}$,

(iii) the operators $(a(s+0)P + a(s-0)Q) + (b(s+0)P + b(s-0)Q)T^0(\rho^{F,G})$ are Fredholm on $l_k^p(\alpha)$ for $s \in \{0, 1\}$,

(iv) the operators $QT^0(a(s - 0) + b(s - 0)\rho^{F,G})Q + P$ and $PT^0(a(s + 0) + b(s + 0)\rho^{F,G})P + Q$ are invertible on $l_k^p(0)$ for $s \in \{0, 1\}$,

(v) the operators

$$\chi_{[-1,1]}I + \chi_{R \setminus [-1,1]}((a(s+0)\chi_{R^+} + a(s-0)\chi_{R^-})I +$$
$$+ (b(s+0)\chi_{R^+} + b(s-0)\chi_{R^-})S_R)\chi_{R \setminus [-1,1]}I$$

are invertible on $L_R^p(\alpha)$ for $s \in \{0, 1\}$.

Proof   The operator under (i) is just $W^1((A_{n,i})_{(n,i)\in T})$. If $s \in \mathbf{R} \setminus \{0,1\}$ then $W_s((A_{n,i})_{(n,i)\in T})$ is a constant sequence which is stable if and only if its generating element $a(s)I + b(s)T^0(\rho^{F,G})$ is invertible on $l_k^p(0)$ or, equivalently, if the matrix function in (ii) is invertible. Similarly, the coset

$$W_\infty((A_{n,i})_{(n,i)\in T}) = \pi(a(\infty)I + b(\infty)T^0(\rho^{F,G}))$$

is invertible in $L(l_k^p(\alpha))/K(l_k^p(\alpha))$ if and only if the matrix function $a(\infty) + b(\infty)\rho^{F,G}$ is invertible.

Finally, the invertibility of the cosets $W_s((A_{n,i})_{(n,i)\in T})$ for $s \in \{0,1\}$ is equivalent to the stability of the infinite section method

$$(Q_i((a(s+0)P+a(s-0)Q)+(b(s+0)P+b(s-0)Q)T^0(\rho^{F,G})+K_s)Q_i+P_i)\,.$$

In case $k = 1$ we have seen in Example 4.6 in 4.1.7 that this method is stable if and only if conditions (iii)-(v) are satisfied; the analogous assertion holds in case $k > 1$ (compare Proposition 4.5 as well as the discussion in Section 4.1.4 concerning finite sections). ∎

The above theorem remains valid for matrix valued coefficients, too. If the coefficients are scalar valued and if $k = 1$ then the conditions (ii)-(v) are effective. This is evident for (ii), and for (iii)-(v) we refer to Example 4.6 in Section 4.1.7.

4.3.2   Continuous coefficients

Suppose now the coefficients a and b in the singular integral operator $aI + bS_R + K$ to be continuous on $\dot{\mathbf{R}}$. Then conditions (iii)-(v) in Theorem 4.7 can be reduced to:

(iii)' the operators $T^0(a(s) + b(s)\rho^{F,G})$ are Fredholm for $s \in \{0,1\}$,

(iv)' the operators $QT^0(a(s) + b(s)\rho^{F,G})Q + P$ and $PT^0(a(s) + b(s)\rho^{F,G})P + Q$ are invertible for $s \in \{0,1\}$,

(v)' the operators $\chi_{R\setminus[-1,1]}(a(s) + b(s)S_R)\chi_{R\setminus[-1,1]} + \chi_{[-1,1]}$ are invertible.

Condition (iii)' is clearly a consequence of (iv)', but (iv)' is essentially stronger than (iii)' (indeed, in the scalar case, the operator in (iii)' is Fredholm if and only if the point 0 does not belong to the curve

$$\Gamma = \{a(s) + b(s)\rho^{F,G}(z) , z \in \mathbf{T} \setminus \{1\}\} \cup \{a(s) \pm b(s)\} \,,$$

whereas, in order to check invertibility of $PT^0(a(s) + b(s)\rho^{F,G})P + Q$, one has to add a certain circular arc to Γ to make Γ to a closed curve, and one has additionally to require that this new curve has winding number 0).

On the other hand, (iii)' is just the local condition at s for the stability of the Galerkin method without cutting off. So the following warning seems to be reasonable: *Do not cut off non-existing singularities!*

4.3.3 Other projections methods

Let now G be a k-tupel of premeasures which is related, for example, to the collocation or qualocation method (as in Sections 3.8.2-3.8.4), and suppose (F, G) to be a canonical pair. Then the local stability conditions for the sequence $(L_n^{F,G}(aI + bS_R + K)|S_n^F)$ have the same structure independent of G (compare Theorems 3.12 and 3.14) and, thus, Theorem 4.7 remains verbatim valid for any "canonical" projection method, which belongs to $\mathcal{A}_{\{1\},\epsilon}^{F,T}$.

4.3.4 A modified cut-off-method

Let F be a one-elementic canonical prebasis throughout this section, and let (F, G) be a canonical pair again. Instead of the cutting-off-sequences

$$(E_n^F \Lambda V_{\{yn+r\}} \Lambda^{-1} Q_i \Lambda V_{-\{yn+r\}} \Lambda^{-1} E_{-n}^F)_{(n,i)\in T} \tag{1}$$

we consider the following ones:

$$(E_n^F T^0(\lambda^{F,G})^{-1} \Lambda V_{\{yn+r\}} \Lambda^{-1} Q_i \Lambda V_{-\{yn+r\}} \Lambda^{-1} T^0(\lambda^{F,G}) E_{-n}^F)_{(n,i)\in T} . \tag{2}$$

These sequences can be written as

$$(E_n^F T^0(\lambda^{F,G})^{-1} E_{-n}^F)(E_n^F \Lambda V_{\{yn+r\}} \Lambda^{-1} Q_i \Lambda V_{-\{yn+r\}} \Lambda^{-1} E_{-n}^F)(E_n^F T^0(\lambda^{F,G}) E_{-n}^F)$$

$$\tag{3}$$

and belong therefore to the algebra $\mathcal{A}_{\Omega,r}^{F,T}$ for all $\Omega \subseteq \mathbf{T}$. (Here we need the restriction $k = 1$ since otherwise the inclusion $(E_n^F T^0(\lambda^{F,G})^{\pm 1} E_{-n}^F) \in \mathcal{A}_{\Omega,r}^F$ cannot be guaranteed.)

Although the sequences (2) seem to be somewhat more complicated than (1), the corresponding approximation methods will be of a simpler structure. To illustrate this we let $y = r = 0$ and consider the singular integral equation $Au = f$ whose solution by a Galerkin method requires to solve the approximating equations

$$L_n^{F,G} Au_n = L_n^{F,G} f \quad , u_n \in S_n^F .$$

Cutting off the (possible) singularity at 0 by (1) gives

$$E_n^F Q_i E_{-n}^F L_n^{F,G} Au_{n,i} = E_n^F Q_i E_{-n}^F L_n^{F,G} f$$

with $u_{n,i} \in E_n^F Q_i E_{-n}^F S_n^F$ or, equivalently,

$$Q_i T^0(\lambda^{F,G})^{-1} P_n^G Au_{n,i} = Q_i T^0(\lambda^{F,G})^{-1} P_n^G f$$

whereas use of (2) leads to

$$E_n^F T^0(\lambda^{F,G})^{-1} Q_i T^0(\lambda^{F,G}) E_{-n}^F L_n^{F,G} Au_{n,i} =$$
$$= E_n^F T^0(\lambda^{F,G})^{-1} Q_i T^0(\lambda^{F,G}) E_{-n}^F L_n^{F,G} f$$

or, equivalently,

$$Q_i P_n^G A u_{n,i} = Q_i P_n^G f \quad , \quad u_{n,i} \in E_n^F Q_i E_{-n}^F S_n^F \; . \tag{4}$$

In detail, (4) is the infinite system

$$\sum_{m \in Z \setminus \{-i, \ldots, i-1\}} \int_R (A\phi_{mn})(\frac{t+l}{n})\overline{\gamma(t)} \, dt = \int_R f(\frac{t+l}{n})\overline{\gamma(t)} \, dt \; ,$$

with $l \in \mathbf{Z} \setminus \{-i, \ldots, i-1\}$, which originates from the full system 3.8.1(2) by removing the "central" $2i$ equations and $2i$ unknowns.

For an application of the cutting-off-sequences (2) we need their images under the homomorphisms W_s and W^t. For brevity, we write $< Q_i, y >$ in place of the sequence (2).

Proposition 4.10 *The sequence $< Q_i, y >$ belongs to $\mathcal{A}_{\Omega,r}^{F,T}$, and*

$$W^t(< Q_i, y >) = I \quad \text{for all } t \, ,$$

$$W_s(< Q_i, y >) = \begin{cases} (T^0(\lambda^{F,G})^{-1} Q_i T^0(\lambda^{F,G}))_{i \in Z^+} + \mathcal{G}(l_1^p(\alpha(s))) & \text{if } s = y \\ (I)_{i \in Z^+} + \mathcal{G}(l_1^p(\alpha(s))) & \text{if } s \neq y \end{cases}$$

$$W_\infty(< Q_i, y >) = I + K(l_1^p(\alpha)) \, .$$

Proof Since W^t and W_s are homomorphisms, and due to representation (3), the assertion follows by combining Propositions 3.13, 3.14, and 4.9(d). ∎

As an example we consider the Galerkin method for the singular integral operator $aI + bS_R + K$ with a, b, K as in 4.3.1 again.

Theorem 4.8 *The modified Galerkin sequence*

$$< Q_i, 0 >< Q_i, 1 > (L_n^{F,G}(aI + bS_R + K)|Q_{n,i}S_n^F) \tag{5}$$

is stable if and only if conditions (i)-(v) in Theorem 4.7 are satisfied.

Indeed, the only difference between the sequences (5) and 4.3.1(1) appears in computing the images of the homomorphisms W_s for $s \in \{0, 1\}$. Their invertibility is now equivalent to the stability of the modified infinite section method

$$(T^0(\lambda^{F,G})^{-1} Q_i T^0(\lambda^{F,G}))[(a(s+0)P + a(s-0)Q) +$$
$$+ (b(s+0)P + b(s-0)Q)T^0(\rho^{F,G}) + K_s] \, T^0(\lambda^{F,G})^{-1} Q_i T^0(\lambda^{F,G}))_{i \in Z^+} . \tag{6}$$

Removing the outer factors $T^0(\lambda^{F,G})^{\pm 1}$ and employing the compactness of the commutator $T^0(\lambda^{F,G})P - PT^0(\lambda^{F,G})$, we can rewrite (6) as

$$(Q_i[(a(s+0)P + a(s-0)Q) + (b(s+0)P + b(s-0)Q)T^0(\rho^{F,G}) + K_s']Q_i)$$

with a new compact operator K_s', and now Example 4.6 in 4.1.7 states that stability of (6) is equivalent to (iii)-(v) in Theorem 4.7. ∎

Clearly, this equivalence is *not* constrained by the special form of the approximation operators in our example. Since $T^0(\lambda^{F,G})$ commutes with each operator in
the Toeplitz algebra $\mathcal{T}_1^p(\alpha)$ modulo a compact operator (Proposition 2.8), the two
methods of cutting off are always simultaneously stable or not.

4.3.5 Projection methods for operators with arbitrary discontinuities in their coefficients

As already announced, the idea of "cutting off" also opens a way for tackling projection methods for singular integral operators with arbitrary piecewise continuous
coefficients.

Recall that the question whether the sequence $(L_n^{F,G} aI|S_n^F)$, $a \in PC(\mathbf{R})$, be-
longs to one of the algebras $\mathcal{A}_{\Omega,r}^F$, can be reduced by localization to the con-
sideration of sequences of the form $(L_n^{F,G} \chi_{[t,\infty)} I|S_n^F)$ (compare Section 3.8.1). In
Proposition 3.24 we have already verified that this sequence is always in $\mathcal{A}_{\Omega,r}^F$ if
t is integer, whereas Exercise 3.26 shows that this sequence does, in general, not
belong to any $\mathcal{A}_{\Omega,r}^F$ if $t \notin \mathbf{Z}$. Here we study the modified sequence

$$(E_n^F T^0(\lambda^{F,G})^{-1} V_{\{tn\}} Q_i V_{-\{tn\}} T^0(\lambda^{F,G}) E_{-n}^F) (L_n^{F,G} \chi_{[t,\infty)} I|S_n^F) \,, \tag{1}$$

where we limit ourselves to the unweighted case.

Proposition 4.11 *The sequence* (1) *is in* $\mathcal{A}_{\Omega,0}^F$ *for each one-elemento prebasis F
and each subset $\Omega \subseteq \mathbf{T}$ whenever i is a largely enough fixed integer.*

Proof Introduce besides (1) the sequence

$$(E_n^F T^0(\lambda^{F,G})^{-1} V_{\{tn\}} Q_i V_{-\{tn\}} T^0(\lambda^{F,G}) E_{-n}^F L_n^{F,G} \chi_{[\{tn\}/n,\infty)} I|S_n^F) \,. \tag{2}$$

One readily checks that this sequence is in $\mathcal{A}_{\Omega,0}^F$, and for the difference of (1) and
(2) one finds

$$(E_n^F T^0(\lambda^{F,G})^{-1} V_{\{tn\}} Q_i V_{-\{tn\}} P_n^G (\chi_{[t,\infty)} - \chi_{[\{tn\}/n,\infty)}) I|S_n^F)$$
$$= (E_n^F T^0(\lambda^{F,G})^{-1} V_{\{tn\}} Q_i P_n^G (\chi_{[t-\{tn\}/n,\infty)} - \chi_{[0,\infty)}) U_{-\frac{\{tn\}}{n}}|S_n^F) \,,$$

and now it is not hard to verify that

$$Q_i P_n^G (\chi_{[t-\{tn\}/n,\infty)} - \chi_{[0,\infty)}) I = 0$$

for large i (if $G = \{\gamma\}$ and $\mathrm{supp}\, \gamma \subseteq [-m, m]$ with an integer m then one can take
$i = m + 1$ since the function $\chi_{[t-\{tn\}/n,\infty)} - \chi_{[0,\infty)}$ vanishes outside the intervall
$[\frac{-1}{n}, 0]$). ∎

If a is a piecewise continuous coefficient with a finite set $\Sigma = \{\sigma_1, \ldots, \sigma_p\}$ of
discontinuities then this proposition yields that the sequence

$$\prod_{s=1}^{p} (E_n^F T^0(\lambda^{F,G})^{-1} V_{\{\sigma_s n\}} Q_i V_{-\{\sigma_s n\}} T^0(\lambda^{F,G}) E_{-n}^F)(L_n^{F,G} aI|S_n^F)$$

belongs to the algebra $\mathcal{A}_{\Omega,0}^F$, and this can be used to derive stability criteria for projection methods for singular integral operators with discontinuities in their coefficients at non-integer points. The treatise of the details will be left to the readers.

4.4 Approximation of compound Mellin operators

4.4.1 Compound Mellin operators

In this section we are going to study spline Galerkin methods for operators of the form

$$\chi_{R^+} \sum_{j=1}^{r} a_j A_j b_j \chi_{R^+} I + \chi_{R^-} I \tag{1}$$

where a_j and b_j are piecewise continuous on $\mathbf{R}^+$ and continuous on $\mathbf{R}^+ \setminus \{1\}$ (for simplicity we suppose the coefficients to have only one jump; the case of finitely many jumps at integer points can be treated analogously), and where the A_j's are operators in the algebra $\Sigma^p(\alpha)$, generated by the singular integral operator $S = S_{R^+}$. By Costabel's decomposition 2.1.2(d), there are complex numbers α_j and β_j and continuous functions c_j with $M^0(c_j) \in N^p(\alpha)$ such that (1) is equal to

$$\chi_{R^+} \sum_{j=1}^{r} a_j (\alpha_j I + \beta_j S_{R^+} + M^0(c_j)) b_j \chi_{R^+} I + \chi_{R^-} I \,.$$

Further we set for brevity

$$a(s) = \sum_{j=1}^{r} a_j(s) \alpha_j b_j(s) \,, \; b(s) = \sum_{j=1}^{r} a_j(s) \beta_j b_j(s) \,, \; c(s) = \sum_{j=1}^{r} a_j(s) c_j b_j(s) \,,$$

and, as in Section 3.3.3(1), we let $L : L(l_1^p(\alpha)) \mapsto L(l_k^p(\alpha))$ stand for the mapping

$$A \mapsto (\int_R \phi_s(x) \, dx \, \alpha_r^F A)_{r,s=0}^{k-1} \,.$$

The functions a and b are complex-valued and piecewise continuous, whereas the values of c are again functions with the property $M^0(c(s)) \in N^p(\alpha)$.

Let, finally, (F,G) be a canonical pair consisting of the k-elementic prebases F and G, and set

$$Q_{n,i} = E_n^F Q_i V_n Q_i V_{-n} E_{-n}^F \,.$$

Theorem 4.9 *The modified Galerkin sequence*

$$(A_{n,i})_{(n,i)\in T} := (Q_{n,i} L_n^{F,G} (\chi_{R^+} \sum_{j=1}^{r} a_j A_j b_j \chi_{R^+} I + \chi_{R^-} I) Q_{n,i} | S_n^F)$$

belongs to $\mathcal{A}_{\{1\},0}^{F,T}$. It is stable if and only if

(i) *the operator $\chi_{R^+} \sum_{j=1}^{r} a_j A_j b_j \chi_{R^+} I + \chi_{R^-} I \in L(L_R^p(\alpha))$ is invertible,*

(ii) *the operator*

$$T(a(0+0) + b(0+0)\rho^{F,G}) + L(G(c(0+0))) + Q$$

is Fredholm on $l_k^p(\alpha)$, the operator $T(a(0+0)+b(0+0)\rho^{F,G}) + Q$ is invertible on $l_k^p(0)$, and the operator

$$\chi_{[1,\infty)} (a(0+0)I + b(0+0)S_{R^+} + M^0(c(0+0)))\chi_{[1,\infty)} I + \chi_{(-\infty,1]} I$$

is invertible on $L_R^p(\alpha)$,

(iii) *the operator*

$$\sum_{j=1}^{r} (a_j(1+0)P + a_j(1-0)Q) T^0(\alpha_j + \beta_j \rho^{F,G})(b_j(1+0)P + b_j(1-0)Q)$$

is Fredholm on $l_k^p(0)$, the operators $T(a(1-0) + b(1-0)\tilde{\rho}^{F,G}) + Q$ (with $\tilde{\rho}^{F,G}(z) = \rho^{F,G}(1/z))$ and $T(a(1+0) + b(1+0)\rho^{F,G}) + Q$ are invertible on $l_k^p(0)$, and the operator

$$\chi_{R\setminus[-1,1]} (\sum_{j=1}^{r} (a_j(1+0)\chi_{R^+} + a_j(1-0)\chi_{R^-})(\alpha_j I + \beta_j S_R)$$

$$(b_j(1+0)\chi_{R^+} + b_j(1-0)\chi_{R^-}))\chi_{R\setminus[-1,1]} I + \chi_{[-1,1]} I$$

is invertible on $L_R^p(0)$,

(iv) *the operators $T^0(a(s) + b(s)\rho^{F,G}) \in L(l_k^p(0))$ are invertible for all $s \in \mathbf{R}^+ \setminus \{0,1\}$,*

(v) *the operator $T(a(\infty) + b(\infty)\rho^{F,G}) + L(G(c(\infty)))$ is Fredholm on $l_k^p(\alpha)$.*

Proof For the inclusion

$$(L_n^{F,G} (\chi_{R^+} \sum_{j=1}^{r} a_j A_j b_j \chi_{R^+} I + \chi_{R^-} I) | S_n^F) \in \mathcal{A}_{\{1\},0}^F$$

one shows that this sequence is of local type and that it belongs locally to $\mathcal{A}^F_{\{1\},0}$ (just in the same way as in Proposition 3.24). The computation of the images of the relevant homomorphisms yields

$$W^1((A_{n,i})_{(n,i)\in T}) = \chi_{R^+} \sum_{j=1}^{r} a_j A_j b_j \chi_{R^+} I + \chi_{R^-} I \,, \tag{2}$$

$$W_0((A_{n,i})_{(n,i)\in T})$$
$$= \ (Q_i(\sum_{j=1}^{r} a_j(0+0)(PT^0(\alpha_j + \beta_j \rho^{F,G})P + L(G(c_j)))b_j(0+0)P +$$
$$+ Q + K_0)Q_i + P_i)_{i\in Z^+} + \mathcal{G}(l_k^p(\alpha))$$
$$= \ (Q_i(T(a(0+0) + b(0+0)\rho^{F,G}) + L(G(c(0+0)))) + Q + K_0)Q_i +$$
$$+ P_i)_{i\in Z^+} + \mathcal{G}(l_k^p(\alpha)) \tag{3}$$

with a certain compact operator K_0,

$$W_1((A_{n,i})_{(n,i)\in T})$$
$$= \ (Q_i(\sum_{j=1}^{r}(a_j(1+0)P + a_j(1-0)Q)T^0(\alpha_j + \beta_j \rho^{F,G}) \times$$
$$\times (b_j(1+0)P + b_j(1-0)Q) + K_1)Q_i + P_i)_{i\in Z^+} + \mathcal{G}(l_k^p(0)) \tag{4}$$

with a compact operator K_1; further, for $s \in \mathbf{R}^+ \setminus \{0,1\}$,

$$W_s((A_{n,i})_{(n,i)\in T})$$
$$= \sum_{j=1}^{r} a_j(s)T^0(\alpha_j + \beta_j \rho^{F,G})b_j(s) = T^0(a(s) + b(s)\rho^{F,G}) \tag{5}$$

and, finally,

$$W_\infty((A_{n,i})_{(n,i)\in T})$$
$$= \sum_{j=1}^{r} a_j(\infty)(PT^0(\alpha_j + \beta_j \rho^{F,G})P + L(G(c_j)))b_j(\infty) + K(l_k^p(\alpha))$$
$$= T(a(\infty) + b(\infty)\rho^{F,G}) \ + \ L(G(c(\infty))) + K(l_k^p(\alpha)) \,. \tag{6}$$

The invertibility of (2), (5) and (6) is just the topic of (i),(iv) and (v), respectively. Further, as in Example 4.7 in 4.1.7, the infinite section sequence (3) is stable if and only if (ii) holds, and similarly one checks that (4) is a stable sequence if and only if (iii) is satisfied. ∎

The theorem is again valid for matrix-valued coefficients. In case the coefficients are scalar-valued, and if $k = 1$, then the conditions (ii), (iv) and (v) are effectively verifiable. They are actually equivalent to *(ii)'* the point 0 has winding number 0 with respect to the curves $\Gamma_1 \cup \Gamma_2$ and $\Gamma_2 \cup \Gamma_3$ with

$$\Gamma_1 = \{a(0+0) + b(0+0)\rho^{F,G}(t) \, , \, t \in \mathbf{T}\} \, ,$$

$$\Gamma_2 = \{a(0+0) + b(0+0)\coth \pi(z+i/p) \, , \, z \in \overline{\mathbf{R}}\} \, ,$$

$$\Gamma_3 = \{a(0+0)+b(0+0)\coth \pi(z+i(1/p+\alpha)) + c(0+0)(z) \, , \, z \in \overline{\mathbf{R}}\} \, ,$$

(iv)' the function $a(s) + b(s)\rho^{F,G}$ has no zero for all $s \in \mathbf{R}^+ \setminus \{0,1\}$,

(v)' the point 0 does not belong to the curve $\Gamma_4 \cup \Gamma_5$ with

$$\Gamma_4 = \{a(\infty) + b(\infty)\rho^{F,G}(t) \, , \, t \in \mathbf{T}\} \, ,$$

$$\Gamma_5 = \{a(\infty) + b(\infty)\coth \pi(z+i(1/p+\alpha)) + c(\infty)(z) \, , \, z \in \overline{\mathbf{R}}\} \, .$$

The proof bases on the symbol calculi for elements of the Toeplitz algebra (Section 2.5.2) and of the algebra generated by singular integral operators (Section 3.1.1) as well as on Coburn's theorem for Toeplitz operators and singular integral operators (see Theorems 2.4 and 3.2).

Note further that the condition $0 \notin \Gamma_5$ is automatically satisfied if the operator (i) is Fredholm (compute the local symbol of (i) at infinity).

The verification of (iii) is less obvious since, even in the scalar case, the invertibility of the singular integral operator in (iii) is equivalent to the invertibility of a Wiener-Hopf operator with piecewise continuous 2×2-matrix-valued generating function. We are going to mention two special situations where an effective verification is possible.

For the first case we suppose all functions b_j to be continuous at 1. Then (iii) of Theorem 4.9 can be written as follows:

(iii)' the operator

$$PT^0(a(1+0) + b(1+0)\rho^{F,G}) + QT^0(a(1-0) + b(1-0)\rho^{F,G})$$

is Fredholm on $l_1^p(0)$, the operators $T(a(1-0).+ b(1-0)\tilde{\rho}^{F,G}) + Q$ and $T(a(1+0) + b(1+0)\rho^{F,G}) + Q$ are invertible on $l_1^p(0)$, and the operator

$$\chi_{R\setminus[-1,1]}((a(1+0)\chi_{R^+} + a(1-0)\chi_{R^-})I+$$
$$+ (b(1+0)\chi_{R^+} + b(1-0)\chi_{R^-})S_R)\chi_{R\setminus[-1,1]}I + \chi_{[-1,1]}I$$

is invertible on $L_R^p(0)$.

These conditions are effective: The paired Toeplitz operator is Fredholm if and only if

$$a(1+0) + b(1+0)\rho^{F,G}(t) \neq 0 \quad \text{for } t \in \mathbf{T} ,$$

$$a(1-0) + b(1-0)\rho^{F,G}(t) \neq 0 \quad \text{for } t \in \mathbf{T} ,$$

$$(a(1-0)-b(1-0))/(a(1-0)+b(1-0))\,(1+\coth\pi(z+i/p))+$$
$$+(a(1+0)-b(1+0))/(a(1+0)+b(1+0))\,(1-\coth\pi(z+i/p)) \neq 0$$

for $z \in \overline{\mathbf{R}}$,
the Toeplitz operators are invertible if and only if the closed curves $\Lambda_1 \cup \Lambda_2$ and $\Lambda_3 \cup \Lambda_4$ with

$$\begin{aligned}
\Lambda_1 &= \{a(1-0) + b(1-0)\tilde{\rho}^{F,G}(t) , \ t \in \mathbf{T}\} , \\
\Lambda_2 &= \{a(1-0) + b(1-0)\coth\pi(-z+i/p) , \ z \in \overline{\mathbf{R}}\} , \\
\Lambda_3 &= \{a(1+0) + b(1+0)\rho^{F,G}(t) , \ t \in \mathbf{T}\} , \\
\Lambda_4 &= \{a(1+0) + b(1+0)\coth\pi(-z+i/p) , \ z \in \overline{\mathbf{R}}\}
\end{aligned}$$

have winding number 0 with respect to the origin, and the singular integral operator is invertible if and only the closed curve $\Lambda_5 \cup \Lambda_6 \cup \Lambda_7$ with

$$\Lambda_5 = \left\{ \frac{a(1+0) - b(1+0)}{a(1+0) + b(1+0)} \frac{1+\coth\pi(z+i/p)}{2} + \frac{1-\coth\pi(z+i/p)}{2} , \ z \in \overline{\mathbf{R}} \right\} ,$$

$$\Lambda_6 = \left\{ \frac{1+\coth\pi(z+i/p)}{2} + \frac{a(1-0) - b(1-0)}{a(1-0) + b(1-0)} \frac{1-\coth\pi(z+i/p)}{2} , \ z \in \overline{\mathbf{R}} \right\} ,$$

$$\Lambda_7 = \left\{ \frac{a(1-0) - b(1-0)}{a(1-0) + b(1-0)} \frac{1+\coth\pi(z+i/q)}{2} + \right.$$
$$\left. + \frac{a(1+0) - b(1+0)}{a(1+0) + b(1+0)} \frac{1-\coth\pi(z+i/q)}{2} , \ z \in \overline{\mathbf{R}} \right\}$$

(where $1/p + 1/q = 1$) has winding number zero with respect to the origin.

For a second situation where condition (iii) is well verifiable we suppose that the coefficients a_j vanish outside the interval $[0,1]$ and that $b_j = \chi_{[0,1]}$ for $j = 1, \ldots, r-1$, and that $a_r = b_r = \chi_{[1,\infty)}$ and $A_r = I$. Then the operator under consideration is of the form

$$\chi_{[0,1]} \sum_{j=1}^{r-1} (a_j\alpha_j I + a_j\beta_j S_R + a_j M^0(c_j))\chi_{[0,1]} I + \chi_{R\setminus[0,1]} I ,$$

and (iii) changes into

(iii)' the operators $T(a(1-0)+b(1-0)\tilde{\rho}^{F,G})+Q$ *and* $T(a(1+0)+b(1+0)\rho^{F,G})+Q$
are invertible on $l_1^p(0)$, *and the operator*

$$a(1-0)\chi_{(-\infty,-1]}I + b(1-0)\chi_{(-\infty,-1]}S_{(-\infty,-1]} + \chi_{[-1,\infty)}I$$

is invertible on $L_R^p(0)$.

The singular integral operator appearing in (iii)' is invertible if and only if the curve $\Lambda_6 \cup \Lambda_8$ with Λ_6 as above and

$$\Lambda_8 = \left\{ \frac{a(1-0)-b(1-0)}{a(1-0)+b(1-0)} \frac{1+\coth\pi(z+i/q)}{2} + \frac{1-\coth\pi(z+i/q)}{2} , z \in \overline{\mathbf{R}} \right\}$$

has winding number zero with respect to the origin.

4.4.2 Operators in $\Sigma^p(\alpha)$

Next we study spline Galerkin methods for operators in the algebra $\Sigma^p(\alpha)$, that is, for operators of the form

$$A = a\chi_{R^+}I + bS_{R^+} + M^0(c) + \chi_{R^-}I. \tag{1}$$

with complex numbers a and b and a continuous function c. The only singular point is $0 \in \mathbf{R}$, and so we modify the usual Galerkin sequences by cutting off just this point.

Theorem 4.10 *Let* (F,G) *be a canonical pair consisting of the* k-*elemetric prebases* F *and* G. *Then the modified Galerkin sequence* $(A_{n,i})_{(n,i)\in T}$ *with*

$$A_{n,i} = E_n^F Q_i E_{-n}^F A E_n^F Q_i E_{-n}^F | S_n^F$$

where A *is defined as in* (1) *is stable if and only if*

(i) *the operator* A *is invertible on* $L_R^p(\alpha)$,

(ii) *the operator* $T(a+b\rho^{F,G}) + L(G(c)) + Q$ *is Fredholm on* $l_k^p(\alpha)$, *the operator* $T(a+b\rho^{F,G}) + Q$ *is invertible on* $l_k^p(0)$, *and the operator*

$$\chi_{[1,\infty)}(aI + bS_{R^+} + M^0(c))\chi_{[1,\infty)}I + \chi_{(-\infty,1]}I$$

is invertible on $L_R^p(\alpha)$.

Proof Condition (i) is just the invertibility of $W^1((A_{n,i})_{(n,i)\in T})$, whereas (ii) is equivalent to the stability of the infinite section sequence

$$(Q_i(T(a+b\rho^{F,G}) + L(G(c)))Q_i)_{i\in Z^+}$$

(compare Example 4.6). Further one has for $s \in \mathbf{R} \setminus \{0\}$

$$W_s((A_{n,i})_{(n,i)\in T}) = PT^0(a+b\rho^{F,G})P + Q$$

and for $s = \infty$

$$W_\infty((A_{n,i})_{(n,i)\in T}) = PT^0(a + b\rho^{F,G})P + L(G(c)) + Q + K(l_k^p(\alpha)) ,$$

and the demand of invertibility of these operators resp. cosets is already part of the condition (ii). ∎

If $k = 1$ then all conditions are verifiable: The operator A in (i) is invertible if and only if its Mellin symbol (depending on p and α) does not degenerate, and the invertibility and Fredholmness of the operators in (ii) was discussed above.

4.4.3   A special case: pure Mellin convolutions

Specifying Theorems 4.10 and 4.9 to operators of the form

$$\chi_{R^+}(aI + M^0(c))\chi_{R^+}I + \chi_{R^-}I \quad \text{and} \quad \chi_{[0,1]}(aI + M^0(c))\chi_{[0,1]}I + \chi_{R\setminus[0,1]}I ,$$

respectively, we obtain

Proposition 4.12 *Let (F, G) be a canonical pair consisting of the k-elementic prebases F and G.*
(a) The modified Galerkin sequence

$$(E_n^F Q_i E_{-n}^F L_n^{F,G}(\chi_{R^+}(aI + M^0(c))\chi_{R^+}I + \chi_{R^-}I)E_n^F Q_i E_{-n}^F|S_n^F)$$

is stable if and only if

   (i) the function $a + c(\cdot)$ does not vanish on $\dot{\mathbf{R}}$,

   (ii) the winding number of this function with respect to the origin is equal to 0.

(b) The modified Galerkin sequence

$$(E_n^F Q_i V_n Q_i V_{-n} E_{-n}^F L_n^{F,G}$$
$$(\chi_{[0,1]}(aI + M^0(c))\chi_{[0,1]}I + \chi_{R\setminus[0,1]}I)E_n^F Q_i V_n Q_i V_{-n} E_{-n}^F|S_n^F)$$

is stable if and only if conditions (i) and (ii) of part (a) are satisfied.

Proof   (a) Condition (i) of Theorem 4.10 is equivalent to the invertibility of the Mellin symbol $a + c$ of the operator $aI + M^0(c)$. Since $c(\infty) = 0$ this involves in particular that $a \neq 0$.

Condition (ii) of Theorem 4.10 reduces to the Fredholmness of the operator $aP + L(G(c))$ and to the invertibility of the singular integral operator

$$\chi_{[1,\infty)}(aI + M^0(c))\chi_{[1,\infty)} + \chi_{(-\infty,1]}I .$$

The latter one is equivalent to a Wiener-Hopf operator with continuous generating function $a+c$, thus, its invertibility is equivalent to condition (ii). Further, Exercise 3.29 entails that $aP + L(G(c))$ is Fredholm whenever $aP + G(c)$ is so, and this is clearly equivalent to the invertibility of $a + c$. The proof of part (b) proceeds analogously. ∎

Remark Projection methods for pure Mellin convolutions on $\mathbf{R}^+$ and $[0, 1]$ show some remarkable pecularities: First, this is one of the rather seldom situations where effective criteria even for k-elementic prebases F and G are derivable. The second one is even more impressive: the stability conditions for (F, G)-Galerkin methods with cutting-off-factor are completely independent of the choice of the prebases F and G! Moreover, since the above results remain valid for arbitrary projection methods for which $(L_n^{F,G} M^0(c)|S_n^F)$ belongs to the algebra $\mathcal{A}_{\{1\},0}^F$ (such as collocation and qualocation), we conclude that if one of the projection methods in Proposition 4.12 is stable (for special F, G), then all of these projection methods are stable (with arbitrary F, G).

The same effect of independence of stability of the concrete choice of the projection method appears for double layer potential operators on simple closed curves with corners if the "geometric singularity" at each corner is cut off (for first results into this direction we refer to [DRS] and Section 6).

4.4.4 Operators with two fixed singularities

As a last application of the general stability theorem 4.5 we consider operators with two fixed singularities at 0 and 1 such as the following one:

$$(Af)(t) = f(t) + \frac{1}{\pi i} \int_0^1 \frac{f(s)}{s + t} ds + \frac{1}{\pi i} \int_0^1 \frac{f(s)}{2 - s - t} ds, \ t \in [0, 1]. \tag{1}$$

We continue this operator by I onto $\mathbf{R} \setminus [0, 1]$ to get an operator with the same mapping properties as (1) which acts on the whole axis, and we analyze the stability of projection methods which cut off the points 0 and 1.

More generally, we consider operators of the form

$$A = a_1 I + a_2 S_R b_2 I + a_3 M^0(c_3) b_3 I + a_4 U_1 J M^0(c_4) J U_{-1} b_4 I + K \tag{2}$$

where $U_{\pm 1}$ are the usual shift operators, the flip J is defined by $(Jf)(t) = f(-t)$, and K is a compact operator. The coefficients a_j, b_j in (2) are supposed to be piecewise continuous on $\mathbf{R}$ and continuous on $\mathbf{R} \setminus \{0, 1\}$, and we further assume that $a_1(t) = 1$ for $t \in \mathbf{R} \setminus [0, 1]$ and $a_j(t) = b_j(t) = 0$ for $t \in \mathbf{R} \setminus [0, 1]$ and $j > 1$. Finally, c_3 and c_4 are functions such that $M^0(c_3), M^0(c_4) \in N^p(0)$. The operator (2) acts boundedly on $L_R^p(0)$ under these restrictions.

It is easy to check that the operator (1) results from (2) by setting $a_1 = 1, a_2 = 0, a_3 = a_4 = b_3 = b_4 = \chi_{[0,1]}$, and $c_3 = c_4 = n_\pi$ with $n_\pi(z) = 1/\sinh \pi(z + i/p)$.

For the approximate solution of the equation $Au = f$ by a spline Galerkin method which cuts off the singularities at 0 and 1, we let F and G be k-elementic prebases which form a canonical pair (F, G), and we consider the approximation equations

$$A_{n,i}u_{n,i} = E_n^F Q_i V_n Q_i V_{-n} E_{-n}^F L_n^{F,G} f$$

where $u_{n,i}$ is seeked in $E_n^F Q_i V_n Q_i V_{-n} E_{-n}^F S_n^F$,

$$A_{n,i} = E_n^F Q_i V_n Q_i V_{-n} E_{-n}^F L_n^{F,G} A E_n^F Q_i V_n Q_i V_{-n} E_{-n}^F | S_n^F , \qquad (3)$$

and A is the operator (2).

For brevity we set

$$
\begin{aligned}
a(s) &:= a_1(s) + a_2(s)b_2(s)(\lambda^{F,G})^{-1}\sigma^{F,G} , \\
b &:= a_3(0+0)b_3(0+0)c_3 , \\
c &:= a_4(1-0)b_4(1-0)c_4 .
\end{aligned}
$$

Theorem 4.11 *The sequence $(A_{n,i})_{(n,i)\in T}$, where $A_{n,i}$ is given by (3) and T is as in Section 4.2.4, is stable (with respect to $L_R^p(0)$) if and only if*

(i) *the operator A is invertible on $L_R^p(0)$,*

(ii) *the operator $T(a(0+0))+L(G(b))$ is Fredholm on $l_k^p(0)$, the operator $T(a(0+0))$ is invertible on $l_k^p(0)$, and the operator*

$$\chi_{[1,\infty)}(a_1(0+0)I + a_2(0+0)b_2(0+0)S_{R^+} + M^0(b))\chi_{[1,\infty)}I + \chi_{(-\infty,1]}I$$

is invertible on $L_R^p(0)$,

(iii) *the functions $a(s)$ are invertible for all $s \in (0,1)$,*

(iv) *the operator $T(\hat{a}(1-0)) + L(G(c))$ (where $\hat{a}(1-0)(t) := a(1-0)(1/t)$) is Fredholm on $l_k^p(0)$, the operator $T(\hat{a}(1-0))$ is invertible on $l_k^p(0)$, and the operator*

$$\chi_{[1,\infty)}(a_1(1-0)I - a_2(1-0)b_2(1-0)S_{R^+} + M^0(c))\chi_{[1,\infty)}I + \chi_{(-\infty,1]}I$$

is invertible on $L_R^p(0)$.

Proof It is not hard to see that $JM^0(c_4)J$ belongs to the algebra $\Sigma_R^p(0)$ generated by the singular integral operator S_R and the operators of multiplication by the characteristic functions $\chi_{R^\pm}$ (this algebra is actually invariant under the mapping $B \mapsto JBJ$) and that

$$E_{-n}^F JM^0(c_4)JE_n^F - L(\tilde{J}G(c_4)\tilde{J}) \in K(l_k^p(0))$$

where

$$\tilde{J} : l_1^p(0) \to l_1^p(0) , \quad (x_r)_{r\in Z} \mapsto (x_{-r-1})_{r\in Z}$$

is the discrete flip operator. The operator $\tilde{J}G(c_4)\tilde{J}$ belongs to the Toeplitz algebra $\mathcal{T}^p_{1,1,\{1\}}(0)$ (observe that this algebra is invariant under the mapping $C \mapsto \tilde{J}C\tilde{J}$), and now it is easy to derive that the sequence $(L_n^{F,G}A|S_n^F)$ is an element of the algebra $\mathcal{A}^F_{\{1\},0}$. Consequently, $(A_{n,i})_{(n,i)\in T} \in \mathcal{A}^{F,T}_{\{1\},0}$.

It remains to compute the images of the sequence $(A_{n,i})$ under the homomorphisms W^1 and W_s, $s \in [0,1]$ (the homomorphisms W^t, $t \neq 1$, and W_s, $s \notin [0,1]$, are not relevant here since $\Omega = \{1\}$ and $W_s((A_{n,i})) = I$ whenever $s < 0$ or $s > 1$). Clearly, $W^1((A_{n,i})) = A$, which gives condition (i). Further,

$$W_0((A_{n,i})) = (Q_i(T(a(0+0)) + L(G(b)) + K_0 + Q)Q_i)_{i\in Z} + \mathcal{G}(l_k^p(0)),$$

$$W_s((A_{n,i})) = T^0(a(s)) \quad \text{for} \quad s \in (0,1),$$

and

$$W_1((A_{n,i})) = (Q_i(QT^0(a(1-0))Q + L(\tilde{J}G(c)\tilde{J}) + K_1 + P)Q_i)_{i\in Z} + \mathcal{G}(l_k^p(0))$$

with certain compact operators K_0 and K_1. The coset $W_0((A_{n,i}))$ is invertible if and only if condition (ii) is satisfied (compare Example 4.7), and the invertibility of $W_s((A_{n,i}))$ for $s \in (0,1)$ is equivalent to (iii).

In order to explain (iv), we abbreviate the $k \times k$ diagonal operator $\mathrm{diag}\,(\tilde{J}, \ldots, \tilde{J})$ to $\tilde{J}$ again. Then $Q_i = \tilde{J}Q_i\tilde{J}$ and

$$
\begin{aligned}
L(\tilde{J}G(c)\tilde{J}) &= \tilde{J}L(G(c))\tilde{J}, \\
QT^0(a(1-0))Q &= \tilde{J}PT^0(\hat{a}(1-0))P\tilde{J}.
\end{aligned}
$$

Since $\tilde{J} : l_k^p(0) \to l_k^p(0)$ is invertible, we conclude that the sequence

$$(Q_i(QT^0(a(1-0))Q + L(\tilde{J}G(c)\tilde{J}) + K_1 + P)Q_i)_{i\in Z}$$

is stable if and only if the sequence

$$(Q_i(PT^0(\hat{a}(1-0))P + L(G(c)) + \tilde{K}_1 + Q)Q_i)_{i\in Z} \tag{4}$$

with the compact operator $\tilde{K}_1 = \tilde{J}K_1\tilde{J}$ is stable. By Example 4.7, the condition (iv) is necessary and sufficient for the stability of (4). $\blacksquare$

After a substitution of $(\lambda^{F,G})^{-1}\sigma^{F,G}$ in the definition of $a(s)$ by the corresponding function for the discretized singular integral operator, this theorem holds for collocation and qualocation as well as for the quadrature and quadrocation methods considered in the second chapter, too.

4.5 Approximation of operators over unbounded domains

4.5.1 Methods cutting off the infinity

If A is a singular integral operator which acts on the whole axis, or on an unbounded part of it, then the formal application of a spline approximation method naturally leads to infinite discrete systems. For an effective solution of these systems we introduce an additional parameter i which cuts off 'the singularity at infinity' and so reduces the infinite system to a finite one. In what follows we consider a cutting-off-factor of the form $(E_n^F R_i E_{-n}^F | S_n^F)$ where R_i, as above, is given by

$$R_i : l_1^p(\alpha) \to l_1^p(\alpha), \ (x_r) \mapsto (\ldots, 0, x_{-i}, \ldots, x_{i-1}, 0, \ldots),$$

and where the same letter is also used for the diagonal operator

$$\mathrm{diag}\,(R_i, \ldots, R_i) \ : \ l_k^p(\alpha) \to l_k^p(\alpha) \,.$$

In order to guarantee convergence of the cut-off-projections $E_n^F R_i E_{-n}^F L_n^{F,G}$, we choose an index set $T \subseteq \mathbf{Z}_+ \times \mathbf{Z}_+$ of the form

$$T = \{(n, i) \in \mathbf{Z}_+^2 \quad \text{with } i \geq n^{1+\delta} \text{ for some fixed } \delta > 0\}$$

or

$$T = \{(n, i) \in \mathbf{Z}_+^2 \quad \text{with } i \geq n \ln n\}$$

or any other subset which forces the convergence of the sequence $(i/n)_{(n,i) \in T}$ to infinity as $(n, i) \to \infty$. As for spline approximation methods which cut off singularities in finite points, the concrete choice of T is not essential.

If T is fixed anyhow we let $\mathcal{F}^{F,\infty}$ stand for the collection of all sequences $(A_{n,i})_{(n,i) \in T}$ of operators $A_{n,i} : S_n^F \to S_n^F$ such that

$$\|(A_{n,i})\| := \sup_{(n,i) \in T} \|A_{n,i} L_n^{F,F}\| < \infty \,.$$

Introducing elementwise operations we can make $\mathcal{F}^{F,\infty}$ to become a Banach algebra. Further we write $\mathcal{K}^\infty$ for the subset of $\mathcal{F}^{F,\infty}$ of all sequences tending to zero in the operator norm, and we let $\mathcal{A}_{\Omega,r}^{F,\infty}$ denote the smallest closed subalgebra of $\mathcal{F}^{F,\infty}$ which contains

(i) all sequences $(A_n)_{(n,i) \in T}$ with $(A_n)_{n \in Z^+} \in \mathcal{A}_{\Omega,r}^F$,

(ii) the sequence $(E_n^F V_{\{r\}} R_i V_{-\{r\}} E_{-n}^F | S_n^F)_{(n,i) \in T}$.

The following lemma prepares the definition of the homomorphisms W^t and W_s.

Lemma 4.1 *Let* $(A_{n,i}) \in \mathcal{A}_{\Omega,r}^{F,\infty}$. *Then*

(a) *the strong limits* s–$\lim_{i \to \infty} A_{n,i}$ *exist for each fixed* n, *and the sequence* (s–$\lim_{i \to \infty} A_{n,i})_{n \in Z^+}$ *belongs to* $\mathcal{A}_{\Omega,r}^{F}$,

(b) *the sequence* $(E_{-1}^{F} A_{1,i} E_{1}^{F})_{i \in Z^+}$ *belongs to* $\mathcal{F}(T_{k,\kappa,\Omega}^{p}(\alpha))$.

Recall that the algebra in (b) is generated by the constant sequences (A), where A runs through $T_{k,\kappa,\Omega}^{p}(\alpha)$, and by the sequence $(R_i)_{i \in Z^+}$; see the 7 th step in 4.2.6.

Proof The assertion (a) is evident for sequences of the form (i), and for (ii) one easily finds

$$\text{s} - \lim_{i \to \infty} E_{n}^{F} V_{\{r\}} R_i V_{-\{r\}} E_{-n}^{F} |S_{n}^{F} = I|S_{n}^{F}. \tag{1}$$

Assertion (b) is a consequence of Proposition 3.14 for sequences of the form (i), and for (ii) one has

$$(E_{-1} A_{1,i} E_{1}^{F}) = (V_{\{r\}} R_i V_{-\{r\}}) = (V_{\{r\}})(R_i)(V_{-\{r\}})$$

which is in $\mathcal{F}(T_{k,\kappa,\Omega}^{p}(\alpha))$ since the constant sequences $(V_{\pm\{r\}})$ as well as the sequence (R_i) belong to this algebra by definition. Now standard arguments yield the assertions for arbitrary sequences in $\mathcal{A}_{\Omega,r}^{F,\infty}$. ∎

The following definitions are correct by the above lemma: For $(A_{n,i})_{(n,i) \in T}$, $t \in \Omega$, and $s \in \mathbf{R}$ we set

$$W^{t}((A_{n,i})_{(n,i) \in T}) := W^{t}((\text{s} - \lim_{i \to \infty} A_{n,i})_{n \in Z^+}) \tag{2}$$

and

$$W_s((A_{n,i})_{(n,i) \in T}) := W_s((\text{s} - \lim_{i \to \infty} A_{n,i})_{n \in Z^+}) \tag{3}$$

where the homomorphisms W^{t} and W_s on the right hand sides of (2) and (3) are those known from Section 3.4.3. The definition is somewhat more complicated for $s = \infty$. Here we have to introduce the smallest closed two-sided ideal $\mathcal{J}(l_{k}^{p}(\alpha))$ of $\mathcal{F}(T_{k,\kappa,\Omega}^{p}(\alpha))$ which contains all sequences of the form $(K + G_i)$ where K is compact on $l_{k}^{p}(\alpha)$ and the sequence $(G_i)_{i \in Z^+}$ belongs to the ideal $\mathcal{G}(l_{k}^{p}(\alpha))$, that is, the G_i's are compact, and $\|G_i\| \to 0$ as $i \to \infty$. Now we set

$$W_\infty((A_{n,i})_{(n,i) \in T}) := (E_{-1}^{F} A_{1,i} E_{1}^{F}) + \mathcal{J}(l_{k}^{p}(\alpha)) \in \mathcal{F}(T_{k,\kappa,\Omega}^{p}(\alpha))/\mathcal{J}(l_{k}^{p}(\alpha)). \tag{4}$$

The definitions (2)-(4) are natural generalizations of the definitions of the homomorphisms W^{t} and W_s related with the algebra $\mathcal{A}_{\Omega,r}^{F}$ in Chapter 3. Indeed, if $(A_{n,i})_{(n,i) \in T}$ is the sequence (i) then

$$W^{t}((A_{n,i})_{(n,i) \in T}) = W^{t}(A_n), \ t \in \Omega,$$
$$W_s((A_{n,i})_{(n,i) \in T}) = W_s(A_n), \ s \in \mathbf{R},$$

where the operators W^t and W_s on the left are associated with $\mathcal{A}_{\Omega,r}^{F,\infty}$ and those on the right with $\mathcal{A}_{\Omega,r}^F$. Further, since $(E_{-1}^F A_1 E_1^F)$ is a constant sequence, the invertibility of the coset

$$W_\infty((A_{n,i})_{(n,i)\in T}) = (E_{-1}^F A_1 E_1^F) + \mathcal{J}(l_k^p(\alpha))$$

is equivalent to the Fredholmness of $E_{-1}^F A_1 E_1^F$, or, again equivalently, to the invertibility of

$$W_\infty(A_n) = \pi(E_{-1}^F A_1 E_1^F).$$

In that sense, the stability theorem 3.5 is a special case of the following general stability theorem for sequences in $\mathcal{A}_{\Omega,r}^{F,\infty}$.

4.5.2   The stability theorem

Theorem 4.12 *Let* $(A_{n,i})_{(n,i)\in T} \in \mathcal{A}_{\Omega,r}^{F,\infty}$. *This sequence is stable if and only if*

(a) *the operators* $W^t((A_{n,i})_{(n,i)\in T})$ *are invertible for all* $t \in \Omega$,

(b) *the operators* $W_s((A_{n,i})_{(n,i)\in T})$ *are invertible for all* $s \in \mathbf{R}$,

(c) *the coset* $W_\infty((A_{n,i})_{(n,i)\in T}) \in \mathcal{F}(\mathcal{T}_{k,\kappa,\Omega}^p(\alpha))/\mathcal{J}(l_k^p(\alpha))$ *is invertible.*

Proof The proof of this theorem proceeds parallel to that of Theorem 4.5. We only give some hints concerning the sequence 4.5.1(ii). First we compute the image of this sequence under the homomorphisms 4.5.1 (2)-(4). From 4.5.1(1) we conclude that

$$W^t((E_n^F V_{\{r\}} R_i V_{-\{r\}} E_{-n}^F)_{(n,i)\in T}) = I, \ t \in \Omega,$$

and

$$W_s((E_n^F V_{\{r\}} R_i V_{-\{r\}} E_{-n}^F)_{(n,i)\in T}) = I, \ s \in \mathbf{R},$$

and it is clear that

$$W_\infty((E_n^F V_{\{r\}} R_i V_{-\{r\}} E_{-n}^F)_{(n,i)\in T}) = (V_{\{r\}} R_i V_{-\{r\}}) + \mathcal{J}(l_k^p(\alpha)).$$

We set $\mathcal{G}^\infty := \mathcal{A}_{\Omega,r}^{F,\infty} \cap \mathcal{K}^\infty$ in analogy with 4.2.6 (second step), and let $\mathcal{J}_{\Omega,r}^{F,\infty}$ refer to the ideal of $\mathcal{A}_{\Omega,r}^{F,\infty}$ consisting of all sequences $(K_{n,i})_{(n,i)\in T}$ of the form

$$K_{n,i} = K_n + G_{n,i}, \ (K_n) \in \mathcal{J}_{\Omega,r}^F, \ (G_{n,i}) \in \mathcal{G}^\infty,$$

(compare the third step in 4.2.6). The general lifting theorem 1.8 leads us to the problem of studying invertibility in the quotient algebra $\mathcal{A}_{\Omega,r}^{F,\infty}/\mathcal{J}_{\Omega,r}^{F,\infty}$, and this can be done by localization, since the cosets

$$(L_n^{F,F} fI|S_n^F)_{(n,i)\in T} + \mathcal{J}_{\Omega,r}^{F,\infty} \tag{1}$$

belong to the center of this algebra for each function $f \in C(\dot{\mathbf{R}})$. The maximal ideal space of the smallest closed subalgebra of $\mathcal{A}_{\Omega,r}^{F,\infty}/\mathcal{J}_{\Omega,r}^{F,\infty}$ generated by all sequences (1) is homeomorphic to $\dot{\mathbf{R}}$, and by $\mathcal{A}_{\Omega,r,s}^{F,\infty}$ and $\Phi_{s,\infty}^{J}$ we denote the local algebra and the local homomorphism from $\mathcal{A}_{\Omega,r}^{F,\infty}$ onto $\mathcal{A}_{\Omega,r,s}^{F,\infty}$ associated with $s \in \dot{\mathbf{R}}$, respectively.

The value of the homomorphisms W_s at the sequence $(A_{n,i}) \in \mathcal{A}_{\Omega,r}^{F,\infty}$ depends on the coset $\Phi_{s,\infty}^{J}((A_{n,i})_{(n,i)\in T})$ only, and so it is correct to think of W_s as acting on the local algebras $\mathcal{A}_{\Omega,r,s}^{F,\infty}$. The proof of this fact bases on two observations:

1° If $(G_{n,i}) \in \mathcal{G}^{\infty}$ then $(\text{s--}\lim_{i\to\infty} G_{n,i})_{n\in Z^+} \in \mathcal{G}$ where $\mathcal{G}$ was introduced in Proposition 3.15.

2° If $(A_{n,i}) \in \mathcal{A}_{\Omega,r}^{F,\infty}$ and n is fixed then

$$(E_{-1}^{F} A_{1,i} E_{1}^{F})_{i\in Z^+} - (E_{-n}^{F} A_{n,i} E_{n}^{F})_{i\in Z^+} \in \mathcal{J}(l_k^p(\alpha))$$

whence follows $(E_{-1}^{F} G_{1,i} E_{1}^{F})_{i\in Z^+} \in \mathcal{J}(l_k^p(\alpha))$, compare Proposition 3.14.

The so-defined quotient homomorphisms

$$\Phi_{s,\infty}^{J}((A_{n,i})_{(n,i)\in T}) \mapsto W_s((A_{n,i})_{(n,i)\in T})$$

are just locally equivalent representations of the local algebras $\mathcal{A}_{\Omega,r,s}^{F,\infty}$. In fact, one readily finds

$$\operatorname{Im} W_s = \begin{cases} \mathcal{T}_{k,\kappa,\Omega}^{p}(\alpha) & \text{if} \quad s = 0 \\ \mathcal{T}_{k,\kappa,\Omega}^{p}(0) & \text{if} \quad s \in \mathbf{R} \setminus \{0\} \\ \mathcal{F}(\mathcal{T}_{k,\kappa,\Omega}^{p}(\alpha))/\mathcal{J}(l_k^p(\alpha)) & \text{if} \quad s = \infty, \end{cases}$$

and all quoted algebras are inverse closed in the algebras $L(l_k^p(\alpha))$, $L(l_k^p(0))$, and $\mathcal{F}_s(l_k^p(\alpha))/\mathcal{J}(l_k^p(\alpha))$, respectively, where $\mathcal{F}_s(l_k^p(\alpha))$ refers to the subalgebra of the algebra $\mathcal{F}(l_k^p(\alpha))$ containing all sequences $(A_i)_{i\in Z^+}$ for which the strong limits $\text{s--}\lim A_i$ and $\text{s--}\lim A_i^*$ exist. (The set $\mathcal{J}(l_k^p(\alpha))$ is actually a closed two-sided ideal in this algebra as one easily checks.)

The inverse homomorphisms W_s' are given by

$$W_0' : \mathcal{T}_{k,\kappa,\Omega}^{p}(\alpha) \to \mathcal{A}_{\Omega,r,0}^{F,\infty},$$
$$A \mapsto \Phi_{0,\infty}^{J}((E_n^F \Lambda V_{\{r\}} \Lambda^{-1} A \Lambda V_{-\{r\}} \Lambda^{-1} E_{-n}^{F})_{(n,i)\in T}),$$

$$W_s' : \mathcal{T}_{k,\kappa,\Omega}^{p}(0) \to \mathcal{A}_{\Omega,r,s}^{F,\infty},$$
$$A \mapsto \Phi_{s,\infty}^{J}((E_n^F \Lambda V_{\{sn+r\}} A V_{-\{sn+r\}} \Lambda^{-1} E_{-n}^{F})_{(n,i)\in T}) \text{ for } s \in \mathbf{R} \setminus \{0\}, \text{ and}$$

$$W_\infty' : \mathcal{F}(\mathcal{T}_{k,\kappa,\Omega}^{p}(\alpha))/\mathcal{J}(l_k^p(\alpha)) \to \mathcal{A}_{\Omega,r,\infty}^{F,\infty},$$
$$(A_i)_{i\in Z^+} + \mathcal{J}(l_k^p(\alpha)) \mapsto \Phi_{\infty,\infty}^{J}((E_n^F A_i E_{-n}^{F})_{(n,i)\in T}).$$

These definitions (in particular, the latter one) are correct, and for all $s \in \dot{\mathbf{R}}$ and $(A_{n,i})_{(n,i)\in T}$ one has

$$\Phi^J_{s,\infty}((A_{n,i})_{(n,i)\in T}) \;=\; W'_s(W_s((A_{n,i})_{(n,i)\in T})). \tag{2}$$

The proof of (2) for sequences of type (i) goes as for Proposition 3.22 whereas the verification of (2) for the sequence (ii) is as in the 7 th step of 4.2.6.

To prove the inverse closedness we set

$$\mathrm{s}-\lim_{i\to\infty} A_{n,i} \;=:\; A_n. \tag{3}$$

It is easy to see that then

$$\mathrm{s}-\lim_{i\to\infty} A^*_{n,i} \;=\; A^*_n, \tag{4}$$

and (3) and (4) can be used to show that the sequence $(A_n) \in \mathcal{A}^F_{\Omega,r}$ is stable whenever the sequence $(A_{n,i}) \in \mathcal{A}^{F,\infty}_{\Omega,r}$ is so. Now the related result for $\mathcal{A}^F_{\Omega,r}$ gives the invertibility of $W^t((A_{n,i})_{(n,i)\in T})$, $t \in \Omega$, and $W_s((A_{n,i})_{(n,i)\in T})$, $s \in \mathbf{R}$, for stable sequences $(A_{n,i})_{(n,i)\in T} \in \mathcal{A}^{F,\infty}_{\Omega,r}$. The invertibility of $W_\infty((A_{n,i})_{(n,i)\in T})$ for stable $(A_{n,i})$ is obvious (remember observation 2° above). ∎

4.5.3   Invertibility in $\mathcal{F}(\mathcal{T}^p_{k,\kappa,\Omega}(\alpha))/\mathcal{J}(l^p_k(\alpha))$

In the formulation of Theorem 4.12 there appears a new object which has not been discussed yet, viz. the algebra $\mathcal{F}(\mathcal{T}^p_{k,\kappa,\Omega}(\alpha))/\mathcal{J}(l^p_k(\alpha))$. The invertibility of an element of this algebra can again be traced back to a stability problem of a sequence in an algebra $\mathcal{A}^F_{\Omega,r}$ which, on its hand, is subject to the stability theorem 3.5. Thereby, the prebasis F must be chosen in accordance with the parameters k and κ, and one can set $r = 0$.

Given a coset $(A_n) + \mathcal{J}(l^p_k(\alpha))$ in $\mathcal{F}(\mathcal{T}^p_{k,\kappa,\Omega}(\alpha))/\mathcal{J}(l^p_k(\alpha))$ we pick any representative (A_n). The sequence $(E^F_n A_n E^F_{-n})$ belongs to $\mathcal{A}^F_{\Omega,r}$ then, and we can form the operators $W^t(E^F_n A_n E^F_{-n})$ with $t \in \Omega$, $W_s(E^F_n A_n E^F_{-n})$ with $s \in \mathbf{R}$, and the coset $W_\infty(E^F_n A_n E^F_{-n})$ with respect to this algebra. If $s \neq 0$ then these operators/cosets depend on the coset $(A_n) + \mathcal{J}(l^p_k(\alpha))$ only, whereas $W_0(E^F_n A_n E^F_{-n})$ actually depends on the choice of the representative. But it turns out that $W_0(E^F_n A_n E^F_{-n})$ and $W_0(E^F_n A'_n E^F_{-n})$ can only differ by a compact operator if (A_n) and (A'_n) are representatives of the same coset. Hence, the coset $W_0(E^F_n A_n E^F_{-n})/K(l^p_k(\alpha))$ depends on $(A_n) + \mathcal{J}(l^p_k(\alpha))$ only.

Now we can formulate a criterion for invertibility of elements in the quotient algebra $\mathcal{F}(\mathcal{T}^p_{k,\kappa,\Omega}(\alpha))/\mathcal{J}(l^p_k(\alpha))$.

Theorem 4.13 *Let $(A_n) \in \mathcal{F}(\mathcal{T}^p_{k,\kappa,\Omega}(\alpha))$. The coset $(A_n) + \mathcal{J}(l^p_k(\alpha))$ is invertible if and only if the operators $W^t(E^F_n A_n E^F_{-n})$ for $t \in \Omega$, the operators $W_s(E^F_n A_n E^F_{-n})$ for $s \in \{-1,1\}$, and the cosets $W_0(E^F_n A_n E^F_{-n}) + K(l^p_k(\alpha))$ and $W_\infty(E^F_n A_n E^F_{-n})$ are invertible.*

Sketch of the proof As in Theorem 4.1 one can check that the only relevant homomorphisms W_s are those with $s \in \{-1, 0, 1, \infty\}$. Further one only needs Fredholmness of $W_0(E_n^F A_n E_{-n}^F)$ since its invertibility would allow to lift the ideal $\mathcal{J}(l_k^p(\alpha))$ to $\mathcal{G}(l_k^p(\alpha))$ by means of the lifting theorem 1.8. ∎

If, in particular, $A_n = R_n A R_n + Q_n$, then this theorem gets a somewhat simpler form:

Corollary 4.1 *If* $A \in \mathcal{T}_{k,\kappa,\Omega}^p(\alpha)$ *then the coset* $(R_n A R_n + Q_n) + \mathcal{J}(l_k^p(\alpha))$ *is invertible in* $\mathcal{F}(\mathcal{T}_{k,\kappa,\Omega}^p(\alpha))/\mathcal{J}(l_k^p(\alpha))$ *if and only if the operators* $W^t(E_n^F(R_n A R_n + Q_n)E_{-n}^F)$, $t \in \Omega$, *and* $W_{\pm 1}(E_n^F(R_n A R_n + Q_n)E_{-n}^F)$ *are invertible and if* A *is Fredholm.*

Compare Theorem 4.1 for a proof.

4.5.4 An example: The Galerkin method

As an illustration we consider the spline Galerkin method for the singular integral operator

$$A = aI + bS_R + K \tag{1}$$

where a, b are continuous on $\mathbf{R}$ and may jump at ∞, and where K is compact. If F and G are k-elementic prebases which form a canonical pair then the formal application of the F, G-Galerkin method to the operator (1) requires to solve the infinite system with system matrix

$$L_n^{F,G}(aI + bS_R + K)|S_n^F . \tag{2}$$

Multiplying both sides of (2) by the factor $(E_n^F R_i E_{-n}^F)$ we cut off the singular point ∞ and obtain a sequence of finite systems with system matrices

$$E_n^F R_i E_{-n}^F L_n^{F,G}(aI + bS_R + K)E_n^F R_i E_{-n}^F|S_n^F . \tag{3}$$

For convenience we change (3) formally into an infinite system again by adding the sequence $(E_n^F Q_i E_{-n}^F|S_n^F)$. This new system has clearly the same stability properties as (3). Finally we choose any index set $T \subseteq \mathbf{Z}^+ \times \mathbf{Z}^+$ which forces the convergence of $(i/n)_{(n,i) \in T}$ to infinity as $(n, i) \to \infty$ (see, for example, Section 4.5.1).

Theorem 4.14 *The modified* (F, G)-*Galerkin method*

$$(A_{n,i}) = (E_n^F R_i E_{-n}^F L_n^{F,G}(aI + bS_R + K)E_n^F R_i E_{-n}^F|S_n^F + E_n^F Q_i E_{-n}^F)$$

is stable if and only if

(i) the operator $aI + bS_R + K$ *is invertible on* $L_R^p(\alpha)$,

(ii) the functions $a(s) + b(s)\rho^{F,G}$ with $\rho^{F,G} = (\lambda^{F,G})^{-1}\sigma^{F,G}$ are invertible for all $s \in \mathbf{R}$,

(iii) the operator

$$(a(+\infty)P + a(-\infty)Q) + (b(+\infty)P + b(-\infty)Q)T^0(\rho^{F,G})$$

is Fredholm on $l_k^p(\alpha)$, the Toeplitz operators

$$PT^0(a(-\infty) + b(-\infty)\rho^{F,G})P + Q \text{ and } QT^0(a(+\infty) + b(+\infty)\rho^{F,G})Q + P$$

are invertible on $l_k^p(0)$, and the singular integral operator

$$\chi_{[-1,1]}[(a(+\infty)\chi_{R^+} + a(-\infty)\chi_{R^-})I + (b(+\infty)\chi_{R^+} + b(-\infty)\chi_{R^-})S_R]\chi_{[-1,1]}I$$

is invertible on $L_{[-1,1]}^p(\alpha)$.

Condition (i) is just the invertibility of $W^1((A_{n,i})_{(n,i)\in T})$, condition (ii) is equivalent to the invertibility of $W_s((A_{n,i})_{(n,i)\in T})$ with $s \in \mathbf{R}$, whereas (iii) contains necessary and and sufficient conditions for the invertibility of the coset $W_\infty((A_{n,i})_{(n,i)\in T})$. For the latter equivalence we refer to Example 4.2. The conditions (ii) and (iii) are effective in scalar case.

Analogously one can study stability of Galerkin methods for compound Mellin operators as well as other projection or quadrature methods for these operators.

4.5.5   Approximation methods cutting off finite and infinite points

It is obvious how to combine cutting-off-methods both for finite points and for infinity. These methods should depend on the mesh parameter n, on the parameter i controlling the size of the cuts at finite points, and on a parameter j for the cut at infinity. As index set we choose a $T \subseteq \mathbf{Z}^+ \times \mathbf{Z}^+ \times \mathbf{Z}^+$ which guarantees the convergences $(i/n) \to 0$ and $(j/n) \to \infty$ as $(n,i,j) \to \infty$. The system matrices result from multiplying the original sequence by the cutting-off-factors both for finite points and for infinity; for instance

$$(E_n^F Q_i E_{-n}^F)(E_n^F R_j E_{-n}^F)(L_n^{F,G} A | S_n^F)(E_n^F Q_i E_{-n}^F)(E_n^F R_j E_{-n}^F | S_n^F) \tag{1}$$

is a cutting-off-method for the points 0 and ∞ which substitutes the infinite sytem matrix

$$(L_n^{F,G} A | S_n^F) . \tag{2}$$

The associated "global" stability conditions, i.e. those related with W^t, coincide for (1) and (2), whereas the "local" stability conditions for (1) (related with the homomorphisms W_s) coincide with those of (2) if $s \neq 0, s \neq \infty$, are as in Theorem 4.5(c) if $s = 0$, are as in Theorem 4.12(c) if $s = \infty$.

4.6 Exercises

E 4.1 Let $a_i \in PC_{p,a}$ for $i = 1, \ldots, r$, and set $\tilde{a}_i(t) := a_i(1/t)$. Prove that the finite section method $(R_n A R_n + Q_n)$ applies to the operator $A = \prod_{i=1}^{r} PT^0(a_i)P + Q$ if and only if the operators $A \in L(l_Z^p(\alpha))$, $PT^0(\prod_{i=1}^{r} \tilde{a}_i)P + Q \in L(l_Z^p(0))$, and $\chi_{[0,1]}(\prod_{i=1}^{r}(a_i(t-0)(I+S_R)/2 + a_i(t+0)(I-S_R)/2))\chi_{[0,1]}I + \chi_{R\backslash[0,1]}I \in L(L_R^p(\alpha))$, $t \in \mathbf{T}$, are invertible.

E 4.2 Show that the *modified* finite section method

$$(\prod_{i=1}^{r} R_n(PT^0(a_i)P + Q)R_n + Q_n)_{n \geq 1}$$

applies to the operator A in Exercise 4.1 if and only if the operators $A \in L(l_Z^p(\alpha))$ and $\prod_{i=1}^{r}(PT^0(\tilde{a}_i) P + Q) \in L(l_Z^p(0))$ are invertible.

E 4.3 Let $\mathcal{F}$ stand for the Banach algebra of all bounded sequences (A_n) of operators $A_n \in L(l_{Z+}^2(0))$, denote the smallest closed subalgebra of $\mathcal{F}$ containing all sequences $(P_nT(a)P_n + Q_n)$ with $P_n(x_k) = (x_0, \ldots, x_n, 0, 0, \ldots)$ and $Q_n = I - P_n$ and with $a \in PC$ (= the set of piecewise continuous functions) by $\mathcal{A}$, and write $\mathcal{G}$ for the ideal in $\mathcal{A}$ containing all sequences tending to zero in the norm. Further, let $\mathcal{B}$ stand for the smallest closed subalgebra of $L(l_{Z+}^2(0)) \times L(l_{Z+}^2(0))$ (with element-wise operations and norm $\|(A, B)\| = \sup\{\|A\|, \|B\|\}$) containing all ordered pairs $(T(a), T(\tilde{a}))$ with $a \in PC$. Show that the algebras $\mathcal{A}/\mathcal{G}$ and $\mathcal{B}$ are isometrically isomorphic.
(Hint: Exercises 4.2 and 1.9.)

E 4.4 Let $a, b \in PC_{p,\alpha}$. Show that the modified finite section method

$$(R_nPT^0(a)PR_{kn}PT^0(b)PR_n + Q_n)$$

applies to the product $A := PT^0(a)PT^0(b)P + Q$ if and only if the operators $A \in L(l_Z^p(\alpha))$, $PT^0(\tilde{a}\tilde{b})P + Q \in L(l_Z^p(0))$, and

$$\chi_{[0,1]}(a(t+0)(I-S_R)/2 + a(t-0)(I+S_R)/2)\chi_{[0,k]} \times$$
$$\times (b(t+0)(I-S_R)/2 + b(t-0)(I+S_R)/2)\chi_{[0,1]}I + \chi_{R\backslash[0,1]}I \in L(l_R^p(\alpha))$$

are invertible for all $t \in \mathbf{T}$.

E 4.5 Let $a, b \in PC_{p,\alpha}$. (a) Prove that the modified finite section method

$$(R_nT^0(a)PR_n + PQ_n + R_{kn}T^0(b)QR_{kn} + QQ_{kn})$$

applies to the operator $A = T^0(a)P + T^0(b)Q \in L(l_Z^p(\alpha))$ if and only if the operators A, $PT^0(\tilde{a})P + Q \in L(l_Z^p(0))$, $P + QT^0(b)Q \in L(l_Z^p(0))$, and

$$\chi_{[-1,1]}(a(t+0)(I-S_R)/2 + a(t-0)(I+S_R)/2)\chi_{[0,1]}I + \chi_{(1,\infty)}I +$$
$$+\chi_{[-k,k]}(b(t+0)(I-S_R)/2 + b(t-0)(I+S_R)/2)\chi_{[-k,0]}I +$$
$$+\chi_{(-\infty,-k)}I \in L(L_R^p(\alpha)), \quad t \in \mathbf{T},$$

are invertible.

(b) Prove that the modified finite section method

$$(R_{kn}T^0(a)PR_n + PQ_n + R_{kn}T^0(b)QR_{kn} + QQ_{kn})$$

applies to the operator $A = T^0(a)P + T^0(b)Q \in L(l_Z^p(\alpha))$ if and only if the operators A, $T^0(a)Q + P \in L(l_Z^p(0))$, $P + QT^0(\tilde{b})Q \in L(l_Z^p(0))$, and

$$(I + S_R)/2(\chi_{(-\infty,-k)} + b(t-0)\chi_{[-k,0]} + a(t-0)\chi_{[0,1]} + \chi_{(1,\infty)})I +$$
$$+(I - S_R)/2(\chi_{(-\infty,-k)} + b(t+0)\chi_{[-k,0]} + a(t+0)\chi_{[0,1]} + \chi_{(1,\infty)})I$$

$\in L(L_R^p(\alpha))$, $t \in \mathbf{T}$, are invertible.

E 4.6 Let $a \in PC_{p,0}$ (or any other multiplier on $l_Z^p(0)$) such that $A = PT^0(a)P + Q \in L(l_Z^p(0))$ is an invertible operator. Show that the finite section method $(R_n A^{-1} P_n + Q_n)$ applies to A^{-1} then.

(Hint: Use Kozak's identity; compare [BS 2], Corallary 7.18.)

Remark: It is still an open problem whether this assertion remains true for weighted spaces $l_Z^p(\alpha)$ with $\alpha \neq 0$.

E 4.7 Let $A \in L(l_Z^p(\alpha))$ and $K \in K(l_Z^p(\alpha))$. Prove that the infinite section method $(R_n + Q_n A Q_n)$ is stable if and only if the infinite section method $(R_n + Q_n(A + K)Q_n)$ is stable. (That is, the stability of $(R_n + Q_n A Q_n)$ depends only on the coset $A + K(l_Z^p(\alpha))$.)

E 4.8 Derive necessary and sufficient conditions for the stability of the infinite section method $(R_n + Q_n A Q_n)$ if A is the operator $\prod_{i=1}^r PT^0(a_i)P + Q$ and $a_i \in PC_{p,\alpha}$.

E 4.9 Derive Verbitski's conditions in Examples 4.1 and 4.2 by means of the symbol calculus for singular integral operators and of Coburn's theorem.

E 4.10 Let $T \subseteq \mathbf{Z}^+ \times \mathbf{Z}^+$ be an index set as in Section 4.2.2.

(a) Prove that the ideal $\mathcal{G}^T = \mathcal{A}_{\Omega,r}^{F,T} \cap \mathcal{K}^T$ introduced in the proof of Theorem 4.5 contains all sequences $(G_{n,i})_{(n,i)\in T}$ where $E_{-n_0}^F G_{n_0,i_0} E_{n_0}^F$ is compact for a certain index pair $(n_0, i_0) \in T$ and where $G_{n,i} = 0$ for $(n,i) \neq (n_0, i_0)$.

(b) Let $\tilde{\mathcal{G}}^T$ denote the closure in $\mathcal{A}_{\Omega,r}^{F,T}$ of the collection of all sequences of the form described in (a). Show that the inclusion $\tilde{\mathcal{G}}^T \subseteq \mathcal{G}^T$ is proper.

(Compare these results with Proposition 3.15 .)

E 4.11 Consider the modified Galerkin method in Theorem 4.9 where both the points 0 and $1 \in \mathbf{R}$ are cut of. Modify this method once more by introducing independent cutting-off-parameters, say i at 0 and i' at 1, construct a suitable index set $T \subseteq \mathbf{Z}^+ \times \mathbf{Z}^+$ such that the associated projection sequence

$$\left(E_n^F Q_i V_n Q_{i'} V_{-n} L_n^{F,F}\right)_{(n,i,i')\in T}$$

converges strongly to I as $(n, i, i') \to \infty$, and show that the method modified twice is stable if and only if the method in Theorem 4.9 is stable.

4.7 Comments and references

4.1 Here is the continuation of the brief history of the finite section method for operators in $\mathcal{T}^p_{1,1,T}(\alpha)$. Gohberg and Feldman ([GF 1], Theorem 4.1) showed that the finite section method applies to the Toeplitz operator $T(a)$ on $l^2_{Z+}(0)$ with a having a finite number of discontinuities if and only if this operator is invertible. Verbitski and Krupnik [KV] derived necessary and sufficient conditions for the stability of the finite section method for $T(a)$ on $l^p_{Z+}(\alpha)$ in case the generating function a has exactly one discontinuity and, using separation techniques, Böttcher and Silbermann [BS 4] generalized their results to the case of Toeplitz operators on $l^p_{Z+}(0)$ with finitely many discontinuities in their symbols. The general case, i.e. operators $T(a) + K$ on $l^p_{Z+}(0)$ where a has infinitely many discontinuities and K is compact, was successfully tackled by Silbermann [S 2] who developed for this goal a completely new and universally applicable approach based on a Banach algebra language and on localization techniques (see also Chapter 1). The matrix case is also treated in [S 2]. Silbermann's results were generalized by Böttcher to weighted l^p-spaces (the proof is published in [RS 9]).

The problem of applicability of the finite section method to local representatives of powers resp. products of Toeplitz operators was solved by Verbitski in [V 1] resp. Roch in [R 7]. But there is a mistake in [R 7] in applying the local principle, and so the general statement of this paper concerning arbitrary products of Toeplitz operators is false. This mistake was pointed out and corrected by Rathsfeld (private communication). The first complete proof of stability conditions for arbitrary operators in the algebra generated by Toeplitz operators on $l^2_{Z+}(0)$ with piecewise continuous coefficients was given in the book [BS 2], Theorem 7.72. An extension of these results to the matrix case and to operators on $l^p_{Z+}(0)$ is in [R 2].

The pioneering paper in studying the finite section method for operators $T^0(a)P + T^0(b)Q$ is Verbitski [V 2]. He showed in case of opeators on $l^2_Z(0)$ and of coefficients a and b having finitely many discontinuities that the finite section method applies to $T^0(a)P + T^0(b)Q$ if the operators $T^0(a)P + Q$ and $P + T^0(b)Q$ are invertible and if a certain local condition (which is now usually referred to as Verbitski's condition, compare Example 4.1) is satisfied. Silbermann [S 3] showed the sufficiency of Verbitski's condition in case a and b are arbitrary piecewise continuous functions. It has been an open problem for a long time whether Verbitski's condition is also necessary for the applicability of the finite section method, and how the adequate formulation of this condition in case of matrix coefficients and operators on $l^p_Z(0)$ looks like. These problems were solved by Rathsfeld [Ra 1]. The general case pertaining to arbitrary operators in $(\mathcal{T}^p_{1,1,T}(0))_{k\times k}$ is treated in Roch [R 2] (see also [RS 6] for a detailed proof).

The results for $(\mathcal{T}^p_{1,1,T}(\alpha))_{k\times k}$ being subject to Chapter 3 and Section 4.1 (compare also the comments and references to 3.3) are new.

Theorem 4.2 is due to Spitkovski and Tashbaev [ST] who successfully applied factorization theory to make some of our stability conditions more explicit. The simple but surprisingly useful identity in Section 4.1.5 goes back to Kozak (private communication).

The consideration of the infinite section methods for operators in the Toeplitz algebra was suggested by the cutting-off-techniques studied in the following sections. The results in Section 4.1.7 are new.

4.2-4.4 The principle of cutting off "perturbing" or "singular" points is certainly not new.

As W. L. Wendland told us, already Fredholm used related techniques to "regularize" determinants of infinite matrices. Among the recent papers, in particular Chandler and Graham [CGr 1] and [CGr 2] had a great influence on numerical applications. In the paper [CGr 2], they proved stability of spline collocation methods for Mellin operators on $L^\infty_{[0,1]}$ for spline spaces modified in a neighbourhood of the "singular" point 0 by taking piecewise constant splines in place of higher order ones. In [CGr 1] the spline spaces are modified by ommiting the basis spline functions in a neighbourhood of the critical points in order to get stability. These methods were employed by Elschner [E] for Mellin and Wiener-Hopf operators with continuous generating functions on L^p-spaces, and by Prößdorf and Rathsfeld [PR 1], Remark 34, for quadrature methods for Mellin operators.

The results of 4.2-4.4 (in particular the definition of the algebra of approximation sequences cutting off singularities, the derivation of stability conditions, and the investigation of many of the examples) are taken from Roch [R 8]. In particular, the material presented in 4.2.7, 4.3.4 and 4.3.5 is new.

4.5 These results are published here for the first time.

Chapter 5

Manifolds

The fifth chapter is the quintessence of the whole. Here we are going to deal with spline approximation methods for singular integral operators over curves in the complex plane, and for this, we shall need all things which had been placed at disposal in the preceding chapters. Starting with defining spline spaces and spline projections on curves as in Chapter 2, we then introduce a Banach algebra of spline approximation sequences for operators over curves. The study of stability of these sequences follows again the general scheme presented in Chapter 1. Thereby, the sequences considered in Chapter 3, will serve as local representatives in the following sense: a smooth curve behaves locally, at some given point, as its tangent at this point. Similarly, an approximation sequence over a smooth curve behaves locally as a certain sequence over the tangent, i.e. over the real axis, and these are just the sequences we dealt with in the third chapter.

If the curve is more complicated and possesses corners, intersections, or endpoints, then a new type of singularities appears which we call geometric ones. As we shall point out, geometric singularities can be handled by the same cutting-off-technique as discussed in Chapter 4.

5.1   Algebras of singular integral operators over composed curves

5.1.1   Composed curves

We start with giving a brief account on Fredholm theory of singular integral operators over composed curves.

The simplest composed curves are the *simple arcs*. A simple arc Γ is an oriented and rectifiable curve in the complex plane which is homeomorphic to a closed interval, and which satisfies the so-called *Lyapunov condition*: If $t : [0, l] \to \Gamma$ is the natural parameter representation of Γ by its arc length then t' is Hölder continuous, that is, $|t'(s_1) - t'(s_2)| \leq C |s_1 - s_2|^\alpha$ with some constants $C > 0$ and $\alpha \in (0, 1)$. We shall always suppose that t preserves the orientation if $[0, l]$ is oriented in the common manner. The points $t(0)$ and $t(l)$ are the *end points* of Γ; the other points of Γ are called the *inner* ones.

The *oriented tangent* of the simple arc Γ at an inner point $t(s) \in \Gamma$ is the line $t(s) + t'(s)\mathbf{R}$ provided with the natural orientation of $\mathbf{R}$. At the end points of Γ we define *oriented semi-tangents* by $t(0) + t'(0)\mathbf{R}^+$ and $t(l) + t'(l)\mathbf{R}^-$ and provide them with the natural orientations of $\mathbf{R}^+$ and $\mathbf{R}^-$, respectively. Observe that a simple arc always forms a peak with each of its semi-tangents. For technical reasons we prefer to write sometimes $\mathbf{R}^{+1}$ and $\mathbf{R}^{-1}$ in place of $\mathbf{R}^+$ and $\mathbf{R}^-$. A pair (Γ_1, Γ_2)

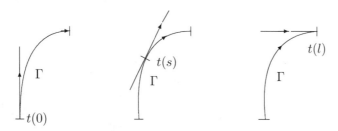

of simple arcs is called *compatible* if either $\Gamma_1 \cap \Gamma_2 = \emptyset$ or if Γ_1 and Γ_2 have exactly one point in common, if this point is end point both of Γ_1 and of Γ_2, and if the one-sided tangents of Γ_1 and Γ_2 at this point do not coincide.

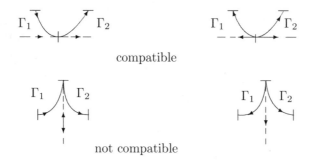

A *composed curve* is the union of a finite number of pairwise compatible simple arcs. Let Γ be a composed curve. A point $x \in \Gamma$ has *order* $w(x)$ if, for each sufficiently small neighbourhood $U \subseteq \mathbf{C}$ of x, the set $(\Gamma \cap U) \setminus \{x\}$ consists of $w(x)$ connected components. A point $x \in \Gamma$ is called *(geometrically) regular* if there is a closed neighbourhood of x such that $\Gamma \cap U$ is a simple arc and if the order of x is 2; otherwise x is *(geometrically) singular*.

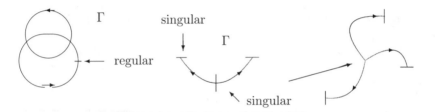

5.1.2 Lebesgue spaces on curves

Let $\alpha : \Gamma \to \mathbf{R}$ be a function on Γ which vanishes everywhere with exception of finitely many points, and let $p \in (1, \infty)$. By $L^p_\Gamma(\alpha)$ we denote the weighted Lebesgue space on Γ provided with norm

$$\|f\| = \left(\int_\Gamma |f(x)|^p \omega(x)^p \, |dx| \right)^{1/p} \tag{1}$$

where ω is the weight function $\omega(x) = \prod_{y \in \Gamma} |x - y|^{\alpha(y)}$ and where $|dx|$ refers to the arc length differential.

The dual space of $L^p_\Gamma(\alpha)$ is isomorphic to $L^q_\Gamma(-\alpha)$ where $1/p + 1/q = 1$, and the action of $g \in L^q_\Gamma(-\alpha)$ at $f \in L^p_\Gamma(\alpha)$ is given by

$$\int_\Gamma f(x) \, g(x) \, |dx| \, . \tag{2}$$

Remember that the notion $L^p_\Gamma(\alpha)$ has a twofold meaning in case Γ is an interval I. If α is a number then the norm on $L^p_I(\alpha)$ is defined by 2.1.1(1), and if it is a function, then by (1). If the number α is identified with the function

$$\alpha : \mathbf{I} \to \mathbf{R} \, , \; t \mapsto \begin{cases} 0 & \text{if} \quad t \neq 0 \\ \alpha & \text{if} \quad t = 0 \end{cases}$$

then the notations coincide. In the sequel we shall often make use of this identification without further comments.

5.1.3 Singular integral operators on Γ

The singular integral operator S_Γ over Γ is defined by

$$(S_\Gamma f)(x) = \frac{1}{\pi i} \int_\Gamma \frac{f(y)}{y - x} \, dy \, , \; x \in \Gamma \, , \tag{1}$$

which has to be understood as Cauchy principal value integral. If Γ is a composed curve, and if $0 < \alpha + 1/p < 1$, then this operator is bounded on $L^p_\Gamma(\alpha)$. Henceforth we shall always suppose these conditions to be satisfied.

A function $a : \Gamma \to \mathbf{C}$ is called *piecewise continuous* if it possesses finite limits at each point $x \in \Gamma$ along each simple arc ending in x. A piecewise continuous function is necessarily bounded, hence, the operators of multiplication by piecewise continuous functions are bounded on $L_\Gamma^p(\alpha)$. By $\mathcal{O}_\Gamma^p(\alpha)$ we denote the smallest closed subalgebra of $L(L_\Gamma^p(\alpha))$ which contains the operator S_Γ and all multiplication operators with piecewise continuous functions. In case $\Gamma = \mathbf{R}$ we have two definitions of $\mathcal{O}_\Gamma^p(\alpha)$ depending on whether α is a number or a function; compare the remark in 5.1.2.

The following analogue of Properties 3.1.1(i) and (ii) are basic:

(i)' The algebra $\mathcal{O}_\Gamma^p(\alpha)$ contains the ideal $K(L_\Gamma^p(\alpha))$ of the compact operators on $L_\Gamma^p(\alpha)$.

(ii)' If f is continuous on Γ then the operator $fS_\Gamma - S_\Gamma fI$ is compact.

On the other hand, the homogenization technique 3.1.1(iii) does clearly not apply directly to arbitrary composed curves. But it applies at least locally as we are going to point out now.

5.1.4   Local homogeneous curves

Let Γ be a composed curve. The union of all $w(x)$ one-sided oriented tangents at a given point $x \in \Gamma$ will be denoted by $\hat{\Gamma}_x$, and we set $\Gamma_x = \hat{\Gamma}_x - x$ (= the algebraic difference). Provide Γ_x with the orientation inherited by those of the one-sided tangents at x. Henceforth we shall always think of Γ_x as being given in the form

$$\Gamma_x = \cup_{j=1}^{w(x)} e^{i\beta_j} \mathbf{R}^{\delta_j}$$

where $\beta_j = \beta_j(x) \in [0, 2\pi)$ and $\delta_j = \delta_j(x) \in \{+1, -1\}$. These numbers are chosen in such a way that $t(x) + e^{i\beta_j}\mathbf{R}^{\delta_j}$ is just a one-sided tangent of Γ at x. Thus, β_j measures the angle between the tangents and the real line, and δ_j indicates the orientation of the tangents.

Obviously, the curves Γ_x and $\hat{\Gamma}_x$ have the same behaviour in a neighbourhood of x but, moreover, Γ_x is a homogeneous curve in the sense that, given $z \in \Gamma_x$ and $s \in \mathbf{R}^+$, then $sz \in \Gamma_x$. If we further define the local weight function $\alpha_x : \Gamma_x \to \mathbf{R}$ by

$$\alpha_x(z) = \begin{cases} \alpha(x) & \text{if} \quad z = 0 \\ 0 & \text{if} \quad z \neq 0 \end{cases}$$

then we can introduce the algebra $\mathcal{O}_{\Gamma_x}^p(\alpha_x)$ of singular integral operators on $L_{\Gamma_x}^p(\alpha_x)$, and we have the following analogue of Property 3.1.1(iii):

(iii)' For $s > 0$, define the operator $Z_s : L_{\Gamma_x}^p(\alpha_x) \to L_{\Gamma_x}^p(\alpha_x)$ by $(Z_s a)(z) = a(z/s)$. Then the strong limit

$$\text{s--}\lim_{s \searrow 0} Z_s^{-1} A Z_s =: W_0(A)$$

exists for each $A \in \mathcal{O}^p_{\Gamma_x}(\alpha_x)$, and $W_0(A)$ is a homogeneous operator, that is, $Z_s^{-1}W_0(A)Z_s = W_0(A)$. Moreover, the mapping $W_0 : \mathcal{O}^p_{\Gamma_x}(\alpha_x) \to L(L^p_{\Gamma_x}(\alpha_x))$ is a continuous algebra homomorphism, which acts as follows:

$$W_0(S_{\Gamma_x}) = S_{\Gamma_x} \tag{1}$$

$$W_0(aI) = \sum_{j=1}^{w(x)} a_j(0)\chi_j I \tag{2}$$

$$W_0(K) = 0 \tag{3}$$

In (2), χ_j refers to the characteristic function of $e^{i\beta_j}\mathbf{R}^{\delta_j}$, and $a_j(0)$ stands for the one-sided limit of the piecewise continuous function a at 0 along $e^{i\beta_j}\mathbf{R}^{\delta_j}$. (The existence of the strong limits s–$\lim_{s\to\infty} Z_s^{-1}AZ_s$ can also be verified, but these limits will not be needed in the sequel.)

5.1.5 Fredholmness of operators in $\mathcal{O}^p_\Gamma(\alpha)$

Let $\mathcal{O}^p_\Gamma(\alpha)^\pi$ denote the quotient algebra $\mathcal{O}^p_\Gamma(\alpha)/K(L^p_\Gamma(\alpha))$, and write π for the canonical homomorphism from $\mathcal{O}^p_\Gamma(\alpha)$ onto $\mathcal{O}^p_\Gamma(\alpha)^\pi$. The following analogue of Theorem 3.1 holds:

Theorem 5.1 *Let Γ be a composed curve.*

(a) *For each $x \in \Gamma$ there is a continuous algebra homomorphism $W_x : \mathcal{O}^p_\Gamma(\alpha)^\pi \to L(L^p_{\Gamma_x}(\alpha_x))$ such that*

$$W_x(\pi(S_\Gamma)) = S_{\Gamma_x}, \tag{1}$$

$$W_x(\pi(aI)) = \sum_{j=1}^{w(x)} a_j(x)\chi_j I, \tag{2}$$

where χ_j is the characteristic function of $e^{i\beta_j}\mathbf{R}^{\delta_j}$, and $a_j(x)$ is the one-sided limit of the piecewise continuous function a at x along the arc having $x + e^{i\beta_j}\mathbf{R}^{\delta_j}$ as its one-sided tangent.

(b) *An operator $A \in \mathcal{O}^p_\Gamma(\alpha)$ is Fredholm if and only if all operators $W_x(\pi(A))$, $x \in \Gamma$, are invertible. If A is Fredholm then it has a regularizer in $\mathcal{O}^p_\Gamma(\alpha)$. The algebras $\mathcal{O}^p_\Gamma(\alpha)^\pi$ and $\mathcal{O}^p_\Gamma(\alpha)$ are inverse closed in $L(L^p_\Gamma(\alpha))/K(L^p_\Gamma(\alpha))$ and $L(L^p_\Gamma(\alpha))$, respectively.*

We complete this theorem in the following subsection by an invertibility criterion for the operators $W_x(\pi(A))$.

5.1.6 Homogeneous singular operators

Let $A \in \mathcal{O}_\Gamma^p(\alpha)$. From 5.1.5(1) and 5.1.5(2) we conclude that $W_x(\pi(A))$ belongs to the smallest closed subalgebra $\Sigma_{\Gamma_x}^p(\alpha_x)$ of $L_{\Gamma_x}^p(\alpha_x)$ which contains the singular integral operator S_{Γ_x} as well as all operators of multiplication by the characteristic functions χ_j, $j = 1, \ldots, w(x)$. The algebra $\Sigma_{\Gamma_x}^p(\alpha_x)$ consists of homogeneous operators only, and it admits a similar characterization as that given in Proposition 2.1 for $\Sigma_R^p(\alpha)$.

Let $L_{R^+}^p(\alpha_x)_{w(x)}$ stand for the Banach space of all vectors $(f_1, \ldots, f_{w(x)})^T$ with $f_j \in L_{R^+}^p(\alpha_x)$, endowed with the norm

$$\|(f_1, \ldots, f_{w(x)})^T\| := \left(\sum_{j=1}^{w(x)} \|f_j\|_{L_{R^+}^p(\alpha_x)}^p \right)^{1/p}.$$

Proposition 5.1 *The algebra $\Sigma_{\Gamma_x}^p(\alpha_x)$ is isometrically isomorphic to the algebra of all operators $(B_{ij})_{i,j=1}^{w(x)} \in L(L_{R^+}^p(\alpha_x)_{w(x)})$ where $B_{ij} \in \Sigma^p(\alpha(x))$ for all i,j and $B_{ij} \in N^p(\alpha(x))$ for all $i \neq j$. The isomorphism takes the operator S_{Γ_x} into the matrix (B_{ij}) with*

$$B_{ij} = \begin{cases} \delta_j S & \text{if} \quad i = j \\ \delta_j N_{\beta_{ij}} & \text{if} \quad i \neq j \end{cases}$$

where the numbers $\beta_{ij} \in (0, 2\pi)$ are determined by $e^{i\beta_{ij}} = \delta_i \delta_j e^{i(\beta_i - \beta_j)}$, and the image of the operator $\sum_{j=1}^{w(x)} a_j \chi_j I$ is the diagonal matrix $\operatorname{diag}(a_1, \ldots, a_{w(x)})$.

If $\eta : L_{\Gamma_x}^p(\alpha_x) \to L_{R^+}^p(\alpha_x)_{w(x)}$ denotes the mapping

$$\eta : f \mapsto (\eta_1 f, \ldots, \eta_{w(x)} f) \quad \text{with} \quad (\eta_j f)(z) = f(\delta_j e^{i\beta_j} z) \tag{1}$$

then the desired isomorphism is just given by $A \mapsto \eta A \eta^{-1}$. The proof proceeds exactly in the same way as that of Proposition 2.1.

Observe that this result provides us with an effective criterion for invertibility of the operators $W_x(\pi(A))$. Indeed, all operators in $\eta \Sigma_{\Gamma_x}^p(\alpha_x) \eta^{-1}$ are Mellin convolution operators with matrix-valued generating functions, and, thus, the operator $W_x(\pi(A))$ is invertible if and only if the associated matrix function is invertible.

5.1.7 Proof of Theorem 5.1

The proof follows the same steps as that of Theorem 3.1. Employing property 5.1.3(ii') we can localize $\mathcal{O}_\Gamma^p(\alpha)^\pi$ over Γ via Allan's local principle. The result is local algebras $\mathcal{O}_\Gamma^p(\alpha)_x^\pi$ with canonical homomorphisms $\Phi_x : \mathcal{O}_\Gamma^p(\alpha)^\pi \to \mathcal{O}_\Gamma^p(\alpha)_x^\pi$, $x \in \Gamma$. Analogously, we localize $\mathcal{O}_{\Gamma_x}^p(\alpha_x)^\pi$ over Γ_x to get local algebras $\mathcal{O}_{\Gamma_x}^p(\alpha_x)_y^\pi$ with canonical homomorphisms $\Psi_y : \mathcal{O}_{\Gamma_x}^p(\alpha_x)^\pi \to \mathcal{O}_{\Gamma_x}^p(\alpha_x)_y^\pi$, $y \in \Gamma_x$. Suppose for a moment we would be aware of the following:

Proposition 5.2 *The algebras* $\mathcal{O}_\Gamma^p(\alpha)_x^\pi$ *and* $\mathcal{O}_{\Gamma_x}^p(\alpha_x)_0^\pi$ *are topologically isomorphic, and the isomorphism sends* $\Phi_x(\pi(S_\Gamma))$ *into* $\Psi_0(\pi(S_{\Gamma_x}))$ *and* $\Phi_x(\pi(aI))$ *into* $\Psi_0(\pi(\sum_j a_j(x)\chi_j I))$ *for each piecewise continuous function a.*

Then we could apply the homogenization property 5.1.4(iii)' in the very same manner as when proving Theorem 3.1 to construct locally equivalent representations of $\mathcal{O}_{\Gamma_x}^p(\alpha_x)_0^\pi$ and to verify the conditions of the concretization step. So we are left on showing the proposition.

For its proof, we pick an open (with respect to the topology of Γ inherited by that one of **C**) and connected subcurve U of Γ which is subject to the following conditions:

(a) $x \in U$,

(b) if $\cup_{j=1}^k \Gamma_j(x)$ is the decomposition of $U \setminus \{x\}$ into its connected components then $k = w(x)$, and the closures of $\Gamma_j(x)$ are simple arcs,

(c) if $y \in \operatorname{clos} U$ and $\alpha(y) \neq 0$ then $y = x$,

(d) the one-sided tangent of $\operatorname{clos} \Gamma_j(x)$ at x is $x + e^{i\beta_j} \mathbf{R}^{\delta_j}$.

If $l_j = l_j(x)$ refers to the arc length of $\operatorname{clos} \Gamma_j(x)$ then we let V denote the subcurve $\cup_{j=1}^w (x)e^{i\beta_j}[0, \delta_j l_j)$ of Γ_x, and we define a mapping $\nu : V \to U$ by

$$\nu(t) = \begin{cases} x & \text{if} \quad t = 0 \\ \text{the point on } \Gamma_j(x) & \\ \text{with arc distance } s \text{ from } x & \text{if} \quad t = e^{i\beta_j}s \text{ with } s \in (0, \delta_j l_j). \end{cases}$$

The homomorphism ν induces via $(\psi f)(t) = f(\nu(t))$, $t \in V$, a continuous and one-to-one mapping ψ from $L_U^p(\alpha|U)$ onto $L_V^p(\alpha_x|V)$, which on its hand defines a continuous algebra isomorphism from $L(L_U^p(\alpha|U))$ into $L(L_V^p(\alpha_x|V))$ by $A \mapsto \psi A \psi^{-1}$.

The following result is fundamental in the theory of singular integral operators over curves.

Lemma 5.1 *(Straightening lemma) The operator*

$$\psi S_U \psi^{-1} - S_V \tag{1}$$

is compact on $L_V^p(\alpha_x|V)$.

The compactness of (1) together with the obvious fact that $\psi a \psi^{-1}$ is again an operator of multiplication by a piecewise continuous function whenever aI is so, involve that the Banach algebras $\mathcal{O}_U^p(\alpha|U)$ and $\mathcal{O}_V^p(\alpha_x|V)$ are topologically isomorphic. Since both algebras contain the compact operators, and since $\psi K \psi^{-1}$ is compact for each compact K, we conclude that the quotient algebras $\mathcal{O}_U^p(\alpha|U)^\pi$ and $\mathcal{O}_V^p(\alpha_x|V)^\pi$ are topologically isomorphic, too. Finally, if f is a continuous function on U which vanishes at x then $\psi f \psi^{-1}$ is the operator of multiplication by a continuous function which vanishes at 0. This can be employed to verify that the isomorphism from $\mathcal{O}_U^p(\alpha|U)^\pi$ onto $\mathcal{O}_V^p(\alpha_x|V)^\pi$ sends the local ideal associated with x to the local ideal at 0. Thus, the quotient algebras $\mathcal{O}_U^p(\alpha|U)_x^\pi$ and $\mathcal{O}_V^p(\alpha_x|V)_0^\pi$ are topologically isomorphic.

It remains to observe that the algebra $\mathcal{O}_\Gamma^p(\alpha)_x^\pi$ is isomorphic to $\mathcal{O}_U^p(\alpha|U)_x^\pi$, and that $\mathcal{O}_V^p(\alpha_x|V)_0^\pi$ and $\mathcal{O}_{\Gamma_x}^p(\alpha_x)_0^\pi$ are isomorphic, which is left as an exercise. ∎

Remark 1 The local algebras $\mathcal{O}_\Gamma^p(\alpha)_x^\pi$ and $\mathcal{O}_{\Gamma_x}^p(\alpha_x)_0^\pi$ (and, consequently, the algebras $\mathcal{O}_\Gamma^p(\alpha)_x^\pi$ and $\Sigma_{\Gamma_x}^p(\alpha_x)$) are even isometrically isomorphic (cp. [RS 4]).

Remark 2 Our proof of Theorem 5.1 is purely local. It is also possible (and seems to be much easier at first glance) to give a more global proof which goes as follows. We break the composed curve Γ into simple arcs Γ_j, $j = 1, \ldots, m$, which are parametrized by the same interval, say $[0, 1]$. Then we think of $L_\Gamma^p(\alpha)$ as a space of vectors $(f_1, \ldots, f_m)^T$ with certain functions $f_j \in L_{[0,1]}^p(\alpha_j)$ with certain weight functions α_j depending on α. If we do this cleverly enough then operators of multiplication over Γ will change into matrices of multiplication operators, and the singular integral operator over Γ turns to a matrix of singular integral and Mellin operators. One "clever" choice of simple arcs is the following: We break Γ into simple arcs $\Gamma_1, \ldots, \Gamma_m$ such that the weight function α vanishes at all inner points of Γ_j, and then we break each Γ_j into two simple arcs Γ_j^+ and Γ_j^-. These will be parametrized by Hölder continuously differentiable mappings $\nu_j^\pm : [0, 1] \to \Gamma_j^\pm$ in such a way that $\nu_j^+(0)$ and $\nu_j^-(0)$ are just the end points of Γ_j and that $\nu_j^+(1) = \nu_j^-(1)$ (possibly, this parametrization changes the orientation).

If we proceed so, the algebra $\mathcal{O}_\Gamma^p(\alpha)$ changes into an algebra of matrix operators the entries of which belong to algebras of singular integral operators over $[0, 1]$, and so a matrix version of Theorem 3.1 could be applied to give Fredholm criteria for operators in $\mathcal{O}_\Gamma^p(\alpha)$. This requires some care since the matrix entries act on spaces with varying weight functions, and it is also not obvious to discover the exact structure of the associated local algebras (locally seen, many entries of the matrix operators vanish!). But the main disadvantage is that this global approach hardly applies to general manifolds whereas the local one works well also in the multidimensional context (see [Hop]).

5.2 Spline approximation methods over homogeneous curves

5.2.1 Seeley's definition

The introduction of singular integral operators over plane curves is fairly simple since 5.1.3(1) allows to define these operators *globally*, that is, at once on the whole curve.

In case of spatial curves, or of higher-dimensional manifolds, there is in general no way to define singular integral operators globally. In 1959, Seeley proposed a *local* definition of singular integral operators over manifolds without boundary. We are going to present a version of Seeley's definition which is specified to the situation of a composed plane curve Γ. For, we indicate the dependence on $x \in \Gamma$ of the open sets U, V and of the homomorphism ψ introduced in 5.1.7 by writing U_x, V_x, and ψ_x, respectively.

Further, if fI is the operator of multiplication by f over U_x then we agree to write f_x for the multiplication operator $\psi f \psi^{-1}$ over V_x.

Definition *Let Γ be a composed curve. An operator $A \in L_{\Gamma}^p(\alpha)$ is called a singular integral operator over Γ if*

(a) *the operators $\psi_x f A g \psi_x^{-1} \chi_{V_x} I - f_x S_{\Gamma_x} g_x I$ are compact on $L_{\Gamma_x}^p(\alpha_x)$ for all $x \in \Gamma$ and for all continuous functions f, g with $\operatorname{supp} f \cup \operatorname{supp} g \subseteq U_x$,*

(b) *the operators $\psi_{x_2} f A g \psi_{x_1}^{-1} \chi_{V_{x_1}} I$ are compact from $L_{\Gamma_{x_1}}^p(\alpha_{x_1})$ into $L_{\Gamma_{x_2}}^p(\alpha_{x_2})$ whenever $x_1 \neq x_2$ and f and g are continuous functions with $\operatorname{supp} f \subseteq U_{x_2}$, $\operatorname{supp} g \subseteq U_{x_1}$, and $\operatorname{supp} f \cap \operatorname{supp} g = \emptyset$.*

For arbitrary manifolds one has to replace the sets U_x by suitable chosen charts, and the ψ_x by smooth mappings between the charts and $\mathbf{C}^n$.

The singular integral operator S_Γ introduced by 5.1.3(1) is clearly a singular operator in Seeley's sense: condition (a) is a reformulation of the straightening lemma, and (b) is obvious.

5.2.2 Fredholm theory versus stability theory

The same local point of view will prove to be advantageous both to introduce and to investigate algebras of approximation sequences for singular integral operators over composed curves (whereas any concrete numerical computation will be undoubtedly done globally).

So our further steps will be as follows: we start with introducing spline spaces and spline approximation methods over homogeneous curves, and then we define approximation sequences over arbitrary composed curves by requiring that, locally seen, the related methods coincide with certain model methods over the local homogeneous curves. The so-defined sequences form an algebra which will be studied by the general scheme of Chapter 1.

Before proceeding with this programme, we shall detail some differences between our approaches to Fredholm theory above and stability theory below which lie in the nature of the matter.

The first thing is that we shall no longer rely on the arc length as only parametrization, since the admission of other parametrizations will provide us more flexibility in defining spline spaces. The second point is that the integers $0, \pm 1, \ldots$ play a distinguished role for spline spaces. Thus it is more natural to map neighbourhoods of inner points of the simple arcs forming our curve to neighbourhoods of inner points of the interval $[0, 1]$ (and not to neighbourhoods of 0). Thirdly, singular integral operators over homogeneous curves and approximation sequences for these operators behave differently under transformation of the curve into a system of semi-axes as in Proposition 5.1.

To explain this effect let, for simplicity, $\Gamma = \mathbf{R}$ and consider the singular integral operator S_R on $L_R^2(0)$. The mapping $\eta : L_R^p(0) \to L_{R^+}^p(0)_2$ defined by

$$\eta : \ f \mapsto (f_1, f_2) \text{ with } f_1(z) = f(z) \, , \ f_2(z) = f(-z) \, , \tag{1}$$

(which is an old crony; see Sections 2.1.3 and 5.1.6) induces an isomorphism between $L(L_R^2(0))$ and $L(L_{R^+}^2(0)_2) \cong L(L_{R^+}^2(0))_{2 \times 2}$ which sends the singular integral operator S_R into the matrix

$$\begin{pmatrix} \chi_{R^+} S_R \chi_{R^+} I & \chi_{R^+} S_R J \chi_{R^+} I \\ \chi_{R^+} J S_R \chi_{R^+} I & \chi_{R^+} J S_R J \chi_{R^+} I \end{pmatrix} \tag{2}$$

where J denotes the flip $(Jf)(t) = f(-t)$. The operator (2) is a pleasant object since it can be rewritten as

$$\begin{pmatrix} S & -N \\ N & -S \end{pmatrix} , \tag{3}$$

and all entries of (3) belong again to an algebra of singular integral operators (now over $\mathbf{R}^+$), see Section 2.1.2.

Now choose $F = \{\chi_{[0,1)}\}$ and consider a spline approximation sequence of the form $(E_n^F T^0(a) E_{-n}^F)$ for S_R. If a is invertible then this sequence is stable, and if $a(1 \pm 0) = \mp 1$ then the sequence $E_n^F T^0(a) E_{-n}^F L_n^{F,F}$ converges to S_R strongly. Applying the transformation (1) again we find

$$\eta(E_n^F T^0(a) E_{-n}^F L_n^{F,F}) \eta^{-1} =$$

$$= \begin{pmatrix} \chi_{R^+} E_n^F T^0(a) E_{-n}^F L_n^{F,F} \chi_{R^+} I & \chi_{R^+} E_n^F T^0(a) E_{-n}^F L_n^{F,F} J \chi_{R^+} I \\ \chi_{R^+} J E_n^F T^0(a) E_{-n}^F L_n^{F,F} \chi_{R^+} I & \chi_{R^+} J E_n^F T^0(a) E_{-n}^F L_n^{F,F} J \chi_{R^+} I \end{pmatrix}$$

which can be rewritten as

$$\begin{pmatrix} E_n^F P T^0(a) P E_{-n}^F L_n^{F,F} & E_n^F P T^0(a) \hat{J} P E_{-n}^F L_n^{F,F} \\ E_n^F P \hat{J} T^0(a) P E_{-n}^F L_n^{F,F} & E_n^F P \hat{J} T^0(a) \hat{J} P E_{-n}^F L_n^{F,F} \end{pmatrix}$$

$$= \begin{pmatrix} E_n^F T(a) E_{-n}^F L_n^{F,F} & E_n^F H(a) E_{-n}^F L_n^{F,F} \\ E_n^F H(\tilde{a}) E_{-n}^F L_n^{F,F} & E_n^F T(\tilde{a}) E_{-n}^F L_n^{F,F} \end{pmatrix} , \tag{4}$$

where $\hat{J}$ refers to the discrete flip $\hat{J}(x_j) = (x_{-j-1})$, $\tilde{a}$ is given by $\tilde{a}(z) = a(1/z)$, and $T(a)$ and $H(a)$ denote the Toeplitz and Hankel operator with generating function a, respectively. The operator (4) is unpleasant since Hankel operators do in general not belong to the Toeplitz algebra $\mathcal{T}^2_{1,1,T}(0)$ (compare Exercise 2.11), and therefore the sequences $(E^F_n H(a) E^F_{-n})$ are not subject to the general stability theorem.

The expedient is evident: we only replace the mapping η in (1) by

$$\eta : L^2_R(0) \to L^2_{R+}(0) \times L^2_{R-}(0) , f \mapsto (f_1, f_2)^T$$

with $f_1(z) = f(z)$ for $z \geq 0$ and $f_2(z) = f(z)$ for $z \leq 0$, and get then

$$\eta(E^F_n T^0(a) E^F_{-n} L^{F,F}_n)\eta^{-1} =$$
$$= \left(\begin{array}{cc} E^F_n PT^0(a)PE^F_{-n}L^{F,F}_n & E^F_n PT^0(a)QE^F_{-n}L^{F,F}_n \\ E^F_n QT^0(a)PE^F_{-n}L^{F,F}_n & E^F_n QT^0(a)QE^F_{-n}L^{F,F}_n \end{array} \right) .$$

But this matrix belongs to $(\mathcal{A}^F_{T,0})_{2\times 2}$.

5.2.3 Spline spaces and Galerkin projections over the semi-axis

We start with the simplest homogeneous curve, the semi-axis $\mathbf{R}^+$. There are a lot of possibilities of introducing a spline space over $\mathbf{R}^+$, e.g. one could simply take the space $\chi_{R^+} S^F_n$ where $F = \{\phi_0, \ldots, \phi_{k-1}\}$ is a canonical prebasis. But for our purposes it is more convenient to have a spline space over $\mathbf{R}^+$ which is naturally embedded into the space S^F_n over the whole axis, and that's why we define $S^F_n(\mathbf{R}^+)$ as the smallest closed subspace of $S^F_n(\subseteq L^p_R(\alpha))$ which contains all basis functions ϕ_{rln} $(r = 0, \ldots, k-1, l \in \mathbf{Z})$ whose support belongs to $\mathbf{R}^+$. If we let $\hat{\chi}^{(n)}_{R^+}$ stand for the operator acting on S^F_n by the rule

$$\hat{\chi}^{(n)}_{R^+} : \sum_{r,l} x_{rl} \phi_{rln} \mapsto \sum_{\text{supp}\phi_{rln}\subseteq R^+} x_{rl} \phi_{rln}$$

then we can simply write $S^F_n(\mathbf{R}^+) = \hat{\chi}^{(n)}_{R^+} S^F_n$.

The *Galerkin projection* from $L^p_{R+}(\alpha)$ onto $S^F_n(\mathbf{R}^+)$ is then the operator $L^{F,F}_{n,R^+}$ which satisfies

$$(L^{F,F}_{n,R^+} f , \phi_{rln}) = (f, \phi_{rln}) \tag{1}$$

for all $f \in L^p_{R+}(\alpha)$ and $\phi_{rln} \in S^F_n(\mathbf{R}^+)$.

With the ansatz $L^{F,F}_{n,R^+} f = \sum_{s,m} x_{sm} \phi_{smn}$, the condition (1) proves to be equivalent to the algebraic system

$$\sum_{s,m} (\phi_{smn}, \phi_{rln}) x_{sm} = (f, \phi_{rln}) . \tag{2}$$

The infiniteness of this system involves that it is by no means clear whether the Galerkin projection actually exists.

Theorem 5.2 *If F is a canonical prebasis then the system matrix of (2) is invertible, and thus the Galerkin projection $L_{n,R^+}^{F,F}$ exists and is uniquely determined.*

Proof   A little thought shows that the operator $\hat{\chi}_{R^+}^{(n)}$ can be rewritten as

$$\hat{\chi}_{R^+}^{(n)} = E_n^F \operatorname{diag}(V_{c_0} P V_{-c_0}, \ldots, V_{c_{k-1}} P V_{-c_{k-1}}) E_{-n}^F, \qquad (3)$$

where the c_r are integers satisfying

$$\operatorname{supp}\phi_{rc_r n} \subseteq \mathbf{R}^+ \quad \text{but} \quad \operatorname{supp}\phi_{r,c_r-1,n} \not\subseteq \mathbf{R}^+.$$

Let us first suppose that $c_0 = c_1 = \ldots = c_{k-1} = 0$. Then the system matrix of (2) is just the block Toeplitz operator

$$T(\lambda^{F,F}) = \operatorname{diag}(P, \ldots, P) T^0(\lambda^{F,F}) \operatorname{diag}(P, \ldots, P),$$

and the assertion of the theorem is equivalent to the following implication

$$T^0(\lambda^{F,F}) \text{ invertible} \Rightarrow T(\lambda^{F,F}) \text{ invertible}. \qquad (4)$$

For a proof of (4) we approximate the basis functions $\phi_r \in F$ by functions $\phi_r^{(n)} \in S(\mathbf{R})$ in the norm of $L_R^2(0)$ such that $\|P_n^F E_n^F - P_n^{F_n} E_n^{F_n}\|_{L_R^2(0)} < \frac{1}{n}$ with F_n standing for $(\phi_0^{(n)}, \ldots, \phi_{k-1}^{(n)})$ (compare the proof of Theorem 2.10). In the same proof we have shown that $\lambda^{F_n, F_n}(t)$ is a Gram matrix for each $t \in \mathbf{T}$, which implies that

$$(\lambda^{F_n, F_n}(t) z, z) \geq 0 \quad \text{for each} \quad z \in \mathbf{C}^k$$

(a Gram matrix is always non-negative). Letting n go to infinity we obtain

$$(\lambda^{F,F}(t) z, z) \geq 0 \quad \text{for all} \quad t \in \mathbf{T}, \ z \in \mathbf{C}^k.$$

Thus the matrices $\lambda^{F,F}(t)$ are non-negative, too, and so their spectra $\sigma(\lambda^{F,F}(t))$ are contained in $[0, \infty)$. Let $\sigma(\lambda^{F,F})$ be the spectrum of the matrix function $\lambda^{F,F}$ viewed as an element of the Banach algebra of all continuous $k \times k$-matrix functions on the unit circle. Clearly,

$$\sigma(\lambda^{F,F}) = \cup_{t \in T} \sigma(\lambda^{F,F}(t)),$$

hence, $\sigma(\lambda^{F,F}) \subseteq [0, \infty)$. Our hypothesis entails that $0 \notin \sigma(\lambda^{F,F})$. Since, on the other hand, $\sigma(\lambda^{F,F})$ is compact, there are strictly positive numbers a and b such that $\sigma(\lambda^{F,F}) \subseteq [a, b]$ and, consequently, $\sigma(\lambda^{F,F}(t)) \subseteq [a, b]$ for all $t \in \mathbf{T}$. But this implies that

$$(\lambda^{F,F}(t) z, z) \geq a \|z\|^2 \quad \text{for all} \quad t \in \mathbf{T}, \ z \in \mathbf{C}^k,$$

or, in other words, the matrix function $\lambda^{F,F}$ is analytically sectorial (see [BS 2], 3.1). Now ([BS 2], Corollary 4.2) gives the invertibility of $T(\lambda^{F,F})$ in $L(l_{\mathbf{Z}^+}^2(\alpha)_k)$. It remains to show its invertibility in $L(l_{\mathbf{Z}^+}^p(\alpha)_k)$.

First observe that $T(\lambda^{F,F})$ is a bounded operator on every $l^r_{Z^+}(\gamma)_k$-space with $r > 1$ and $0 < 1/r + \gamma < 1$ (since $\lambda^{F,F}$ is a polynomial), that $T(\lambda^{F,F})$ is Fredholm on every $l^r_{Z^+}(\gamma)_k$-space if and only if it is Fredholm on $l^2_{Z^+}(0)_k$ (compare [BS 2], Theorem 6.12(c)), and that the index of $T(\lambda^{F,F})$ is independent of r and γ ([BS 2], Theorem 6.12 again). Since $T(\lambda^{F,F})$ is invertible on $l^2_{Z^+}(0)_k$, we so conclude that $T(\lambda^{F,F})$ is Fredholm on every $l^r_{Z^+}(\gamma)_k$, and that $\operatorname{Ind} T(\lambda^{F,F}) = 0$.

Now we invoke a result on Fredholm operators acting on different spaces (see [GF 1], Appendix,1°):

Let the Banach space X_1 be continuously and densely embedded into a Banach space X_2, and let A_1, A_2 be bounded Fredholm operators on X_1 and X_2, respectively, having the same index. If A_2 is an extension of A_1 then $\operatorname{Ker} A_1 = \operatorname{Ker} A_2$.

If $r_1 < 2 < r_2$ then $l^{r_2}_{Z^+}(0) \subseteq l^2_{Z^+}(0) \subseteq l^{r_1}_{Z^+}(0)$, and all embeddings are continuous and dense. So the cited result gives, in combination with the invertibility of $T(\lambda^{F,F})$ on $l^2_{Z^+}(0)_k$, the invertibility of $T(\lambda^{F,F})$ on $l^r_{Z^+}(0)_k$ for all $r > 1$. Further, if $\gamma_1 < 0 < \gamma_2$ then $l^r_{Z^+}(\gamma_2) \subseteq l^r_{Z^+}(0) \subseteq l^r_{Z^+}(\gamma_1)$; the embeddings being continuous and dense again. Applying the same proposition once more we get the invertibility of $T(\lambda^{F,F})$ on every $l^p_{Z^+}(\alpha)_k$-space.

Let us return to the general case of (3) where, possibly, $c_r \neq 0$. Then the system matrix of (2) is

$$\operatorname{diag}(V_{c_0}PV_{-c_0}, \ldots, V_{c_{k-1}}PV_{-c_{k-1}})T^0(\lambda^{F,F})\operatorname{diag}(V_{c_0}PV_{-c_0}, \ldots, V_{c_{k-1}}PV_{-c_{k-1}}),$$
$$(5)$$

and this matrix is invertible if and only if

$$\operatorname{diag}(P,\ldots,P)\operatorname{diag}(V_{-c_0},\ldots,V_{-c_{k-1}})T^0(\lambda^{F,F})\operatorname{diag}(V_{c_0},\ldots,V_{c_{k-1}})\operatorname{diag}(P,\ldots,P)$$

is invertible. Set $F' = \{U_{-c_0}\phi_0, \ldots, U_{-c_{k-1}}\phi_{k-1}\}$. A little thought shows that F' is a canonical prebasis again, and that

$$\operatorname{diag}(V_{-c_0}, \ldots, V_{-c_{k-1}})T^0(\lambda^{F,F})\operatorname{diag}(V_{c_0}, \ldots, V_{c_{k-1}}) = T^0(\lambda^{F',F'}).$$

This reduces all things to showing the implication

$$T^0(\lambda^{F',F'}) \quad invertible \Rightarrow T(\lambda^{F',F'}) \quad invertible,$$

which has already been done. ∎

After establishing the existence of the Galerkin projection onto $S^F_n(\mathbf{R}^+)$ it is now easy to derive its main properties. For this reason, we abbreviate the matrix operator (5) by T for a moment, and write $L^{F,F}_{n,R^+}$ as

$$L^{F,F}_{n,R^+} = E^F_n T^{-1} P^F_n = E^F_n T^{-1} T^0(\lambda^{F,F}) E^F_{-n} E^F_n T^0(\lambda^{F,F})^{-1} P^F_n,$$

and since $T^{-1}T^0(\lambda^{F,F}) = \text{diag}\,(P,\dots,P) + K$ with a certain compact operator K (see Proposition 2.8(a)), we finally get

$$L_{n,R^+}^{F,F} = E_n^F(\text{diag}\,(P,\dots,P) + K)E_{-n}^F L_n^{F,F}\,. \tag{6}$$

Observing that $(E_n^F(\text{diag}\,(P,\dots,P) + K)E_{-n}^F)$ is in $\mathcal{A}_{\Omega,r}^F$ by definition, one can straightforwardly show that

- $L_{n,R^+}^{F,F}$ commutes with $f \in C(\overline{\mathbf{R}^+})$ modulo zero sequences,

- $L_{n,R^+}^{F,F}$ is asymptotically quasi-local,

- $L_{n,R^+}^{F,F}$ converges strongly to $\chi_{R^+}I$ as $n \to \infty$.

5.2.4 Spline spaces over homogeneous curves

Given $w \in \mathbf{Z}^+$, $\beta = (\beta_1,\dots,\beta_w)$ with $\beta_j \in [0,2\pi)$ and $\delta = (\delta_1,\dots,\delta_w)$ with $\delta_j \in \{-1,1\}$ such that $e^{i\beta_j}\mathbf{R}^{\delta_j} \neq e^{i\beta_k}\mathbf{R}^{\delta_k}$ whenever $j \neq k$, we let $\Gamma = \Gamma(\beta,\delta)$ denote the homogeneous curve $\cup_{j=1}^w e^{i\beta_j}\mathbf{R}^{\delta_j}$. Further we choose and fix a vector $\xi = (\xi_1,\dots,\xi_w)$ of positive real numbers and define, in analogy and generalization of the η-mapping in 5.1.6 (1),

$$\eta : L_\Gamma^p(\alpha) \to L_R^p(\alpha)_w\,, \quad \eta f = (\eta_1 f,\dots,\eta_w f)^T \tag{1}$$

with

$$(\eta_j f)(s) = \begin{cases} f(e^{i\beta_j}\xi_j s) & \text{if} \quad s \in \mathbf{R}^{\delta_j} \\ 0 & \text{if} \quad s \in \mathbf{R}^{-\delta_j}\,. \end{cases}$$

Finally, given a subset M of $\mathbf{R}$, we write $\hat{\chi}_M^{(n)}$ for the mapping of S_n^F into itself which sends the spline function $\sum_{r,l} x_{rl}\,\phi_{rln}$ into $\sum_{\text{supp}\,\phi_{rln} \subseteq \overline{M}} x_{rl}\,\phi_{rln}$. The operators $\hat{\chi}_M^{(n)}$ are analogues of the operators of multiplication by characteristic functions χ_M; as the latter ones they are projections, but observe that, for example, $\chi_{R^+}I + \chi_{R^-}I = I$ whereas $\hat{\chi}_{R^+}^{(n)} + \hat{\chi}_{R^-}^{(n)} \neq I|S_n^F$ in general.

Now define the spline space $S_n^F(\Gamma) = S_n^F(\Gamma,\xi)$ over Γ as the smallest closed subspace of $L_\Gamma^p(\alpha)$ which contains all functions f with $\eta_j f \in \hat{\chi}_{R^{\delta_j}}^{(n)} S_n^F$ for $j = 1,\dots,w$. The vector ξ just controlls the size of the equidistant partition on each of the semi-axis forming Γ. Further, one clearly has an isomorphy

$$S_n^F(\Gamma,\xi) \cong \text{diag}\,(\hat{\chi}_{R^{\delta_1}}^{(n)},\dots,\hat{\chi}_{R^{\delta_n}}^{(n)})(S_n^F)_w\,.$$

Let us emphasize once more that $S_n^F(\mathbf{R},(1,1))$ is, in general, a proper subspace of S_n^F since $S_n^F(\mathbf{R},(1,1))$ contains only basis splines ϕ_{rln} whose supports are entirely contained in either $\mathbf{R}^+$ or $\mathbf{R}^-$.

5.2.5 Galerkin projections over homogeneous curves

If $g \in L_\Gamma^q(-\alpha)$ is a linear functional over $L_\Gamma^p(\alpha)$ then one usually thinks of g as acting by

$$(f, g) = \int_\Gamma f(t)g(t) \, |dt| . \tag{1}$$

For our purposes it is more convenient to introduce another action of g by setting

$$(f, g)_\Gamma := (\eta f, \eta g)_{L_R^p(\alpha)_w} = \sum_{j=1}^w \int_R (\eta_j f)(s)(\eta_j g)(s) \, ds , \tag{2}$$

or, what is the same,

$$(f, g)_\Gamma = \sum_{j=1}^w \int_{e^{i\beta_j} R^{\delta_j}} f(t)g(t)\xi_j^{-1} \, |dt| .$$

Now we define the Galerkin projection $L_{n,\Gamma}^{F,F}$ mapping $L_\Gamma^p(\alpha)$ onto $S_n^F(\Gamma, \xi)$ by requiring that

$$(L_{n,\Gamma}^{F,F} f, g)_\Gamma = (f, g)_\Gamma$$

for all $f \in L_\Gamma^p(\alpha)$ and $g \in S_n^F(\Gamma, \xi)$. The advantage of (2) in comparision with (1) is that $\eta L_{n,\Gamma}^{F,F} \eta^{-1}$ is just the Galerkin projection of $L_R^p(\alpha)_w$ onto $\hat{\chi}_n^\delta(S_n^F)_w$, where $\hat{\chi}_n^\delta$ abbreviates the diagonal operator

$$\mathrm{diag}\,(\hat{\chi}_{R^{\delta_1}}^{(n)}, \ldots, \hat{\chi}_{R^{\delta_w}}^{(n)}) : (S_n^F)_w \to (S_n^F)_w .$$

Thus, if we define $L_{n,R^-}^{F,F}$ in analogy to $L_{n,R^+}^{F,F}$ (see Section 5.2.3), then

$$\eta L_{n,\Gamma}^{F,F} \eta^{-1} = \mathrm{diag}\,(L_{n,R^{\delta_1}}^{F,F}, \ldots, L_{n,R^{\delta_w}}^{F,F}) . \tag{3}$$

Having (3) in mind it is easy to derive both the existence of the Galerkin projection over Γ and its main properties (commutator property, asymptotic quasi-locality, strong convergence to the identity operator on Γ).

5.2.6 Approximation sequences over homogeneous curves

The approximation sequences over Γ which we will take into consideration are those sequences (A_n), $A_n : S_n^F(\Gamma, \xi) \to S_n^F(\Gamma, \xi)$ for which

$$(\eta A_n \eta^{-1}|\hat{\chi}_n^\delta(S_n^F)_w) \in (\hat{\chi}_n^\delta)(\mathcal{A}_{\Omega,r}^F)_{w \times w}(\hat{\chi}_n^\delta) .$$

The collection of all of these sequences will be denoted by $\mathcal{A}_{\Omega,r}^F(\Gamma) = \mathcal{A}_{\Omega,r}^F(\Gamma, \xi)$. The stability of a sequence $(A_n) \in \mathcal{A}_{\Omega,r}^F(\Gamma)$ is equivalent to the stability of the sequence $(\eta A_n \eta^{-1}|\hat{\chi}_n^\delta(S_n^F)_w)$ or, what is the same, to the stability of the sequence

$$(\eta A_n \eta^{-1}|\hat{\chi}_n^\delta(S_n^F)_w) + ((I - \hat{\chi}_n^\delta)|(S_n^F)_w) .$$

This sequence is in $(\mathcal{A}_{\Omega,r}^F)_{w \times w}$ since $(\hat{\chi}_n^\delta)$ is in $(\mathcal{A}_{\Omega,r}^F)_{w \times w}$ (see Exercise 5.2), and so it is subject to the the matrix version of the general stability theorem.

We conclude this section with some examples of sequences in $\mathcal{A}_{\Omega,r}^F(\Gamma, \xi)$.

Example 5.1 *If K is compact on $L_\Gamma^p(\alpha)$ and if $1 \in \Omega$ then $(L_{n,\Gamma}^{F,F} K | S_n^F(\Gamma, \xi)) \in$* $\mathcal{A}_{\Omega,r}^F(\Gamma, \xi)$.

This is immediate from the identity

$$\eta L_{n,\Gamma}^{F,F} K \eta^{-1} = \eta L_{n,\Gamma}^{F,F} \eta^{-1} \eta K \eta^{-1} = \mathrm{diag}\,(L_{n,R^{\delta_1}}^{F,F}, \dots, L_{n,R^{\delta_w}}^{F,F})\, K'$$

with $K' \in K(L_R^p(\alpha)_w)$, from the representation 5.2.3(6) of the projections $L_{n,R^\pm}^{F,F}$, and from the inclusion $(L_n^{F,F} K | S_n^F) \in \mathcal{A}_{\Omega,r}^F$ established in Proposition 3.17.

Example 5.2 *If a is continuous on $\Gamma \setminus \{0\}$ and has one-sided limits at 0 and ∞ along each semi-axis forming Γ then $(L_{n,\Gamma}^{F,F} aI | S_n^F(\Gamma, \xi)) \in \mathcal{A}_{\Omega,r}^F(\Gamma, \xi)$.*

The proof proceeds as in Example 5.1 and rests on the facts that $(L_n^{F,F} fI | S_n^F) \in \mathcal{A}_{\Omega,r}^F$ for each $f \in C(\dot{\mathbf{R}})$ (Proposition 3.16) and for $f = \chi_{R^+}$ (see the proofs of Propositions 3.16 and 3.24 and the definition of $\mathcal{A}_{\Omega,r}^F$).

Example 5.3 $(L_{n,\Gamma}^{F,F} S_\Gamma | S_n^F(\Gamma, \xi)) \in \mathcal{A}_{\{1\},r}^F(\Gamma, \xi)$.

The proof of this fact is more complicated and requires some preparations. Let J denote the flip operator on the real axis, $(Jf)(t) = f(-t)$, set $J_+ = I$ and $J_- = J$, and define $J_\delta := \mathrm{diag}\,(J_{\delta_1}, \dots, J_{\delta_w})$. Further we let the operators Z_s on $\mathbf{R}$ be defined as in Section 3.1.1 by $(Z_s f)(x) = f(x/s)$, and set $Z_\xi := \mathrm{diag}\,(Z_{\xi_1}, \dots, Z_{\xi_w})$. For η being defined by 5.2.4(1) it is now easy to check that $Z_\xi J_\delta \eta$ is just the mapping 5.1.6(1). Hence,

$$\eta S_\Gamma \eta^{-1} = J_\delta Z_\xi^{-1}(B_{ij}) Z_\xi J_\delta \tag{1}$$

where the matrix (B_{ij}) is given as in Proposition 5.1 by

$$B_{ij} = \begin{cases} \delta_j S & \text{if} \quad i = j \\ \delta_j N_{\beta_{ij}} & \text{if} \quad i \neq j \end{cases}$$

with $\beta_{ij} \in (0, 2\pi)$ being determined by $e^{i\beta_{ij}} = \delta_i \delta_j e^{i(\beta_i - \beta_j)}$. Let (C_{ij}) denote the matrix (1). We are going to show that the entries of this matrix are in $\mathcal{O}_R^p(\alpha)$.

If $i = j$ then

$$C_{jj} = J_{\delta_j} Z_{\xi_j}^{-1} \delta_j S_{R^+} Z_{\xi_j} J_{\delta_j} = S_{R^{\delta_j}}$$

since S is homogeneous and $JS_R J = -S_R$. In case $i \neq j$ we have

$$C_{ij} = J_{\delta_i} Z_{\xi_i}^{-1} \delta_j N_{\beta_{ij}} Z_{\xi_j} J_{\delta_j} = \delta_j J_{\delta_i} N_{\beta_{ij}} Z_{\xi_j / \xi_i} J_{\delta_j} ,$$

and our claim is a consequence of the following proposition where $N^p(\alpha)$ refers to the ideal of $\Sigma^p(\alpha)$ introduced in Section 2.1.2.

Proposition 5.3 *(a)* $JN^p(\alpha) \subseteq \Sigma^p_R(\alpha)$, $N^p(\alpha)J \subseteq \Sigma^p_R(\alpha)$.
(b) For all $s > 0$, $N^p(\alpha)Z_s = Z_s N^p(\alpha) = N^p(\alpha)$.

(*Warning*: Neither the operators J and Z_s (unless $s = 1$) nor the operators JS_{R^+} and $Z_s S_{R^+}$ are in $\Sigma^p_R(\alpha)$ again! The proof is left as an exercise.)

Proof Part (a) has already been established by means of Costabel's decomposition of $\Sigma^p_R(\alpha)$ in Section 2.1.3. It is also not too hard to give a direct proof; see Exercise 2.3(a). For a proof of part (b) we first show that $NZ_s \in N^p(\alpha)$ for all $s \in \mathbf{R}^+$. Let $k(t) = 1/(1+t)$ be the kernel function of the Hankel operator N and $k_s(t) = k(t/s) = 1/(1+t/s)$ be the kernel function of the (homogeneous) operator NZ_s. Then

$$(Mk_s)(z) = \int_0^\infty x^{1/p+\alpha-zi-1} k(x/s)\,dx = s^{1/p+\alpha-zi}(Mk)(z) \tag{2}$$

whence in particular follows that the Mellin symbol of NZ_s equals

$$b(z) = \frac{s^{1/p+\alpha-zi}}{\sinh(z+i(1/p+\alpha))\pi} \ . \tag{3}$$

(Observe that identity (2) remains valid if k and k_s are the kernel functions of $M^0(c)$ and $M^0(c)Z_s$ with arbitrary $c \in \mathcal{N}$, respectively.) The function b in (3) is continuous on $\dot{\mathbf{R}}$ and vanishes at $\pm\infty$. We are going to show that its total variation is finite whence follows that $M^0(b) = NZ_s \in N^p(\alpha)$ (compare 2.1.2(b)).

Indeed, abbreviate the function $z \mapsto s^{1/p+\alpha-zi}$ by f, write b as fn_π, and suppose for definiteness that $s > 1$. Then

$$\begin{aligned}
V^\infty_{-\infty}(fn_\pi) &= \sum_{k\in Z} V^{2(k+1)\pi/\ln s}_{2k\pi/\ln s}(fn_\pi) \\
&\leq \sum_{k\in Z}(V^{2(k+1)\pi/\ln s}_{2k\pi/\ln s}(n_\pi)\,\|f\|_{C[2k\pi/\ln s,2(k+1)\pi/\ln s]} + \\
&\quad + V^{2(k+1)\pi/\ln s}_{2k\pi/\ln s}(f)\,\|n_\pi\|_{C[2k\pi/\ln s,2(k+1)\pi/\ln s]}) \\
&\leq \sup_{t\in R}|f(t)|\,V^\infty_{-\infty}(n_\pi) + \\
&\quad + 2\pi s^{1/p+\alpha}\sum_{k\in Z}\|n_\pi\|_{C[2k\pi/\ln s,2(k+1)\pi/\ln s]} \ ,
\end{aligned} \tag{4}$$

and since $\sup_{t\in R}|f(t)| = s^{1/p+\alpha}$, n_π has a finite total variation, and the series in (4) has a finite sum, we actually find $V^\infty_{-\infty}(fn_\pi) < \infty$. Thus, $NZ_s \in N^p(\alpha)$.

Since any operator in $N^p(\alpha)$ can be approximated as closely as desired by a polynomial in N and S, this yields the assertion in the general case. ∎

In particular, the entries C_{ij} of the matrix (1) are for $i \neq j$ of the form

$$C_{ij} = J_{\delta_i}\,M^0(b_{ij})\,J_{\delta_j} \quad \text{with} \quad M^0(b_{ij}) \in N^p(\alpha)\ .$$

To complete the proof of Example 5.3 we have to show that $(L_n^{F,F} C_{ij} | S_n^F) \in \mathcal{A}_{\Omega,r}^F$ for all i, j and then we could argue as in Example 5.1. The desired inclusion has already been shown in Section 2.11.5 for $i = j$. Its validity for $i \neq j$ is stated in the following proposition.

Proposition 5.4 *Let $M^0(b) \in N^p(\alpha)$. Then the sequences*

$$(L_n^{F,F} M^0(b) | S_n^F), \ (L_n^{F,F} M^0(b) J | S_n^F), \ (L_n^{F,F} J M^0(b) | S_n^F), \ (L_n^{F,F} J M^0(b) J | S_n^F)$$

are in $\mathcal{A}_{\Omega,r}^F$.

Proof   We have already seen that the first sequence is in $\mathcal{A}_{\Omega,r}^F$. So let us show for example, that the second one is in $\mathcal{A}_{\Omega,r}^F$, too. Let $\hat{J}$ stand for the discrete flip $(x_k) \mapsto (x_{-k})$, and denote the diagonal operator diag $(\hat{J}, \ldots, \hat{J})$ by the same symbol $\hat{J}$. Further, given k-tupels $F = (\phi_0, \ldots, \phi_{k-1})$ and $G = (\gamma_0, \ldots, \gamma_{k-1})$ of presplines and premeasures, respectively, we set

$$JF := (J\phi_0, \ldots, J\phi_{k-1}) \quad \text{and} \quad JG := (J\gamma_0, \ldots, J\gamma_{k-1}) \,.$$

Our first claim is that

$$JE_n^F = E_n^{JF} \hat{J} \quad \text{and} \quad P_n^G J = \hat{J} P_n^{JG} \,. \tag{5}$$

For the first assertion in (5), let $x = (x_{rl}) \in l_k^p(\alpha)$. Then

$$JE_n^F x = J \sum_r \sum_l x_{rl} \, \phi_{rln} = \sum_r \sum_l x_{rl} \, J\phi_{rln} \,,$$

and since

$$(J\phi_{rln})(t) = \phi_{rln}(-t) = \phi_r(-tn - l) = (J\phi_r)(tn + l)$$

we immediately get our claim. Similarly the second assertion can be checked.
  Now we have

$$E_{-n}^F L_n^{F,F} M^0(b) J E_n^F = T^0 (\lambda^{F,F})^{-1} P_n^F M^0(b) J E_n^F = T^0 (\lambda^{F,F})^{-1} P_n^F M^0(b) E_n^{JF} \hat{J},$$

and since JF is a canonical prebasis, we conclude from Proposition 2.21 that

$$E_{-n}^F L_n^{F,F} M^0(b) J E_n^F = (\alpha_r^{JF} \int (J\phi_s)(x) \, dx \, G(b) \hat{J})_{r,s=0}^{k-1} + K$$

with a compact operator K. It remains to show that $G(b)\hat{J}$ is in the Toeplitz algebra $\mathcal{T}_{1,1,\{1\}}^p(\alpha)$. But this has already been accomplished in Section 2.5.2, implication (9). ∎

5.3 Spline approximation methods on composed curves

5.3.1 Spline spaces

We start with defining a global parametrization of the composed plane curve Γ which is well suited for introducing a spline space over Γ.

Let Γ' be a finite subset of Γ which contains all singular points (but which is allowed to contain regular points, too), and suppose that the weight function α vanishes on $\Gamma \setminus \Gamma'$. Further we let $\Gamma \setminus \Gamma' = \cup_{j=1}^m \Gamma_j$ be the decomposition of $\Gamma \setminus \Gamma'$ into its connected components.

Each component is subject to one of the following types: either the closure $\operatorname{clos} \Gamma_j$ of Γ_j is homeomorphic to an interval (type I), or $\operatorname{clos} \Gamma_j$ is homeomorphic to a circle, but Γ_j is not closed (type II), or Γ_j is homeomorphic to a circle (type III).

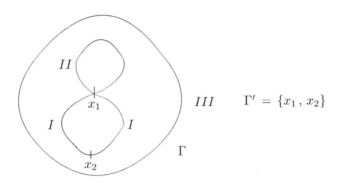

Now we parametrize the closure of each component by a mapping $\zeta_j : [0,1] \to \operatorname{clos} \Gamma_j$ which preserves the orientation and has a Hölder continuous and non-vanishing derivative. Additionally we require that $\zeta_j'(0) = \zeta_j'(1)$ for Γ_j being of type III. If Γ_j is of type I then ζ_j is one-to-one, and $\zeta_j(0)$ and $\zeta_j(1)$ are just the end-points of $\operatorname{clos} \Gamma_j$. If Γ_j is of type II or III then ζ_j is one-to-one on $(0,1)$, but $\zeta_j(0) = \zeta_j(1)$, and the point $\zeta_j(0)$ belongs to Γ' for type II and can be arbitrarily chosen for type III. For example, one can take

$$\zeta_j(s) = t_j(l_j s)$$

where l_j is the arc length of Γ_j and $t_j : [0, l_j] \to \operatorname{clos} \Gamma_j$ is the natural parametrization of $\operatorname{clos} \Gamma_j$ by its arc length.

Further we need some more notation: by per S_n^F we denote the subspace of S_n^F which contains all functions f with

$$f(s-1) = f(s) = f(s+1) \quad \text{for} \quad s \in [0,1] .$$

These functions are "periodic in a neighbourhood of $[0,1]$", and they are intended to introduce spline functions over curves of type III which have no favoured points.

Abbreviating the vector $(\zeta_1, \ldots, \zeta_m)$ by ζ we now introduce the spline space $S_n^F(\Gamma) = S_n^F(\Gamma, \Gamma', \zeta)$ over Γ as the smallest subspace of $L_\Gamma^p(\alpha)$ which contains all functions f with

$$f(z) = \begin{cases} g(\zeta_j^{-1}(z)) & \text{if} \quad z \in \Gamma_j \\ 0 & \text{else} \end{cases}$$

where g runs through $\hat{\chi}_{[0,1]}^{(n)} S_n^F$ if Γ_j is type I or II, and where $g \in \text{per} S_n^F$ if Γ_j is type III.

Example 5.4 *The spline space $S_n^F([0,1], \{0,1\}, \zeta)$ with $\zeta(s) = s$ coincides with $\hat{\chi}_{[0,1]}^{(n)} S_n^F$.*

Example 5.5 *Let $F = F_{d,e}$ be as in Section 2.9.4, and set $\zeta : [0,1] \to \mathbf{T}$, $z \mapsto e^{2\pi i z}$. Then $S_n^F(\mathbf{T}, \emptyset, \zeta)$ is the space of piecewise polynomial splines with defect over the equidistant partition of the unit circle $\mathbf{T}$. This space is clearly isomorphic to $\chi_{[0,1]} \text{per} S_n^F$.*

Example 5.6 *Choose F and ζ as in the previous example. The space $S_n^F(\mathbf{T}, \{1\}, \zeta)$ contains just those basis functions of $S_n^F(\mathbf{T}, \emptyset, \zeta)$ whose supports do not contain 1 as inner point. Thus, this space is isomorphic to $\hat{\chi}_{[0,1]}^{(n)} S_n^F$.*

Example 5.7 *Let Γ be the union of the semi circle $\{z \in \mathbf{T} : \text{Im} z \geq 0\}$ and $[-1,1]$, $\Gamma' = \{-1,1\}$, and set $\zeta = (\zeta_1, \zeta_2)$ with $\zeta_1 : [0,1] \to [-1,1]$, $z \mapsto 2z - 1$, $\zeta_2 : [0,1] \to \{z \in \mathbf{T} : \text{Im} z \geq 0\}$, $z \mapsto e^{i\pi z}$. Then $S_n^F(\Gamma, \Gamma', \zeta)$ is isomorphic to the direct sum of two spaces $\hat{\chi}_{[0,1]}^{(n)} S_n^F$.*

Example 5.8 *Let Γ, Γ', and ζ_2 be as in Example 5.7, but choose a mapping $\zeta_1 : z \mapsto (2\pi - 4)z^3 + (6 - 3\pi)z^2 + \pi z - 1$. The resulting spline space is isomorphic to that one in Example 5.7, but now one has $|\zeta_1'(0)| = |\zeta_2'(1)| = \pi$ and $|\zeta_1'(1)| = |\zeta_2'(0)| = \pi$, that is, the associated partition is "almost" equidistant in neighbourhoods $U_{\pm 1}$ of ± 1, and the typical mesh wide is π/n (whereas in Example 5.7 $|\zeta_1'(0)| = 2 \neq |\zeta_2'(1)|$ which involves that the typical mesh wides are $2/n$ over $U_{-1} \cap [-1,1]$ and π/n over $U_{-1} \cap \{z \in \mathbf{T} : \text{Im} z \geq 0\}$).*

5.3.2   Spline projections on composed curves

As for homogeneous curves, we modify the standard action of a functional $g \in L_\Gamma^q(-\alpha)$ at a function $f \in L_\Gamma^p(\alpha)$ given by 5.1.2(2) by defining a new action by

$$
\begin{aligned}
(f,g)_\Gamma &= \sum_{j=1}^m ((\chi_{\Gamma_j} f) \circ \zeta_j, (\chi_{\Gamma_j} g) \circ \zeta_j) \\
&= \sum_{j=1}^m \int_0^1 f(\zeta_j(s)) \, g(\zeta_j(s)) \, ds .
\end{aligned}
\tag{1}
$$

If we abbreviate the numbers $|\zeta_j'(\zeta_j^{-1}(t))|$ for $t \in \Gamma_j$ by $\xi_j(t)$ then (1) can also be written as

$$(f,g)_\Gamma = \sum_j \int_{\Gamma_j} (\chi_{\Gamma_j} f)(t) \, (\chi_{\Gamma_j} g)(t) \, \frac{1}{\xi_j(t)} \, |dt| \, .$$

The Galerkin projection $L_{n,\Gamma}^{F,F}$ from $L_\Gamma^p(\alpha)$ onto $S_n^F(\Gamma,\Gamma',\zeta)$ is now defined by

$$(L_{n,\Gamma}^{F,F} f \, , g)_\Gamma \; = \; (f,g)_\Gamma$$

for all $f \in L_\Gamma^p(\alpha)$ and $g \in S_n^F(\Gamma,\Gamma',\zeta)$.

Observing that the spline space $S_n^F(\Gamma,\Gamma',\zeta)$ is the direct sum of the spline spaces over the connected components Γ_j of $\Gamma \setminus \Gamma'$, and taking into account the special form (1) of the action of functionals, one can readily verify the main properties of the Galerkin projections $L_{n,\Gamma}^{F,F}$.

Proposition 5.5 *(a) (Commutator property) If $f \in C(\Gamma)$ then*

$$\|L_{n,\Gamma}^{F,F} f I - f L_{n,\Gamma}^{F,F}\|_{L(L_\Gamma^p(\alpha))} \to 0 \quad \text{as} \quad n \to \infty \, .$$

(b) $L_{n,\Gamma}^{F,F}$ is asymptotically quasi-local.

(c) $L_{n,\Gamma}^{F,F}$ converges strongly to the identity operator over Γ.

5.3.3 Local parametrizations

The local parametrizations of Γ which we are going to define now are adapted to the local introduction and investigation of spline approximation methods over Γ. As in 5.1.7, we associate to each point $x \in \Gamma$ a connected open subcurve U_x such that

(a) $x \in U_x$,

(b) if $\cup_{j=1}^k \Gamma_j(x)$ is the decomposition of $U_x \setminus \{x\}$ into its connected components then $k = w(x)$, and $\operatorname{clos}\Gamma_j(x)$ are simple arcs,

(c) $(\operatorname{clos} U_x \setminus \{x\}) \cap \Gamma' = \emptyset$,

(d) the one-sided tangent of $\operatorname{clos}\Gamma_j(x)$ at x is $x + e^{i\beta_j(x)}\mathbf{R}^{\delta_j(x)}$,

(e) the intersection $U_x \cap U_y$ is connected for $x, y \in \Gamma$ and empty for $x, y \in \Gamma'$.

A consequence of (e) is that, for all $x, y \in \Gamma$, the intersection $U_x \cap U_y$ is contained in one component Γ_j (recall that $\Gamma \setminus \Gamma' = \cup\Gamma_j$ is the decomposition of $\Gamma \setminus \Gamma'$ into its connected components and that ζ_j is the global parametrization of $\operatorname{clos}\Gamma_j$).

We start our definition with the simplest case, viz. $x \in \Gamma_j$ but $x \neq \zeta_j(0)$ (the latter condition is automatically satisfied if Γ_j is type I or II). Then we set

$$V_x = \zeta_j^{-1}(U_x) \, , \quad \nu_x : V_x \to U_x \, , \quad s \mapsto \zeta_j(s) \, .$$

If $x \in \Gamma_j$ but $x = \zeta_j(0)$ (which only can happen if Γ_j is type III) then $\zeta_j^{-1}(U_x)$ consists of two connected components $V_{x,0}$ and $V_{x,1}$ which contain 0 and 1, respectively, and we set

$$V_x = V_{x,0} \cup (V_{x,1} - 1)$$

(the algebraic difference!), and

$$\nu_x : V_x \to U_x, \ s \mapsto \begin{cases} \zeta_j(s) & \text{if } s \geq 0 \\ \zeta_j(1+s) & \text{if } s < 0 . \end{cases}$$

Let, finally, $x \in \Gamma'$, and suppose $\Gamma_j(x) \subseteq \operatorname{clos}\Gamma_{k(j)}$. Let us first introduce positive numbers $\xi_j(x)$ by $\xi_j(x) = |\zeta_j'(\zeta_j^{-1}(x))|$ for $j = 1, \dots, w(x)$. These numbers measure the local distortions of the spline spaces nearby x. Then we set, again for $j = 1, \dots, w(x)$,

$$V_{x,j} = \begin{cases} e^{i\beta_j(x)}\, \xi_j(x)\, \zeta_j^{-1}(\Gamma_j(x)) & \text{if } \delta_j(x) = 1 \\ e^{i\beta_j(x)}\, \xi_j(x)\, (\zeta_j^{-1}(\Gamma_j(x)) - 1) & \text{if } \delta_j(x) = -1 \end{cases}$$

and define

$$V_x = V_{x,1} \cup \dots \cup V_{x,w(x)} \tag{1}$$

and

$$\nu_x : V_x \to U_x, \ s \mapsto \begin{cases} \zeta_j\big(\frac{e^{-i\beta_j(x)}}{\xi_j(x)}s\big) & \text{if } \delta_j(x) = 1 \\ \zeta_j\big(1 + \frac{e^{-i\beta_j(x)}}{\xi_j(x)}s\big) & \text{if } \delta_j(x) = -1 \end{cases} \tag{2}$$

where s is supposed to belong to the component $V_{x,j}$. The consideration of the distortion factors in (1) and (2) is necessary to guarantee that ν_x has a derivative which is continuous on all of V_x (whereas the one-sided derivatives of ζ_j at $\zeta_j^{-1}(x)$ can differ from each other). The continuity of ν_x' on its hand will be needed to prove an analogon of the straightening lemma (see the subsequent section).

5.3.4 Compatibility

The local parametrizations ν_x are compatible with the Lebesgue spaces, the spline spaces, the associated Galerkin projections, and the singular integral operators. Precisely, if we introduce operators ψ_x by $(\psi_x f)(s) = f(\nu_x(s))$ for $s \in V_x$, if we abbreviate the vector $(\xi_1(x), \dots, \xi_{w(x)}(x))$ to ξ_x, and if we define, in analogy to 5.2.3, $\hat{\chi}_M^{(n)}$ for subsets M of the composed curve Γ, then we have

Proposition 5.6 (a) The mapping ψ_x is a continuous and invertible mapping from $L^p_{U_x}(\alpha|U_x)$ onto $L^p_{V_x}(\alpha_x|V_x)$. (The latter is a subspace of $L^p_R(0)$ in case $x \in \Gamma \setminus \Gamma'$ and of $L^p_{\Gamma_x}(\alpha_x)$ in case $x \in \Gamma'$.)
(b) The operator ψ_x maps the spline space $\hat{\chi}_{U_x}^{(n)} S_n^F(\Gamma, \Gamma', \zeta)$ one-to-one onto $\hat{\chi}_{V_x}^{(n)} S_n^F$ in case $x \in \Gamma \setminus \Gamma'$ and onto $\hat{\chi}_{V_x}^{(n)} S_n^F(\Gamma_x, \xi_x)$ in case $x \in \Gamma'$.

(c) One has

$$\psi_x \chi_{U_x} L_{n,\Gamma}^{F,F} \psi_x^{-1} \chi_{V_x} I = \begin{cases} \chi_{V_x} L_n^{F,F} \chi_{V_x} I & \text{if} \quad x \in \Gamma \setminus \Gamma' \\ \chi_{V_x} L_{n,\Gamma_x}^{F,F} \chi_{V_x} I & \text{if} \quad x \in \Gamma' . \end{cases}$$

(d) (General straightening lemma) The operator $\psi_x S_{U_x} \psi_x^{-1} \chi_{V_x} I - S_{V_x}$ is compact on $L_{V_x}^p (\alpha_x | V_x)$.

The proof is either immediate from the definitions (parts (a)-(c)) or well known (part (d), see, e.g.([GK 1], Lemma 3.2)).

5.3.5 An algebra of approximation methods

As already announced, we define the approximation methods under consideration in full analogy to Seeley's definition of singular integral operators. Due to the different qualities of the spline spaces S_n^F and $S_n^F(\Gamma_x)$ at the origin, this requires to distinguish some cases. Let us further agree to indicate the dependence on x of the mapping $\eta : L_{\Gamma_x}^p(\alpha) \to L_R^p(\alpha)_{w(x)}$ introduced in 5.2.4(1) by writing η_x, and recall that f_x abbreviates the function $\psi_x f \psi_x^{-1}$.

Definition *Let $\mathcal{A}_{\Omega,r}^F(\Gamma,\Gamma',\zeta)$ stand for the collection of all sequences (A_n), $A_n : S_n^F(\Gamma,\Gamma',\zeta) \to S_n^F(\Gamma,\Gamma',\zeta)$, which are subject to the following conditions:*
(a) For all $x \in \Gamma$, and for all continuous functions f and g on Γ with $\operatorname{supp} f \cup \operatorname{supp} g \subseteq U_x$,

$$\psi_x f A_n L_{n,\Gamma}^{F,F} g \psi_x^{-1} \chi_{V_x} I = \begin{cases} f_x B_n L_n^{F,F} g_x I & \text{if} \quad x \in \Gamma \setminus \Gamma' \\ f_x B_n L_{n,\Gamma_x}^{F,F} g_x I & \text{if} \quad x \in \Gamma' \end{cases} \tag{1}$$

where

$$(B_n) \in \begin{cases} \mathcal{A}_{\Omega,r}^F & \text{if} \quad x \in \Gamma \setminus \Gamma' \\ \mathcal{A}_{\Omega,r}^F(\Gamma_x, \xi_x) & \text{if} \quad x \in \Gamma' . \end{cases} \tag{2}$$

(b) For all $x, y \in \Gamma$, and for all continuous functions f and g on Γ with $\operatorname{supp} f \subseteq U_x$, $\operatorname{supp} g \subseteq U_y$, and $\operatorname{supp} f \cap \operatorname{supp} g = \emptyset$,

$$\psi_y f A_n L_{n,\Gamma}^{F,F} g \psi_x^{-1} \chi_{V_x} I =$$

$$= \begin{cases} f_y J_n L_n^{F,F} g_x I & \text{if} \quad x, y \in \Gamma \setminus \Gamma' \\ f_y \eta_y^{-1} J_n L_n^{F,F} g_x I & \text{if} \quad x \in \Gamma \setminus \Gamma', y \in \Gamma' \\ f_y J_n \eta_x L_{n,\Gamma_x}^{F,F} g_x I & \text{if} \quad x \in \Gamma', y \in \Gamma \setminus \Gamma' \\ f_y \eta_y^{-1} J_n \eta_x L_{n,\Gamma_x}^{F,F} g_x I & \text{if} \quad x, y \in \Gamma' \end{cases} \tag{3}$$

where

$$(J_n) \in \begin{cases} \mathcal{J}_{\Omega,r}^F & \text{if} \quad x, y \in \Gamma \setminus \Gamma' \\ (\hat{\chi}_n^{\delta_y})(\mathcal{J}_{\Omega,r}^F)_{w(y) \times 1} & \text{if} \quad x \in \Gamma \setminus \Gamma', y \in \Gamma' \\ (\mathcal{J}_{\Omega,r}^F)_{1 \times w(x)}(\hat{\chi}_n^{\delta_x}) & \text{if} \quad x \in \Gamma', y \in \Gamma \setminus \Gamma' \\ (\hat{\chi}_n^{\delta_y})(\mathcal{J}_{\Omega,r}^F)_{w(y) \times w(x)}(\hat{\chi}_n^{\delta_x}) & \text{if} \quad x, y \in \Gamma' . \end{cases} \tag{4}$$

Similarly we define the analogue of the ideal $\mathcal{J}_{\Omega,r}^F$:

Definition *Let* $\mathcal{J}_{\Omega,r}^{F}(\Gamma,\Gamma',\zeta)$ *stand for the subset of* $\mathcal{A}_{\Omega,r}^{F}(\Gamma,\Gamma',\zeta)$ *which contains all sequences which satisfy besides (2) additionally*

$$(B_n) \in \begin{cases} \mathcal{J}_{\Omega,r}^{F} & \text{if} \quad x \in \Gamma \setminus \Gamma' \\ \eta_x^{-1}(\hat{\chi}_n^{\delta_x})(\mathcal{J}_{\Omega,r}^{F})_{w(x)\times w(x)}(\hat{\chi}_n^{\delta_x})\eta_x & \text{if} \quad x \in \Gamma' \, . \end{cases}$$

Sometimes we shall simply write $\mathcal{A}_{\Omega,r}^{F}(\Gamma)$ and $\mathcal{J}_{\Omega,r}^{F}(\Gamma)$ in place of $\mathcal{A}_{\Omega,r}^{F}(\Gamma,\Gamma',\zeta)$ and $\mathcal{J}_{\Omega,r}^{F}(\Gamma,\Gamma',\zeta)$.

The sets $\mathcal{A}_{\Omega,r}^{F}(\Gamma)$ and $\mathcal{J}_{\Omega,r}^{F}(\Gamma)$ are certainly not empty. Here are some examples of sequences in it.

Example 5.9 *The set of all sequences* (G_n), $G_n : S_n^F(\Gamma) \to S_n^F(\Gamma)$ *with* $\|G_n L_{n,\Gamma}^{F,F}\| \to 0$ *as* $n \to \infty$, *belongs to* $\mathcal{J}_{\Omega,r}^{F}(\Gamma)$.

It is evident that $\|\psi_x f G_n L_{n,\Gamma}^{F,F} g\psi_x^{-1}\chi_{V_x} I\| \to 0$ and, thus, $\|f_x B_n L_{n,\Gamma}^{F,F} g_x I\| \to 0$ for all $x \in \Gamma \setminus \Gamma'$, and a little thought shows that the operators B_n can be chosen so that $E_{-n}^{F} B_n E_n^{F}$ are compact (and even finite rank since f_x and g_x have bounded support). Similarly one verifies the other conditions.

Example 5.10 *If K is compact on $L_\Gamma^p(\alpha)$ then $(L_{n,\Gamma}^{F,F} K | S_n^F(\Gamma))$ belongs to $\mathcal{J}_{\Omega,r}^{F}(\Gamma)$.*

The proof of this fact is an immediate consequence of part (c) of Proposition 5.6: Let, for instance, $x \in \Gamma'$. Then, up to a sequence tending to zero in the norm,

$$\psi_x f L_{n,\Gamma}^{F,F} K L_{n,\Gamma}^{F,F} g\psi_x^{-1}\chi_{V_x} I =$$
$$= \psi_x f L_{n,\Gamma}^{F,F}\chi_{U_x} K\chi_{U_x} L_{n,\Gamma}^{F,F} g\psi_x^{-1}\chi_{V_x} I + C_n$$
$$= f_x L_{n,\Gamma_x}^{F,F}\psi_x\chi_{U_x} K\chi_{U_x}\psi_x^{-1} L_{n,\Gamma_x}^{F,F} g_x I + C_n$$

with a certain compact operator $\psi_x\chi_{U_x} K\chi_{U_x}\psi_x^{-1}$, and with a sequence (C_n) tending to zero in the norm by Proposition 5.5. Now recall Example 5.1 where we have seen that

$$\eta_x L_{n,\Gamma_x}^{F,F} K' L_{n,\Gamma_x}^{F,F}\eta_x^{-1} \in (\hat{\chi}_n^{\delta_x})(\mathcal{J}_{\Omega,r}^{F})_{w(x)\times w(x)}(\hat{\chi}_n^{\delta_x}) \, ,$$

for all compact operators K' on $L_{\Gamma_x}^p(\alpha_x)$. This verifies condition (a) of the definition, and the proof of (b) proceeds analogously.

Example 5.11 *If the function a is piecewise continuous over Γ and continuous over $\Gamma \setminus \Gamma'$ then the sequence $(L_{n,\Gamma}^{F,F} aI | S_n^F(\Gamma))$ is in $\mathcal{A}_{\Omega,r}^{F}(\Gamma)$.*

Indeed, since the spline space $S_n^F(\Gamma)$ is the direct sum of the spline spaces $S_n^F(\Gamma_j)$ over the connected components Γ_j of $\Gamma \setminus \Gamma'$, and since a is continuous on each component, we get parallely to the previous example

$$\psi_x f L_{n,\Gamma}^{F,F} a L_{n,\Gamma}^{F,F} g\psi_x^{-1}\chi_{V_x} I = f_x L_{n,\Gamma_x}^{F,F} a_x L_{n,\Gamma_x}^{F,F} g_x I + C_n$$

with a sequence (C_n) tending to zero in the norm. Now it remains to recall Example 5.2 as well as the inclusion $(L_n^{F,F} bI | S_n^F) \in \mathcal{A}_{\Omega,r}^{F}$ for b being continuous on $\dot{\mathbf{R}} \setminus \{0\}$ and piecewise continuous on $\mathbf{R}$ to get the assertion for the part (a) of the definition. That this sequence is also subject to condition (b) can be seen in the same manner.

Example 5.12 $(L_{n,\Gamma}^{F,F} S_\Gamma | S_n^F(\Gamma, \Gamma', \varsigma)) \in \mathcal{A}_{\{1\},0}^F(\Gamma, \Gamma', \varsigma).$

Here, the same reasoning as above yields

$$\psi_x f L_{n,\Gamma}^{F,F} S_\Gamma L_{n,\Gamma}^{F,F} g \psi_x^{-1} \chi_{V_x} I = f_x L_{n,\Gamma_x}^{F,F} (S_{\Gamma_x} + K) L_{n,\Gamma_x}^{F,F} g_x I + C_n$$

with a sequence (C_n) having norm limit 0 and with a compact operator K. Now remember Example 5.3 and Section 2.11.1, where the inclusions

$$(L_{n,\Gamma_x}^{F,F} S_{\Gamma_x} | S_n^F(\Gamma_x)) \in \mathcal{A}_{\{1\},0}^F(\Gamma_x) \quad \text{and} \quad (L_n^{F,F} S_R | S_n^F) \in \mathcal{A}_{\{1\},0}^F$$

have been shown. The other conditions can be verified similarly.

We conclude this section by showing the following result:

Theorem 5.3 *Provided with elementwise operations the set* $\mathcal{A}_{\Omega,r}^F(\Gamma)$ *becomes a Banach algebra, and* $\mathcal{J}_{\Omega,r}^F(\Gamma)$ *its closed two-sided ideal.*

Proof It is evident that the sum of two sequences $(A_n), (B_n) \in \mathcal{A}_{\Omega,r}^F(\Gamma)$ is in $\mathcal{A}_{\Omega,r}^F(\Gamma)$ again. Let us show that the product $(A_n)(B_n) = (A_n B_n)$ is in $\mathcal{A}_{\Omega,r}^F(\Gamma)$, too.

Let, for definiteness, $x \in \Gamma'$, and let f and g be continuous functions as in part (a) of the definition in this section. Choose a finite covering $(U_{x_j})_{j=1}^k$ of Γ with $x_1 = x$ (such a covering exists because Γ is compact, note also that all neighbourhoods U_y with $y \in \Gamma'$ necessarily belong to $(U_{x_j})_{j=1}^k$), and let $1 = f_1 + \ldots + f_k$ be a nonnegative and continuous partition of unity subordinate to the selected covering such that

$$(\operatorname{supp} f \cup \operatorname{supp} g) \cap \operatorname{supp} f_j = \emptyset, \quad j = 2, \ldots, k.$$

Set $h_j := \sqrt{f_j}$. Then

$$\psi_x f A_n B_n L_{n,\Gamma}^{F,F} g \psi_x^{-1} \chi_{V_x} I$$

$$= \psi_x f A_n L_{n,\Gamma}^{F,F} B_n L_{n,\Gamma}^{F,F} g \psi_x^{-1} \chi_{V_x} I$$

$$= \sum_{j=1}^k \psi_x f A_n L_{n,\Gamma}^{F,F} h_j^2 B_n L_{n,\Gamma}^{F,F} g \psi_x^{-1} \chi_{V_x} I$$

$$= \sum_{j=1}^k (\psi_x f A_n L_{n,\Gamma}^{F,F} h_j \psi_{x_j}^{-1} \chi_{V_j})(\psi_{x_j} h_j B_n L_{n,\Gamma}^{F,F} g \psi_x^{-1} \chi_{V_x} I). \tag{5}$$

The summand with $j = 1$ is by condition (1) of the form

$$f_x A_n' L_{n,\Gamma_x}^{F,F} (h_1)_x (h_1)_x B_n' L_{n,\Gamma_x}^{F,F} g_x I = f_x A_n' L_{n,\Gamma_x}^{F,F} (f_1)_x B_n' L_{n,\Gamma_x}^{F,F} g_x I \tag{6}$$

with $(A_n'), (B_n') \in \mathcal{A}_{\Omega,r}^F(\Gamma_x, \xi_x)$. Since $(L_{n,\Gamma_x}(f_1)_x | S_n^F(\Gamma_x)) \in \mathcal{A}_{\Omega,r}^F(\Gamma_x, \xi_x)$ (compare Example 5.2), we conclude that (6) is of the form $f_x D_n^{(1)} L_{n,\Gamma_x}^{F,F} g_x I$ with a sequence $(D_n^{(1)})$ in $\mathcal{A}_{\Omega,r}^F(\Gamma_x, \xi_x)$.

Let now $j = 2$ and suppose for example that $x_2 \in \Gamma \setminus \Gamma'$. By condition (3), the second term can be rewritten as

$$f_x \eta_x^{-1} J_n' L_n^{F,F}(h_2)_{x_2}(h_2)_{x_2} J_n'' \eta_x L_{n,\Gamma_x}^{F,F} g_x I = f_x \eta_x^{-1} J_n' L_n^{F,F}(f_2)_{x_2} J_n'' \eta_x L_{n,\Gamma_x}^{F,F} g_x I \tag{7}$$

with sequences (J_n') and (J_n'') in $(\mathcal{J}_{\Omega,r}^F)_{w(x_2) \times 1}$ and $(\mathcal{J}_{\Omega,r}^F)_{1 \times w(x_1)}$, respectively. Since $(L_n^{F,F}(f_2)_{x_2} I | S_n^F)$ is in $\mathcal{A}_{\Omega,r}^F$ (Proposition 3.16) and $\mathcal{J}_{\Omega,r}^F$ is an ideal in $\mathcal{A}_{\Omega,r}^F$, we get $(J_n' L_n^{F,F}(f_2)_{x_2} J_n'') \in (\mathcal{J}_{\Omega,r}^F)_{w(x_2) \times w(x_1)}$. Hence, the term in (7) is of the form $f_x D_n^{(2)} L_{n,\Gamma_x}^{F,F} g_x I$ with a sequence $(D_n^{(2)})$ in $\mathcal{A}_{\Omega,r}^F(\Gamma_x, \xi_x)$ again. Summarizing these facts we see that

$$\psi_x f A_n B_n L_{n,\Gamma}^{F,F} g \psi_x^{-1} \chi_{V_x} I = f_x D_n L_{n,\Gamma_x}^{F,F} g_x I$$

with a sequence $(D_n) = \sum_{j=1}^k (D_n^{(j)})$ in $\mathcal{A}_{\Omega,r}^F(\Gamma_x, \xi_x)$, which shows conditions (1) and (2) for the product $(A_n B_n)$. The verification of part (b) goes similarly: Given functions f supported on U_{x_1} and g supported on U_{x_2} with $\operatorname{supp} f \cap \operatorname{supp} g = \emptyset$ we pick a covering (U_{x_j}) (including U_{x_1} and U_{x_2}) and choose a continuous and non-negative partition $1 = f_1 + \ldots + f_k$ subordinate to this covering such that

$$\operatorname{supp} f \cap \operatorname{supp} f_j = \emptyset \quad \text{for} \quad j \neq 1, \quad \operatorname{supp} g \cap \operatorname{supp} f_j = \emptyset \quad \text{for} \quad j \neq 2,$$

and then we proceed as in part (a).

So we have seen that $\mathcal{A}_{\Omega,r}^F(\Gamma)$ is actually an algebra. The same arguments yield moreover that $\mathcal{J}_{\Omega,r}^F(\Gamma)$ is a two-sided ideal of this algebra.

Finally, the proofs of the completeness of $\mathcal{A}_{\Omega,r}^F(\Gamma)$ as well as of the closedness of $\mathcal{J}_{\Omega,r}^F(\Gamma)$ are standard. ∎

5.4   The stability theorem

5.4.1   The homomorphisms $W_{s,\Gamma}$

The stability theorem aimed at will be of the same structure as its predecessors Theorem 3.5, Theorem 4.5 and Theorem 4.12: a sequence in $\mathcal{A}_{\Omega,r}^F(\Gamma)$ is stable if the images of this sequence under two families $(W_{s,\Gamma})_{s \in \Gamma}$ and $(W^{t,\Gamma})_{t \in \Omega}$ of homomorphisms are invertible. We are going to introduce these homomorphisms.

Given $s \in \Gamma$ we let f and g be continuous functions as in part (a) of the definition in 5.3.5 which additionally satisfy $f(s) = g(s) = 1$, and we define for $(A_n) \in \mathcal{A}_{\Omega,r}^F(\Gamma)$:

$$W_{s,\Gamma}(A_n) = \begin{cases} W_{\nu_s^{-1}(s)}(L_n^{F,F} \psi_s f A_n L_{n,\Gamma}^{F,F} g \psi_s^{-1} \chi_{V_s} I | S_n^F) & \text{if } s \in \Gamma \setminus \Gamma' \\ W_0(\eta_s L_{n,\Gamma_s}^{F,F} \psi_s f A_n L_{n,\Gamma}^{F,F} g \psi_s^{-1} \eta_s^{-1} |\hat{\chi}_n^{\delta(s)}(S_n^F)_{w(s)}) & \text{if } s \in \Gamma'. \end{cases}$$

This definition is correct. Indeed, in case $s \in \Gamma \setminus \Gamma'$ we have to show that the sequence

$$(L_n^{F,F} \psi_s f A_n L_{n,\Gamma}^{F,F} g \psi_s^{-1} | S_n^F) \tag{1}$$

is in $\mathcal{A}_{\Omega,r}^F$ (and thus the application of $W_{\nu_s^{-1}(s)}$ to this sequence is justified), and that $W_{s,\Gamma}(A_n)$ is independent of the choice of f and g.

For the first point, write (1) as

$$(L_n^{F,F} f_s B_n L_n^{F,F} g_s I | S_n^F) = (L_n^{F,F} f_s I | S_n^F)(B_n)(L_n^{F,F} g_s I | S_n^F) \tag{2}$$

which shows via Proposition 3.16 that (1) is in fact in $\mathcal{A}_{\Omega,r}^F$. Moreover, (2) shows that $W_{s,\Gamma}(A_n) = W_{\nu_s^{-1}(s)}(B_n)$ since

$$W_{\nu_s^{-1}(s)}(L_n^{F,F} f_s I | S_n^F) = W_{\nu_s^{-1}(s)}(L_n^{F,F} g_s I | S_n^F) = 1.$$

Now suppose $\tilde{f}$ and $\tilde{g}$ to be other functions, and consider the sequence (C_n) with

$$C_n = L_n^{F,F} \psi_s f A_n L_{n,\Gamma}^{F,F} g \psi_s^{-1} \chi_{V_s} I | S_n^F - L_n^{F,F} \psi_s \tilde{f} A_n L_{n,\Gamma}^{F,F} \tilde{g} \psi_s^{-1} \chi_{V_s} I | S_n^F.$$

This sequence is in $\mathcal{A}_{\Omega,r}^F$, and rewriting it as

$$
\begin{aligned}
C_n &= L_n^{F,F} \psi_s (f - \tilde{f}) A_n L_{n,\Gamma}^{F,F} g \psi_s^{-1} \chi_{V_s} I | S_n^F - \\
&\quad - L_n^{F,F} \psi_s \tilde{f} A_n L_{n,\Gamma}^{F,F} (\tilde{g} - g) \psi_s^{-1} \chi_{V_s} I | S_n^F \\
&= L_n^{F,F} (f - \tilde{f})_s B_n L_n^{F,F} g_s I | S_n^F - L_n^{F,F} \tilde{f}_s \tilde{B}_n L_n^{F,F} (\tilde{g} - g)_s I | S_n^F
\end{aligned}
$$

with $(B_n), (\tilde{B}_n) \in \mathcal{A}_{\Omega,r}^F$, and observing that

$$(f - \tilde{f})_s(\nu_s^{-1}(s)) = (\tilde{g} - g)_s(\nu_s^{-1}(s)) = 0$$

we easily get $W_s(C_n) = 0$.

The proof of the correctness is similar for $s \in \Gamma'$; the only difference is that $(\eta_s L_{n,\Gamma_s}^{F,F} \psi_s f A_n L_{n,\Gamma}^{F,F} g \psi_s^{-1} \eta_s^{-1} | \hat{\chi}_n^{\delta(s)} (S_n^F)_{w(s)})$ is now a sequence in $(\mathcal{A}_{\Omega,r}^F)_{w(s) \times w(s)}$, and thus, the mapping W_0 appearing in the definition of $W_{s,\Gamma}$ has to be interpreted as the operator

$$W_0 : (\mathcal{A}_{\Omega,r}^F)_{w(s) \times w(s)} \to L(l_{kw(s)}^p(\alpha(s))) \quad (A_{ij})_{i,j=1}^{w(s)} \mapsto (W_0(A_{ij}))_{i,j=1}^{w(s)}$$

where the latter W_0 is that one associated with the algebra $\mathcal{A}_{\Omega,r}^F$, and k is the number of basis splines in F.

Proposition 5.7 *The mappings $W_{s,\Gamma}$ ($s \in \Gamma$) are continuous algebra homomorphisms mapping $\mathcal{A}_{\Omega,r}^F(\Gamma)$ into $L(l_k^p(0))$ if $s \in \Gamma \setminus \Gamma'$ and into $L(l_{kw(s)}^p(\alpha(s)))$ if $s \in \Gamma'$.*

Proof The continuity of $W_{s,\Gamma}$ for $s \in \Gamma \setminus \Gamma'$ is a consequence of the estimates

$$
\begin{aligned}
\|W_{s,\Gamma}(A_n)\| &= \|W_{\nu_s^{-1}(s)}(L_n^{F,F}\psi_s f A_n L_{n,\Gamma}^{F,F} g\psi_s^{-1}\chi_{V_s} I|S_n^F)\| \\
&\leq C_1 \sup_n \|L_n^{F,F}\psi_s f A_n L_{n,\Gamma}^{F,F} g\psi_s^{-1}\chi_{V_s} I|S_n^F\| \\
&\leq C_2 \sup_n \|A_n\| = C_2\|(A_n)\|
\end{aligned}
$$

with certain constants C_1, C_2, and the proof for $s \in \Gamma'$ is the same. It is further evident that $W_{s,\Gamma}$ is an additive mapping. So we are left on showing that

$$ W_{s,\Gamma}(A_n B_n) = W_{s,\Gamma}(A_n)W_{s,\Gamma}(B_n) . $$

Again, we let for example $s \in \Gamma \setminus \Gamma'$. Choose and fix a covering $(U_{x_j})_{j=1}^l$ of Γ with $x_1 = s$, and let $1 = f_1 + \ldots + f_l$ be a partition of unity subordinate to this covering just as in the first part of the proof of Theorem 5.3. Then, proceeding as in 5.3.5 (5), we find

$$
\begin{aligned}
W_{s,\Gamma}(A_n B_n) &= \\
&= W_{\nu_s^{-1}(s)}(L_n^{F,F}\psi_x f A_n B_n L_{n,\Gamma}^{F,F} g\psi_x^{-1}\chi_{V_x} I|S_n^F) \\
&= \sum_{j=1}^l W_{\nu_s^{-1}(s)}(L_n^{F,F} f_s \tilde{A}_n L_n^{F,F}(h_j)_{x_j} I|S_n^F) W_{\nu_s^{-1}(s)}(L_n^{F,F}(h_j)_{x_j} \tilde{B}_n L_n^{F,F} g_s I|S_n^F)
\end{aligned}
$$

with sequences $(\tilde{A}_n), (\tilde{B}_n) \in \mathcal{A}_{\Omega,r}^F$. All terms with $j > 1$ vanish since the sequences $(L_n^{F,F} f_s \tilde{A}_n L_n^{F,F}(h_j)_{x_j} I|S_n^F)$ are in $\mathcal{J}_{\Omega,r}^F$. Thus,

$$
\begin{aligned}
W_{s,\Gamma}(A_n B_n) &= W_{\nu_s^{-1}(s)}(L_n^{F,F} f_s \tilde{A}_n L_n^{F,F}(h_1)_s I|S_n^F) \times \\
&\quad \times W_{\nu_s^{-1}(s)}(L_n^{F,F}(h_1)_s \tilde{B}_n L_n^{F,F} g_s I|S_n^F) ,
\end{aligned}
\tag{3}
$$

and now use that $h_1(s) = 1$ (which is evident from the construction of the partition) and that the definition of $W_{s,\Gamma}$ is independent of the choice of the "cutting-off-functions" to find that the right hand side of (3) equals $W_{s,\Gamma}(A_n)W_{s,\Gamma}(B_n)$ as desired. ■

5.4.2   The homomorphisms $W^{t,\Gamma}$

The definition of the homomorphisms $W^{t,\Gamma}$ for $t \in \Omega$ is less explicit than those of $W_{s,\Gamma}$. Only in case $t = 1$ (being without doubt of extraordinary importance) we shall give an explicit definition of $W^{t,\Gamma}$ later on (simply by setting $W^{1,\Gamma}(A_n) = $ s-$\lim_{n\to\infty} A_n L_{n,\Gamma}^{F,F}$). But let us start with the general case. The basis for introducing $W^{t,\Gamma}$ is the following lemma.

Lemma 5.2 Let $(A_n) \in \mathcal{A}_{\Omega,r}^F(\Gamma)$ and $t \in \Omega$. Then there exists a uniquely determined operator $W^{t,\Gamma}(A_n)$ in $\mathcal{O}_\Gamma^p(\alpha)$ such that

(a) for all $x \in \Gamma$, and for all continuous functions f and g on Γ with $\operatorname{supp} f \cup \operatorname{supp} g \subseteq U_x$,

$$\psi_x f W^{t,\Gamma}(A_n) g \psi_x^{-1} \chi_{V_x} I =$$
$$= \begin{cases} W^t(L_n^{F,F} \psi_x f A_n L_{n,\Gamma}^{F,F} g \psi_x^{-1} \chi_{V_x} I | S_n^F) & \text{if } x \in \Gamma \setminus \Gamma' \\ \eta_x^{-1} W^t(\eta_x L_{n,\Gamma_x}^{F,F} \psi_x f A_n L_{n,\Gamma}^{F,F} g \psi_x^{-1} \eta_x^{-1} | \hat{\chi}_n^{\delta(x)}(S_n^F)_{w(x)}) \eta_x & \text{if } x \in \Gamma' \end{cases} \tag{1}$$

(b) for all $x, y \in \Gamma$, and for all continuous functions f and g on Γ with $\operatorname{supp} f \subseteq U_x$, $\operatorname{supp} g \subseteq U_y$, and $\operatorname{supp} f \cap \operatorname{supp} g = \emptyset$,

$$\psi_y f W^{t,\Gamma}(A_n) g \psi_x^{-1} \chi_{V_x} I =$$
$$= \begin{cases} W^t(L_n^{F,F} \psi_y f A_n L_{n,\Gamma}^{F,F} g \psi_x^{-1} \chi_{V_x} I | S_n^F) & \text{if } x, y \in \Gamma \setminus \Gamma' \\ \eta_y^{-1} W^t(\eta_y L_{n,\Gamma_y}^{F,F} \psi_y f A_n L_{n,\Gamma}^{F,F} g \psi_x^{-1} \chi_{V_x} I | S_n^F) & \text{if } x \in \Gamma \setminus \Gamma', y \in \Gamma' \\ W^t(L_n^{F,F} \psi_y f A_n L_{n,\Gamma}^{F,F} g \psi_x^{-1} | \hat{\chi}_n^{\delta(x)}(S_n^F)_{w(x)}) \eta_x & \text{if } x \in \Gamma', y \in \Gamma \setminus \Gamma' \\ \eta_y^{-1} W^t(\eta_y L_{n,\Gamma_y}^{F,F} \psi_y f A_n \times \\ \quad \times L_{n,\Gamma}^{F,F} g \psi_x^{-1} \eta_x^{-1} | \hat{\chi}_n^{\delta(x)}(S_n^F)_{w(x)}) \eta_x & \text{if } x, y \in \Gamma' \end{cases} \tag{2}$$

Let us for example explain the first identity in (1). If (B_n) is a sequence in $\mathcal{A}_{\Omega,r}^F$ such that

$$\psi_x f A_n L_{n,\Gamma}^{F,F} g \psi_x^{-1} \chi_{V_x} I = f_x B_n L_n^{F,F} g_x I$$

(such a sequence exists by definition) then

$$(L_n^{F,F} \psi_x f A_n L_{n,\Gamma}^{F,F} g \psi_x^{-1} \chi_{V_x} I | S_n^F) = (L_n^{F,F} f_x I | S_n^F)(B_n)(L_n^{F,F} g_x I | S_n^F) . \tag{3}$$

From Proposition 3.16 we infer that $(L_n^{F,F} f_x I | S_n^F)$ and $(L_n^{F,F} g_x I | S_n^F)$ are in $\mathcal{A}_{\Omega,r}^F$, and so it is correct to apply W^t to both sides of (3). Moreover, since

$$W^t(L_n^{F,F} f_x I | S_n^F) = f_x I \quad \text{for } t \in \Omega$$

(again by Proposition 3.16), we conclude from (3) that

$$W^t(L_n^{F,F} f_x B_n L_n^{F,F} g_x I | S_n^F) = f_x W^t(B_n) g_x I .$$

Similarly, if (B_n) and (J_n) denote the sequences introduced in 5.3.5(2) and 5.3.5(4) then (1) and (2) can be equivalently rewritten as

$$\psi_x f W^{t,\Gamma}(A_n) g \psi_x^{-1} \chi_{V_x} I =$$
$$= \begin{cases} f_x W^t(B_n) g_x I & \text{if } x \in \Gamma \setminus \Gamma' \\ f_x \eta_x^{-1} W^t(\eta_x B_n \psi_x^{-1} \eta_x^{-1} | \hat{\chi}_n^{\delta(x)}(S_n^F)_{w(x)}) \eta_x g_x & \text{if } x \in \Gamma'. \end{cases} \tag{4}$$

and

$$\psi_y f W^{t,\Gamma}(A_n) g \psi_x^{-1} \chi_{V_x} I =$$

$$= \begin{cases} f_y W^t(J_n) g_x I & \text{if } x,y \in \Gamma \setminus \Gamma' \\ f_y \eta_y^{-1} W^t(J_n) g_x I & \text{if } x \in \Gamma \setminus \Gamma', y \in \Gamma' \\ f_y W^t(J_n) \eta_x g_x & \text{if } x \in \Gamma', y \in \Gamma \setminus \Gamma' \\ f_y \eta_y^{-1} W^t(J_n) \eta_x g_x & \text{if } x,y \in \Gamma'. \end{cases} \tag{5}$$

Clearly, in case $x \in \Gamma'$ (resp. $x \in \Gamma'$ or $y \in \Gamma'$), the homomorphism W^t in (4) (resp. (5)) is thought of as acting on matrix operators.

Proof of the lemma: First we show the *uniqueness* of the operator $W^{t,\Gamma}(A_n)$ satisfying (1) and (2). Suppose, for contrary, that there are operators A and B satisfying (1) and (2) in place of $W^{t,\Gamma}(A_n)$. Then we choose a covering $\Gamma = \cup_{j=1}^{l} U_{x_j}$ and a continuous partition of unity $1 = f_1 + \ldots + f_l$ subordinate to this covering, and for each $i = 1, \ldots, l$ we pick a further continuous partition of unity $1 = g_{i1} + \ldots + g_{il}$ subordinate to the same covering such that

$$\operatorname{supp} f_i \cap \operatorname{supp} g_{ij} = \emptyset \quad \text{whenever} \quad i \neq j .$$

Clearly, $A = \sum_i \sum_j f_i A g_{ij} I$ and $B = \sum_i \sum_j f_i B g_{ij} I$, and from (1) and (2) we conclude that

$$f_i A g_{ij} I = f_i B g_{ij} I$$

for all i, j. Consequently, $A = B$.

Let us now verify the *existence* of an operator $W^{t,\Gamma}(A_n)$ subject to (1) and (2). For, we abbreviate the operator on the right hand side of (1) by $W^t(x, f, A_n, g, x)$, and the operator on the right hand side of (2) by $W^t(y, f, A_n, g, x)$. We claim that the operator

$$W^{t,\Gamma}(A_n) := \sum_i \sum_j \psi_{x_i}^{-1} W^t(x_i, f_i, A_n, g_{ij}, x_j) \psi_{x_j} \chi_{U_j} I \tag{6}$$

actually satisfies (1) and (2). We only accomplish part (a) in case $x \in \Gamma \setminus \Gamma'$. Let $f, g \in C(\Gamma)$ with $\operatorname{supp} f \cup \operatorname{supp} g \subseteq U_x$. Our goal is to show the identity

$$\sum_{ij} \psi_x f \psi_{x_i}^{-1} W^t(x_i, f_i, A_n, g_{ij}, x_j) \psi_{x_j} g \psi_x^{-1} \chi_{V_x} I =$$

$$= W^t(L_n^{F,F} \psi_x f A_n L_{n,\Gamma}^{F,F} g \psi_x^{-1} \chi_{V_x} I | S_n^F) . \tag{7}$$

For brevity we will do the following: we focus our attention on the part of (7) which contains the functions f_i and f, and we leave the remaining part of (7) invariant. To avoid unnecessary notations we simply indicate this invariant part by dots. If x_i is in $\Gamma \setminus \Gamma'$ then the ij th summand on the left hand side of (7) is

$$\psi_x f \psi_{x_i}^{-1} W^t(x_i, f_i, A_n, g_{ij}, x_j) \psi_{x_j} g \psi_x^{-1} \chi_{V_x} I =$$

$$= \psi_x f \psi_{x_i}^{-1} W^t(L_n^{F,F} \psi_{x_i} f_i A_n \ldots$$

$$= f_x \psi_x \chi_{U_x \cap U_{x_i}} \psi_{x_i}^{-1} W^t(L_n^{F,F} \psi_{x_i} f_i A_n \ldots \tag{8}$$

This is clearly zero if $U_x \cap U_{x_i} = \emptyset$. If $U_x \cap U_{x_i} \neq \emptyset$ then U_x and U_{x_i} belong to the very same component of $\Gamma \setminus \Gamma'$, say to Γ_r, and so the local parametrizations are defined by means of the same global parametrization, ζ_r. So, in this case,

$$\psi_x \chi_{U_x \cap U_{x_i}} \psi_{x_i}^{-1} = \chi_{V_x \cap V_{x_i}} \tag{9}$$

and this involves that (8) is equal to

$$f_x \chi_{V_x \cap V_{x_i}} W^t(L_n^{F,F} \psi_{x_i} f_i A_n \ldots \quad = f_x W^t(L_n^{F,F} \psi_{x_i} f_i A_n \ldots . \tag{10}$$

Now, write $f_x = W^t(L_n^{F,F} f_x I | S_n^F)$ and use the multiplicativity of W^t to find that (10) is the same as

$$W^t(L_n^{F,F} f_x I | S_n^F) W^t(L_n^{F,F} \psi_{x_i} f_i A_n \ldots \quad = W^t(L_n^{F,F} f_x L_n^{F,F} \psi_{x_i} f_i A_n \ldots ,$$

or, by the commutator property, as $W^t(L_n^{F,F} f_x \psi_{x_i} f_i A_n \ldots .$ Further, we have

$$f_x \psi_{x_i} f_i = \psi_x f \psi_x^{-1} \psi_{x_i} f_i = \psi_x f f_i$$

again by (9), and this finally gives

$$\psi_x f \psi_{x_i}^{-1} W^t(L_n^{F,F} \psi_{x_i} f_i A_n \ldots \quad = W^t(L_n^{F,F} \psi_x f f_i A_n \ldots . \tag{11}$$

Of course, (11) remains valid if $U_x \cap U_{x_i} = \emptyset$.

We are going to show that the same holds true for x_i being in Γ'. In this case, the ij th summand on the left hand side of (7) equals

$$\psi_x f \psi_{x_i}^{-1} \eta_{x_i}^{-1} W^t(\eta_{x_i} L_{n,\Gamma_{x_i}}^{F,F} \psi_{x_i} f_i A_n \ldots$$

$$= f_x \psi_x \psi_{x_i}^{-1} \eta_{x_i}^{-1} W^t(\eta_{x_i} L_{n,\Gamma_{x_i}}^{F,F} \psi_{x_i} f_i A_n \ldots . \tag{12}$$

Again, either $U_x \cap U_{x_i} = \emptyset$ or $U_x \cap U_{x_i} \neq \emptyset$, and in the latter case the intersection is completely contained in one component of $U_{x_i} \setminus \{x_i\}$, say in $\Gamma_r(x_i)$. Let, for definiteness, $\Gamma_r(x_i)$ be oriented away from x_i. Examining the local representations for this case we conclude that

$$\psi_x \psi_{x_i}^{-1} \eta_{x_i}^{-1} = (0, \ldots, 0, 1, 0, \ldots, 0)$$

on $\eta_{x_i} \psi_{x_i}(U_{x_i} \cap U_x)$ where the 1 stands at the r th place. Writing moreover f_x as $W^t(L_n^{F,F} f_x I | S_n^F)$ we get that (12) is

$$W^t(L_n^{F,F} f_x I | S_n^F)(0, \ldots, 0, 1, 0, \ldots, 0) W^t(\eta_{x_i} L_{n,\Gamma_{x_i}}^{F,F} \psi_{x_i} f_i A_n \ldots$$

$$= W^t(L_n^{F,F} f_x I(0, \ldots, 0, 1, 0, \ldots, 0) \eta_{x_i} L_{n,\Gamma_{x_i}}^{F,F} \psi_{x_i} f_i A_n \ldots$$

$$= W^t(L_n^{F,F} f_x I(0, \ldots, 0, 1, 0, \ldots, 0) \operatorname{diag}(L_{n,R^{\delta_1(x_i)}}^{F,F}, \ldots, L_{n,R^{\delta_{w(x_i)}(x_i)}}^{F,F}) \times$$

$$\times \eta_{x_i} \psi_{x_i} f_i A_n \ldots \qquad \text{(by 5.2.5(3))}$$

$$= W^t(L_n^{F,F} f_x L_{n,R^{\delta_r(x_i)}}^{F,F} \eta_{x_i} \psi_{x_i} f_i A_n \ldots$$

$$= W^t(L_n^{F,F} f_x \eta_{x_i} \psi_{x_i} f_i A_n \ldots$$

by the commutator property. The same arguments as above show that this is just

$$W^t(L_n^{F,F}\psi_x f f_i A_n \ldots$$

which again holds for $U_x \cap U_{x_i} = \emptyset$, too. This gives (11), and treating the other possible cases analogously we arrive at the identity

$$\psi_x f \psi_{x_i}^{-1} W^t(x_i, f_i, A_n, g_{ij}, x_j)\psi_{x_j} g \psi_x^{-1} \chi_{V_x} I =$$
$$= W^t(L_n^{F,F}\psi_x f f_i A_n L_{n,\Gamma}^{F,F} g_{ij} g \psi_x^{-1} \chi_{V_x} I | S_n^F)$$

holding for all i and j. Summarizing these identities with respect to i and j we finally obtain (7). The verification of part (b) of the lemma proceeds analogously.

It remains to show that $W^{t,\Gamma}(A_n)$ is always in $\mathcal{O}_\Gamma^p(\alpha)$. This is immediate from formula (6), from the fact that W^t maps $\mathcal{A}_{\Omega,r}^F$ into $\mathcal{O}_R^p(\alpha)$, and from the behaviour of the operators aI (with a piecewise continuous) and S_R under the mappings ψ_x and η_x; see the straightening lemma and Proposition 5.6(d). ∎

Now it is not hard to prove the following

Proposition 5.8 *The mapping $W^{t,\Gamma}$ is a continuous algebra homomorphism from $\mathcal{A}_{\Omega,r}^F(\Gamma)$ into $\mathcal{O}_\Gamma^p(\alpha)$.*

Proof   We use the same covering and the same partition of unity as in the preceding proof again. The continuity of $W^{t,\Gamma}$ is easy to check: from the continuity of $W^t : \mathcal{A}_{\Omega,r}^F \to \mathcal{O}_R^p(\alpha)$ (see Section 3.4.2 and Proposition 3.12) we infer that

$$\|W^t(x_i, f_i, A_n, g_{ij}, x_j)\| \leq C_1 \|(A_n)\|$$

with a constant C_1, whence via formula (6) follows

$$\|W^{t,\Gamma}(A_n)\| \leq C_2 \, l^2 \, \|(A_n)\|, \tag{13}$$

where l is the number of subcurves in the covering. (The estimate (13) is not the best possible one; finer estimates can be proved by employing KMS-techniques similarly to Sections 3.6.3-3.6.4.)

It is further immediate that $W^{t,\Gamma}$ is additive. For proving its multiplicativity we associate with each pair $(i,j) \in \{1,\ldots,l\} \times \{1,\ldots,l\}$ a non-negative continuous partition of unity $1 = h_{ij1} + \ldots + h_{ijl}$ with respect to the covering $(U_{x_j})_{j=1}^l$ and with the property that $\mathrm{supp}\, f_i \cap \mathrm{supp}\, h_{ijk} = \emptyset$ if $i \neq k$ and $\mathrm{supp}\, g_{ij} \cap \mathrm{supp}\, h_{ijk} = \emptyset$ for $j \neq k$. Set $m_{ijk} = \sqrt{h_{ijk}}$. Then

$$W^{t,\Gamma}(A_n B_n) =$$
$$= \sum_{ij} \psi_{x_i}^{-1} W^t(x_i, f_i, A_n B_n, g_{ij}, x_j)\psi_{x_j}\chi_{U_j} I$$
$$= \sum_{ijk} \psi_{x_i}^{-1} W^t(x_i, f_i, A_n L_{n,\Gamma}^{F,F} m_{ijk} m_{ijk} B_n, g_{ij}, x_j)\psi_{x_j}\chi_{U_j} I$$

$$= \sum_{ijk} \psi_{x_i}^{-1} W^t(x_i, f_i, A_n, m_{ijk}, x_k) \psi_{x_k} \chi_{U_k} \times$$

$$\times \psi_{x_k}^{-1} W^t(x_k, m_{ijk}, B_n, g_{ij}, x_j) \psi_{x_j} \chi_{U_j} I$$

$$= \sum_{ijk} f_i W^{t,\Gamma}(A_n) m_{ijk} I \, m_{ijk} W^{t,\Gamma}(B_n) g_{ij} I$$

$$= W^{t,\Gamma}(A_n) \, W^{t,\Gamma}(B_n) . \blacksquare$$

We conclude this section by stating the announced equivalent definition of $W^{1,\Gamma}$.

Proposition 5.9 (a) If $(A_n) \in \mathcal{A}_{\Omega,r}^F(\Gamma)$ then the strong limit $\operatorname*{s-lim}_{n\to\infty} A_n L_{n,\Gamma}^{F,F}$ exists.

(b) $W^{1,\Gamma}(A_n) = \operatorname*{s-lim}_{n\to\infty} A_n L_{n,\Gamma}^{F,F}$.

Proof (a) Choose a covering of Γ and corresponding partitions as in the proof of Lemma 5.2. Clearly,

$$A_n L_{n,\Gamma}^{F,F} = \sum_{ij} f_i A_n L_{n,\Gamma}^{F,F} g_{ij} I , \tag{14}$$

and (14) converges strongly if only each summand $f_i A_n L_{n,\Gamma}^{F,F} g_{ij} I$ converges strongly. But, if for example $x_i, x_j \in \Gamma \setminus \Gamma'$, then

$$f_i A_n L_{n,\Gamma}^{F,F} g_{ij} I = \psi_{x_i}^{-1} (f_i)_{x_i} B_n L_n^{F,F} (g_{ij})_{x_j} \psi_{x_j} \chi_{U_j} I \tag{15}$$

with a sequence (B_n) in $\mathcal{A}_{\Omega,r}^F$ whence the strong convergence for this special case follows. The other cases can be treated analogously.
(b) From (15) we conclude

$$f_i W^{t,\Gamma}(A_n) g_{ij} I = \psi_{x_i}^{-1} W^t (L_n^{F,F} \psi_{x_i} f_i A_n L_{n,\Gamma}^{F,F} g_{ij} \psi_{x_j}^{-1} \chi_{V_{x_j}} I | S_n^F) \psi_{x_j}$$

in case $x_i, x_j \in \Gamma \setminus \Gamma'$. The other conditions can be verified similarly. $\blacksquare$

5.4.3 The stability theorem

After finishing all preparations we are now going to present the main result of this chapter.

Theorem 5.4 (Stability theorem) A sequence (A_n) in $\mathcal{A}_{\Omega,r}^F(\Gamma)$ is stable if and only if all operators $W_{s,\Gamma}(A_n)$ and $W^{t,\Gamma}(A_n)$ with $s \in \Gamma$ and $t \in \Omega$ are invertible.

Its proof follows again the scheme pointed out in the first chapter. We start with preparing the application of the lifting theorem 1.8.

For $t \in \Omega$ we let $\mathcal{J}^t$ stand for the subset 3.6.2(1) of the algebra $\mathcal{A}_{\Omega,r}^F$, and write $\mathcal{J}^t(\Gamma)$ for the collection of all sequences in $\mathcal{A}_{\Omega,r}^F(\Gamma)$ which satisfy the definitions in 5.3.5 instead of $\mathcal{J}_{\Omega,r}^F$. Evidently,

$$\mathcal{G}(\Gamma) \subseteq \mathcal{J}^t(\Gamma) \subseteq \mathcal{J}_{\Omega,r}^F(\Gamma) ,$$

and repeating the arguments of the proof of Theorem 5.3 one can see that $\mathcal{J}^t(\Gamma)$ is a closed two-sided ideal of $\mathcal{A}^F_{\Omega,r}(\Gamma)$, and that $\mathcal{J}^F_{\Omega,r}(\Gamma)$ is the smallest closed two-sided ideal of $\mathcal{A}^F_{\Omega,r}(\Gamma)$ which encloses all $\mathcal{J}^t(\Gamma)$ with t running through Ω.

Proposition 5.10 *Let $t \in \Omega$. Then the mapping*

$$\Pi^t: \ \mathcal{J}^t(\Gamma)/\mathcal{G}(\Gamma) \to L(L^p_\Gamma(\alpha)), \quad (K_n) + \mathcal{G}(\Gamma) \mapsto W^{t,\Gamma}(K_n) \tag{1}$$

is correctly defined, and it is a continuous isomorphism from $\mathcal{J}^t(\Gamma)/\mathcal{G}(\Gamma)$ onto the ideal $K(L^p_\Gamma(\alpha))$ of all compact operators.

Proof The correctness of the definition (1) is immediate from the obvious identity $W^{t,\Gamma}(G_n) = 0$ for all $(G_n) \in \mathcal{G}(\Gamma)$. Furthermore, the formula 5.4.2(6) for $W^{t,\Gamma}$ involves that Π^t maps $\mathcal{J}^t(\Gamma)/\mathcal{G}(\Gamma)$ *into* the ideal $K(L^p_\Gamma(\alpha))$.

It remains to show that (1) is *onto* $K(L^p_\Gamma(\alpha))$ and *one-to-one*. Let K be compact on $L^p_\Gamma(\alpha)$, and let f_i and g_{ij} be functions defining partitions of unity over Γ as in the proof of Lemma 5.2. Then $K = \sum_{ij} f_i K g_{ij} I$, and so it suffices to prove that all operators $f_i K g_{ij} I$ lie in the image of $\mathcal{J}^t(\Gamma)$ under the homomorphism $W^{t,\Gamma}$. Suppose, for definiteness, $x_i, x_j \in \Gamma'$. Then

$$\psi_{x_i} f_i K g_{ij} \psi^{-1}_{x_j} \chi_{V_j} I = (f_i)_{x_i} \eta^{-1}_{x_i} K^{ij} \eta_{x_j} (g_{ij})_{x_j} I$$

with a compact $w(x_i) \times w(x_j)$-matrix operator K^{ij}. Since W^t is an isomorphism from $\mathcal{J}^t$ onto $K(L^p_R(\alpha))$, there is a sequence (K^{ij}_n) in $(\mathcal{J}^t)_{w(x_i) \times w(x_j)}$ such that $W^t(K^{ij}_n) = K^{ij}$. Set

$$K_n = L^{F,F}_{n,\Gamma} \psi^{-1}_{x_i} (f_i)_{x_i} \eta^{-1}_{x_i} K^{ij}_n \eta_{x_j} L^{F,F}_{n,\Gamma_{x_j}} (g_{ij})_{x_j} \psi_{x_j} \chi_{U_j} |S^F_n \ .$$

Proceeding in an analogous manner as in the second part of the proof of Lemma 5.2 one finds that $(K_n) \in \mathcal{J}^t(\Gamma)$ and $W^{t,\Gamma}(K_n) = f_i K g_{ij} I$ as desired.

To verify the injectivity of $W^{t,\Gamma}$ on $\mathcal{J}^t(\Gamma)$, we let $(A_n), (B_n) \in \mathcal{J}^t(\Gamma)$ have the same images under $W^{t,\Gamma}$. If, again for definiteness, $x_i, x_j \in \Gamma'$, then

$$\begin{aligned}
0 &= \psi_{x_i} f_i W^{t,\Gamma}(A_n - B_n) g_{ij} \psi^{-1}_{x_j} \chi_{V_j} I \\
&= \eta^{-1}_{x_i} W^t(\eta_{x_i} L_{n,\Gamma_{x_i}} \psi_{x_i} f_i (A_n - B_n) L^{F,F}_{n,\Gamma} g_{ij} \psi^{-1}_{x_j} \eta^{-1}_{x_j}) \eta_{x_j}
\end{aligned}$$

and since the sequence

$$(\eta_{x_i} L_{n,\Gamma_{x_i}} \psi_{x_i} f_i (A_n - B_n) L^{F,F}_{n,\Gamma} g_{ij} \psi^{-1}_{x_j} \eta^{-1}_{x_j} |\hat{\chi}^{\delta(x_j)}_n| (S^F_n)_{w(x_j)}) \tag{2}$$

is in $(\mathcal{J}^t)_{w(x_i) \times w(x_j)}$ and W^t acts injectively on this ideal we conclude that (2) tends to zero in the norm. Consequently, for all i, j,

$$\|L^{F,F}_{n,\Gamma} f_i (A_n - B_n) L^{F,F}_{n,\Gamma} g_{ij} |S^F_n\| \to 0$$

whence, of course, $\|(A_n - B_n) L^{F,F}_{n,\Gamma}\| \to 0$ or, equivalently, $(A_n - B_n) \in \mathcal{G}(\Gamma)$. $\blacksquare$

The general lifting theorem gives now:

Proposition 5.11 *Let* $(A_n) \in \mathcal{A}^F_{\Omega,r}(\Gamma)$. *The coset* $(A_n) + \mathcal{G}(\Gamma)$ *is invertible if and only if all operators* $W^{t,\Gamma}(A_n)$, $t \in \Omega$, *are invertible, and if the coset* $(A_n) + \mathcal{J}^F_{\Omega,r}(\Gamma)$ *is invertible in* $\mathcal{A}^F_{\Omega,r}(\Gamma)/\mathcal{J}^F_{\Omega,r}(\Gamma)$.

The coset $(A_n) + \mathcal{J}^F_{\Omega,r}(\Gamma)$ will be abbreviated to $\Phi^J(A_n)$. The following proposition indicates that the invertibility of $\Phi^J(A_n)$ is again a matter of localizing via Allan's local principle.

Proposition 5.12 *(a) If* $f \in C(\Gamma)$ *then the coset* $\Phi^J(L^{F,F}_{n,\Gamma} fI | S^F_n(\Gamma))$ *belongs to the center of the quotient algebra* $\mathcal{A}^F_{\Omega,r}(\Gamma)/\mathcal{J}^F_{\Omega,r}(\Gamma)$.
(b) The set

$$\mathcal{B}^F_{\Omega,r}(\Gamma) := \{\Phi^J(L^{F,F}_{n,\Gamma} fI | S^F_n(\Gamma)) \text{ with } f \in C(\Gamma)\}$$

is a closed subalgebra of $\mathcal{A}^F_{\Omega,r}(\Gamma)/\mathcal{J}^F_{\Omega,r}(\Gamma)$. *The maximal ideal space of this algebra is homeomorphic to* Γ, *and the maximal ideal associated with* $s \in \Gamma$ *is*

$$\{\Phi^J(L^{F,F}_{n,\Gamma} fI | S^F_n(\Gamma)) \text{ with } f \in C(\Gamma) \text{ and } f(s) = 0\}. \tag{3}$$

(c) There are constants $C_1, C_2 > 0$ *such that*

$$C_1 \|f\|_\infty \leq \|\Phi^J(L^{F,F}_{n,\Gamma} fI | S^F_n(\Gamma))\| \leq C_2 \|f\|_\infty$$

for all $f \in C(\Gamma)$.

To settle the proof of part (a) one chooses suitable coverings of Γ and reduces all things to sequences over the axis or over a system of axes, which are subject to Proposition 3.18. Assertions (b) and (c) can be verified by repeating parts of the proof of the same proposition; compare also Sections 3.1.2 and 3.1.5 for the Fredholm case. ∎

Proposition 5.12 offers the applicability of Allan's local principle to study the invertibility in $\mathcal{A}^F_{\Omega,r}(\Gamma)/\mathcal{J}^F_{\Omega,r}(\Gamma)$. Given $s \in \Gamma$ we denote by $\mathcal{J}^F_{\Omega,r,s}(\Gamma)$ the smallest closed two-sided ideal in $\mathcal{A}^F_{\Omega,r}(\Gamma)/\mathcal{J}^F_{\Omega,r}(\Gamma)$ containing the maximal ideal (3) of $\mathcal{B}^F_{\Omega,r}(\Gamma)$. We write $\mathcal{A}^F_{\Omega,r,s}(\Gamma)$ for the quotient algebra $(\mathcal{A}^F_{\Omega,r}(\Gamma)/\mathcal{J}^F_{\Omega,r}(\Gamma))/\mathcal{J}^F_{\Omega,r,s}(\Gamma)$ and Φ^J_s for the canonical homomorphism from $\mathcal{A}^F_{\Omega,r}(\Gamma)$ onto $\mathcal{A}^F_{\Omega,r,s}(\Gamma)$.

Proposition 5.13 *The mapping*

$$\Phi^J_s(A_n) \mapsto W_{s,\Gamma}(A_n) \tag{4}$$

is correctly defined, and it is a locally equivalent representation of the local algebra $\mathcal{A}^F_{\Omega,r,s}(\Gamma)$ *for all* $s \in \Gamma$.

Proof   The correctness of the mapping (4) follows from the identities

$$W_{s,\Gamma}(K_n) = 0 \quad \text{for all} \quad (K_n) \in \mathcal{J}_{\Omega,r}^F(\Gamma)$$

and

$$W_{s,\Gamma}(L_{n,\Gamma} f I | S_n^F) = f(s)I \quad \text{for} \quad f \in C(\Gamma)$$

which are readily verifiable by using Propositions 3.16 and 3.17. Let us now characterize the images of $\mathcal{A}_{\Omega,r}^F(\Gamma)$ under $W_{s,\Gamma}$. Given $\delta \in \{-1,1\}$ we denote the projection operator $E_{-n}^F \hat{\chi}_{R^\delta}^{(n)} E_n^F \in L(l_k^p(\alpha))$ (which is actually independent of n) by Π_δ. Using the notations in 5.2.3 we have

$$\Pi_1 = \text{diag}\,(V_{c_0} P V_{-c_0}, \dots, V_{c_{k-1}} P V_{-c_{k-1}})$$

and, thus, $\Pi_1 = \text{diag}\,(P, \dots, P) + \text{compact}$. Similarly, $\Pi_{-1} = \text{diag}\,(I - P, \dots, I - P) + \text{compact}$. If now $\delta = (\delta_1, \dots, \delta_w) \in \{-1,1\}^w$ then we define

$$\Pi_\delta = \text{diag}\,(\Pi_{\delta_1}, \dots, \Pi_{\delta_w})\,.$$

The Π_δ are projections acting on $l_{wk}^p(\alpha)$, and $E_n^F \Pi_\delta E_{-n}^F = \hat{\chi}_n^\delta$.
   We claim that

$$\text{Im}\,W_{s,\Gamma} = \begin{cases} \mathcal{T}_{k,\kappa,\Omega}^p(0) & \text{if} \quad s \in \Gamma \setminus \Gamma' \\ \Pi_{\delta(s)}(\mathcal{T}_{k,\kappa,\Omega}^p(\alpha_s))_{w(s) \times w(s)} \Pi_{\delta(s)} & \text{if} \quad s \in \Gamma'\,. \end{cases}$$

Indeed, this is immediate from the definitions of $\mathcal{A}_{\Omega,r}^F(\Gamma)$ and $W_{s,\Gamma}$ and from the mapping properties of W_s established in Section 3.6.4.
   The inverse closedness of $\mathcal{T}_{k,\kappa,\Omega}^p(\alpha_s)$ in $L(l_k^p(\alpha_s))$ has already been mentioned (Corollary 3.2), and similar arguments show that $(\mathcal{T}_{k,\kappa,\Omega}^p(\alpha_s))_{w(s) \times w(s)}$ is inverse closed in $L(l_k^p(\alpha_s))_{w(s) \times w(s)}$. Exercise 1.12 entails the inverse closedness of the algebra $\Pi_{\delta(s)}(\mathcal{T}_{k,\kappa,\Omega}^p(\alpha_s))_{w(s) \times w(s)} \Pi_{\delta(s)}$. This shows condition 1.6.4(v) of local equivalence, and the existence of the mappings $W_{s,\Gamma}'$ subject to 1.6.4(vi) can be seen by composing the mappings W_s' introduced in 3.6.1 (or their matrix versions) with the local parametrizations to "heave all onto the curve". For example, if $s \in \Gamma \setminus \Gamma'$, one chooses a continuous function f supported on U_s and taking the value 1 at s and defines $W_{s,\Gamma}' : \mathcal{T}_{k,\kappa,\Omega}^p(0) \to \mathcal{A}_{\Omega,r,s}^F(\Gamma)$ by

$$A \mapsto \Phi_s^J(L_{n,\Gamma}^{F,F} \psi_s^{-1} f_s(A, \nu_s^{-1}(s)) L_n^{F,F} \psi_s f I | S_n^F(\Gamma))$$

where, as in 3.5.1,

$$(A, \nu_s^{-1}(s)) = (E_n^F V_{\{\nu_s^{-1}(s)n+r\}} A V_{-\{\nu_s^{-1}(s)n+r\}} E_{-n}^F)\,. \quad \blacksquare$$

To finish the proof of Theorem 5.4 we must show that stability of a sequence $(A_n) \in \mathcal{A}_{\Omega,r}^F(\Gamma)$ implies invertibility of $W_{s,\Gamma}(A_n)$ and $W^{t,\Gamma}(A_n)$.
   Let $(A_n) \in \mathcal{A}_{\Omega,r}^F(\Gamma)$ be stable. Then there is a bounded sequence (B_n) of operators $B_n : S_n^F(\Gamma) \to S_n^F(\Gamma)$ such that

$$B_n A_n = A_n B_n = I | S_n^F(\Gamma) \quad \text{for all} \quad n \geq n_0\,. \tag{5}$$

For the first family of homomorphisms, we let $s \in \Gamma \setminus \Gamma'$ for definiteness, and we fix continuous functions f, g supported on U_s and taking the value 1 at s. If we abbreviate $\nu_s^{-1}(s)$ by σ then the homomorphisms $W_{s,\Gamma}$ are of the form

$$W_{s,\Gamma}(C_n) = \operatorname*{s-lim}_{n \to \infty} V_{-\{\sigma n + r\}} E_{-n}^F L_n^{F,F} \psi_s f C_n L_{n,\Gamma}^{F,F} g \psi_s^{-1} \chi_{V_s} E_n^F V_{\{\sigma n + r\}} \ .$$

(Recall that the weight function vanishes on $\Gamma \setminus \Gamma'$.)

From (5) we immediately conclude that $W_{s,\Gamma}(B_n A_n) = I$. Now we pick a non-negative and continuous partition of unity, $1 = f_1 + \ldots + f_k$, such that $(\operatorname{supp} f \cup \operatorname{supp} g) \cap \operatorname{supp} f_j = \emptyset$ for $j = 2, \ldots, k$ (compare 5.3.5) and claim that

$$W_{s,\Gamma}(B_n L_{n,\Gamma}^{F,F} f A_n) = I \ . \tag{6}$$

Clearly, for (6) it is sufficient to verify that

$$W_{s,\Gamma}(B_n L_{n,\Gamma}^{F,F} f_j A_n) = 0 \ , \quad j = 2, \ldots, k \ . \tag{7}$$

If we set $h_i = \sqrt{f_i}$ then

$$W_{s,\Gamma}(B_n L_{n,\Gamma}^{F,F} f_j A_n) =$$

$$= \operatorname*{s-lim}_{n \to \infty} V_{-\{\sigma n + r\}} E_{-n}^F L_n^{F,F} \psi_s f B_n L_{n,\Gamma}^{F,F} h_j h_j A_n L_{n,\Gamma}^{F,F} g \psi_s^{-1} \chi_{V_s} E_n^F V_{\{\sigma n + r\}}$$

$$= \operatorname*{s-lim}_{n \to \infty} (V_{-\{\sigma n + r\}} E_{-n}^F L_n^{F,F} \psi_s f B_n L_{n,\Gamma}^{F,F} h_j \psi_{x_j}^{-1} \chi_{V_{x_j}} E_n^F V_{\{\sigma n + r\}}) \times$$

$$\times (V_{-\{\sigma n + r\}} E_{-n}^F L_n^{F,F} \psi_{x_j} h_j A_n L_{n,\Gamma}^{F,F} g \psi_s^{-1} \chi_{V_s} E_n^F V_{\{\sigma n + r\}}) \ , \tag{8}$$

where we used the compatibility of the projections with the local parametrizations, see 5.3.4(c). Since $g h_j = 0$ one easily finds that the second factor in (8) goes strongly to zero. Then the uniform boundedness of the first factor gives our claim (7). Thus, (6) holds, and rewriting this identity in a similar manner as (8) we obtain

$$I = W_{s,\Gamma}(B_n L_{n,\Gamma}^{F,F} f A_n) =$$

$$= \operatorname*{s-lim}_{n \to \infty} (V_{-\{\sigma n + r\}} E_{-n}^F L_n^{F,F} \psi_s f B_n L_{n,\Gamma}^{F,F} h_1 \psi_s^{-1} \chi_{V_s} E_n^F V_{\{\sigma n + r\}}) \times$$

$$\times (V_{-\{\sigma n + r\}} E_{-n}^F L_n^{F,F} \psi_s h_1 A_n L_{n,\Gamma}^{F,F} g \psi_s^{-1} \chi_{V_s} E_n^F V_{\{\sigma n + r\}}) \ . \tag{9}$$

Abbreviate the first factor on the right hand side of (9) by B_n', and the second one by A_n' for a moment. Then the operators B_n' are uniformly bounded with respect to n whereas the operators A_n' converge strongly to $W_{s,\Gamma}(A_n)$ by the definition of the homomorphism $W_{s,\Gamma}$. Set $C_n' = B_n' A_n' - I$ and let $x \in l_k^p(0)$. Then

$$\|x + C_n' x\| = \|B_n' A_n' x\| \leq \sup_n \|B_n'\| \, \|A_n' x\| \leq C \|A_n' x\|$$

and letting n go to infinity we obtain

$$\|x\| \leq C \|W_{s,\Gamma}(A_n) x\| \ .$$

Taking adjoints (which is possible because $W_s(C_n^*) = W_s(C_n)^*$ for all sequences $(C_n) \in \mathcal{A}_{\Omega,r}^F$; see also 3.6.2 for this symmetry) one gets analogously

$$\|y\| \le C \, \|W_{s,\Gamma}(A_n)^* y\|$$

for all $y \in l_k^p(0)^*$. Hence, both the operator $W_{s,\Gamma}(A_n)$ and its adjoint are of regular type which implies the invertibility of $W_{s,\Gamma}(A_n)$ (see Section 1.2.2).

It is the concrete form of the homomorphisms $W^{t,\Gamma}$ which makes it more complicated to prove the triviality of their kernels. To start with, let $\Gamma \setminus \Gamma' = \cup_{j=1}^m \Gamma_j$ be the decomposition of $\Gamma \setminus \Gamma'$ into its open and connected components, and suppose Γ' to consist of l points, say x_j with $j = m+1, \ldots, m+l$. To each $x_j \in \Gamma'$ we choose a (very small) neighbourhood $U_{x_j} \subseteq \Gamma$ (later on we shall see how small the U_{x_j} must actually be), and we set $U = \cup_{x_j \in \Gamma'} U_{x_j}$. Further we introduce projections by

$$\tilde{\chi}_{\Gamma_j}^{(n)} = \hat{\chi}_{\Gamma_j}^{(n)} - \chi_{\Gamma_j} \hat{\chi}_U^{(n)} \quad (j = 1, \ldots, m)$$

and

$$\tilde{\chi}_{\Gamma \setminus U}^{(n)} = \sum_{j=1}^m \tilde{\chi}_{\Gamma_j}^{(n)}$$

(that is, $\tilde{\chi}_{\Gamma_j}^{(n)}$ maps $S_n^F(\Gamma)$ onto its subspace consisting of all basis functions which are completely supported on Γ_j but not completely supported on U). Clearly,

$$\hat{\chi}_U^{(n)} + \tilde{\chi}_{\Gamma \setminus U}^{(n)} = I | S_n^F(\Gamma) \, .$$

Next we choose and fix a continuous partition of unity, $1 = \sum_{j=1}^{m+l} f_j$, where

$$\operatorname{supp} f_j \subseteq \begin{cases} \Gamma_j & \text{for} \quad j = 1, \ldots, m \\ U_{x_j} & \text{for} \quad j = m+1, \ldots, m+l \, , \end{cases}$$

and we suppose moreover that $f_j \tilde{\chi}_{\Gamma_j}^{(n)} = \tilde{\chi}_{\Gamma_j}^{(n)}$ for $j = 1, \ldots, m$ which is possible if n is large enough. Let (A_n) be an arbitrary sequence in $\mathcal{A}_{\Omega,r}^F(\Gamma)$. Then, clearly,

$$W^{t,\Gamma}(\tilde{\chi}_{\Gamma \setminus U}^{(n)} A_n \tilde{\chi}_{\Gamma \setminus U}^{(n)}) = \chi_{\Gamma \setminus U} W^{t,\Gamma}(A_n) \chi_{\Gamma \setminus U} I \, , \tag{10}$$

and our first goal is to explain how the operator on the left hand side of (10) can be computed by means of a unique strong limit (just as in the case $\Gamma = \mathbf{R}$).

For, let ξ_j $(j = 1, \ldots, m)$ stand again for the global parametrization of Γ_j, and define

$$(\psi_j f)(s) = f(\xi_j(s)) \, , \ s \in [0,1] \, .$$

Given $t \in \mathbf{T}$ and $j \in \{1, \ldots, m\}$ we introduce operators $Y_{t,j}^{(n)} : S_n^F(\Gamma) \to S_n^F(\Gamma)$ by

$$Y_{t,j}^{(n)} = \psi_j^{-1} E_n^F Y_t E_{-n}^F \psi_j \hat{\chi}_{\Gamma_j}^{(n)}$$

and set $Y_t^{(n)} = Y_{t,1}^{(n)} + \ldots + Y_{t,m}^{(n)}$. Evidently, $(Y_t^{(n)})^{-1} = Y_{t-1}^{(n)}$. Now our precise goal is to prove that

$$W^{t,\Gamma}(\tilde{\chi}_{\Gamma\backslash U}^{(n)} A_n \tilde{\chi}_{\Gamma\backslash U}^{(n)}) = \operatorname*{s-lim}_{n\to\infty} Y_{t-1}^{(n)} \tilde{\chi}_{\Gamma\backslash U}^{(n)} A_n \tilde{\chi}_{\Gamma\backslash U}^{(n)} Y_t^{(n)} L_{n,\Gamma}^{F,F} .$$

Indeed, taking into account that

$$f_j \tilde{\chi}_{\Gamma\backslash U}^{(n)} = \begin{cases} \tilde{\chi}_{\Gamma_j}^{(n)} & \text{if } j = 1,\ldots,m \\ 0 & \text{if } j = m+1,\ldots,m+l , \end{cases}$$

one easily gets from the definition of $W^{t,\Gamma}$ that

$$
\begin{aligned}
& W^{t,\Gamma}(\tilde{\chi}_{\Gamma\backslash U}^{(n)} A_n \tilde{\chi}_{\Gamma\backslash U}^{(n)}) \\
&= \sum_{i,j=1}^{m+l} f_i W^{t,\Gamma}(\tilde{\chi}_{\Gamma\backslash U}^{(n)} A_n \tilde{\chi}_{\Gamma\backslash U}^{(n)}) f_j I \\
&= \sum_{i,j=1}^{m} \psi_i^{-1} W^t(L_n^{F,F} \psi_i f_i \tilde{\chi}_{\Gamma\backslash U}^{(n)} A_n \tilde{\chi}_{\Gamma\backslash U}^{(n)} L_{n,\Gamma}^{F,F} f_j \psi_j^{-1} \hat{\chi}_{[0,1]}^{(n)} I | S_n^F) \psi_j \\
&= \sum_{i,j=1}^{m} \psi_i^{-1} W^t(L_n^{F,F} \psi_i \tilde{\chi}_{\Gamma_i}^{(n)} A_n \tilde{\chi}_{\Gamma_j}^{(n)} \psi_j^{-1} \hat{\chi}_{[0,1]}^{(n)} I | S_n^F) \psi_j \\
&= \operatorname*{s-lim}_{n\to\infty} \sum_{i,j=1}^{m} \psi_i^{-1} E_n^F Y_{t-1} E_{-n}^F \psi_i \tilde{\chi}_{\Gamma_i}^{(n)} A_n \tilde{\chi}_{\Gamma_j}^{(n)} \psi_j^{-1} E_n^F Y_t E_{-n}^F \psi_j \hat{\chi}_{\Gamma_j}^{(n)} L_{n,\Gamma}^{F,F} \\
&= \operatorname*{s-lim}_{n\to\infty} Y_{t-1}^{(n)} \tilde{\chi}_{\Gamma_i}^{(n)} A_n \tilde{\chi}_{\Gamma_j}^{(n)} Y_t^{(n)} L_{n,\Gamma}^{F,F} .
\end{aligned}
$$

as desired.

Now let (A_n) be a stable sequence in $\mathcal{A}_{\Omega,r}^F(\Gamma)$, i.e. there is a bounded sequence (B_n) of operators $B_n : S_n^F(\Gamma) \to S_n^F(\Gamma)$ such that $I = B_n A_n = A_n B_n$, and suppose there is a non-zero function g in the kernel of $W^{t,\Gamma}(A_n)$. Then write

$$\tilde{\chi}_{\Gamma\backslash U}^{(n)} = \tilde{\chi}_{\Gamma\backslash U}^{(n)} B_n A_n \tilde{\chi}_{\Gamma\backslash U}^{(n)} = Y_{t-1}^{(n)} \tilde{\chi}_{\Gamma\backslash U}^{(n)} B_n A_n \tilde{\chi}_{\Gamma\backslash U}^{(n)} Y_t^{(n)}$$

to obtain

$$
\begin{aligned}
\tilde{\chi}_{\Gamma\backslash U}^{(n)} L_{n,\Gamma}^{F,F} g &= Y_{t-1}^{(n)} \tilde{\chi}_{\Gamma\backslash U}^{(n)} B_n \hat{\chi}_U^{(n)} A_n \tilde{\chi}_{\Gamma\backslash U}^{(n)} Y_t^{(n)} L_{n,\Gamma}^{F,F} g + \\
&\quad + Y_{t-1}^{(n)} \tilde{\chi}_{\Gamma\backslash U}^{(n)} B_n \tilde{\chi}_{\Gamma\backslash U}^{(n)} Y_t^{(n)} Y_{t-1}^{(n)} \tilde{\chi}_{\Gamma\backslash U}^{(n)} A_n \tilde{\chi}_{\Gamma\backslash U}^{(n)} Y_t^{(n)} L_{n,\Gamma}^{F,F} g .
\end{aligned}
\tag{11}
$$

The left hand side of (11) goes strongly to $\chi_{\Gamma\backslash U} g$. Thus, if n is sufficiently large and U is sufficiently small, then the norm of $\tilde{\chi}_{\Gamma\backslash U}^{(n)} L_{n,\Gamma}^{F,F} g$ is arbitrarily close to $\|g\|$ (say, greater than $\frac{2}{3}\|g\|$). We claim that, on the other side, the norm of the function on the right-hand side of (11) can be made arbitrarily small by choosing n large and U small (say, less than $\frac{1}{3}\|g\|$). This contradiction will yield $g \equiv 0$, thus, the kernel of $W^{t,\Gamma}(A_n)$ is trivial, and so the proof can be finished as in Section 3.6.5.

The first summand of (11) can be written as

$$\sum_{i,j=1}^{m} \sum_{k=m+1}^{m+l} Y_{t^{-1}}^{(n)} \tilde{\chi}_{\Gamma_i}^{(n)} B_n \hat{\chi}_{U_{x_k}}^{(n)} A_n \tilde{\chi}_{\Gamma_j}^{(n)} Y_t^{(n)} L_{n,\Gamma}^{F,F} g \, , \tag{12}$$

and the ijk th summand in (12) is just

$$[\psi_i^{-1} E_n^F b^F Y_{t^{-1}} E_{-n}^F \psi_i \tilde{\chi}_{\Gamma_i}^{(n)} B_n \hat{\chi}_{U_{x_k}}^{(n)} \psi_{x_k}^{-1} \eta_{x_k}^{-1} \mathrm{diag}\,(E_n^F Y_t E_{-n}^F, \dots, E_n^F Y_t E_{-n}^F)] \times$$
$$\times \; [\mathrm{diag}\,(E_n^F Y_{t^{-1}} E_{-n}^F, \dots, E_n^F Y_{t^{-1}} E_{-n}^F) \times$$
$$\times \eta_{x_k} \psi_{x_k} \hat{\chi}_{U_{x_k}}^{(n)} A_n \tilde{\chi}_{\Gamma_j}^{(n)} \psi_j^{-1} E_n^F Y_t E_{-n}^F L_n^{F,F} \psi_j g] \, . \tag{13}$$

Herein the second factor converges strongly to

$$\chi_{U_{x_k}} W^{t,\Gamma}(A_n) \chi_{\Gamma_j} g \, ,$$

and this can be made as small as desired by choosing U_{x_k} small. Since the first factor in (13) is uniformly bounded, and since (12) consists of ml factors only, this shows our claim for the first term in (11). For its second one we recall that

$$Y_{t^{-1}}^{(n)} \tilde{\chi}_{\Gamma\setminus U}^{(n)} A_n \tilde{\chi}_{\Gamma\setminus U}^{(n)} Y_t^{(n)} L_{n,\Gamma}^{F,F} g \to \chi_{\Gamma\setminus U} W^{t,\Gamma}(A_n) \chi_{\Gamma\setminus U} g \quad \text{strongly}$$

by what has already been proved. But

$$\chi_{\Gamma\setminus U} W^{t,\Gamma}(A_n) \chi_{\Gamma\setminus U} g = -\chi_{\Gamma\setminus U} W^{t,\Gamma}(A_n) \chi_U g$$

(because $g \in \mathrm{Ker}\, W^{t,\Gamma}(A_n)$), and $\|\chi_U g\|$ becomes arbitrarily small for small U. The uniform boundedness with respect to n of the operators $Y_{t^{-1}}^{(n)} \tilde{\chi}_{\Gamma\setminus U}^{(n)} B_n \tilde{\chi}_{\Gamma\setminus U}^{(n)} Y_t^{(n)}$ yields our claim for the first item in (11). ∎

5.5 A Galerkin method

5.5.1 Description of the curve

We are going to illustrate the application of the stability theorem 5.4 by an example. Let Γ be a curve in the complex plane which is composed by the simple arcs

$$\begin{aligned}
\Gamma^{(1)} &= \{z \in \mathbf{C} : \mathrm{Re}\, z = 0, \, \mathrm{Im}\, z \in [-1,0]\} \, , \\
\Gamma^{(2)} &= \{z \in \mathbf{C} : \mathrm{Im}\, z = 0, \, \mathrm{Re}\, z \in [0,1]\} \, , \\
\Gamma^{(3)} &= \{z \in \mathbf{C} : \mathrm{Im}\, z \geq 0, \, |z| = 1\} \\
\Gamma^{(4)} &= \{z \in \mathbf{C} : \mathrm{Im}\, z = 0, \, \mathrm{Re}\, z \in [-1,0]\} \, ,
\end{aligned}$$

with orientations as pictured. This "umbrella curve" exhibits both corners, end points, and intersections.

Choose $\Gamma' = \{-1, 0, 1, -i\}$. The connected components of $\Gamma \setminus \Gamma'$ are just the arcs $\Gamma^{(j)}$ without their endpoints.

We examine two versions of global parametrizations of Γ. The first one is given by

$$
\begin{aligned}
\zeta_1 &: [0,1] \to \Gamma^{(1)}, \; s \mapsto i(s-1), \\
\zeta_2 &: [0,1] \to \Gamma^{(2)}, \; s \mapsto s, \\
\zeta_3 &: [0,1] \to \Gamma^{(3)}, \; s \mapsto e^{i\pi s}, \\
\zeta_4 &: [0,1] \to \Gamma^{(4)}, \; s \mapsto s-1,
\end{aligned} \tag{1}
$$

and the second one by

$$
\begin{aligned}
\zeta_1 &: [0,1] \to \Gamma^{(1)}, \; s \mapsto i(s-1), \\
\zeta_2 &: [0,1] \to \Gamma^{(2)}, \; s \mapsto (\pi-1)s^3 + (1-\pi)s^2 + s, \\
\zeta_3 &: [0,1] \to \Gamma^{(3)}, \; s \mapsto e^{i\pi s}, \\
\zeta_4 &: [0,1] \to \Gamma^{(4)}, \; s \mapsto (\pi-1)s^3 + 2(1-\pi)s^2 + \pi s - 1.
\end{aligned} \tag{2}
$$

Both parametrizations preserve the orientation; the second one has, moreover, a derivative whose absolute value is continuous on Γ.

Given a k-elementic canonical prebasis F we define the spline space $S_n^F(\Gamma)$ in accordance with 5.3.1. The essential difference between the two versions is that, for the second version, the spline partitions are "almost equidistant" in a neighbourhood of each point, whereas in the first version, the characteristic mesh wide jumps at -1 and 1; compare also Examples 5.7 and 5.8.

5.5.2 The Galerkin method

Given a weight function $\alpha : \Gamma \to \mathbf{R}$ which vanishes on $\Gamma \setminus \Gamma'$ and satisfies $0 < 1/p + \alpha < 1$, we let $L_\Gamma^p(\alpha)$ denote the Lebesgue space over Γ, and we shall write $L_{n,\Gamma}^{F,F}$ for the Galerkin projection operator from $L_\Gamma^p(\alpha)$ onto $S_n^F(\Gamma)$.

The Galerkin method with projection $L_{n,\Gamma}^{F,F}$ (i.e. with $S_n^F(\Gamma)$ both as trial and test space) will be applied to solve the singular integral equation

$$
(aI + bS_\Gamma + T)\, u = f,
$$

where $u, f \in L_\Gamma^p(\alpha)$, and with a compact T and piecewise continuous coefficients a and b which are supposed to be continuous on $\Gamma \setminus \Gamma'$.

The following proposition shows that the theory established in the preceding sections applies to this approximation method.

Proposition 5.14 *The sequence* $(L_{n,\Gamma}^{F,F}(aI + bS_\Gamma + T)|S_n^F(\Gamma))$ *belongs to* $\mathcal{A}_{\{1\},0}^F(\Gamma)$ *under the above restrictions.*

The proof can easily be given by invoking Proposition 3.24 (where the corresponding result for $\mathbf{R}$ was stated). Thus, the general stability theorem applies, and we are recommended to compute the images of the sequence

$$(A_n) = (L_{n,\Gamma}^{F,F}(aI + bS_\Gamma + T)|S_n^F(\Gamma))$$

under the homomorphisms $W^{t,\Gamma}$ and $W_{s,\Gamma}$.

5.5.3   Computation of the homomorphisms

The only relevant homomorphism in the family $(W^{t,\Gamma})_{t\in\Omega}$ is $W^{1,\Gamma}$ (since $\Omega = \{1\}$ can be chosen), and Proposition 5.9 in combination with the strong convergence $L_{n,\Gamma}^{F,F} \to I$ immediately gives

$$W^{1,\Gamma}(A_n) = aI + bS_\Gamma + T .$$

For the homomorphisms $W_{s,\Gamma}$ we let first $s \in \Gamma \setminus \Gamma'$. Then

$$L_n^{F,F}\psi_s f A_n L_{n,\Gamma}^{F,F} g\psi_s^{-1}\chi_{V_s} I|S_n^F$$
$$= L_n^{F,F}\psi_s f(aI + bS_\Gamma + T)L_{n,\Gamma}^{F,F} g\psi_s^{-1}\chi_{V_s} I|S_n^F$$
$$= L_n^{F,F}\psi_s f(aI + bS_\Gamma + T)g L_{n,\Gamma}^{F,F} \psi_s^{-1}\chi_{V_s} I|S_n^F \;+\; C_n$$
$$= L_n^{F,F}\psi_s f(aI + bS_\Gamma + T)g\psi_s^{-1}\chi_{V_s} L_n^{F,F}\chi_{V_s} I|S_n^F \;+\; C_n$$

where the sequence (C_n) is in $\mathcal{G}$ due to the commutator property, and where we used Proposition 5.6(c) for the last step. The function fag is continuous and

$$(\psi_s fag\psi_s^{-1})(\nu_s^{-1}(s)) = f(s)a(s)g(s) = a(s) .$$

Further, by Proposition 5.6(d),

$$\psi_s fbS_\Gamma g\psi_s^{-1} = \psi_s fbS_{U_s} g\psi_s^{-1} = f_s b_s(S_{V_s} + K)g_s$$

with a compact operator K and with continuous functions f_s, b_s, g_s satisfying

$$f_s(\nu_s^{-1}(s)) = g_s(\nu_s^{-1}(s)) = 1 \quad\text{and}\quad b_s(\nu_s^{-1}(s)) = b(s) .$$

Finally, the operator $\psi_s fTg\psi_s^{-1}$ is compact again. Computing for each of these cases the strong limits W_s and summarizing the results we find for $s \in \Gamma \setminus \Gamma'$

$$W_{s,\Gamma}(A_n) = a(s)I + b(s)T^0((\lambda^{F,F})^{-1}\sigma^{F,F}) , \tag{1}$$

and this is valid both for version 1 and for version 2. Let now $s \in \Gamma'$. Then the local (homogeneous) curves are

$$\begin{array}{rlll}
\Gamma_{-i} = & i\mathbf{R}^+ & = \Gamma((\pi/2),(1)) , \\
\Gamma_{-1} = & \mathbf{R}^+ \cup (-i)\mathbf{R}^- & = \Gamma((0,3\pi/2),(1,-1)) , \\
\Gamma_1 = & \mathbf{R}^- \cup i\mathbf{R}^+ & = \Gamma((0,\pi/2),(-1,1)) , \\
\Gamma_0 = & \mathbf{R}^- \cup \mathbf{R}^+ \cup i\mathbf{R}^- & = \Gamma((0,0,\pi/2),(-1,1,-1)) ,
\end{array}$$

where we have used the notations of 5.2.4, and the spline spaces over Γ_s, $s \in \{-i, -1, 1, 0\}$, are the spaces $S_n^F(\Gamma_s, \xi_s)$ where

$$\xi_{-i} = (1) , \ \xi_{-1} = (1, \pi) , \ \xi_1 = (1, \pi) , \ \xi_0 = (1, 1, 1)$$

for version 1 and

$$\xi_{-i} = (1) , \ \xi_{-1} = (\pi, \pi) , \ \xi_1 = (\pi, \pi) , \ \xi_0 = (1, 1, 1)$$

for version 2.

If we introduce the notation $a_s I = \psi_s a \psi_s^{-1} \chi_{V_s} I$ for discontinuous functions a, too, then the general straightening lemma gives, exactly as for $s \in \Gamma \setminus \Gamma'$, in any case

$$L_{n,\Gamma_s}^{F,F} \psi_s f A_n L_{n,\Gamma}^{F,F} g \psi_s^{-1} \chi_{V_s} I | S_n^F(\Gamma_s, \xi_s)$$
$$= L_{n,\Gamma_s}^{F,F} \psi_s f L_{n,\Gamma}^{F,F} (aI + bS_\Gamma + T) g \psi_s^{-1} \chi_{V_s} I | S_n^F(\Gamma_s, \xi_s)$$
$$= L_{n,\Gamma_s}^{F,F} (f_s(a_s I + b_s S_{\Gamma_s}) g_s I + T_s) I | S_n^F(\Gamma_s, \xi_s) + C_n^s \qquad (2)$$

with compact operators T_s and sequences (C_n^s) of operators on $S_n^F(\Gamma_s, \xi_s)$ tending to zero in the norm. These sequences do not influence the values of the homomorphisms $W_{s,\Gamma}$ and will be neglected.

Applying the mappings η_s, η_s^{-1} to both sides of the first sequence in (2) we obtain, again up to a sequence going to zero in the norm,

$$\eta_s L_{n,\Gamma_s}^{F,F} (f_s(a_s I + b_s S_{\Gamma_s}) g_s I + T_s) \eta_s^{-1} | \hat{\chi}_n^{\delta(s)}(S_n^F)_{w(s)}$$
$$= \eta_s L_{n,\Gamma_s}^{F,F} f_s a_s g_s \eta_s^{-1} | \hat{\chi}_n^{\delta(s)}(S_n^F)_{w(s)} +$$
$$+ \eta_s L_{n,\Gamma_s}^{F,F} f_s b_s \eta_s^{-1} \eta_s S_{\Gamma_s} \eta_s^{-1} \eta_s g_s \eta_s^{-1} | \hat{\chi}_n^{\delta(s)}(S_n^F)_{w(s)} +$$
$$+ \eta_s L_{n,\Gamma_s}^{F,F} T_s \eta_s^{-1} | \hat{\chi}_n^{\delta(s)}(S_n^F)_{w(s)} . \qquad (3)$$

Since W_0 is an algebra homomorphism, it suffices to apply this operator to each of the sequences in (3).

Recalling that

$$\eta_s L_{n,\Gamma_s}^{F,F} \eta_s^{-1} = \text{diag}\,(L_{n,R^{\delta_1(s)}}^{F,F}, \ldots, L_{n,R^{\delta_{w(s)}(s)}}^{F,F})$$

by 5.2.5(3) and taking into account that the system matrix associated with $L_{n,R^+}^{F,F}$ is, by 5.2.3(5),

$$\text{diag}(V_{c_0} P V_{-c_0}, \ldots, V_{c_{k-1}} P V_{-c_{k-1}}) T^0(\lambda^{F,F}) \text{diag}(V_{c_0} P V_{-c_0}, \ldots, V_{c_{k-1}} P V_{-c_{k-1}}),$$

whence follows that its inverse is of the form

$$\text{diag}\,(P, \ldots, P) T^0(\lambda^{F,F})^{-1} \text{diag}\,(P, \ldots, P) + K$$

with a compact operator K, one easily gets that

$$W_0(\eta_s L_{n,\Gamma_s}^{F,F} f_s a_s g_s \eta_s^{-1} |\hat{\chi}_n^{\delta(s)} (S_n^F)_{w(s)}) =$$
$$= \operatorname{diag}(a_s^{(1)} \operatorname{diag}(P_{\delta_1(s)}, \dots, P_{\delta_1(s)}), \dots$$
$$\dots, a_s^{(w(s))} \operatorname{diag}(P_{\delta_{w(s)}(s)}, \dots, P_{\delta_{w(s)}(s)})) + K,$$

where $a_s^{(j)}$ denotes the limit of a_s at 0 along $e^{i\beta_j(s)} \mathbf{R}^{\delta_j(s)}$, and where we wrote for brevity $P_1 = P$, $P_{-1} = I - P$. The operator K is again compact. An analogous formula holds for the first factor of the second summand in (3). For its central factor we have to take into account identity 5.2.6(1) as well as Proposition 5.4 with its proof. The third summand in (3) belongs to the kernel of W_0 by Proposition 3.17. To write down the final results we let, for simplicity, $F = \{\phi\}$, i.e. $k = 1$. In the general case all operators appearing below have to be replaced by their matrix-analogues. Then we obtain

$$W_{-i,\Gamma}(A_n) = a_{-i}^{(1)} P + b_{-i}^{(1)} PT^0(\rho^{F,F}) P + K_{-i}, \tag{4}$$

$$W_{-1,\Gamma}(A_n) = \begin{pmatrix} a_{-1}^{(1)} P & 0 \\ 0 & a_{-1}^{(2)} Q \end{pmatrix} +$$
$$+ \begin{pmatrix} b_{-1}^{(1)} P & 0 \\ 0 & b_{-1}^{(2)} Q \end{pmatrix} \begin{pmatrix} T^0(\rho^{F,F}) & -G(d_{-1})\hat{J} \\ \hat{J}G(c_{-1}) & T^0(\rho^{F,F}) \end{pmatrix} \begin{pmatrix} P & 0 \\ 0 & Q \end{pmatrix} + K_{-1}, \tag{5}$$

$$W_{1,\Gamma}(A_n) = \begin{pmatrix} a_1^{(1)} Q & 0 \\ 0 & a_1^{(2)} P \end{pmatrix} +$$
$$+ \begin{pmatrix} b_1^{(1)} Q & 0 \\ 0 & b_1^{(2)} P \end{pmatrix} \begin{pmatrix} T^0(\rho^{F,F}) & \hat{J}G(d_1) \\ -G(c_1)\hat{J} & T^0(\rho^{F,F}) \end{pmatrix} \begin{pmatrix} Q & 0 \\ 0 & P \end{pmatrix} + K_1 \tag{6}$$

and

$$W_{0,\Gamma}(A_n) = \begin{pmatrix} a_0^{(1)} Q & 0 & 0 \\ 0 & a_0^{(2)} P & 0 \\ 0 & 0 & a_0^{(3)} Q \end{pmatrix} + \begin{pmatrix} b_0^{(1)} Q & 0 & 0 \\ 0 & b_0^{(2)} P & 0 \\ 0 & 0 & b_0^{(3)} Q \end{pmatrix} \times$$
$$\times \begin{pmatrix} T^0(\rho^{F,F}) & \hat{J}G(d_0) & -\hat{J}G(f_0)\hat{J} \\ -G(c_0)\hat{J} & T^0(\rho^{F,F}) & -G(g_0)\hat{J} \\ -\hat{J}G(h_0)\hat{J} & \hat{J}G(k_0) & T^0(\rho^{F,F}) \end{pmatrix} \begin{pmatrix} Q & 0 & 0 \\ 0 & P & 0 \\ 0 & 0 & Q \end{pmatrix} + K_0. \tag{7}$$

Herein, the K_s are certain compact operators, the $\hat{J}$ marks the discrete flip $\hat{J}(x_k) = (x_{-k})$, and the generating functions $c_{-1}, d_{-1}, \dots$ of the discretized Mellin convolutions are given by

$$c_{-1}(z) = \begin{cases} (1/\pi)^{1/p+\alpha-zi} n_{\pi/2}(z) & \text{for version 1} \\ n_{\pi/2}(z) & \text{for version 2} \end{cases},$$

$$d_{-1}(z) = \begin{cases} \pi^{1/p+\alpha-zi}n_{3\pi/2}(z) & \text{for version 1} \\ n_{3\pi/2}(z) & \text{for version 2} \end{cases},$$

$$c_1(z) = \begin{cases} \pi^{1/p+\alpha-zi}n_{3\pi/2}(z) & \text{for version 1} \\ n_{3\pi/2}(z) & \text{for version 2} \end{cases},$$

$$d_1(z) = \begin{cases} (1/\pi)^{1/p+\alpha-zi}n_{\pi/2}(z) & \text{for version 1} \\ n_{\pi/2}(z) & \text{for version 2} \end{cases},$$

and

$$c_0(z) = n_\pi(z) \,, \ d_0(z) = n_\pi(z) \,, \ f_0(z) = n_{3\pi/2}(z)$$

$$g_0(z) = n_{\pi/2}(z) \,, \ h_0(z) = n_{\pi/2}(z) \,, \ k_0(z) = n_{3\pi/2}(z) \,.$$

Theorem 5.5 *The Galerkin method with $S_n^F(\Gamma)$ both as trial and test space applies to the singular integral operator $aI + bS_\Gamma + T$ with a, b, T being specified as above if and only the operator $aI + bS_\Gamma + T$ as well as all operators $W_{s,\Gamma}(A_n)$ given by (1) for $s \in \Gamma \setminus \Gamma'$ and by (4)-(7) for $s \in \Gamma'$ are invertible.*

5.5.4 Cutting off singularities

To obtain verifiable conditions for the Galerkin method we apply at each point of Γ' a cutting-off-technique. The corresponding modified Galerkin scheme is stable if and only if the operator $aI + bS_\Gamma + T$ itself as well as all operators $W_{s,\Gamma}(A_n)$ for $s \in \Gamma \setminus \Gamma'$ (given by 5.5.3(1)) are invertible and if the infinite section method for the operators $W_{s,\Gamma}(A_n)$, $s \in \Gamma'$ (given by 5.5.3(4)-(7)) applies.

Let us analyse the latter in detail. This can be most easily done for $W_{-i,\Gamma}(A_n) \in l_k^p(\alpha(-i))$. As in Example 4.7 one immediately obtains that the *infinite section method for this operator is stable if and only if*

(a) *the Toeplitz operator $T(a_{-i}^{(1)} + b_{-i}^{(1)}\rho^{F,F})$ is Fredholm on $(l_{Z+}^p(\alpha(-i)))_k$,*

(b) *the same Toeplitz operator is invertible on $(l_{Z+}^p(0))_k$,*

(c) *the singular integral operator $\chi_{[1,\infty)}(a_{-i}^{(1)}I + b_{-i}^{(1)}S)\chi_{[1,\infty)}I + \chi_{[0,1]}I$ is invertible on $L_{R+}^p(\alpha(-i))$.*

Evidently, in case $k = 1$, all conditions are effective.

Next we consider $W_{-1,\Gamma}(A_n) \in l_k^p(\alpha(-1))$. For technical reasons it is more convenient to work with the operator

$$\mathrm{diag}\,(I, \hat{J})W_{-1,\Gamma}(A_n)\mathrm{diag}\,(I, \hat{J}) \tag{1}$$

in place of $W_{-1,\Gamma}$, and to apply the infinite section method to this operator (the equivalence of these two problems results from the commutativity of the flip operator $\hat{J}$ with the cutting-off-operators Q_i in the infinite section method). Obviously, (1) is equal to

$$\begin{pmatrix} T(a_{-1}^{(1)} + b_{-1}^{(1)}\rho^{F,F}) + K_{11} & -b_{-1}^{(1)}G(d_{-1}) + K_{12} \\ b_{-1}^{(2)}G(c_{-1}) + K_{21} & T(a_{-1}^{(2)} + b_{-1}^{(2)}\tilde{\rho}^{F,F}) + K_{22} \end{pmatrix} \in (l_{Z+}^p(\alpha(-1)))_{2k} \,,$$

where $\tilde{\rho}^{F,F}(z) = \rho^{F,F}(z^{-1})$, and where the K_{ij} refer to compact operators. *The infinite section method applies to this operator if and only if*

(d) the operator

$$
\begin{pmatrix}
T(a_{-1}^{(1)} + b_{-1}^{(1)}\rho^{F,F}) & -b_{-1}^{(1)}G(d_{-1}) \\
b_{-1}^{(2)}G(c_{-1}) & T(a_{-1}^{(2)} + b_{-1}^{(2)}\tilde{\rho}^{F,F})
\end{pmatrix}
$$

is Fredholm on $(l_{Z^+}^p(\alpha(-1)))_{2k}$,

(e)' the operator

$$
\begin{pmatrix}
T(a_{-1}^{(1)} + b_{-1}^{(1)}\rho^{F,F}) & 0 \\
0 & T(a_{-1}^{(2)} + b_{-1}^{(2)}\tilde{\rho}^{F,F})
\end{pmatrix}
$$

is invertible on $(l_{Z^+}^p(0))_{2k}$, and

(f)' the operator

$$
\begin{pmatrix}
\chi_{[0,1]}I & 0 \\
0 & \chi_{[0,1]}I
\end{pmatrix}
+
\begin{pmatrix}
\chi_{[1,\infty)}I & 0 \\
0 & \chi_{[1,\infty)}I
\end{pmatrix}
\times
$$

$$
\times
\begin{pmatrix}
a_{-1}^{(1)}I + b_{-1}^{(1)}S & -b_{-1}^{(1)}M^0(d_{-1}) \\
b_{-1}^{(2)}M^0(c_{-1}) & a_{-1}^{(2)}I + b_{-1}^{(2)}S
\end{pmatrix}
\begin{pmatrix}
\chi_{[1,\infty)}I & 0 \\
0 & \chi_{[1,\infty)}I
\end{pmatrix}
$$

is invertible on $(L_{R^+}^p(\alpha(-1)))_{2k}$.

Again, all conditions are effective for $k = 1$. This is clear for (d) since the quotient algebra alg $T(PC_{p,\alpha(-1)})/K(l^p(\alpha(-1)))$ is commutative, and almost clear for (e)' since this is equivalent to

(e) the Toeplitz operators $T(a_{-1}^{(1)} + b_{-1}^{(1)}\rho^{F,F})$ and $T(a_{-1}^{(2)} + b_{-1}^{(2)}\tilde{\rho}^{F,F})$ are invertible on $(l_{Z^+}^p(0))_k$.

For (f)' we restrict ourselves to version 2, in which case this operator can be written as

$$
\begin{pmatrix}
\chi_{[0,1]}I & 0 \\
0 & \chi_{[0,1]}I
\end{pmatrix}
+
\begin{pmatrix}
a_{-1}^{(1)}\chi_{[1,\infty)}I & 0 \\
0 & a_{-1}^{(2)}\chi_{[1,\infty)}I
\end{pmatrix}
+
$$

$$
+
\begin{pmatrix}
b_{-1}^{(1)} & 0 \\
0 & b_{-1}^{(2)}
\end{pmatrix}
\begin{pmatrix}
S_{[1,\infty)} & -\chi_{[1,\infty)}N_{3\pi/2}\chi_{[1,\infty)}I \\
\chi_{[1,\infty)}N_{\pi/2} & -S_{[1,\infty)}
\end{pmatrix}
$$

and *this operator is invertible if and only if*

(f) the singular integral operator

$$\chi_{\Gamma_{-1}\backslash\tilde{\Gamma}_{-1}}(a_{-1}I+b_{-1}S_{\Gamma_{-1}})\chi_{\Gamma_{-1}\backslash\tilde{\Gamma}_{-1}}I+\chi_{\tilde{\Gamma}_{-1}}\,, \tag{2}$$

where $\Gamma_{-1}=\mathbf{R}^{+}\cup(-i)\mathbf{R}^{-}$, $\tilde{\Gamma}_{-1}=[0,1]\cup(-i)[-1,0]$,

$$a_{-1}=\begin{cases} a_{-1}^{(1)} & \text{on } \mathbf{R}^{+} \\ a_{-1}^{(2)} & \text{on } (-i)\mathbf{R}^{-} \end{cases}, \quad b_{-1}=\begin{cases} b_{-1}^{(1)} & \text{on } \mathbf{R}^{+} \\ b_{-1}^{(2)} & \text{on } (-i)\mathbf{R}^{-} \end{cases},$$

is invertible on $L_{\Gamma_{-1}}^{p}(\alpha(-1))$.

The operator (2) is subject to Coburn's theorem in case $k=1$, i.e. its invertibility depends on the winding number of the symbol.

In a completely analogous manner one obtains (also for the second version) : *The infinite section method applies to the operator* $W_{1,\Gamma}(A_n)$ *if and only if*

(g) the operator

$$\begin{pmatrix} T(a_1^{(1)}+b_1^{(1)}\tilde{\rho}^{F,F}) & b_1^{(1)}G(d_1) \\ -b_1^{(2)}G(c_1) & T(a_1^{(2)}+b_1^{(2)}\rho^{F,F}) \end{pmatrix}$$

is Fredholm on $(l_{Z^+}^p(\alpha(1)))_{2k}$,

(h) the Toeplitz operators $T(a_1^{(1)}+b_1^{(1)}\tilde{\rho}^{F,F})$ *and* $T(a_1^{(2)}+b_1^{(2)}\rho^{F,F})$ *are invertible on* $(l_{Z^+}^p(0))_k$,

(i) the singular integral operator

$$\chi_{\Gamma_1\backslash\tilde{\Gamma}_1}(a_1I+b_1S_{\Gamma_1})\chi_{\Gamma_1\backslash\tilde{\Gamma}_1}I+\chi_{\tilde{\Gamma}_1}\,,$$

where $\Gamma_1=\mathbf{R}^{-}\cup i\mathbf{R}^{+}$, $\tilde{\Gamma}_1=[-1,0]\cup i[0,1]$,

$$a_1=\begin{cases} a_1^{(1)} & \text{on } \mathbf{R}^{-} \\ a_1^{(2)} & \text{on } i\mathbf{R}^{+} \end{cases}, \quad b_1=\begin{cases} b_1^{(1)} & \text{on } \mathbf{R}^{-} \\ b_1^{(2)} & \text{on } i\mathbf{R}^{+} \end{cases},$$

is invertible on $L_{\Gamma_1}^{p}(\alpha(1))$.

The infinite section method applies to the operator $W_{0,\Gamma}(A_n)$ *if and only if*

(j) the operator

$$\begin{pmatrix} T(a_0^{(1)}+b_0^{(1)}\tilde{\rho}^{F,F}) & b_0^{(1)}G(d_0) & -b_0^{(1)}G(f_0) \\ -b_0^{(2)}G(c_0) & T(a_0^{(2)}+b_0^{(2)}\rho^{F,F}) & -b_0^{(2)}G(g_0) \\ -b_0^{(3)}G(h_0) & b_0^{(3)}G(k_0) & T(a_0^{(3)}+b_0^{(3)}\tilde{\rho}^{F,F}) \end{pmatrix}$$

is Fredholm on $(l_{Z^+}^p(\alpha(0)))_{3k}$,

(k) the Toeplitz operators

$$T(a_0^{(1)} + b_0^{(1)} \tilde{\rho}^{F,F}) \,,\; T(a_0^{(2)} + b_0^{(2)} \rho^{F,F}) \,,\; T(a_0^{(3)} + b_0^{(3)} \tilde{\rho}^{F,F})$$

are invertible on $(l_{Z^+}^p(0))_k$,

(l) the singular integral operator

$$\chi_{\Gamma_0 \backslash \tilde{\Gamma}_0} (a_0 I + b_0 S_{\Gamma_0}) \chi_{\Gamma_0 \backslash \tilde{\Gamma}_0} I + \chi_{\tilde{\Gamma}_0} \,,$$

where $\Gamma_0 = \mathbf{R}^- \cup \mathbf{R}^+ \cup i\,\mathbf{R}^-$, $\tilde{\Gamma}_0 = [-1,0] \cup [0,1] \cup (-i)[-1,0]$,

$$a_0 = \begin{cases} a_0^{(1)} & \text{on} & \mathbf{R}^- \\ a_0^{(2)} & \text{on} & \mathbf{R}^+ \\ a_0^{(3)} & \text{on} & i\,\mathbf{R}^- \end{cases} \,,\; b_0 = \begin{cases} b_0^{(1)} & \text{on} & \mathbf{R}^- \\ b_0^{(2)} & \text{on} & \mathbf{R}^+ \\ b_0^{(3)} & \text{on} & i\,\mathbf{R}^- \end{cases} \,,$$

is invertible on $L_{\Gamma_0}^p(\alpha(0))$.

Let us remark that conditions (a)-(l) indicate in an obvious manner how the conditions for the applicability of a modified Galerkin method for singular integral operators over an arbitrarily complicated curve actually look like.

5.5.5   Other projection methods

If another projection method (collocation or qualocation, say) is defined on a complicated curve in such a way that local compatibility in sense of 5.3.4 is guaranteed then the corresponding stability theorem can be derived in an analogous manner as for Galerkin methods. For our concrete example examined in 5.5.1-5.5.4, the final result for other projection methods follows by replacing the function $\rho^{F,F}$ by $(\lambda^{F,G})^{-1}\sigma^{F,G}$ at each occurance.

In particular, the results in Section 2.9.5 concerning compactly supported wavelets involve stability assertions for wavelet approximation methods for singular integral operators over complicated curves.

Compare also the next section where some numerical results for the collocation method for singular integral equations on the "umbrella curve" are presented.

5.5.6   Numerical experiments

For some first numerical experiments we consider the singular integral equation

$$(2I + S_\Gamma)u = f \tag{1}$$

on the "umbrella curve" introduced in 5.5.1. The right-hand side f is given by either

$$f(t) = t\,s(t) + 1/\pi \tag{2}$$

or

$$f(t) = (t^5 - t)\,s(t) + \frac{1}{\pi}\left(t^4 - \frac{i}{2}t^3 - \frac{1}{3}t^2 + \frac{i}{4}t - \frac{4}{5}\right) \tag{3}$$

where s is the function with

$$\operatorname{Re} s(t) = \begin{cases} 1 - \frac{1}{\pi}\arctan\frac{2|t|}{1-|t|^2} & t \in \Gamma^{(1)} \\ 1 & t \in \Gamma^{(2)},\, \Gamma^{(4)} \\ 1/2 & t \in \Gamma^{(3)} \end{cases}$$

and

$$\operatorname{Im} s(t) = \frac{1}{\pi}\ln\left|\frac{i+t}{t}\right|, \quad t \in \Gamma.$$

If f is given by (2) or (3) then the solutions of equation (1) are just the functions

$$u(t) = t \tag{4}$$

or

$$u(t) = t^5 - t, \tag{5}$$

respectively. Observe that the solution (5) vanishes at all singular points of the curve Γ.

Besides the global parametrizations 5.5.1(1) and (2) of Γ we consider the following one:

$$\zeta_1: \quad [0,1] \to \Gamma^{(1)}, \quad s \mapsto i(q(s) - 1)$$
$$\zeta_2: \quad [0,1] \to \Gamma^{(2)}, \quad s \mapsto q(s)$$
$$\zeta_3: \quad [0,1] \to \Gamma^{(3)}, \quad s \mapsto e^{\pi i q(s)}$$
$$\zeta_4: \quad [0,1] \to \Gamma^{(4)}, \quad s \mapsto q(s) - 1$$

with q standing for the polynomial $q(s) = 3s^2 - 2s^3$. The latter choice involves a certain condensation of the knots in neighbourhoods of the singular points. Later on, we refer to these parametrizations by e (for equidistant), a (for adapted), and c (for cubic), respectively.

The trial spaces consist of piecewise constant splines or piecewise linear and continuous splines with $i_0 = 0$ or $i_0 = 1$. The functionals are specified as δ_ϵ-functionals (i.e. ϵ-collocation) with $\epsilon = 0$ or $\epsilon = 1/2$, and the resulting integrals are evaluted by a Simpson rule with 40 nodes. The curves $\Gamma^{(1)}$, $\Gamma^{(2)}$ and $\Gamma^{(4)}$ are subdivided into the same number, n, of intervals, whereas $\Gamma^{(3)}$ is divided into kn parts with $k = 1, 2, 3$ in order to get a better approximation of the geometry of $\Gamma^{(3)}$. The actual choice of the factor k is indicated in the tables below.

Example 1   Exact solution: $u(t) = t$
            Degree of splines: 0
            Method: collocation, $\epsilon = 1/2$

n	dim S_n	parametrization	i_0	k	error in L^2
15	60	e	0	1	0.12357797
30	120	e	0	1	0.06122160
12	60	e	0	2	0.08846480
24	120	e	0	2	0.04399998
15	52	e	1	1	0.83135671
30	112	e	1	1	0.58833155
12	52	e	1	2	0.74111744
24	112	e	1	2	0.52707331
15	60	a	0	1	0.13359056
30	120	a	0	1	0.06601450
12	60	a	0	2	0.10889629
24	120	a	0	2	0.05383925
15	52	a	1	1	0.94626427
30	112	a	1	1	0.68638889
12	52	a	1	2	0.89694533
24	112	a	1	2	0.66380884

Example 2   Exact solution: $u(t) = t^5 - t$
            Degree of splines: 0
            Method: collocation, $\epsilon = 1/2$

n	dim S_n	parametrization	i_0	k	error in L^2
15	60	e	0	1	0.60307626
30	120	e	0	1	0.29233359
12	60	e	0	2	0.37248184
24	120	e	0	2	0.18287055
15	52	e	1	1	0.65976903
30	112	e	1	1	0.30672011
12	52	e	1	2	0.40064611
24	112	e	1	2	0.18974039
15	60	a	0	1	0.61135331
30	120	a	0	1	0.29698494
12	60	a	0	2	0.39102693
24	120	a	0	2	0.19372008
15	52	a	1	1	0.68957769
30	112	a	1	1	0.31928604
12	52	a	1	2	0.46859450
24	112	a	1	2	0.22153955

Example 3   Exact solution: $u(t) = t$
            Degree of splines: 1
            Method: collocation, $\epsilon = 0$

n	dim S_n	parametrization	i_0	k	error in L^2
15	56	e	0	1	0.49267895
30	116	e	0	1	0.34839455
12	56	e	0	2	0.44096445
24	116	e	0	2	0.31203391
15	56	a	0	1	0.57944316
30	116	a	0	1	0.41205268
12	56	a	0	2	0.56871150
24	116	a	0	2	0.40657872
15	56	c	0	1	0.15830879
30	116	c	0	1	0.07932539
12	56	c	0	2	0.13608711
24	116	c	0	2	0.06809960
15	48	e	1	1	0.95922662
30	108	e	1	1	0.68653546
12	48	e	1	2	0.85338029
24	108	e	1	2	0.61397467
15	48	a	1	1	1.06953968
30	108	a	1	1	0.79038843
12	48	a	1	2	1.00539995
24	108	a	1	2	0.76200692
15	48	c	1	1	0.49624695
30	108	c	1	1	0.25324753
12	48	c	1	2	0.42671213
24	108	c	1	2	0.21815981

Remark: The exact solution does not vanish at the singular points $-i$ and ± 1, but all spline functions in the trial space S_n^F (piecewise linear and continuous splines) vanish at these points (compare the definitions of the trial spaces in 5.2.3, 5.2.4, and 5.3.1). This leads to "local " errors in the neighbourhoods of the singular points $-i$ and ± 1 (see the pictures below).

Example 4   Exact solution: $u(t) = t^5 - t$
            Degree of splines: 1
            Method: collocation, $\epsilon = 0$

n	$\dim S_n$	parametrization	i_0	k	error in L^2
15	56	e	0	1	0.12685965
30	116	e	0	1	0.03168408
12	56	e	0	2	0.05106834
24	116	e	0	2	0.01275197
15	56	a	0	1	0.13586109
30	116	a	0	1	0.03489310
12	56	a	0	2	0.08487744
24	116	a	0	2	0.02341189
15	56	c	0	1	0.20659673
30	116	c	0	1	0.05107600
12	56	c	0	2	0.08074626
24	116	c	0	2	0.02008321
15	48	e	1	1	0.52253029
30	108	e	1	1	0.18237137
12	48	e	1	2	0.28391272
24	108	e	1	2	0.10130937
15	48	a	1	1	0.58891277
30	108	a	1	1	0.22568782
12	48	a	1	2	0.43395826
24	108	a	1	2	0.19650076
15	48	c	1	1	0.21324935
30	108	c	1	1	0.05144638
12	48	c	1	2	0.08408387
24	108	c	1	2	0.02024621

The behaviour of the solution in the situation considered in the first line of the table in Example 3 is illustrated by the following pictures.

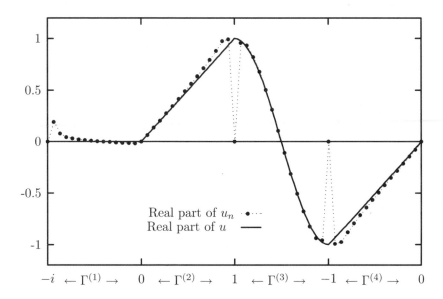

The large errors at the points -1 and 1 are caused by the circumstance that all spline functions in the corresponding spline space vanish at these singular points of the curve Γ.

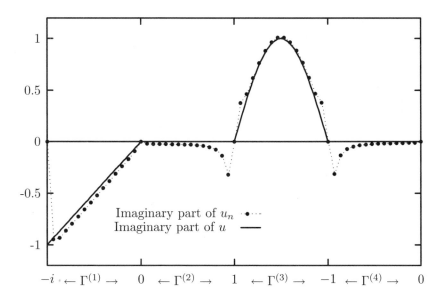

The large error at the point $-i$ is also caused by the fact that the spline functions in the corresponding spline space vanish at this singular point of the curve Γ.

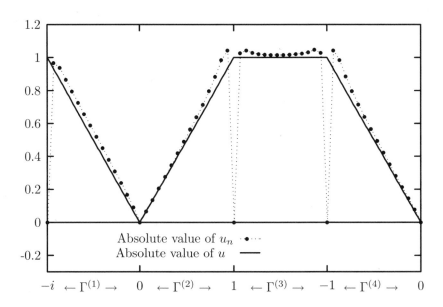

These experimental values confirm the convergence results established above. The relatively large errors in examples with spline degree 0 and $i_0 = 1$ and with spline degree 1 are caused by the circumstance that all spline functions in the corresponding spaces vanish at the singular points of the curve Γ. This leads to a "natural" error of order $\|\phi_{k,n}^{F,F}\|_{L^2} \approx n^{-1/2}$ which is also confirmed by the numerical results. In case the solution vanishes at the singular points, the error rates are much better. In these situations our computations indicate a convergence rate of order n^{-1} for piecewise constant and n^{-2} for piecewise linear splines, respectively. Thus, for practical purposes, the cutting-off-technique should be combined with a regularization of the solution in the neighbourhoods of the singular points.

5.6 Exercises

E 5.1 Let $L_{n,\Gamma}^{F,F}$ be the Galerkin projection over the homogeneous curve Γ introduced in 5.2.6. Prove that $\eta L_{n,\Gamma}^{F,F} \eta^{-1}$ is the Galerkin projection of $L_{R^{\delta_j}}^p(\alpha)+, \ldots$
$\ldots, +L_{R^{\delta_w}}^p(\alpha)$ onto $\hat{\chi}_{R^{\delta_1}} S_n^F+, \ldots, +\hat{\chi}_{R^{\delta_w}} S_n^F$.

E 5.2 Show that the sequence $(\hat{\chi}_{R^+}^{(n)})$ belongs to each of the algebras $\mathcal{A}_{\Omega,r}^F$.

E 5.3 Let Γ be a composed curve. An operator $A \in L(L_\Gamma^p(\alpha))$ is called operator of local type if the commutator $fA - AfI$ is compact for all functions f in $C(\Gamma)$. The collection of all operators of local type will be denoted by OLT_Γ, and we let OLT_Γ^π refer to the quotient $OLT_\Gamma / K(L_\Gamma^p(\alpha))$.

1. Prove that OLT_Γ^π is KMS with respect to the algebra $\pi(C(\Gamma)I)$.

2. Formulate and prove a local inclusion theorem for the algebra $\mathcal{O}_\Gamma^p(\alpha)$.

(Compare Theorems 3.6 and 3.9.)

E 5.4 Let C denote the operator of conjugation, $(Cf)(z) = \overline{f(z)}$. Using the results of the previous exercise, show that the operator $CS_\Gamma C$ belongs to $\mathcal{O}_\Gamma^p(\alpha)$, and that

$$\eta W_x(\pi(CS_\Gamma C))\eta^{-1} =$$

$$-\begin{pmatrix} S & N_{\beta_2-\beta_1} & \cdots & N_{\beta_k-\beta_1} \\ N_{2\pi+\beta_1-\beta_2} & S & \cdots & N_{\beta_k-\beta_2} \\ \vdots & \vdots & \vdots & \vdots \\ N_{2\pi+\beta_1-\beta_k} & N_{2\pi+\beta_2-\beta_k} & \cdots & S \end{pmatrix} \begin{pmatrix} \delta_1(x) & & \\ & \ddots & \\ & & \ddots \\ & & & \delta_{w(x)}(x) \end{pmatrix}$$

where W_x is as in Theorem 5.1 and η denotes the mapping 5.1.6(1). Combine this result with Proposition 5.1 to show that the operator $S_\Gamma + CS_\Gamma C$ (which is, up to a compact summand, nothing else than the double layer potential operator over Γ) is compact if and only if Γ is a composed curve where each point has a closed neighbourhood which is a simple arc.

(In particular, each point of Γ is of order 1 or 2, and Γ has no corners.)

E 5.5 Prove that neither Z_s nor $Z_s S_{R^+}$ belong to $\Sigma^p(\alpha)$ if $s \neq 1$.

5.7   Comments and references

The basic Fredholm theory for singular integral operators with piecewise continuous coefficients on complicated curves can be found in the monographs by Gohberg/Krupnik
[GK 1] (Chapter 10) or Michlin/Prößdorf [MP] (Chapter IV). The pioneering paper on
this topic is Gohberg/Krupnik [GK 3] (see also [GK 2] for the special case of continuous coefficients), where moreover the (non-closed) algebra generated by singular integral operators is introduced and investigated. The closure of this algebra was studied in
Roch/Silbermann [RS 1], [RS 4], and the approach developed in these papers is the one
presented here. For another approach see also Duduchava [Du 2], [Du 3] and Costabel
[Co 2], [Co 3].

   The intensive study of spline approximation methods for singular integral operators
on curves with corners was initiated by Costabel and Stephan in their works [CS 1]
(Galerkin methods) and [CS 2] (collocation). Essential progress is also due to Prößdorf
and Rathsfeld [PR 1], [PR 3] who considered quadrature and collocation methods for
singular integral operators on curves with corners (but without intersections).

   The definition and analysis of the algebra $\mathcal{A}_{\Omega,r}^{F}(\Gamma)$ containing a bulk of concrete spline
approximation methods is new, and it goes back to the authors (see Hagen and Silber-
mann [HS 2] for the case of smooth curves and periodic splines, and also Roch [R 1] for
the special case of spline spaces constituted by smoothest piecewise polynomial splines).
In their recent paper [DRS], Didenko, Roch and Silbermann considered generalizations
of these results for singular integral operators with conjugation.

Chapter 6

Around finite sections of integral operators

6.1 Finite sections of singular integrals

6.1.1 Finite sections of singular integral operators on **R**

The finite section method is primarily intended to make integral equations over unbounded domains (the axis) practically solvable by reducing them to equations over bounded domains (intervals). For the singular integral equation

$$(aP_R + bQ_R)u = f \tag{1}$$

with $P_R = (I + S_R)/2$, $Q_R = I - P_R$, $u, f \in L_R^p(\alpha)$, and piecewise continuous coefficients a and b, this requires to solve instead of (1) equations of the form

$$P_t(aP_R + bQ_R)P_t u_t = P_t f, \tag{2}$$

where the projections P_t are given by

$$(P_t f)(s) = \begin{cases} f(s) & \text{if} \quad |s| \le t \\ 0 & \text{if} \quad |s| > t, \end{cases}$$

and where t is positive (and large). The *finite section method* is said to apply to equation (1) if the equations (2) are uniquely solvable for all right sides f and for all t large enough (say, $t \ge t_0$), and if their solutions u_t converge to a solution u of (1) as $t \to \infty$ in the norm of $L_R^p(\alpha)$.

The solvability of equation (2) is readily verified. Indeed, it is not hard to show that the operator in (2) is invertible if and only if the operator

$$a_t P_R + b_t Q_R, \tag{3}$$

with

$$a_t(s) = \begin{cases} a(s) & \text{if } |s| \le t \\ 1 & \text{if } |s| > t, \end{cases} \qquad b_t(s) = \begin{cases} b(s) & \text{if } |s| \le t \\ 1 & \text{if } |s| > t \end{cases}$$

is invertible (compare [MP], IV1.1). The invertibility of the operators (1) and (3) can be described in terms of their symbols. For simplicity, we let $p = 2$ and $\alpha = 0$, and we abbreviate a/b to g. Then the symbol of $aP_R + bQ_R$ corresponds to a curve G, which is obtained from the range of g by filling in line segments into each gap $[g(s-0), g(s+0)]$, and the operator $aP_R + bQ_R$ is invertible if and only if the winding number of this curve with respect to the point 0 is zero. Analogously, the invertibility of $a_t P_R + b_t Q_R$ is equivalent to a vanishing winding number of a curve G_t which results from G as follows: from $g(-t+0)$ to $g(t-0)$ let G_t coincide with G, and to this curve add the line segments joining the points $g(-t+0)$ and $g(t-0)$ with 1.

This indicates that the applicability of the finite section method only depends on the invertibility of the operator $aP_R + bQ_R$ and on the behaviour of the coefficients a,b at infinity. We are going to point out that this assertion remains true for a large variety of operators.

Let $\mathcal{F}$ stand for the set of all (generalized) bounded sequences $(A_t)_{t>0}$ of operators A_t on $L_R^p(\alpha)$. Provided with elementwise operations and with the norm $\|(A_t)\| = \sup_t \|A_t\|$, the set $\mathcal{F}$ becomes a Banach algebra, and the set $\mathcal{G}$ of all bounded sequences (G_t) with $\|G_t\| \to 0$ as $t \to \infty$ forms a closed two-sided ideal in $\mathcal{F}$. Furthermore, we denote by $\mathcal{C}$ the smallest closed subalgebra of $\mathcal{F}$ containing the constant sequences (I), (P_R), $(\chi_{R^+} I)$, as well as the sequence (P_t) and the ideal $\mathcal{G}$. Finally, we write $\mathcal{A}$ for the subalgebra of F which is generated by $\mathcal{C}$ and by the constant sequences (aI) with a running through PC. Note that this algebra contains all sequences (K) with K being compact (see Section 3.1.1).

The sequences in $\mathcal{A}$ can be "homogenized" in the following sense: if Z_t again refers to the operator $(Z_t f)(s) = f(s/t)$ then the following proposition holds.

Proposition 6.1 *Let $(A_t) \in \mathcal{A}$. Then there exists an operator $W(A_t) : L_R^p(\alpha) \to L_R^p(\alpha)$ such that $Z_{t^{-1}} A_t Z_t \to W(A_t)$ strongly as $t \to \infty$. The mapping $W : \mathcal{A} \to L(L_R^p(\alpha))$ is a continuous algebra homomorphism, and*

$$\begin{aligned} W(P_R) &= P_R, \\ W(P_t) &= P_1, \\ W(aI) &= a(+\infty)\chi_{R^+} I + a(-\infty)\chi_{R^-} I \quad \text{for} \quad a \in PC, \\ W(K) &= 0 \quad \text{for} \quad K \text{ compact}. \end{aligned}$$

The main results concerning stability of sequences in $\mathcal{A}$ are as follows:

Theorem 6.1 *Let $(A_t) \in \mathcal{A}$, $A = \text{s}-\lim_{t\to\infty} A_t$, and set $B_t := Z_t W(A_t) Z_{t^{-1}}$ for $t > 0$. Then $(B_t) \in \mathcal{C}$, and the sequence (A_t) is stable if and only if the operator A is invertible and the sequence (B_t) is stable.*

Theorem 6.2 *The following assertions are equivalent for* $(B_t) \in \mathcal{C}$:

(i) *the sequence* (B_t) *is stable,*

(ii) *one of the operators* B_t *is invertible,*

(iii) B_1 *is invertible.*

The proof of Theorem 6.2 is almost evident since $Z_t^{-1} B_t Z_t = B_1$. Let us outline the proof of Theorem 6.1.

The sequence (A_t) is stable if and only if the coset $(A_t) + \mathcal{G}$ is invertible in $\mathcal{F}/\mathcal{G}$. To start with we ask for whether this coset is invertible in $\mathcal{A}/\mathcal{G}$. Set

$$\mathcal{J} = \{(K) + (G_t) \quad \text{with} \quad (G_t) \in \mathcal{G}, \; K \in K(L^p_R(\alpha))\} \; .$$

Clearly, $\mathcal{J}$ is a closed two-sided ideal of $\mathcal{A}$, and to lift this ideal in sense of 1.6.2 only one homomorphism is needed, viz. $W^0(A_t) := \text{s} - \lim_{t \to \infty} A_t$. The general lifting theorem states that $(A_t) \in \mathcal{A}$ is stable if and only if the operator $A = W^0(A_t)$ is invertible and if the coset $\Phi^{\mathcal{J}}(A_t) := (A_t) + \mathcal{J}$ is invertible in $\mathcal{A}/\mathcal{J}$.

If f is in $C(\dot{\mathbf{R}})$ then the coset $\Phi^{\mathcal{J}}(fI)$ is obviously in the center of $\mathcal{A}/\mathcal{J}$ ((fI) commutes with (P_t) exactly and with (A), $A \in \mathcal{O}^p_R(\alpha)$, modulo sequences (K) with K compact). The collection of all cosets $\Phi^{\mathcal{J}}(fI)$ generates a subalgebra of the center of $\mathcal{A}/\mathcal{J}$ which has $\dot{\mathbf{R}}$ as its maximal ideal space, and localizing via Allan leads to local ideals I_x ($x \in \dot{\mathbf{R}}$), local algebras $\mathcal{A}_x := (\mathcal{A}/\mathcal{J})/I_x$, and to associated homomorphisms $\Phi^{\mathcal{J}}_x : \mathcal{A} \to \mathcal{A}_x$, with the property that $\Phi^{\mathcal{J}}(A_t)$ is invertible if and only if all cosets $\Phi^{\mathcal{J}}_x(A_t)$ are invertible.

First let $x \neq \infty$. Then

$$\Phi^{\mathcal{J}}_x(A_t) = \Phi^{\mathcal{J}}_x(A) , \tag{4}$$

and thus, invertibility of A entails invertibility of $\Phi^{\mathcal{J}}_x(A_t)$. To verify (4) observe that this is obvious for (A_t) being a constant sequence. So we are left on showing (4) for $(A_t) = (P_t)$. Choose $y \in \mathbf{R}$ with $y > |x|$ and define for $t > 0$

$$P'_t = \begin{cases} P_t & \text{if} \quad t \geq y \\ P_y & \text{if} \quad t < y . \end{cases}$$

Since $(P'_t) - (P_t) \in \mathcal{G}$ we conclude that $\Phi^{\mathcal{J}}_x(P'_t) = \Phi^{\mathcal{J}}_x(P_t)$. Now pick a function $f \in C(\dot{\mathbf{R}})$ with $f(x) = 1$ and with support in $(\frac{-|x|-y}{2}, \frac{|x|+y}{2})$. Then $fP'_t = P'_t fI = fI$ for all t which leads to

$$\Phi^{\mathcal{J}}_x(P'_t) = \Phi^{\mathcal{J}}_x(fP'_t) = \Phi^{\mathcal{J}}_x(fI) = \Phi^{\mathcal{J}}_x(I)$$

as desired.

In case $x = \infty$ one similarly shows that

$$\Phi^{\mathcal{J}}_\infty(A_t) = \Phi^{\mathcal{J}}_\infty(B_t) .$$

Hence, stability of (B_t) yields invertibility of $\Phi^J_\infty(A_t)$ and, conversely, one can prove in a standard way that

$$\Phi^J_\infty(A_t) \mapsto (B_t) \in \mathcal{C}$$

is a correctly defined locally equivalent representation of the local algebra at infinity. Thus, $(A_t) + \mathcal{G}$ is invertible in $\mathcal{A}/\mathcal{G}$ if and only if A is invertible and (B_t) is stable. The concluding concretization step does not involve any problems since all occuring homomorphisms are strong limits (compare Section 1.6.5). ∎

Let us complete these results by the following remarks.

1. The assertions of Theorems 6.1 and 6.2 remain valid (and the proofs are verbatim the same) if the class of piecewise continuous coefficients is replaced by the space $L^\infty_c(\mathbf{R})$ consisting of all functions f which behave as L^∞-functions on each finite interval and as PC-function at infinity. More precisely, a bounded measurable function f is in $L^\infty_c(\mathbf{R})$ if there are numbers $f(\pm\infty)$ (= the essential limits of f at infinity) such that

$$\mathrm{ess}\lim_{\pm t \geq t_0} |f(t) - f(\pm\infty)| \to 0 \quad \text{as} \quad t_0 \to \infty .$$

2. The assertions remain valid for matrix-valued coefficients (again with the same proofs).

3. In the scalar case, and if (A_t) is a sequence of the form

$$(P_t \chi_{R^+} A \chi_{R^+} P_t + \chi_{R^-} P_t + (I - P_t)) \quad \text{with} \quad A \in \mathcal{O}^p_R(\alpha)$$

then the invertibility of B_1 is effectively verifiable. Indeed, the operator B_1 takes the form $\chi_{[0,1]} M^0(b) \chi_{[0,1]} I + \chi_{R \setminus [0,1]}$ in this setting where b is a continuous function on $\overline{\mathbf{R}}$, and this operator is equivalent to a Wiener-Hopf operator whose invertibility is subject to Coburn's theorem.

6.1.2   Finite section methods cutting off singularities

The methods used in Section 6.1 to cut off the "singularity" at infinity also apply to free ourselves from the geometric singularities of complicated curves. This is, in a sense, the analytic counterpart to the cutting-off-techniques working on the spline level which had been presented in the fourth chapter.

Let, as in Chapter 5, Γ be a composed curve, and let Γ' be the set of its singular points (or any larger finite subset of Γ). Given $\epsilon > 0$ we denote by Γ^ϵ the curve resulting from Γ by removing all simple subarcs of arc length ϵ which have a point of Γ' as one of its end points.

The *finite setion method* we are going to explain here consits in substituting the original equation

$$Au = f \quad , \quad f, u \in L^p_\Gamma(\alpha) , \tag{1}$$

by the sequence of the equations

$$(\chi_{\Gamma^\epsilon} A \chi_{\Gamma^\epsilon} I + \chi_{\Gamma \backslash \Gamma^\epsilon} I) u_\epsilon = f \quad , \quad \epsilon > 0 \, . \tag{2}$$

This method is called *stable* if the equations (2) are uniquely solvable for all right sides f and for all ϵ small enough, say $\epsilon < \epsilon_0$, and if the sequence (u_ϵ) tends to a solution u of (1) as $\epsilon \to 0$ in the norm of $L^p_\Gamma(\alpha)$. Since Γ^ϵ consits of a finite number of pairwise disjoint simple arcs one can often think of (2) as a system of equations over an interval.

We need some notations in order to formulate the results. For $x \in \Gamma'$ we let Γ_x stand for the local curve $\Gamma_x = \cup_{j=1}^{w(x)} e^{i\beta_j} \mathbf{R}^{\delta_j}$ introduced in 5.1.4, and we set

$$\Gamma^1_x = \cup_{j=1}^{w(x)} e^{i\beta_j} (\mathbf{R}^{\delta_j} \backslash [0, \delta_j]) \, ,$$

that is, Γ^1_x originates from Γ_x by removing all intervals of length 1 having 0 as one of its endpoints. Finally, given a piecewise continuous function a on Γ, we denote its one-sided limits at x along the subcurve, having $x + e^{i\beta_j} \mathbf{R}^{\delta_j}$ as its one-sided tangent at x, by $a_j(x)$, we write $\chi_j(x)$ for the characteristic function of $e^{i\beta_j} \mathbf{R}^{\delta_j}$, and we set $a(x) := \sum_{j=1}^{w(x)} a_j(x) \chi_j(x)$.

Theorem 6.3 *The finite section method (2) applies to the singular integral operator*

$$A = aI + bS_\Gamma + T \tag{3}$$

with piecewise continuous coefficients a and b and a compact perturbation T if and only if the operator A is invertible and if the operators

$$\chi_{\Gamma^1_x} (a(x)I + b(x)S_{\Gamma_x}) \chi_{\Gamma^1_x} I + \chi_{\Gamma_x \backslash \Gamma^1_x} I \tag{4}$$

are invertible for all $x \in \Gamma'$.

The proof goes as in Section 1: Introduce an algebra $\mathcal{A}$ containing all constant sequences (A) with A as in (3), the sequence $(\chi_{\Gamma_\epsilon} I)_{\epsilon > 0}$, and all bounded sequences $(G_\epsilon)_{\epsilon > 0}$ with $\|G_\epsilon\| \to 0$ as $\epsilon \to 0$. Factorize this algebra by the ideal $\mathcal{J}$ of all sequences $(K) + (G_\epsilon)$ where K is compact and (G_ϵ) as above. The ideal $\mathcal{J}$ can be lifted via the homomorphism $(A_\epsilon) \mapsto s - \lim_{\epsilon \to 0} A_\epsilon$. Localize the algebra $\mathcal{A}/\mathcal{J}$ over the maximal ideal space Γ of its central subalgebra spanned by the cosets $(fI) + \mathcal{J}$ with f continuous. If $x \in \Gamma \backslash \Gamma'$ then the invertibility of A entails local invertibility at x, and if $x \in \Gamma'$ then the local invertibility of $(\chi_{\Gamma^\epsilon} A \chi_{\Gamma^\epsilon} I + \chi_{\Gamma \backslash \Gamma^\epsilon} I)$ at x is equivalent to the invertibility of (4).

Let us emphasize that moreover, one can show that the mapping

$$(S_\Gamma) \mapsto S_{\Gamma_x} \, , \quad (a) \mapsto a(x)I \, , \quad (\chi_{\Gamma^\epsilon} I) \mapsto \chi_{\Gamma^1_x} I$$

extends to a continuous Banach algebra homomorphism, $W_x : \mathcal{A} \to L(L^p_{\Gamma_x}(\alpha_x))$, and that an arbitrary sequence (A_ϵ) in $\mathcal{A}$ is stable if and only if the operators $A := s - \lim_{\epsilon \to 0} A_\epsilon$ and $W_x(A_\epsilon)$ are invertible for all $x \in \Gamma'$.

6.1.3   Pilidi's method

Pilidi [Pi 1] proposed an approximation method for singular integral operators which is aimed to associate with each Cauchy singular integral S_R a sequence of *compact* approximation operators and which is, both in its results and in its proofs, closely related to the finite section methods studied above. We are going to present Pilidi's results for singular integral operators $aI + bS_R + T$ on the real line where a and b are continuous coefficients and T is a compact perturbation.

Given numbers ϵ and N we introduce the operators:

$$(S_\epsilon f)(t) = \frac{1}{\pi i} \int_{|s-t|\geq\epsilon} \frac{f(s)}{s-t}\, ds\,,\ t \in \mathbf{R}\,,$$

$$(S^N f)(t) = \frac{1}{\pi i} \int_{|s-t|\leq N} \frac{f(s)}{s-t}\, ds\,,\ t \in \mathbf{R}\,,$$

$$(S^N_\epsilon f)(t) = \frac{1}{\pi i} \int_{\epsilon\leq|s-t|\leq N} \frac{f(s)}{s-t}\, ds\,,\ t \in \mathbf{R}\,,$$

and consider the following approximation operators for the operator $aI+bS_R+T$:

$$
\begin{aligned}
A_\epsilon &= aI + bS_\epsilon + T\,,\\
A^N &= aI + bS^N + T\,,\\
A^N_\epsilon &= aI + bS^N_\epsilon + T\,.
\end{aligned}
$$

As $\epsilon \to 0$ and $N \to \infty$, the sequences $(A_\epsilon), (A^N)$, and (A^N_ϵ) converge strongly to A (this is due to the definition of the singular integral operator S_R as Cauchy principal value). Finally we set

$$\lambda_0 = \frac{2}{\pi} \int_0^\pi \frac{\sin z}{z}\, dz \quad (\approx 1,17898)\,.$$

Now Pilidi's result reads as follows.

Theorem 6.4 *Let the operator $A = aI + bS_R + T$ be invertible in L^p_R.*

(a) *The method (A_ϵ) is stable if and only if $a(x) + \lambda b(x) \neq 0$ for all $x \in \dot{\mathbf{R}}$ and $\lambda \in [-1,1]$.*

(b) *The method (A^N) is stable if and only if $a(\infty) + \lambda b(\infty) \neq 0$ for all $\lambda \in [-\lambda_0, \lambda_0]$.*

(c) *The method (A^N_ϵ) is stable if and only if $a(x) + \lambda b(x) \neq 0$ for all $x \in \mathbf{R}$ and $\lambda \in [-1,1]$ and if $a(\infty) + \lambda b(\infty) \neq 0$ for all $\lambda \in [-\lambda_0, \lambda_0]$.*

To avoid misunderstandings: the sequences $(A_\epsilon), (A^N)$, and (A^N_ϵ) are *stable* if there are constants ϵ_0, N_0 such that all operators A_ϵ with $\epsilon \leq \epsilon_0$, all operators A^N with $N \geq N_0$, and all operators A^N_ϵ with $\epsilon \leq \epsilon_0$ and $N \geq N_0$, are invertible, respectively, and if the norms of their inverses are uniformly bounded.

Let us give a brief account of the proof. The operator S_ϵ can be written as

$$(S_\epsilon f)(t) = \frac{1}{\pi i} \int_R \frac{(1 - \chi_{[-\epsilon,\epsilon]})(s-t)}{s-t} f(s)\, ds \,,$$

hence, it is a Fourier convolution operator with kernel function

$$k_\epsilon(x) = \frac{1}{\pi i} \frac{(1 - \chi_{[-\epsilon,\epsilon]})(x)}{x} \,,$$

the Fourier transform a_ϵ of which equals $a_\epsilon(\lambda) = \frac{2}{\pi}\mathrm{Si}\,(2\pi\lambda\epsilon) - \mathrm{sign}\lambda$ where Si stands for the integral sinus function $\mathrm{Si}\, x = \int_0^x \frac{\sin y}{y}\, dy$.

Analogously, $S^N = W^0(a^N)$ with $a^N(\lambda) = -\frac{2}{\pi}\mathrm{Si}\,(2\pi\lambda N)$ and

$$S_\epsilon^N = W^0(a_\epsilon^N) \quad \text{with} \quad a_\epsilon^N(\lambda) = \frac{2}{\pi}\mathrm{Si}\,(2\pi\lambda\epsilon) - \frac{2}{\pi}\mathrm{Si}\,(2\pi\lambda N) \,.$$

Now consider, for example, the sequence $(aI + bS_\epsilon + T)_{\epsilon>0}$. Let $\mathcal{F}$ stand for the algebra of all bounded sequences $(A_\epsilon)_{\epsilon>0}$, $\mathcal{G}$ for the ideal of all sequences $(G_\epsilon) \in \mathcal{F}$ with $\|G_\epsilon\| \to 0$ as $\epsilon \to 0$, $\mathcal{J}$ for the subalgebra of all sequences $(K + G_\epsilon)$ where K is compact and $(G_\epsilon) \in \mathcal{G}$, and write $\mathcal{A}$ for the smallest closed subalgebra of $\mathcal{F}$ which contains the constant sequences (fI) with $f \in C(\dot{\mathbf{R}})$, the sequence $(W^0(a_\epsilon))_{\epsilon>0}$, and the algebra $\mathcal{J}$. The algebra $\mathcal{J}$ is even an ideal in $\mathcal{A}$, and the general lifting theorem gives:

> The coset $(aI + bS_\epsilon + T) + \mathcal{G}$ is invertible in $\mathcal{A}/\mathcal{G}$ if and only if the operator $aI + bS_R + T = s - \lim_{\epsilon\to 0} aI + bS_\epsilon + T$ is invertible and if the coset $(aI + bS_\epsilon + T) + \mathcal{J}$ is invertible in $\mathcal{A}/\mathcal{J}$.

Let now $f \in C(\dot{\mathbf{R}})$. It is evident that $(fI) + \mathcal{J}$ commutes with all sequences $(gI) + \mathcal{J}$, $g \in C(\dot{\mathbf{R}})$, and, furthermore, a straightforward estimate shows that $(fI) + \mathcal{J}$ commutes with $(W^0(a_\epsilon) + \mathcal{J})$.

Thus, the algebra $\mathcal{B} := \{(fI) + \mathcal{J}, \ f \in C(\dot{\mathbf{R}})\}$ is central in $\mathcal{A}/\mathcal{J}$, and we can localize over the maximal ideal space $\dot{\mathbf{R}}$ of $\mathcal{B}$. For $x \in \dot{\mathbf{R}}$, we shall write $\mathcal{A}_x$ for the associated local algebra and $\Phi_x : \mathcal{A} \to \mathcal{A}_x$ for the canonical homomorphism.

Let $x \in \mathbf{R}$. Then, clearly,

$$\begin{aligned}
\Phi_x(aI + bW^0(a_\epsilon) + T) &= \Phi_x(a(x)I + b(x)W^0(a_\epsilon)) \\
&= \Phi_x(W^0(a(x) + b(x)a_\epsilon)) \,,
\end{aligned}$$

and invoking the translation invariance of Fourier convolution operators one can easily verify that $\Phi_x(W^0(a(x) + b(x)a_\epsilon))$ is invertible if and only if $\Phi_0(W^0(a(x) + b(x)a_\epsilon))$ is invertible. We claim that the latter is equivalent to the invertibility of the function $a(x) + b(x)a_1$. Indeed, suppose there are sequences $(B_\epsilon) \in \mathcal{A}$, $(J_\epsilon) \in \mathcal{J}$, and (J_ϵ^0) with $(J_\epsilon^0) + \mathcal{J}$ being in the local ideal at zero, such that

$$W^0(a(x) + b(x)a_\epsilon)B_\epsilon = I + J_\epsilon + J_\epsilon^0 \,.$$

Multiplying this by Z_ϵ and Z_ϵ^{-1} (the homogenization operators introduced in Section 3.1.1) and passing to the strong limit as $\epsilon \to 0$ yields

$$W^0(a(x) + b(x)a_1)B = I\,,$$

where $B = s-\lim Z_\epsilon B_\epsilon Z_\epsilon^{-1}$. Here we used that

$$Z_\epsilon W^0(a(x) + b(x)a_\epsilon)Z_\epsilon^{-1} = W^0(a(x) + b(x)a_1)$$

and that $Z_\epsilon(J_\epsilon + J_\epsilon^0)Z_\epsilon^{-1} \to 0$ strongly as $\epsilon \to 0$. Thus, the Fourier convolution operator $W^0(a(x) + b(x)a_1)$ and, consequently, the function $a(x) + b(x)a_1$ must be invertible. The converse direction is evident. ∎

Pilidi studied in [Pi 1] the analog of method (a) for singular integral operators on composed curves which are homeomorphic to the unit circle. As already announced, the approximation operators S_ϵ are compact in this case, and showed that the assertion (a) of the theorem remains valid. Pilidi generalized the results cited here to operators with piecewise continuous coefficients in [Pi 4], to singular integral operators with non-compact perturbations ([Pi 3], [Pi 5]) as well as to bisingular operators with continuous coefficients ([Pi 2]). For further generalizations the reader should also see a recent paper by Duduchava and Prößdorf [DP]. It should be finally mentioned that the harmonic extensions, used in [S 8] and [S 7] (these papers appeared already in 1985 and 1986) to define and compute indices of Toeplitz operators with piecewise quasicontinuous generating functions, provides a similar method of regularization of singular kernels by smooth ones.

6.2 Around finite sections of discrete convolution operators

6.2.1 Finite sections of operators with flip

The results of Sections 4.1.1-4.1.3 can be extended to a larger class of operators including Hankel operators with piecewise continuous generating functions.

Throughout Sections 6.2.1-6.2.3 let $\mathcal{T}$ denote the smallest closed subalgebra of $L(l_Z^p(0))$ which contains the multiplication operators $T^0(a)$ with $a \in PC_{p,0}$, the projection

$$P: (x_k) \mapsto (y_k) \quad \text{with} \quad y_k = \begin{cases} x_k & \text{if} \quad k \geq 0 \\ 0 & \text{if} \quad k < 0\,, \end{cases}$$

and the flip

$$J: (x_k) \mapsto (y_k) \quad \text{with} \quad y_k = x_{-k-1}\,.$$

Prominent elements of this algebra are the singular integral operators with Carleman shift,

$$T^0(a)P + T^0(b)Q + (T^0(c)P + T^0(d)Q)J\,, \tag{1}$$

where we wrote $Q = I - P$, and the Toeplitz + Hankel operators

$$PT^0(a)P + PT^0(b)QJ + Q \tag{2}$$

(or, in earlier notation, $T(a) + H(b) + Q$).

Let us further write $\mathcal{F}$ for the collection of all bounded sequences (A_n) of operators $A_n \in L(l^p_Z(0))$. Introducing elementwise operations and the supremum norm we make $\mathcal{F}$ to become a Banach algebra, and we let $\mathcal{A}$ stand for the smallest closed subalgebra of $\mathcal{F}$ which contains the constant sequences (A), with A running through $\mathcal{T}$, and the sequences $(R_{kn})_{n \in Z^+}$ with k running through the positive integers. Recall that the projections R_n are defined by

$$R_n : (x_k) \mapsto (y_k) \quad \text{with} \quad y_k = \begin{cases} x_k & \text{if} \quad -n \le k \le n-1 \\ 0 & \text{else} \end{cases}.$$

Our stability criterion for sequences in $\mathcal{A}$ again relates stability with invertibility of two families of operators. The following propositions provide us with these families. For, let P_R denote the projection $(I + S_R)/2$, set $Q_R = I - P_R$, and write J_R for the flip

$$J_R : L^p_R(0) \to L^p_R(0) , \quad (J_R f)(t) = f(-t) .$$

Proposition 6.2 *For each $t \in \{z \in \mathbf{T} : \operatorname{Im} z \ge 0\}$, there is a continuous algebra homomorphism W^t from $\mathcal{A}$ into $L(L^p_R(0))$ if $\operatorname{Im} t = 0$ and into $L(L^p_R(0)_2) \cong L(L^p_R(0))_{2 \times 2}$ if $\operatorname{Im} t > 0$. This homomorphism acts on the generating sequences of $\mathcal{A}$ as follows:*

$$W^t(P) = \begin{cases} \chi_{[0,\infty)} I & \text{if} \quad \operatorname{Im} t = 0 \\ \begin{pmatrix} \chi_{[0,\infty)} I & 0 \\ 0 & \chi_{(-\infty,0]} I \end{pmatrix} & \text{if} \quad \operatorname{Im} t > 0 , \end{cases}$$

$$W^t(R_{kn}) = \begin{cases} \chi_{[-k,k]} I & \text{if} \quad \operatorname{Im} t = 0 \\ \begin{pmatrix} \chi_{[-k,k]} I & 0 \\ 0 & \chi_{[-k,k]} I \end{pmatrix} & \text{if} \quad \operatorname{Im} t > 0 , \end{cases}$$

$$W^t(T^0(a)) = \begin{cases} a(t+0)Q_R + a(t-0)P_R & \text{if} \quad \operatorname{Im} t = 0 \\ \begin{pmatrix} a(t+0)Q_R + a(t-0)P_R & 0 \\ 0 & \tilde{a}(t+0)Q_R + \tilde{a}(t-0)P_R \end{pmatrix} \\ \quad \text{if} \quad \operatorname{Im} t > 0 , \end{cases}$$

$$W^t(J) = \begin{cases} \pm J_R & \text{if} \quad t = \pm 1 \\ \begin{pmatrix} 0 & I \\ I & 0 \end{pmatrix} & \text{if} \quad \operatorname{Im} t > 0 . \end{cases}$$

Recall that $\tilde{a}$ is defined by $\tilde{a}(t) = a(1/t)$.

Proposition 6.3 *(a) For each* $s \in \{l \in \mathbf{Z}, l \geq 0\}$*, there is a continuous al-gebra homomorphism* W_s *from* $\mathcal{A}$ *into* $L(l_{\mathbf{Z}}^p(0))$ *if* $s = 0$ *and into* $L(l_{\mathbf{Z}}^p(0)_2) \cong L(l_{\mathbf{Z}}^p(0))_{2 \times 2}$ *if* $s > 0$*. This homomorphism acts on the generating sequences of* $\mathcal{A}$ *as follows:*

$$W_s(P) = \begin{cases} P & \text{if } s = 0 \\ \begin{pmatrix} I & 0 \\ 0 & 0 \end{pmatrix} & \text{if } s > 0, \end{cases}$$

$$W_s(R_{kn}) = \begin{cases} I & \text{if } s = 0 \\ \begin{pmatrix} I & 0 \\ 0 & I \end{pmatrix} & \text{if } 0 < s < k \\ \begin{pmatrix} Q & 0 \\ 0 & Q \end{pmatrix} & \text{if } s = k \\ \begin{pmatrix} 0 & 0 \\ 0 & 0 \end{pmatrix} & \text{if } s > k, \end{cases}$$

$$W_s(J) = \begin{cases} J_R & \text{if } s = 0 \\ \begin{pmatrix} 0 & I \\ I & 0 \end{pmatrix} & \text{if } s > 0. \end{cases}$$

$$W_s(T^0(a)) = \begin{cases} T^0(a) & \text{if } s = 0 \\ \begin{pmatrix} T^0(a) & 0 \\ 0 & T^0(\tilde{a}) \end{pmatrix} & \text{if } s > 0, \end{cases}$$

(b) The mapping $W_\infty : \mathcal{A} \to L(l_{\mathbf{Z}}^p(0))/K(l_{\mathbf{Z}}^p(0))$ *defined by* $(A_n) \mapsto A_1 + K(l_{\mathbf{Z}}^p(0))$ *is a continuous algebra homomorphism.*

Now we have the following generalization of Theorem 4.1:

Theorem 6.5 *A sequence* $(A_n) \in \mathcal{A}$ *is stable if and only if all operators (resp. cosets)* $W^t(A_n)$ *with* $t \in \{z \in \mathbf{T}, \operatorname{Im} z \geq 0\}$ *and* $W_s(A_n)$ *with* $s \in \{l \in \mathbf{Z}, l \geq 0\} \cup \{\infty\}$ *are invertible.*

It is not hard to rediscover the results of Sections 4.1.1 and 4.1.2 from this theorem. Indeed, if we temporarily let $\hat{\mathcal{A}}$ stand for the subalgebra of $\mathcal{A}$ which is generated by (P), $(T^0(a))$, with $a \in PC_{p,0}$, and (R_{kn}) (but which does not contain the sequence (J)), and if we write $\hat{W}^t$ and $\hat{W}_s$ for the homomorphisms W^t and W_s used in 4.1.1 then one can show that, for $(A_n) \in \hat{\mathcal{A}}$,

$$W^t(A_n) = \operatorname{diag}\left(\hat{W}^t(E_n^F A_n E_{-n}^F), J_R \hat{W}^{\bar{t}}(E_n^F A_n E_{-n}^F) J_R\right) \quad \text{if } \operatorname{Im} t > 0,$$

$$W_s(A_n) = \operatorname{diag}\left(\hat{W}_s(E_n^F A_n E_{-n}^F), J_R \hat{W}_{-s}(E_n^F A_n E_{-n}^F) J_R\right) \quad \text{if } s > 0,$$

and

$$W^{\pm 1}(A_n) = \hat{W}^{\pm 1}(E_n^F A_n E_{-n}^F),$$

$$W_0(A_n) = \hat{W}_0(E_n^F A_n E_{-n}^F),$$
$$W_\infty(A_n) = \hat{W}_\infty(E_n^F A_n E_{-n}^F).$$

The proof of this theorem is principally the same as for Theorem 4.1: one chooses a suitable spline space S_n^F (say, $F = \{\chi_{[0,1)}\}$), translates the sequence $(A_n) \in \mathcal{A}$ into the sequence $(E_n^F A_n E_{-n}^F)$ of operators on S_n^F, factorizes the translated algebra by a certain ideal $\mathcal{J}$, and localizes this quotient algebra over the maximal ideal space of the algebra

$$\mathcal{B} := \{(L_n^{F,F} fI|S_n^F) + \mathcal{J} \text{ where } f \in C(\dot{\mathbf{R}}) \text{ and } f(t) = f(-t) \text{ for all } t \in \mathbf{R}\}.$$

This maximal ideal space is just homeomorphic to the compactification of the semi-axis $\mathbf{R}_+$ by the point $\{\infty\}$ (i.e. to a closed interval), and this also makes plain, why only homomorphisms W_s with $s \geq 0$ occur in the theorem.

6.2.2 Finite sections of singular integral operators with Carleman shift

Let us now verify what Theorem 6.5 says about stability of the finite section method for the operator 6.2.1(1).

Proposition 6.4 *Let $a, b, c, d \in PC_{p,0}$ and $A = T^0(a)P + T^0(b)Q + (T^0(c)P + T^0(d)Q)J$. The finite section method $(R_n A R_n + Q_n)$ is stable if and only if*

(i) the operator A is invertible.

(ii) the operator $\begin{pmatrix} QT^0(a)Q + P & QT^0(c)Q \\ QT^0(\tilde{d})Q & QT^0(\tilde{b})Q + P \end{pmatrix}$ *is invertible.*

(iii) the operators

$$\chi_{R\backslash[-1,1]}I + \chi_{[-1,1]}\Big[Q_R(a(t+0)\chi_{R^+}I + b(t+0)\chi_{R^-}I)+$$

$$+P_R(a(t-0)\chi_{R^+}I + b(t-0)\chi_{R^-}I) \pm \Big(Q_R(c(t+0)\chi_{R^+}I + d(t+0)\chi_{R^-}I)$$

$$+P_R(c(t-0)\chi_{R^+}I + d(t-0)\chi_{R^-}I)\Big)J_R\Big]\chi_{[-1,1]}I$$

where $t = \pm 1$ *and*

$$\begin{pmatrix} Q_{[-1,1]} & 0 \\ 0 & Q_{[-1,1]} \end{pmatrix} a_t^+ + \begin{pmatrix} P_{[-1,1]} & 0 \\ 0 & P_{[-1,1]} \end{pmatrix} a_t^- \quad \text{with}$$

$$a_t^\pm(x) = \begin{cases} \begin{pmatrix} a(t\pm 0) & c(t\pm 0) \\ \tilde{d}(t\pm 0) & \tilde{b}(t\pm 0) \end{pmatrix} & \text{if} \quad x \in [-1,0) \\[2ex] \begin{pmatrix} b(t\pm 0) & d(t\pm 0) \\ \tilde{c}(t\pm 0) & \tilde{a}(t\pm 0) \end{pmatrix} & \text{if} \quad x \in [0,1], \end{cases}$$

where $\operatorname{Im} t > 0$, $|t| = 1$, *are invertible on $L_R^p(0)$ and $L_{[-1,1]}^p(0)$, respectively.*

These results hold both for scalar- and for matrix-valued coefficients. In the scalar case, the invertibility of the 2×2 matrix operator in (iii) can be effectively checked due to a result of Spitkovski and Tashbaev ([ST], Corollary 2). For its formulation, we set

$$A = a_t^+(-1/2)(a_t^-(-1/2))^{-1}, \quad B = a_t^+(1/2)(a_t^-(1/2))^{-1},$$

write γ_{0k}, γ_{1k} and γ_{2k} $(k = 1, 2)$, respectively, for the arguments of the eigenvalues of A, $A^{-1}B$ and B divided by 2π, and define

$$l_k = \sum_{j=0}^{2} (\gamma_{jk} + [1/p - \gamma_{jk}]), \quad k = 1, 2,$$

where $[.]$ means the integer part.

Proposition 6.5 *The 2×2-matrix operator in Proposition 6.4(iii) is invertible if and only if the matrices $a_t^+(\pm 1/2)$ and $a_t^-(\pm 1/2)$ are invertible, if none of the numbers $1/p - \gamma_{jk}$ $(j = 0, 1, 2, \ k = 1, 2)$ is an integer, and if at least one of the following conditions is satisfied:*

(i) *A and B have no common eigenvectors, and $l_1 = -l_2$.*

(ii) *A and B do not commute, they possess a common eigenvector, and $l_1 = -l_2 \geq 0$.*

(iii) *A and B commute and $l_1 = l_2 = 0$.*

6.2.3   Finite sections of Toeplitz + Hankel operators

As a further example, we apply Theorem 6.5 to the operator 6.2.1(2).

Proposition 6.6 *Let $a, b \in PC_{p,0}$ and $A = PT^0(a)P + PT^0(b)JP + Q$. The finite section method $(R_n A R_n + Q_n)$ is stable if and only if*

(i) *the operator A is invertible.*

(ii) *the operator $QT^0(a)Q + P$ is invertible.*

(iii) *the operators*

$$\chi_{R \setminus [0,1]} I +$$
$$+ \chi_{[0,1]} \left(a(t+0)Q_R + a(t-0)P_R \pm (b(t+0)Q_R + b(t-0)P_R)J_R \right) \chi_{[0,1]} I$$

where $t = \pm 1$ *and*

$$\begin{pmatrix} \chi_{R\setminus[0,1]} I & 0 \\ 0 & \chi_{R\setminus[-1,0]} I \end{pmatrix} +$$

$$\begin{pmatrix} \chi_{[0,1]}(a_t^+ Q_R + a_t^- P_R)\chi_{[0,1]} I & \chi_{[0,1]}(b_t^+ Q_R + b_t^- P_R)\chi_{[-1,0]} I \\ \chi_{[-1,0]}(\tilde{b}_t^+ Q_R + \tilde{b}_t^- P_R)\chi_{[0,1]} I & \chi_{[-1,0]}(\tilde{a}_t^+ Q_R + \tilde{a}_t^- P_R)\chi_{[-1,0]} I \end{pmatrix}$$

where $t \in \mathbf{T}$, $\operatorname{Im} t > 0$, $a_t^\pm = a(t \pm 0)$, $b_t^\pm = b(t \pm 0)$, *are invertible.*

A detailed analysis using highly non-trivial matrix factorization techniques shows that condition (iii) is effective in case of scalar-valued coefficients a and b. We are going to present the related results of [ST] here.

Given complex numbers x, y, we define the set $\Omega(x, y)$ as follows. Let l_x and l_y denote the rays coming out from the origin and passing through the points $-ix$ and $-iy$, respectively. If $l_x = l_y$ then $\Omega(x, y)$ is the complex plane minus the part of l_x joining $-2ix|y/x|^{1/2}$ (inclusively) to infinity. In case $l_x \neq l_y$, we write γ for the branch of the hyperbola with the asymptotes l_x and l_y which contains the point $-i(x + y)$, and we define $\Omega(x, y)$ as the connected component of $\mathbf{C} \setminus \gamma$ containing the origin.

Proposition 6.7 [ST] *Let a, b, A be as in Proposition 6.6 and suppose a and b to be scalar-valued. The finite section method applies to $A \in L(l_Z^2(0))$ if and only if*

(i) *the operator A is invertible.*

(ii) *the function a is Φ-factorizable in L^2 and its index equals zero.*

(iii) *for all $t \in \mathbf{T}$ with $\operatorname{Im} t > 0$, the roots of the quadratic polynomial*

$$z^2 - \left(\frac{a(t+0)}{a(t-0)} + \frac{a(\bar{t}+0)}{a(\bar{t}-0)} + \frac{b(t+0) - b(t-0)}{a(t-0)} \frac{b(\bar{t}+0) - b(\bar{t}-0)}{a(\bar{t}-0)} \right) z +$$
$$+ \frac{a(t+0)}{a(t-0)} \frac{a(\bar{t}+0)}{a(\bar{t}-0)}$$

satisfy the inequality

$$\left| \arg \frac{a(t+0)}{a(t-0)} + \arg \frac{a(\bar{t}+0)}{a(\bar{t}-0)} - \arg z \right| < \pi .$$

(iv) *the two numbers $\pm(b(\pm 1 - 0) - b(\pm 1 + 0))$ belong to $\Omega(a(\pm 1 - 0), a(\pm 1 + 0))$, respectively.*

6.3   Around spline approximation methods

6.3.1   Non-equidistant partitions

A common method to model equations over regions with complicated geometry numerically is to drop the uniform partitions and to replace them by partitions which are concentrated at the singular points (as corners, intersections, end points). In order to explain some effects we consider the following simple example where the point 0 is supposed to be the critical one.

Example Let $A \in \Sigma^p(\alpha)$ be an operator over the semi-axis $\mathbf{R}^+$. For solving the equation $Au = f$ numerically, we choose and fix a parameter $c \geq 1$ and define collocation points $t^c_{kn} = (\frac{k+\epsilon}{n})^c$ with some ϵ.

We introduce basis functions $\phi^c_{kn}(t) := \chi_{[(k/n)^c,((k+1)/n)^c)}$ and look for an approximation solution $u_n = \sum_k a_{kn} \phi^c_{kn}$ of this equation by requiring

$$(Au_n)(t^c_{kn}) = f(t^c_{kn}) \quad \text{for all} \quad k . \tag{1}$$

The system matrix of (1) is just the infinite matrix

$$((A\phi^c_{kn})(t^c_{kn}))^\infty_{j,k=0} . \tag{2}$$

On defining operators $\hat{X}_c$ by

$$(\hat{X}_c f)(t) := \begin{cases} f(t^{1/c}) & \text{if} \quad t \geq 0 \\ f(-|t|^{1/c}) & \text{if} \quad t < 0 , \end{cases}$$

the matrix (2) can be rewritten as

$$\left((\hat{X}_c^{-1} A \hat{X}_c \phi^1_{kn})(t^1_{kn})\right)^\infty_{j,k=0} ,$$

that is, it coincides with the system matrix for the common ϵ-collocation over uniform partitions (as considered in Section 3.8.2), but applied to the transformed operator $\hat{X}_c^{-1} A \hat{X}_c$. ∎

Generalizing this, we arrive at the following concept of dealing with non-uniform partitions: Instead of solving the operator equation $Au = f$ with $A \in L(L^p_R(\alpha))$ by a certain approximation method which is related to the non-uniform partition

$$\ldots < -(2/n)^c < -(1/n)^c < 0 < (1/n)^c < (2/n)^c < \ldots$$

with a $c \geq 1$, we transform the equation $Au = f$ into $\hat{A}_c u_c = \hat{f}_c$ with $\hat{f}_c = \hat{X}_c^{-1} f$ and $\hat{A}_c = \hat{X}_c^{-1} A \hat{X}_c$, and solve this new equation by a spline approximation method using trial and test spaces over *uniform* partitions.

One disadvantage of this approach is that the transformed operator has to be considered on a different space than the original one since

$$\|\hat{X}_c f\|_{L^p_R((\alpha p+1-c)/pc)} = \sqrt[p]{c} \, \|f\|_{L^p_R(\alpha)} ,$$

and this often involves serious complications when studying approximation methods for the transformed equation. If, for example, the operator A in the above equation is $aI + bS_{R^+}$ then $\hat{X}_c^{-1}I\hat{X}_c = I$ but

$$(\hat{X}_c^{-1}S_{R^+}\hat{X}_c f)(t) = \frac{1}{\pi i}\int_0^\infty \frac{f(s^{1/c})}{s - t^c}\,ds = \frac{c}{\pi i}\int_0^\infty \frac{f(s)}{1 - (\frac{t}{s})^c}\frac{ds}{s}\,,$$

i.e., the operator $\hat{X}_c^{-1}S_{R^+}\hat{X}_c$ can be viewed as Mellin convolution operator with kernel function $k(x) = \frac{c}{\pi i}\frac{1}{1-x^c}$, and this operator acts boundedly on $L^p_{R^+}(\alpha + \frac{c-1}{p})$ if S_{R^+} is thought of acting on $L^p_{R^+}(\alpha)$. Now, if it would be $1 \leq c < (\alpha + 1/p)^{-1}$, then the singular integral operator S_{R^+} is bounded on $L^p_{R^+}(\alpha + \frac{c-1}{p})$, and we could decompose the operator $\hat{X}_c^{-1}S_{R^+}\hat{X}_c$ into $eI + fS_{R^+} + M^0(b)$ with $M^0(b) \in N^p(\alpha + \frac{c-1}{p})$ (Costabel's decomposition, compare 2.1.2(d)), and all results for spline approximation methods derived above apply immediately to this operator. For general c (in particular, $c \geq (\alpha + 1/p)^{-1}$), the analysis of approximation methods for $\hat{X}_c^{-1}S_{R^+}\hat{X}_c$ is rather complicated and requires new techniques; the reader is recommended to consult ([PS 2], 11.18) where certain collocation and quadrature methods were studied.

The situation simplifies substantially if the operators $\hat{X}_c$ are replaced by the following ones:

$$(X_c f)(t) := \sqrt[p]{\frac{1}{c}}\,|t|^{(1/c-1)(\alpha+1/p)}\begin{cases} f(t^{1/c}) & \text{if} \quad t \geq 0 \\ f(-|t|^{1/c}) & \text{if} \quad t < 0\,. \end{cases}$$

Clearly, $X_c^{-1} = X_{c^{-1}}$, and these operators are isometries from $L^p_{R^+}(\alpha)$ onto itself. Thus, the transformed equation $X_c^{-1}AX_cu = X_c^{-1}f$ can always be considered on the very same space as the original equation $Au = f$, and the following proposition shows that, moreover, the transformation $A \mapsto X_c^{-1}AX_c$ leaves the algebra $\Sigma^p(\alpha)$ invariant.

Proposition 6.8 $X_c^{-1}\Sigma^p(\alpha)X_c = \Sigma^p(\alpha)$. *In particular,*

$$X_c^{-1}S_{R^+}X_c = S_{R^+} + M^0(x_c) \quad \text{with} \quad M^0(x_c) \in N^p(\alpha) \quad \text{and}$$

$$x_c(z) = \coth\pi(\frac{z}{c} + i(1/p + \alpha)) - \coth\pi(z + i(1/p + \alpha))$$

and, for $M^0(b) \in N^p(\alpha)$, one has $X_c^{-1}M^0(b)X_c = M^0(b_c)$ with $M^0(b_c) \in N^p(\alpha)$ and $b_c(z) = b(\frac{z}{c})$.

The proof is straightforward.

For illustration, let us once more return to our example. If A is the operator $aI + bS_{R^+}$ with constant coefficients a and b, thought of as acting on $L^p_{R^+}$, then

$$(X_c^{-1}AX_c u)(t) = au(t) + \frac{bc}{\pi i}\int_0^\infty \frac{(t/s)^{(c-1)(\alpha+1/p)}}{1 - (t/s)^c}f(s)\frac{ds}{s}\,, \tag{3}$$

and the common collocation method with spline space S_n^F, $F = \{\chi_{[0,1)}\}$ and with collocation points $\frac{k+\epsilon}{n}$, $\epsilon \in (0,1)$, applied to the operator (3), leads on the n th level to a system matrix with jk th entry

$$a\chi_{[j/n,(j+1)/n)}\left(\frac{k+\epsilon}{n}\right) + \frac{bc}{\pi i}\int_0^\infty \frac{((k+\epsilon)/(ns))^{(c-1)(\alpha+1/p)}}{1 - ((k+\epsilon)/(ns))^c}\chi_{[j/n,(j+1)/n)}(s)\frac{ds}{s}$$

$$= a\delta_{jk} + \frac{b}{\pi i}\left(\frac{k+\epsilon}{n}\right)^{(c-1)(\alpha+1/p)}\int_{(j/n)^c}^{((j+1)/n)^c}\frac{x^{(1/c-1)(\alpha+1/p)}}{x - ((k+\epsilon)/(n))^c}\,dx$$

$$= a\gamma_{kn}(\phi_{jn}) + \frac{b}{\pi i}\gamma_{kn}(S_{R^+}\phi_{jn})$$

with

$$\phi_{jn}(t) = t^{(1/c-1)(\alpha+1/p)}\chi_{[(j/n)^c,((j+1)/n)^c)}(t) \tag{4}$$

and

$$\gamma_{kn}(f) = \left(\frac{k+\epsilon}{n}\right)^{(c-1)(\alpha+1/p)} f\left(\left(\frac{k+\epsilon}{n}\right)^c\right). \tag{5}$$

Thus, it is equivalent to the modified collocation method with trial space generated by the functions (4) and with test functionals (5), but now applied to the original operator A. Now one immediately gets from Sections 2.11.4 and 2.11.5 that this method is stable if and only if the operator

$$T(a + b\sigma^{F,G}) + G(x_c) + K$$

(with a certain compact operator K and with $G = \{\delta_\epsilon\}$) is invertible. The undetermined operator K suggests to modify this method by cutting off the singularity at 0. Doing this in the obvious manner one finally arrives at a method which both has a finer resolution at the point 0 (controlled by the parameter c) and whose stability conditions are well verifiable (see Example 4.7 in 4.1.7).

6.3.2   Approximation methods for additive operators

Hitherto, we have only dealt with approximation methods for linear equations. One of the simplest non-linear operators is the operator C of complex conjugation, $(Cf)(t) = \overline{f(t)}$. In combination with the singular operator S_Γ it involves operators of the form

$$aI + bS_\Gamma + cC + dS_\Gamma C - eCS_\Gamma - fCS_\Gamma C + T_1 + T_2 C, \tag{1}$$

where a, b, c, d, e, f are given functions and T_1 and T_2 are linear compact operators. Operators of this form often occur in elasticity theory, hydrodynamics, acoustics, and deformation theory. As a special case of particular interest which is also covered by (1), we mention the double layer potential operator

$$(Au)(t) = a(t)u(t) + \frac{b(t)}{\pi}\int_\Gamma u(s)\frac{d}{dn_s}\log|t - s|\,d\Gamma_s + (Tu)(t), \tag{2}$$

where n_s refers to the inner normal to Γ at s, and T stands for a compact operator. (Indeed, the operator (2) can be rewritten as

$$aI + \frac{b}{2}S_\Gamma + \frac{b}{2}CS_\Gamma C + T$$

and is obviously of form (1).)

In case Γ is the boundary of a polygonal domain, and if $a = b = 1$ on Γ, Costabel and Stephan verified the stability of the 0-collocation as well as of certain Galerkin methods for operator (2) (see [CS 1], [CS 2]), and Atkinson, Chandler, Graham, and Kress studied stability and convergence of modified quadrature rules ([AG], [CGr 1], [Kr]). The decisive observation widely used in these papers is, that under the above restrictions, the operator (2) can be rewritten as $I + K + T'$, where T' is compact and K is small (i.e. $\|K\| < 1$). But this representation is no longer valid if the coefficients of A are not equal to 1, and thus, the general situation requires other techniques to study approximation methods for (2).

The Banach algebra approach discussed throughout this book is also suitable to attack stability of approximation methods for additive operators of the form (1), but it needs some modifications. The point is that the (anti-linear) operator of conjugation does not belong to a complex Banach algebra (since $C\alpha I = \bar{\alpha}C$ for all $\alpha \in \mathbf{C}$), whereas some techniques (for example, the standard proof of Allan's local principle by using Proposition 1.4) work for Banach algebras over the complex field only.

One modification which has been set forth by Didenko, Roch, Silbermann in [DRS] in detail, consists of a consequent replacement of complex Banach algebra techniques by real ones. In particular, the role of Allan's local principle is taken upon by that due to Gohberg-Krupnik ([GK]). In this way, in [DRS] necessary and sufficient conditions for the stability of quadrature, ϵ-collocation, and qualocation methods for the operator (1) were derived for the case of curves Γ where all points have order 2 (or, in other words, for simple closed curves with corners). Let us further emphasize that Gohberg and Krupnik's local principle also applies to derive Allan's local principle for Banach algebras over the real field.

For a third approach, one simply writes the operator (1) in the form $X + YC$ where X and Y are linear operators, and then the identity

$$\begin{pmatrix} X+YC & 0 \\ 0 & X-YC \end{pmatrix} = \frac{1}{2}\begin{pmatrix} I & C \\ I & -C \end{pmatrix}\begin{pmatrix} X & Y \\ CYC & CXC \end{pmatrix}\begin{pmatrix} I & I \\ C & -C \end{pmatrix}$$

shows that the operator (1) is, in a sense, equivalent to the (linear) operator $\begin{pmatrix} X & Y \\ CYC & CXC \end{pmatrix}$. (Observe that $1/2\begin{pmatrix} I & C \\ I & -C \end{pmatrix}$ and $\begin{pmatrix} I & I \\ C & -C \end{pmatrix}$ are inverse to each other and that $-i(X + YC)iI = X - YC$.)

One can prove that each entry of the matrix $\begin{pmatrix} X & Y \\ CYC & CXC \end{pmatrix}$ belongs to the algebra generated by the singular integral operator over Γ and by operators of multiplication by piecewise continuous functions (compare [RS 1]), and so a bulk of the results in Sections 3-5 can be applied to operators with conjugation without great effort. For details we refer to the papers [Di 1], [Di 2], [DiM], [DRS] and [DiS].

6.3.3   The double layer potential operator

Because of the practical importance of the double layer potential operator, we are going to cite the results of [DRS] concerning the equation

$$Au = (aI + \frac{b}{2}S_\Gamma + \frac{b}{2}CS_\Gamma C)u = f \,. \tag{1}$$

For, we suppose Γ to be a simple closed curce and $s : [0, 1] \to \Gamma$ to be a continuous parametrization of Γ with $s(0) = s(1)$. Further we assume that s is twice continuously differentiable on each open interval $(j/n_0, (j+1)/n_0)$ where n_0 is a fixed positive integer and $j = 0, \ldots, n_0 - 1$, and that s' and s'' have finite one-sided limits at j/n_0 such that $|s'(\frac{j}{n_0} - 0)| = |s'(\frac{j}{n_0} + 0)|$ but $\arg s'(\frac{j}{n_0} - 0) \neq \arg s'(\frac{j}{n_0} + 0)$.

Now we choose and fix different numbers ϵ and δ in $[0, 1]$, and for each multiple n of n_0 and for each integer k we define

$$t_k^{(n)} = s(\frac{k+\delta}{n}) \,, \quad \tau_k^{(n)} = s(\frac{k+\epsilon}{n}) \,, \quad \Delta t_k^{(n)} = s(\frac{k+1}{n}) - s(\frac{k}{n}) \,.$$

The quadrature method for solving (1) determines approximate values $u_k^{(n)}$ for the exact value of u at $t_k^{(n)}$ by solving the linear system (where $k = 0, \ldots, n-1$)

$$a(\tau_k^{(n)}) + \frac{b(\tau_k^{(n)})}{2\pi i} \sum_{j=0}^{n-1} \left(\frac{\Delta t_j^{(n)}}{t_j^{(n)} - \tau_k^{(n)}} - \frac{\overline{\Delta t_j^{(n)}}}{\overline{t_j^{(n)}} - \overline{\tau_k^{(n)}}} \right) u_j^{(n)} = f(\tau_k^{(n)}) \,. \tag{2}$$

The following notations are needed to formulate stability conditions for (2).

For each $\tau \in \Gamma$ we let w_τ denote the angle formed by the one-sided tangent at τ, and we set

$$B^\tau = (B_{rl}^\tau)_{r,l=1}^2 \,,$$

where $B_{11}^\tau = B_{22}^\tau = aI$ and

$$B_{21}^\tau = \frac{b}{2\pi i} \left(\frac{1}{j + \delta + (\epsilon - k - 1)e^{iw_\tau}} - \frac{1}{j + \delta + (\epsilon - k - 1)e^{i(2\pi - w_\tau)}} \right)_{k,j=0}^\infty \,,$$

$$B_{12}^\tau = \frac{b}{2\pi i} \left(\frac{e^{iw_\tau}}{(\delta - j + 1)e^{iw_\tau} + \epsilon + k} - \frac{e^{i(2\pi - w_\tau)}}{(\delta - j + 1)e^{i(2\pi - w_\tau)} + \epsilon + k} \right)_{k,j=0}^\infty \,.$$

The operator B^τ is Fredholm if and only if the function

$$\phi_\tau(x) = a^2 - \frac{b^2}{4} \left[4x(1 - x) + 2(x^{\delta_\tau}(1 - x)^{2-\delta_\tau} + (1 - x)^{\delta_\tau} x^{2-\delta_\tau}) \cos w_\tau - \right.$$
$$\left. - 2i(x^{\delta_\tau}(1 - x)^{2-\delta_\tau} - (1 - x)^{\delta_\tau} x^{2-\delta_\tau}) \sin w_\tau \right]$$

with $\delta_\tau = w_\tau/2\pi$ does not vanish on $[0,1]$. Let B_τ stand for the curve $\{\phi_\tau(x), \ x \in [0,1]\}$ and, if $\tau_0, \ldots, \tau_{n_0-1}$ are the corner points of Γ, then introduce the following curves:

$$\mathcal{C}_r = \left\{ \frac{(1+x^2)}{4x + 2(x^{\delta_r} + x^{2-\delta_r})\cos w_{\tau_r} - 2i(x^{\delta_r} - x^{2-\delta_r})\sin w_{\tau_r}} \right.$$

$$\left. \text{where} \quad x \in (0,\infty) \text{ and } \delta_r = \frac{w_{\tau_r}}{2\pi} \right\}$$

for $r = 0, \ldots, n_0 - 1$ and

$$\mathcal{C}_{a,b} = \left\{ \frac{b^2(t)}{4a^2(t)}, \quad t \in \Gamma \right\}.$$

Theorem 6.6 *Let $a, b \in C(\Gamma)$ and $a(t) \neq 0$ for all $t \in \Gamma$. The quadrature method (2) is stable if and only if the following conditions are fulfilled:*

(a) *the operator A given in (1) is invertible.*

(b) $\mathcal{C}_{a,b} \cap \cup_{r=0}^{n_0-1} \mathcal{C}_r = \emptyset$.

(c) *the curves B_{τ_r}, $r = 0, \ldots, n_0 - 1$, have winding number zero with respect to the origin.*

(d) $\dim \operatorname{Ker} B^\tau = 0$, $\tau \in \Gamma$.

Let us emphasize that conditions (b) and (c) are well verifiable, but the authors are not aware of an effective algorithm for estimating the kernel dimension of the operators B^τ.

Observe that there are some deciding differences between quadrature methods for singular integral operators and for double layer potential operators:

- For singular integral operators without conjugation, the index of the local operators is always zero, whereas the local operators B^τ can have non-vanishing indices! For example, one can prove that if

$$\frac{a^2(\tau_r)}{b^2(\tau_r)} \in \mathbf{R} \quad \text{and} \quad 0 \notin \left(\frac{4a^2(\tau_r)}{b^2(\tau_r)} - 1 - \cos w_{\tau_r}, \frac{4a^2(\tau_r)}{b^2(\tau_r)} \right)$$

then $\operatorname{Ind} B^{\tau_r} = 0$, but if the point 0 belongs to the indicated interval, then $|\operatorname{Ind} B^{\tau_r}| = 1$.

- The conditions (b) and (c) are completely independent of δ and ϵ, i.e. of the special choice of the quadrature method. The same independence appears for ϵ-collocation and qualocation methods (see [DRS] for details). Thus, if one combines these methods with a cutting-off-technique around the corner points then the stability of these modified methods will only depend on the operator A but not on the choice of the method (if one of these methods is stable then all of them are stable). The reader is recommended to compare these facts with the stability conditions of modified Galerkin methods for Mellin operators (see Section 4.4.1).

Bibliography

1. Monographs, Textbooks, Surveys

1.1. Functional analysis

[Pe]   Pedersen, G.K.: Analysis now.– Graduate Texts in Mathematics 118, Springer Verlag New York Berlin Heidelberg London Paris Tokyo 1989.

[ReS]   Reed, M., and Simon, B.: Methods of modern mathematical physics, vol. I-IV.– Academic Press, New York 1972-79.

[Ru]   Rudin, W.: Functional analysis.– Mc Graw-Hill Book Company, New York 1973.

[Yo]   Yosida, K.: Functional Analysis.– Springer Verlag, Berlin-Göttingen-Heidelberg 1965.

1.2. Banach algebras

[Dix]   Dixmier, J.: Les C^*-algébres et leurs représentations.– Gauthier-Villars Éditeur, Paris 1969.

[D 1]   Douglas, R.G.: Banach algebra techniques in operator theory. – Academic Press, New York 1972.

[D 2]   Douglas, R.G.: Banach algebra techniques in the theory of Toeplitz operators. – CBMS Lecture Notes 15, Amer. Math. Soc., Providence, R.I., 1973.

[GRS]   Gelfand, I.M., Raikov, D.A., and Shilov, G.E.: Kommutative normierte Algebren.– Deutscher Verlag der Wissenschaften, Berlin 1964.

[KR]   Kadison, R.V., and Ringrose, J.R.: Fundamentals of the theory of operator algebras, Volume I: Elementary theory. – Academic press, New York 1983.

[Kh]   Khelemskii, A.Ya.: Banach and polynormed algebras: general theory, rep-
          resentations, homology. – Nauka, Moscow 1989 (Russian).

[Kr]   Krupnik, N.: Banach algebras with symbol and singular integral opera-
          tors. – Operator Theory 26, Birkhäuser Verlag, Basel 1987.

[Na]   Naimark, M.A.: Normierte Algebren.– Deutscher Verlag der Wissenschaf-
          ten, Berlin 1959. (English translation: Hafner, New York 1964).

1.3. Concrete operator theory

[BS 1]   Böttcher, A., and Silbermann, B.: Invertibility and asymptotics of
          Toeplitz matrices. – Akademie-Verlag, Berlin 1983.

[BS 2]   Böttcher, A., and Silbermann, B.: Analysis of Toeplitz operators. –
          Akademie-Verlag, Berlin 1989, and Springer-Verlag, Berlin 1990.

[CG]   Clancey, K.F., and Gohberg, I.: Factorization of matrix functions and
          singular integral operators. – Birkhäuser Verlag, Basel-Boston-Stuttgart
          1981.

[Du 1]   Duduchava, R.: Integral equations with fixed singularities. – B.G. Teub-
          ner Verlagsgesellschaft, Leipzig 1979.

[FS]   Fenyö, S., and Stolle, H.W.: Theorie und Praxis der linearen Integral-
          gleichungen, Bände 1-4.– Deutscher Verlag der Wissenschaften, Berlin
          1982-84.

[GF 1]   Gohberg, I.C., and Feldman, I.A.: Convolution equations and projec-
          tion methods for their solution. – Nauka, Moscow 1971 (Russian);
          Engl. Transl.: Amer. Math. Soc. Transl. of Math. Monographs 41, Prov-
          idence, R.I., 1974; German Transl.: Akademie-Verlag, Berlin 1974.

[GG]   Gohberg I., and Goldberg, S.: Basic operator theory. – Birkhäuser Verlag,
          Basel 1981.

[GGK]   Gohberg, I., Goldberg, S., and Kaashoek M.A.: Classes of linear opera-
          tors, Vol.1.– OT 49, Birkhäuser Verlag, Basel-Boston-Berlin 1990 (Vol. 2
          in preparation).

[GK 1]   Gohberg, I., and Krupnik, N.: One-dimensional linear singular integral
          equations, Vol. I and II.– Operator Theory 53 and 54, Birkhäuser Verlag,
          Basel-Boston-Berlin, 1992.

[GS]   Grenander, U., and Szegö, G.: Toeplitz forms and their applications. –
          University of Califormia Press, Berkeley and Los Angeles 1958.

[L]   Litvinchuk, G.S.: Boundary value problems and singular integral equations with shift. – Nauka, Mosow 1977 (Russian).

[LS]   Litvinchuk, G.S., and Spitkovski, I.M.: Factorization of Measurable Matrix Functions. – OT 25, Birkhäuser Verlag, Basel 1987, and Akademie-Verlag, Berlin 1987.

[MP]   Michlin, S.G., and Prößdorf, S.: Singuläre Integraloperatoren.– Akademie-Verlag, Berlin 1980. (Extended english translation: Singular integral operators).

[P 1]   Prößdorf, S.: Some classes of singular equations.– North-Holland Publ. Comp., Amsterdam, New York, Oxford 1978 (Russian transl. Mir, Moscow 1979).

[RS 1] Roch, S., and Silbermann, B.: Algebras of convolution operators and their image in the Calkin algebra.– Report R-MATH-05/90 des Karl-Weierstraß-Instituts für Mathematik, Berlin 1990.

[SC]   Simonenko, I.B., and Chin Ngok Minh.: A local method in the theory of one-dimensional singular integral equations with piecewise continuous coefficients. Fredholmness.– Izd. Rostovsk. Univ., Rostov-na-Donu 1986 (Russian).

1.4. Numerical analysis for convolution operators

[BL]   Belotserkovsky, S.M., and Lifanov, I.K.: Method of discrete vortices.– CRC Press, Boca Ratoa, Ann Arbor, London, Tokyo, 1993.

[Hop]   Hoppe, W.: Stabilität von Splineapproximationsverfahren für singuläre Integralgleichungen auf kompakten, glatten Mannigfaltigkeiten ohne Rand. – Dissertationsschrift an der TU Chemnitz, Chemnitz 1993.

[PS 1] Prößdorf, S., and Silbermann, B.: Projektionsverfahren und die näherungsweise Lösung singulärer Gleichungen.– B.G. Teubner Verlagsgesellschaft, Leipzig 1977.

[PS 2] Prößdorf, S., and Silbermann, B.: Numerical analysis for integral and related operator equations. – Akademie-Verlag, Berlin, 1991, and Birkhäuser Verlag, Basel-Boston-Stuttgart 1991.

[R 1]   Roch, S.: Nichtkommutative Gelfandtheorien und ihre Anwendungen in Operatortheorie und numerischer Analysis. – Habilitationsschrift an der TU Chemnitz, Chemnitz 1990-92.

[Sc 1] Schmidt, G.: Splines und die näherungsweise Lösung von Pseudodifferentialgleichungen auf geschlossenen Kurven.– Report R-MATH 09/86, Karl-Weierstraß-Inst. f. Mathematik, Berlin 1986.

[S 1]   Silbermann, B.: Asymptotische Invertierung von Faltungsoperatoren.
          – In: Jahrbuch Überblicke Mathematik, 1993, Vieweg-Verlag, Braun-
          schweig 1993.

1.5. Numerical inversion of structured matrices

[GvL]   Golub, G., and van Loan, C.F.: Matrix computations. – John Hopkins
          University Press, 1989.

[HR]   Heinig, G., and Rost, K.: Algebraic methods for Toeplitz-like matrices
          and operators. – Operator Theory 13, Birkhäuser Verlag, Basel 1984; and
          Akademie-Verlag, Berlin 1984.

1.6. Splines and Wavelets

[BCD]   Beylkin, G., Coifmann, R., Daubechies, I., Mallat, S., Meyer, Y., Raphael,
          L., Ruskai, B. (Eds.): Wavelets and their applications. – Jones and Bar-
          lett, Cambridge MA 1992.

[dB]   de Boor, C.: A practical guide to splines.– Springer Verlag, New York-
          Heidelberg-Berlin 1978.

[C 1]   Chui, C.K.: Multivariate splines. – CMBS-NSF Series in Appl. Math. 54,
          SIAM Publ., Philadelphia 1988.

[C 2]   Chui, C.K.: An introduction to wavelets. Wavelet Analysis and its Ap-
          plications, Vol.1. – Academic Press Inc., Boston 1992.

[CGT]   Combes, J.M., Grossmann, A., Tchamitchian, P. (Eds.): Wavelets: Time-
          frequency methods and phase space. – Springer-Verlag, New York 1989.

[Dau 1]   Daubechies, I.: Ten lectures on wavelets.– CMBS-NSF Series in Appl.-
          Math., SIAM Publ., Philadelphia 1992.

[Me]   Meyer, Y.: Ondelettes et Opèrateurs, 1-4. – Hermann, Paris 1990.

1.7. Further Textbooks

[BL]   Bergh, J., and Löfström, J.: Interpolation spaces. An Introduction.–
          Springer Verlag, Berlin-Heidelberg-New York 1976.

[Die]   Dieudonné, J.: Elements d' Analyse Tome II, Chapitres XII à XV.– 2^e
          édition, revue et augmentée Gauthier-Villars, Editeur Paris/Bruxel-
          les/Montréal 1974.

[FF]   Foias, C., and Frazho, A.E.: The commutant lifting approach to interpolation problems.– Birkhäuser Verlag, Basel-Boston-Berlin, 1990

[GR]   Gradstein, I.S, and Ryshik, I.M.: Tables of integrals, sums, series and products.– Nauka, Moscow 1971 (Russian).

[PBM]   Prudnikov, A.P., Brychkov, Y.A., and Marichev, O.I.: Integrals and Series, Vol. 1 : Elementary functions.– Nauka, Moscow 1983, Engl Transl. Gordon and Beach Science Publ. New York-London-Paris-Montreux-Tokyo-Melbourne 1986.

[Re]   Remmert, R.: Theory of complex functions.– Springer Verlag, New York 1991.

2. Research Papers

[A]   Allan, G.R.: Ideals of vector valued functions. – Proc. London Math. Soc. 18(1968), 3, 193-216.

[ABCR]   Alpert, B., Beylkin, G., Coifman, R., Rokhlin, V.: Wavelets for the fast solution of second-kind integral equations. – Research Report 837, Yale University, 1990.

[AG]   Atkinson, K.E., and Graham, I.G.: An iterative variant of the Nyström method for boundary integral equations in domains with corners. – Numer. Math. 58 (1990), 145-161.

[Bat]   Battle, G.: A block spin construction of ondelettes, Part I: Lemarié functions.– Comm. Math. Phys. 1987.

[Ba]   Baxter, G.: A norm inequality for a finite section Wiener-Hopf equation.– Illinois J.Math. 7(1963), 97-103.

[BCR]   Beylkin, G., Coifmann, R., and Rokhlin, V.: Fast wavelet transforms and numerical algorithms I. – Comm. on Pure and Applied Math., Vol. XLIV (1991), 141-183.

[BKS]   Böttcher, A., Krupnik, A., and Silbermann, B.: A general look at local principles with special emphasis on the norm computation effect. – IEOT 11(1988), 4, 455-479.

[BS 3]   Böttcher, A., and Silbermann, B.: The finite section method for Toeplitz operators on the quarter plane with piecewise continuous symbols. – Math. Nachr. 110(1983), 279-291.

[BS 4]   Böttcher, A., and Silbermann, B.: Über das Reduktionsverfahren für diskrete Wiener-Hopf Gleichungen mit unstetigem Symbol.– ZfAA 1:2(1982), 1-5.

[CGr 1]   Chandler, G.A., and Graham, I.G.: High-order methods for linear functionals of solutions of second kind integral equations. – SIAM Jour. Numer. Anal 25(1988), 1118-1179.

[CGr 2]   Chandler, G.A., and Graham, I.G.: Product integration-collocation methods for non-compact integral operator equations.– Math. Comp. 50(1988), 125-138.

[CS]   Chandler, G.A., and Sloan I.H.: Spline Qualocation Methods For Boundary Integral Equations.– Num. Math. 58(1990), 5, 537-567.

[Co 1]   Costabel, M.: An inverse for the Gohberg-Krupnik symbol map.– Proc. Royal Soc. of Edinburgh, 87A(1980), 153-165.

[Co 2]   Costabel, M.: Singular integral operators on curves with corners.– IEOT 3(1980), 323-349.

[Co 3]   Costabel, M.: Singuläre Integralgleichungen mit Carlemanschen Verschiebungen auf Kurven mit Ecken.– Math. Nachr. 109(1982), 29-37.

[CS 1]   Costabel, M., and Stephan, E.P.: Boundary integral equations for mixed boundary value problems in polygonial domains and Galerkin approximations. – Banach Center Publications Vol. 15, PWN, Warsaw 1985, 175-251.

[CS 2]   Costabel, M., and Stephan, E.P.: On the convergence of collocation methods for boundary integral equations on polygons. – Math. Comp. 49(1987), 461-678.

[CS 3]   Costabel, M., and Stephan, E.P.: The method of Mellin transformation for boundary integral equations on curves with corners.– Preprint 761, Fachbereich Mathematik, Technische Hochschule Darmstadt, 1983.

[DKPS]   Dahmen, W., Kleemann, B., Prößdorf, S., and Schneider, R.: A multiscale method for the double layer potential equation on a polyhedron.– Bericht Nr. 91 des Instituts für Geometrie und praktische Mathematik der RWTH Aachen, 1993.

[DPS 1]   Dahmen, W., Prößdorf, S., and Schneider, R.: Wavelet approximation methods for pseudodifferential equations I: Stability and convergence.– Preprint 7/92 of the IAAS, Berlin 1992, to appear in: Mathematische Zeitschrift.

[DPS 2]   Dahmen, W., Prößdorf, S., and Schneider, R.: Wavelet approximation methods for pseudodifferential operators II: Matrix compression and fast solution.– Advances in Computational Mathematics 1(1993), 259-335.

[DPS 3] Dahmen, W., Prößdorf, S., and Schneider, R.: Multiscale methods for pseudodifferential equations.– In: Recent Advances in Wavelet Analysis (eds. L. Schumaker and G. Webb), Academic Press, Inc. 1994, 191-235.

[Dau 2] Daubechies, I.: Orthonormal bases of compactly supported wavelets. – Comm. on Pure and Applied Math. Vol. XLI (1988), 909-996.

[Di 1] Didenko, V.D.: On the approximate solution of singular integral equations with Carleman shift and complex conjugate values of the unknown function. – Ukr. Math. Journ. 32, 2(1980), 378-382 (Russian).

[Di 2] Didenko, V.D.: On a method for the approximate solution of singular integral equations with conjugation on piecewise smooth curves. – Dokl. AN SSSR 318, 6(1991), 1298-1301 (Russian).

[DiM] Didenko, V.D., and Matskul, V.N.: On the approximate solution of singular integral equations with conjugation. – Zh. Vychisl. Mat. i Mat. Fiz. 29, 3(1989), 392-404 (Russian; Engl. transl. in USSR Comput. Math. and Mat. Phys. 29(1989)).

[DRS] Didenko, V.D., Roch, S., and Silbermann, B.: Approximation methods for singular integral equations with conjugation on curves with corners. – SIAM Journ. of Numer. Anal (to appear).

[DiS] Didenko, V.D., and Silbermann, B.: On the stability of some operator sequences and the approximate solution of singular integral equations with conjugation. – IEOT 16(1993), 224-243.

[Du 2] Duduchava, R.V.: On bisingular integral operators with discontinuous coefficients.– Math. USSR Sbornik 30(1976), 4, 515-537.

[Du 3] Duduchava, R.V.: On the index of bisingular integral operators.– Math. Nachr. 91(1979), 431-460, and – Math. Nachr. 92(1979), 289-307 (Russian).

[DP] Duduchava, R.V., and Prößdorf, S.: On the approximation of singular integral equations by equations with smooth kernels.– Preprint, to appear.

[E 1] Elschner, J.: On spline approximation for a class of non-compact integral equations.– Report R-MATH-09/88, Karl-Weierstraß-Inst. Math., Akad. Wiss. DDR, Berlin 1988.

[E 2] Elschner, J.: On spline approximation for singular integral equations on an interval.– Math. Nachr. 139(1988), 309-319.

[GF 2] Gohberg, I.Z., and Feldman, I.A.: On the approximate solution of certain classes of linear equations.– Dokl. AN SSSR 160,4(1965), 750-753 (Russian).

[Go]   Gohberg, I.C.: On an application of the theory of normed rings to singular integral equations.– Uspekhi mat. nauk 7, 2(1952), 149-156 (Russian).

[GK 2]   Gohberg, I.C., and Krupnik, N.Ya.: Singular integral equations with continuous coefficients on a composed curve.– Mat. Issled. Kishinev 5(1970), 2, 89-103 (Russian).

[GK 3]   Gohberg, I.C., and Krupnik, N.Ya.: On the algebra generated by one-dimensional singular integral operators with piecewise continuous coefficients.– Funkc. Analiz i ego prilozh., 4(1970), 3, 26-36.

[GK 4]   Gohberg, I.C., and Krupnik, N.Ya.: On singular integral operators on a non-simple curve.– Soobshzh. AN Gruz. SSR 64(1971), 21-24 (Russian).

[HS 1]   Hagen, R., and Silbermann, B.: Local theory of the collocation method for the approximative solution of singular integral equations, II.– Seminar Analysis, Operator Equat. and Numer. Anal. 1986/1987, Karl-Weierstraß-Institut für Mathematik, Berlin 1987, 41-56.

[HS 2]   Hagen, R., and Silbermann, B.: A Banach algebra approach to the stability of projection methods for singular integral equations.– Math. Nachr. 140(1989), 285-297.

[HS 3]   Hagen, R., and Silbermann, B.: On the stability of the qualocation method.– Seminar Analysis, Operator Equat. and Numer. Anal. 1988/89, Karl-Weierstraß-Institut für Mathematik, Berlin 1989, 43-52.

[HS 4]   Hagen, R., and Silbermann, B.: On the convergence of the qualocation method.– TU Chemnitz, Preprint Nr. 207, 5.Jg., 1991.

[HS 5]   Hagen, R., and Silbermann, B.: A finite element collocation method for bisingular integral equations.– Appl. Anal., 1985, Vol. 19, 117-135.

[HRS]   Hagen, R., Roch, S., and Silbermann, B.: Stability of spline approximation methods for multidimensional pseudodifferential operators.– IEOT 19(1994), 25-64.

[Ho]   Hörmander, L.: Pseudo-differential operators.– Commun. on pure and appl. math. 188(1965), 3, 501-517.

[JS]   Junghanns, P., and Silbermann, B.: Local theory of the collocation method for the approximate solution of singular integral equations, I.– IEOT 7(1984), 791-807.

[KN]   Kohn, J.J., and Nirenberg, L.I.: An algebra of pseudo-differential operators.– Commun. on pure and appl. math. 18(1965), 1/2, 269-305.

[Ko]   Kozak, A.: A local method in the theory of projection methods. – Dokl. AN SSSR 212(1974), 6, 1287-1289 (Russian).

[Kre]   Kress, R.: A Nyström method for boundary integral equations in domains
          with corners.– Numer. Math. 58(1990), 145-161.

[KV]   Krupnik, N. Ya., and Verbitski, I.E.: On the applicability of the reduc-
          tion method to discrete Wiener-Hopf equations with piecewise continuous
          symbol.– In: Spektr. Svoistva Oper. (Mat. issled. 45), Shtiintsa, Kishinev
          1977, 17-28 (Russian).

[Ma]   Mallat, S.G.: Multiresolution approximations and wavelet orthonormal
          bases of $L^2(R)$.– TAMS Vol. 315, Number1, 1989.

[M 1]   Michlin, S.G.: The composition of twofold singular integrals. – Dokl. AN
          SSSR 2(11), 1(87), 1936, 3-6 (Russian).

[M 2]   Michlin, S.G.: Singular integral equation with two independent variables.
          – Mat. Sbornik 1(43), 4, 1936, 535-550 (Russian).

[Pi 1]   Pilidi, V.S.: On the uniform invertibility of regular approximations of one-
          dimensional singular integral operators. – Dokl. AN SSSR 299(1988), 6,
          1317-1320 (Russian).

[Pi 2]   Pilidi, V.S.: Foundation of a method of cutting off singularities for
          bisingular integral operators with continuous coefficients. – Izv. Vys-
          sh. Uchebn. Zaved. 7(1990), 51-60 (Russian).

[Pi 3]   Pilidi, V.S.: The method of excising singularities for a class of singular
          integral operators. – Izv. Vyssh. Uchebn. Zaved. 2(1990), 78-80 (Russian).

[Pi 4]   Pilidi, V.S.: On uniform invertibility of strong approximations of one-
          dimensional singular integral operators with piecewise continuous co-
          effients on the real line. – Izv. Sev.-Kavk. Nauchn. Tsentra Vyssh. Shk.
          Estestv. Nauki 3(71), 1990, 44-46 (Russian).

[Pi 5]   Pilidi, V.S.: On uniform invertibility of regular approximations of one-
          dimensional singular integral operators with non-compact perturbations.
          – Diff. Urav. 12(1990), 2127-2136 (Russian).

[Po]   Polski, N.I.: Projection methods for solving linear equations.– Uspekhi
          mat. Nauk 18(1963), 2, 179-180.

[P 2]   Prößdorf, S.: Ein Lokalisierungsprinzip in der Theorie der Splineapproxi-
          mation und einige seiner Anwendungen. – Math. Nachr. 119(1984), 239-
          255.

[PR 1]   Prößdorf, S., and Rathsfeld, A.: Mellin techniques in the numerical anal-
          ysis for one-dimensional singular integral equations.– Report R-MATH
          06/88, Karl-Weierstraß-Inst., Berlin 1988.

[PR 2]   Prößdorf, S., and Rathsfeld, A.: Stabilitätskriterien für Näherungs-
          verfahren bei singulären Integralgleichungen in L^p.– Z. Anal. Anw.
          6(1987), 6, 539-558.

[PR 3]   Prößdorf, S., and Rathsfeld, A.: Quadrature and collocation methods
          for singular integral equations on curves with corners.– Z. Anal. Anw.
          8(1989), 3, 197-220.

[PR 4]   Prößdorf, S., and Rathsfeld, A.: Quadrature methods for strongly elliptic
          Cauchy singular integral equations on an interval.– In: H. Dym et al.
          (eds.), The Gohberg anniversary collection, Vol.2: Topics in analysis and
          operator theory. Birkhäuser Verlag, Basel-Boston-Berlin 1989, 435-471.

[PSch 1] Prößdorf, S., and Schneider, R.: Spline approximation methods for mul-
          tidimensional pseudodifferential equations. – IEOT 15(1992), 626-672.

[PSch 2] Prößdorf, S., and Schneider, R.: Pseudodifference operators – A sym-
          bolic calculus for approximation methods for periodic pseudodifferential
          equations. Part I: Calculus. – Preprint 1992.

[Ra]   Rathsfeld, A.: Reduktionsverfahren für singuläre Integralgleichungen mit
          stückweise stetigen Koeffizienten.– Math. Nachr. 127(1986), 125-143.

[Re]   Reich, E.: On non-Hermitian Toeplitz matrices.– Math. Scand. 10(1962),
          145-152.

[R 2]   Roch, S.: Finite sections of operators belonging to the closed algebra
          of singular integral operators. – Seminar Analysis: Operator Equat. and
          Numer. Anal., 1986/87, Berlin 1987, 139-148.

[R 3]   Roch, S.: Finite sections of operators generated by singular integrals with
          Carleman shift. – Preprint TU Karl-Marx-Stadt (Chemnitz), 52(1987).

[R 4]   Roch, S.: Finite sections of singular integral operators with measur-
          able coefficients. – Wiss. Zeitschr. der TU Karl-Marx-Stadt (Chemnitz)
          31(1989), 2, 236-242.

[R 5]   Roch, S.: Finite sections of operators generated by convolutions. – Semi-
          nar Analysis: Operator Equat. and Numer. Anal. 1987/88, Berlin 1988,
          118-138.

[R 6]   Roch, S.: Local algebras of Toeplitz operators. – Math. Nachr. 152(1991),
          69-81.

[R 7]   Roch, S.: Das Reduktionsverfahren für Produktsummen von Toeplitz-
          operatoren mit stückweise stetigen Symbolen.– Wiss. Zeitschr. TH Karl-
          Marx-Stadt 26(1984), 2, 265-273.

[R 8]   Roch, S.: Spline approximation methods cutting off singularities.– ZfAA
          (to appear).

[RS 2]   Roch, S., and Silbermann, B.: A symbol calculus for finite sections of
          singular integral operators with flip and piecewise continuous coefficients.
          – Journ. of Funct. Analysis 78(1988), 2, 365-389.

[RS 3]   Roch, S., and Silbermann, B.: Finite sections of singular integral op-
          erators with Carleman shift. – Seminar Analysis: Operator Equat. and
          Numer. Anal. 1986/87, Berlin 1987, 149-180.

[RS 4]   Roch, S., and Silbermann, B.: The Calkin image of algebras of singular
          integral operators. – IEOT, 12(1989), 855-897.

[RS 5]   Roch, S., and Silbermann, B.: The structure of algebras of singular inte-
          gral operators. – Journ. Int. Eq. and Applic. 4(1992), 3, 421-442.

[RS 6]   Roch, S., and Silbermann, B.: Limiting sets of eigenvalues and singular
          values of Toeplitz matrices. – Asymptotic Analysis 8(1994), 293-309.

[RS 7]   Roch, S., and Silbermann, B.: Algebras generated by idempotents and
          the symbol calculus for singular integral operators.– IEOT 11(1988), 3,
          385-419.

[RS 8]   Roch, S., and Silbermann, B.: Representations of non-commutative Ba-
          nach algebras by continuous functions.– Algebra i Analiz, 3(1991), 4,
          171-185.

[RS 9]   Roch, S., and Silbermann, B.: Toeplitz-like operators, quasicommutator
          ideals, numerical analysis. Part I, Math. Nachr. 120(1985), 141-173.

[Sa]   Saranen, J.: On the effect of numerical quadratures in solving boundary
          integral equations.– Panel methods in fluid mechanics with emphasis on
          aerodynamics, Proc. 3rd GAMM-Semin., Kiel/FRG 1987, Notes Numer.
          Fluid Mech. 21, 196-209, (1988).

[Sa]   Saranen, J.: The modified quadrature method for logarithmic-kernel in-
          tegral equations on closed curves.– J. Integral Equat. Appl. 3, No.4, 575-
          600, (1991).

[Sc 2]   Schmidt, G.: On spline collocation methods for boundary integral equa-
          tions in the plane. – Math. Methods Appl. Sci. 7(1985), 74-89.

[Sch]   Schneider, R.: Integral equations with piecewise continuous coefficients
          in L^p-spaces with weight.– Journ. Int. Equat. 9(1985), 135-152.

[Se]   Seeley, R.T.: Singular integrals on compact manifolds. – Amer. J. of
          Math. 81(1959), 3, 658-690.

[S 2]   Silbermann, B.: Lokale Theorie des Reduktionsverfahrens für Toeplitz-
           operatoren. – Math. Nachr. 104(1981), 137-146.

[S 3]   Silbermann, B.: Lokale Theorie des Reduktionsverfahrens für singuläre
           Integralgleichungen. – ZfAA 1(1982), 6, 45-56.

[S 4]   Silbermann, B.: The Banach algebra approach to the reduction method
           for Toeplitz operators. – In "Linear and Complex Analysis Problem
           Book", Lecture Notes in Math., Vol. 1043, Springer-Verlag, New York-
           Berlin 1984, 293-297.

[S 5]   Silbermann, B.: Symbol constructions and numerical analysis. – Preprint
           116 der TU Chemnitz, 1989.

[S 6]   Silbermann, B.: On the limiting set of singular values of Toeplitz matri-
           ces. – Lin. Alg. Appl. 182(1993), 35-43.

[S 7]   Silbermann, B.: Local objects in the theory of Toeplitz operators.– IEOT
           9(1986), 706-738.

[S 8]   Silbermann, B.: Harmonic approximation of Toeplitz operators.– IEOT
           8(1985), 842-853.

[Si]   Simonenko, I.B.: A new general method of studying linear operator equa-
           tions of the type of singular integral equations, Part I.– Izv. Akad. Nauk
           SSSR, Ser. Mat. 29(1965), 3, 567-586; Part II.– Izv. Akad. Nauk SSSR,
           Ser. Mat. 29(1965), 4, 757-782 (Russian).

[Sl]   Sloan, I.H.: A quadrature-based approach to improving the collocation
           method.– Num. Math. 54(1988), 1, 41-56.

[Sl]   Sloan, I.H., and Wendland, W.L.: A quadrature-based approach to im-
           proving the collocation method for splines of even degree.– Z. Anal. An-
           wend. 8, No.4, 361-376, (1989).

[ST]   Spitkovski, I.M., and Tashbaev, A.M.: Factorization of certain piecewise
           constant matrix functions and its applications. – Math. Nachr. 151(1991),
           241-261.

[Sz]   Szyszka, U.: Splinekollokationsmethoden für singuläre Integralgleichun-
           gen auf geschlossenen Kurven in L^2. – In: Seminar Analysis: Operator
           Equat. and Numer. Anal., Berlin 1989, 153-184.

[V 1]   Verbitski, I.E.: On the reduction method for powers of Toeplitz matrices.–
           In: Oper. v. Banach. Prostr. (Mat. Issled. 47), Shtiintsa, Kishinev 1978,
           3-11 (Russian).

[V 2]   Verbitski, I.E.: Projection methods for the solution of singular integral equations with piecewise continuous coefficients.– In: Oper. v. Banach. Prostr. (Mat. Issled. 47), Shtiintsa, Kishinev 1978, 12-24 (Russian).

[W 1]   Widom, H.: Asymptotic behavior of block Toeplitz matrices and determinants, II.– Adv. Math. 21(1976), 1-29.

[W 2]   Widom, H.: On the singular values of Toeplitz matrices. – Zeitschr. Anal. Anwend. 8(1989), 3, 221-229.

[W 3]   Widom, H.: Eigenvalue distribution of nonselfadjoint Toeplitz matrices and the asymptotics of Toeplitz determinants in the case of nonvanishing index. – In: de Branges, L., Gohberg, I., and Rovnyak, J. (Eds.): Topics in Operator Theory. Ernst D. Hellinger Memorial Volume. Operator Theory: Advances and Applications Vol. 48, Birkhäuser Verlag, Basel-Boston-Berlin 1990, 387-421.

Index

algebra
    Banach, 1
    Calkin, 4, 147
    C^*, 3
    KMS-, 32
    semi-simple, 3
algebraization, 24
applicability
    of an approximation method, 7,
        244
approximation method
    cutting off the infinity, 270
    double indicated, 227
    multiindiced, 242

canonical projection, 86
Carleman shift, 344
center, 14
central subalgebra, 14
collocation method
    for equations on $\mathbf{R}$, 211, 213
    non-equidistant partitions, 350
    numerical examples, 329
    on complicated curves, 328
collocation projection, 103
commutator property, 87
compatible pair of arcs, 282
composed curve, 282
compression of F, 92
convergence
    strong, 4
    uniform, 4
    weak, 4
Costabel's decomposition, 50
cut-off-method, 227, 242

cutting off singularities, 325, 340

de Boor-type estimate, 80

essentialization, 25

finite section method, 228
    cutting off singularities, 340
    for block Toeplitz operators, 235
    for operators with Carleman shift,
        347
    for operators with flip, 345
    for paired operators, 229
    for singular integral operators, 337
    for Toeplitz + Hankel operators,
        348
    for Toeplitz operators, 19
Fourier transform, 46

Galerkin method, 225, 260, 321
    cutting off singularities, 268
    for equations on $[0, 1]$, 58, 209
    for equations on $\mathbf{R}$, 56, 108, 206
    for equations on $\mathbf{R}^+$, 58, 114
    for Mellin convolutions, 59, 115
    modified, 255, 258, 260, 265, 275
Galerkin projection, 54, 102, 291, 295,
    301
Gelfand transform, 13

homomorphism, 2

ideal, 3
identification, 26
index
    of an operator, 6

infinite section method, 237
inverse closedness, 2, 29
invertibility
    approximate, 11
    essential, 11
    essential approximate, 11
    usual, 11

Kozak's identity, 237

lifting theorem, 25
local inclusion theorem, 34
local principle
    Allan's, 14
    Douglas', 17
localization, 26
locally equivalent representations, 27

maximal ideal, 3
maximal ideal space, 13
Mellin transform, 47
method of discrete whirls, 119
mother wavelet, 100
multiplier
    continuous, 60
    $l^p(\alpha)$, 59
    piecewise continuous, 60
multiresolution analysis, 100

normally solvable, 6

operator
    conjugation, 335, 352
    discrete flip, 76, 269, 344
    discretized Mellin convolution, 64
    double layer potential, 352, 354
    flip, 52, 267
    Fourier convolution, 46
    Fredholm, 6
    generalized Hankel, 49
    Hankel, 18
    homogeneous, 62
    Mellin convolution, 48
    of local type, 197
    of regular type, 6

projection, 7
    singular integral, 46, 283, 289
    Toeplitz, 18
    weighted generalized Hankel, 49
    weighted singular integral, 49

partition
    non-equidistant, 350
    non-uniform, 350
    uniform, 350
piecewise continuous function, 146, 284
Pilidi's method, 342, 344
point
    regular, 282
    singular, 282
prebasis, 91
premeasure, 85
prespline, 79
    canonical, 82

quadrature method
    for double layer potential opera-
        tors, 354
    for equations on $\mathbf{R}$, 118, 120, 215
    for Mellin convolutions, 121, 122
    for singular integral operators, 215
quadrocation method, 123
    for singular integral operators, 217
qualocation method
    for equations on $\mathbf{R}$, 214
qualocation projection, 104

radical, 3

sequence
    generalized, 242
    infinite-section, 227
    of local type, 199
simple arc, 281
singularity
    analytic, 225, 226
    geometric, 225, 226
    numeric, 226
spectrum, 2

spline, 79
    piecewise constant, 54, 96
    piecewise polynomial with defect, 97
    smoothest piecewise polynomial, 97
spline projection
    cutting-off singularities, 245
spline space, 54, 79, 294, 300
stability
    of a generalized sequence, 244
    of an operator sequence, 7
symbol mapping, 2

wavelet, 100
winding number, 78

Notation index

algebras
    $\mathrm{alg}\,(T^0(PC_{p,\alpha}), P)$, 61
    $\mathrm{alg}\,T(PC_{p,\alpha})$, 61
    OLT, 197
    $SLT^F_{\Omega,r}$, 199
    $\Sigma^p(\alpha)$, 47
    $\Sigma^p_R(\alpha)$, 47
    $\mathcal{A}^{F,T}_{\Omega,r}$, 246
    $\mathcal{A}^{F,\infty}_{\Omega,r}$, 270
    $\mathcal{A}^F_{\Omega,r}$, 173
    $\mathcal{A}^F_{\Omega,r}(\Gamma,\Gamma',\zeta)$, 303
    $\mathcal{A}^F_{\Omega}$, 172
    $\mathcal{O}^p_\Gamma(\alpha)$, 284
    $\mathcal{O}^p_R(\alpha)$, 146
    $\mathcal{T}^p_{1,1,\Omega}(\alpha)$, 163
    $\mathcal{T}^p_{k,\kappa,\Omega}(\alpha)$, 164
function spaces
    $C_{p,\alpha}$, 60
    $L^p_\Gamma(\alpha)$, 283
    $L^\infty(\mathbf{T})$, 17
    $L^p_I(\alpha)$, 46
    PC, 146
    $PC^\Omega_{p,\alpha}$, 163
    $PC_{p,\alpha}$, 60
    $PC^{k,\kappa,\Omega}_{p,\alpha}$, 168
$l^p_k(\alpha)$, 80
$l^p_I(\alpha)$, 54
operators
    E^F_n, 80
    E^F_{-n}, 81
    $G(b)$, 64
    $H(a)$, 18
    $L^{F,G}_n$, 86
    $M^0(b)$, 48
    N_β, 49
    P, 61
    P_n, 19
    P^G_n, 85
    Q, 77, 208
    Q_n, 228
    R_n, 228
    S_Γ, 283
    S_I, 46
    $T(a)$, 18, 61
    $T^0(a)$, 59
    $W^0(b)$, 47
premeasures
    $G = \{\gamma_0, \ldots, \gamma_{k-1}\}$, 85
spline functions
    $F = \{\phi_0, \ldots, \phi_{k-1}\}$, 79
spline spaces
    S^F_n, 79
    $S^F_n(\Gamma,\Gamma',\zeta)$, 300
    $S^F_n(\Gamma,\xi)$, 294
symbol functions
    $\lambda^{F,G}$, 85
    $\rho^{F,G}$, 107
    $\sigma^{F,G}$, 107

OPERATOR THEORY: ADVANCES AND APPLICATIONS
BIRKHÄUSER VERLAG

7. **M.G. Krein:** Topics in Differential and Integral Equations and Operator Theory, 1983, (3-7643-1517-2)

9. **H. Baumgärtel, M. Wollenberg**: Mathematical Scattering Theory, 1983, (3-7643-1519-9)

11. **C. Apostol, C.M. Pearcy, B. Sz.-Nagy, D. Voiculescu, Gr. Arsene** (Eds.): Dilation Theory, Toeplitz Operators and Other Topics, 1983, (3-7643-1516-4)

13. **G. Heinig, K. Rost:** Algebraic Methods for Toeplitz-like Matrices and Operators, 1984, (3-7643-1643-8)

15. **H. Baumgärtel:** Analytic Perturbation Theory for Matrices and Operators, 1984, (3-7643-1664-0)

17. **R.G. Douglas, C.M. Pearcy, B. Sz.-Nagy, F.-H. Vasilescu, D. Voiculescu, Gr. Arsene** (Eds.): Advances in Invariant Subspaces and Other Results of Operator Theory, 1986, (3-7643-1763-9)

18. **I. Gohberg** (Ed.): I. Schur Methods in Operator Theory and Signal Processing, 1986, (3-7643-1776-0)

19. **H. Bart, I. Gohberg, M.A. Kaashoek** (Eds.): Operator Theory and Systems, 1986, (3-7643-1783-3)

20. **D. Amir:** Isometric characterization of Inner Product Spaces, 1986, (3-7643-1774-4)

21. **I. Gohberg, M.A. Kaashoek** (Eds.): Constructive Methods of Wiener-Hopf Factorization, 1986, (3-7643-1826-0)

22. **V.A. Marchenko:** Sturm-Liouville Operators and Applications, 1986, (3-7643-1794-9)

23. **W. Greenberg, C. van der Mee, V. Protopopescu:** Boundary Value Problems in Abstract Kinetic Theory, 1987, (3-7643-1765-5)

24. **H. Helson, B. Sz.-Nagy, F.-H. Vasilescu, D. Voiculescu, Gr. Arsene** (Eds.): Operators in Indefinite Metric Spaces, Scattering Theory and Other Topics, 1987, (3-7643-1843-0)

25. **G.S. Litvinchuk, I.M. Spitkovskii:** Factorization of Measurable Matrix Functions, 1987, (3-7643-1883-X)

26. **N.Y. Krupnik:** Banach Algebras with Symbol and Singular Integral Operators, 1987, (3-7643-1836-8)

27. **A. Bultheel:** Laurent Series and their Pade Approximation, 1987, (3-7643-1940-2)

28. **H. Helson, C.M. Pearcy, F.-H. Vasilescu, D. Voiculescu, Gr. Arsene** (Eds.): Special Classes of Linear Operators and Other Topics, 1988, (3-7643-1970-4)

29. **I. Gohberg** (Ed.): Topics in Operator Theory and Interpolation, 1988, (3-7634-1960-7)

30. **Yu.I. Lyubich:** Introduction to the Theory of Banach Representations of Groups, 1988, (3-7643-2207-1)

31. **E.M. Polishchuk:** Continual Means and Boundary Value Problems in Function Spaces, 1988, (3-7643-2217-9)

32. **I. Gohberg** (Ed.): Topics in Operator Theory. Constantin Apostol Memorial Issue, 1988, (3-7643-2232-2)

33. **I. Gohberg** (Ed.): Topics in Interplation Theory of Rational Matrix-Valued Functions, 1988, (3-7643-2233-0)

34. **I. Gohberg** (Ed.): Orthogonal Matrix-Valued Polynomials and Applications, 1988, (3-7643-2242-X)

35. **I. Gohberg, J.W. Helton, L. Rodman** (Eds.): Contributions to Operator Theory and its Applications, 1988, (3-7643-2221-7)

36. **G.R. Belitskii, Yu.I. Lyubich:** Matrix Norms and their Applications, 1988, (3-7643-2220-9)

37. **K. Schmüdgen:** Unbounded Operator Algebras and Representation Theory, 1990, (3-7643-2321-3)

38. **L. Rodman:** An Introduction to Operator Polynomials, 1989, (3-7643-2324-8)

39. **M. Martin, M. Putinar:** Lectures on Hyponormal Operators, 1989, (3-7643-2329-9)

40. **H. Dym, S. Goldberg, P. Lancaster, M.A. Kaashoek** (Eds.): The Gohberg Anniversary Collection, Volume I, 1989, (3-7643-2307-8)

41. **H. Dym, S. Goldberg, P. Lancaster, M.A. Kaashoek** (Eds.): The Gohberg Anniversary Collection, Volume II, 1989, (3-7643-2308-6)

42. **N.K. Nikolskii** (Ed.): Toeplitz Operators and Spectral Function Theory, 1989, (3-7643-2344-2)

43. **H. Helson, B. Sz.-Nagy, F.-H. Vasilescu, Gr. Arsene** (Eds.): Linear Operators in Function Spaces, 1990, (3-7643-2343-4)

44. **C. Foias, A. Frazho:** The Commutant Lifting Approach to Interpolation Problems, 1990, (3-7643-2461-9)

45. **J.A. Ball, I. Gohberg, L. Rodman:** Interpolation of Rational Matrix Functions, 1990, (3-7643-2476-7)

46. **P. Exner, H. Neidhardt** (Eds.): Order, Disorder and Chaos in Quantum Systems, 1990, (3-7643-2492-9)

47. **I. Gohberg** (Ed.): Extension and Interpolation of Linear Operators and Matrix Functions, 1990, (3-7643-2530-5)

48. **L. de Branges, I. Gohberg, J. Rovnyak** (Eds.): Topics in Operator Theory. Ernst D. Hellinger Memorial Volume, 1990, (3-7643-2532-1)

49. **I. Gohberg, S. Goldberg, M.A. Kaashoek:** Classes of Linear Operators, Volume I, 1990, (3-7643-2531-3)

50. **H. Bart, I. Gohberg, M.A. Kaashoek** (Eds.): Topics in Matrix and Operator Theory, 1991, (3-7643-2570-4)

51. **W. Greenberg, J. Polewczak** (Eds.): Modern Mathematical Methods in Transport Theory, 1991, (3-7643-2571-2)

52. **S. Prössdorf, B. Silbermann:** Numerical Analysis for Integral and Related Operator Equations, 1991, (3-7643-2620-4)

53. **I. Gohberg, N. Krupnik:** One-Dimensional Linear Singular Integral Equations, Volume I, Introduction, 1992, (3-7643-2584-4)

54. **I. Gohberg, N. Krupnik:** One-Dimensional Linear Singular Integral Equations, Volume II, General Theory and Applications, 1992, (3-7643-2796-0)

55. **R.R. Akhmerov, M.I. Kamenskii, A.S. Potapov, A.E. Rodkina, B.N. Sadovskii:** Measures of Noncompactness and Condensing Operators, 1992, (3-7643-2716-2)

56. **I. Gohberg** (Ed.): Time-Variant Systems and Interpolation, 1992, (3-7643-2738-3)

57. **M. Demuth, B. Gramsch, B.W. Schulze** (Eds.): Operator Calculus and Spectral Theory, 1992, (3-7643-2792-8)

58. **I. Gohberg** (Ed.): Continuous and Discrete Fourier Transforms, Extension Problems and Wiener-Hopf Equations, 1992, (3-7643-2809-6)

59. **T. Ando, I. Gohberg** (Eds.): Operator Theory and Complex Analysis, 1992, (3-7643-2824-X)

60. **P.A. Kuchment:** Floquet Theory for Partial Differential Equations, 1993, (3-7643-2901-7)

61. **A. Gheondea, D. Timotin, F.-H. Vasilescu** (Eds.): Operator Extensions, Interpolation of Functions and Related Topics, 1993, (3-7643-2902-5)

62. **T. Furuta, I. Gohberg, T. Nakazi** (Eds.): Contributions to Operator Theory and its Applications. The Tsuyoshi Ando Anniversary Volume, 1993, (3-7643-2928-9)

63. **I. Gohberg, S. Goldberg, M.A. Kaashoek:** Classes of Linear Operators, Volume 2, 1993, (3-7643-2944-0)

64. **I. Gohberg** (Ed.): New Aspects in Interpolation and Completion Theories, 1993, (3-7643-2948-3)

65. **M.M. Djrbashian:** Harmonic Analysis and Boundary Value Problems in the Complex Domain, 1993, (3-7643-2855-X)

66. **V. Khatskevich, D. Shoiykhet:** Differentiable Operators and Nonlinear Equations, 1993, (3-7643-2929-7)

67. **N.V. Govorov †:** Riemann's Boundary Problem with Infinite Index, 1994, (3-7643-2999-8)

68. **A. Halanay, V. Ionescu:** Time-Varying Discrete Linear Systems Input-Output Operators. Riccati Equations. Disturbance Attenuation, 1994, (3-7643-5012-1)

69. **A. Ashyralyev, P.E. Sobolevskii:** Well-Posedness of Parabolic Difference Equations, 1994, (3-7643-5024-5)

70. **M. Demuth, P. Exner, G. Neidhardt, V. Zagrebnov** (Eds): Mathematical Results in Quantum Mechanics. International Conference in Blossin (Germany), May 17-21, 1993, 1994, (3-7643-5025-3)

71. **E.L. Basor, I. Gohberg** (Eds): Toeplitz Operators and Related Topics. The Harold Widom Anniversary Volume. Workshop on Toeplitz and Wiener-Hopf Operators, Santa Cruz, California, September 20–22, 1992, 1994 (3-7643-5068-7)

72. **I. Gohberg, L.A. Sakhnovich** (Eds): Matrix and Operator Valued Functions. The Vladimir Petrovich Potapov Memorial Volume, (3-7643-5091-1)

73. **A. Feintuch, I. Gohberg** (Eds): Nonselfadjoint Operators and Related Topics. Workshop on Operator Theory and Its Applications, Beersheva, February 24–28, 1994, (3-7643-5097-0)

74. **R. Hagen, S. Roch, B. Silbermann:** Spectral Theory of Approximation Methods for Convolution Equations, 1994, (3-7643-5112-8)

G. Kaiser, University of Massachusetts at Lowell, MA, USA

A Friendly Guide to Wavelets

1994. 320 pages. Hardcover
ISBN 3-7643-3711-7

Please order through your bookseller or write to:

Birkhäuser Verlag AG
P.O. Box 133
CH-4010 Basel / Switzerland
FAX: ++41 / 61 / 271 76 66

For orders originating in the USA or Canada:

Birkhäuser
333 Meadowlands Parkway
Secaucus, NJ 07094-2491
USA

Birkhäuser

Birkhäuser Verlag AG
Basel · Boston · Berlin

This volume consists of two parts. Chapters 1–8, Basic Wavelet Analysis, are aimed at graduate students or advanced undergraduates in science, engineering and mathematics designed for an introductory one-semester course on wavelets and time-frequency analysis, and can also be used for self-study or by book for practicing researchers in signal analysis and related areas. The reader is not presumed to have a sophisticated mathematical background; therefore, much of the needed analytical machinery is developed from the beginning. The only prerequisites are matrix theory, Fourier series, and Fourier integral transforms. Notation is introduced that facilitates the formulation of signal analysis in a modern and general mathematical language, and the illustrations should further ease comprehension. Each ends with a set of straightforward exercises designed to drive home the concepts.

Chapters 9–11, Physical Wavelets, are at a more advanced level and represent original research. They can be used as a text for a second-semester course or, when combined with Chapters 1 and 3, as a reference for a research seminar. Whereas the wavelets of Part I can be any functions of "time," physical wavelets are funtions of space-time constrained by differential equations. In Chapter 9, wavelets specifically dedicated to Maxwell's equations are constructed. These wavelets are electromagnetic pulses parameterized by their point and time of emission or absorption, their duration, and the velocity of the emitter or absorber. The duration also acts as a scale parameter. We show that every electromagnetic wave can be composed from such wavelets. This fact is used in Chapter 10 give a new formulation of electromagnetic imaging, such as radar, accompanied by a geometrical model for scattering based on conformal transformations. In Chapter 11 a similar set of wavelets is developed for acoustics. A relation is established at the fundamental level of differential equations between physical waves and time signals.